Molecular and Cellular Approaches to the Control of Proliferation and Differentiation

CELL BIOLOGY: A Series of Monographs

EDITORS

A sampling of published volumes

John Morrow. EUKARYOTIC CELL GENETICS, 1983

John F. Hartmann (editor). MECHANISM AND CONTROL OF ANIMAL FERTILIZATION, 1983

Gary S. Stein and Janet L. Stein (editors). RECOMBINANT DNA AND CELL PROLIFERATION, 1984

Prasad S. Sunkara (editor). NOVEL APPROACHES TO CANCER CHEMOTHERAPY, 1984

B. G. Atkinson and D. B. Walden (editors). CHANGES IN GENE EXPRESSION IN RESPONSE TO ENVIRONMENTAL STRESS, 1984

Reginald M. Gorczynski (editor). RECEPTORS IN CELLULAR RECOGNITION AND DEVELOPMENTAL PROCESSES, 1986

Govindjee, Jan Amesz, and David Charles Fork (editors). LIGHT EMISSION BY PLANTS AND BACTERIA, 1986

Peter B. Moens (editor). MEIOSIS, 1986

Robert A. Schlegel, Margaret S. Halleck, and Potu N. Rao (editors). MOLECULAR REGULATION OF NUCLEAR EVENTS IN MITOSIS AND MEIOSIS, 1987

Monique C. Braude and Arthur M. Zimmerman (editors). GENETIC AND PERINATAL EFFECTS OF ABUSED SUBSTANCES, 1987

E. J. Rauckman and George M. Padilla (editors). THE ISOLATED HEPATOCYTE: USE IN TOXICOLOGY AND XENOBIOTIC BIOTRANSFORMATIONS, 1987

Heide Schatten and Gerald Schatten (editors). THE MOLECULAR BIOLOGY OF FERTILIZATION, 1988

Heide Schatten and Gerald Schatten (editors). THE CELL BIOLOGY OF FERTILIZATION, 1988

Anwar Nasim, Paul Young, and Byron F. Johnson (editors). MOLECULAR BIOLOGY OF THE FISSION YEAST, 1989

Mary P. Moyer and George Poste (editors). COLON CANCER CELLS, 1989

Gary S. Stein and Jane B. Lian (editors). MOLECULAR AND CELLULAR APPROACHES TO THE CONTROL OF PROLIFERATION AND DIFFERENTIATION, 1992

Molecular and Cellular Approaches to the Control of Proliferation and Differentiation

Edited by

Gary S. Stein
Jane B. Lian

Department of Cell Biology
University of Massachusetts Medical Center
Worcester, Massachusetts

ACADEMIC PRESS, INC.
Harcourt Brace Jovanovich, Publishers
San Diego New York Boston London Sydney Tokyo Toronto

This book is printed on acid-free paper. ∞

Academic Press, Inc.
San Diego, California 92101

United Kingdom Edition published by
Academic Press Limited
24–28 Oval Road, London NW1 7DX

Library of Congress Cataloging-in-Publication Data

Molecular and cellular approaches to the control of proliferation and differentiation / edited by Gary S. Stein, Jane B. Lian.
p. cm. --(Cell biology)
Includes index.
ISBN 0-12-664745-3
1. Cell proliferation--Molecular aspects. 2. Cell differentiation--Molecular aspects. I. Stein, Gary S. II. Lian, Jane B. III. Series.
[DNLM: 1. Cell Differentiation. 2. Cell Division. QH 605 M718]
QH604.7.M65 1991
574.87'612--dc
DNLM/DLC
for Library of Congress 91-22590
CIP

PRINTED IN THE UNITED STATES OF AMERICA
91 92 93 94 9 8 7 6 5 4 3 2 1

Contents

3 The Control of Mitotic Division

Potu N. Rao

4 Cell Cycle and Cell-Growth Control

Kenneth J. Soprano and Stephen C. Cosenza

5 Regulation of Gene Expression by Serum Growth Factors

Gregg T. Williams, Andrew S. Abler, and Lester F. Lau

II Cellular, Biochemical, and Molecular Parameters of *in vitro* Model Systems in Which Modifications in Cell-Growth Control Are Functionally Related to the Onset of Differentiation

9 Growth and Differentiation in Melanocytes

Ann Richmond

III Exploring Mechanisms of Control

10 Molecular Mechanisms That Mediate a Functional Relationship between Proliferation and Differentiation

Gary S. Stein, Jane B. Lian, Thomas A. Owen, Joost Holthuis, Rita Bortell, and Andre J. van Wijnen

11 The Nuclear Matrix: Structure and Involvement in Gene Expression

Jeffrey A. Nickerson and Sheldon Penman

12 Histone Modifications Associated with Mitotic Chromosome Condensation

John P. H. Th'ng, Xiao-Wen Guo, and E. Morton Bradbury

Contributors

Numbers in parentheses indicate the pages on which the authors' contributions begin.

Andrew S. Abler (116), Department of Genetics, The University of Illinois College of Medicine, Chicago, Illinois 60612

Michael Aronow (165), Department of Cell Biology, University of Massachusetts Medical Center, Worcester, Massachusetts 01655

Leesa Barone (165), Department of Cell Biology, University of Massachusetts Medical Center, Worcester, Massachusetts 01655

Joseph Bidwell (165), Department of Cell Biology, University of Massachusetts Medical Center, Worcester, Massachusetts 01655

Rita Bortell (302), Department of Cell Biology, University of Massachusetts Medical Center, Worcester, Massachusetts 01655

E. Morton Bradbury (381), Department of Biological Chemistry, School of Medicine, University of California, Davis, California 95616, and Life Sciences Division, Los Alamos National Laboratories, Los Alamos, New Mexico 87545

Bruno Calabretta (29), Department of Microbiology and Immunology and Jefferson Cancer Institute, Thomas Jefferson Medical College, Philadelphia, Pennsylvania 19107

David Collart (165), Department of Cell Biology, University of Massachusetts Medical Center, Worcester, Massachusetts 01655

Stephen C. Cosenza (73), Department of Microbiology and Immunology, Temple University School of Medicine, Philadelphia, Pennsylvania 19140

Vladimir Drozdoff (4), Department of Cell Biology, Vanderbilt University School of Medicine, Nashville, Tennessee 37203

Steven Dworetzky (165), Department of Cell Biology, University of Massachusetts Medical Center, Worcester, Massachusetts 01655

Xiao-Wen Guo (381), Department of Biological Chemistry, School of Medicine, University of California, Davis, California 95616

Joost Holthuis (302), Department of Cell Biology, University of Massachusetts Medical Center, Worcester, Massachusetts 01655

Brent L. Kreider (223), The Wistar Institute of Anatomy and Biology, Philadelphia, Pennsylvania 19104

Lester F. Lau (116), Department of Genetics, The University of Illinois College of Medicine, Chicago, Illinois 60612

Jane B. Lian (165, 302), Department of Cell Biology, University of Massachusetts Medical Center, Worcester, Massachusetts 01655

Paul A. Marks (243), DeWitt Wallace Research Laboratories, Memorial Sloan-Kettering Cancer Center, and the Sloan-Kettering Division of the Graduate School of Medical Sciences, Cornell University, New York, New York 10021

Joseph Michaeli (243), DeWitt Wallace Research Laboratories, Memorial Sloan-Kettering Cancer Center, and the Sloan-Kettering Division of the Graduate School of Medical Sciences, Cornell University, New York, New York 10021

Jeffrey A. Nickerson (343), Department of Biology, Massachusetts Institute of Technology, Cambridge, Massachusetts 02139

Nancy E. Olashaw (4), Department of Cell Biology, Vanderbilt University School of Medicine, Nashville, Tennessee 37203

James E. Olson (4), Department of Cell Biology, Vanderbilt University School of Medicine, Nashville, Tennessee 37203

Thomas A. Owen (165, 302), Department of Cell Biology, University of Massachusetts Medical Center, Worcester, Massachusetts 01655

Sheldon Penman (343), Department of Biology, Massachusetts Institute of Technology, Cambridge, Massachusetts 02139

Scott Peura (165), Department of Cell Biology, University of Massachusetts Medical Center, Worcester, Massachusetts 01655

W. J. Pledger (4), Department of Cell Biology, Vanderbilt University School of Medicine, Nashville, Tennessee 37203

Shirwin Pockwinse (165), Department of Cell Biology, University of Massachusetts Medical Center, Worcester, Massachusetts 01655

Potu N. Rao (49), Department of Medical Oncology, The University of Texas M. D. Anderson Cancer Center, Houston, Texas 77030

Ann Richmond (269), Department of Veterans Affairs, and Departments of Cell Biology and Medicine, Vanderbilt University School of Medicine, Nashville, Tennessee 37232

Victoria M. Richon (243), DeWitt Wallace Research Laboratories, Memorial Sloan-Kettering Cancer Center, and the Sloan-Kettering Division of the Graduate School of Medical Sciences, Cornell University, New York, New York 10021

Richard A. Rifkind (243), DeWitt Wallace Research Laboratories, Memorial Sloan-Kettering Cancer Center, and the Sloan-Kettering Division of the Graduate School of Medical Sciences, Cornell University, New York, New York 10021

Giovanni Rovera (223), The Wistar Institute of Anatomy and Biology, Philadelphia, Pennsylvania 19104

Victoria Shalhoub (165), Department of Cell Biology, University of Massachusetts Medical Center, Worcester, Massachusetts 01655

Kenneth J. Soprano (73), Department of Microbiology and Immunology, Temple University School of Medicine, Philadelphia, Pennsylvania 19140

Gary S. Stein (165, 302), Department of Cell Biology, University of Massachusetts Medical Center, Worcester, Massachusetts 01655

Melissa S. Tassinari (165), Department of Cell Biology, University of Massachusetts Medical Center, Worcester, Massachusetts 01655

John P. H. Th'ng (381), Department of Biological Chemistry, School of Medicine, University of California, Davis, California 95616

Andre J. van Wijnen (302), Department of Cell Biology, University of Massachusetts Medical Center, Worcester, Massachusetts 01655

Donatella Venturelli (29), Department of Microbiology and Immunology and Jefferson Cancer Institute, Thomas Jefferson Medical College, Philadelphia, Pennsylvania 19107

Gregg T. Williams (116), Department of Genetics, The University of Illinois College of Medicine, Chicago, Illinois 60612

Preface

For more than a century, it has been acknowledged that proliferation and differentiation are fundamental biological processes. Equally important, it has been understood that a relationship between cell growth and expression of phenotypic properties characteristic of specialized cells and tissues is associated with key regulatory events in the control of development as well as tissue repair. However, until recently, proliferation and differentiation were experimentally addressed independently.

It would be arbitrary and less than accurate to invoke any single explanation for the convergence of both concepts and experimental approaches that have provided the basis for addressing the integrated relationship between proliferation and differentiation. Indeed, the advances that have been made in molecular biology have played an important role in this context, permitting the assessment of a broad spectrum of biological parameters in single cells and tissue preparations and facilitating identification of cell growth and tissue-specific genes and their regulatory complexes (transcription factors and cognate regulatory elements). But it appears that it has been the combined application of molecular, biochemical, and morphological approaches, together with the development of *in vitro* systems that support differentiation and tissue organization, that has led to significant increments in our ability to define the proliferation–differentiation relationship.

To attempt coverage of all aspects of cell growth and differentiation in a single volume would be unrealistic and, at best, treatment of the principal elements of the developmental process and their control would be descriptive and superficial. Rather, in this volume we restrict our considerations to basic mechanisms involved in cell growth control, emphasizing the coupling of proliferation and the progressive expression of several specific cellular phenotypes. The manner in which cell structure is involved in the selective expression of genes associated with proliferation and differentiation and, in turn, how expression of such genes in response modulates both intracellular (nuclear matrix and cytoskeleton) and extracellular (extracellular matrix) architecture are emerging concepts that are addressed.

Most authors have focused primarily on a single model system or cell phenotype. But collectively these chapters provide information for beginning to assess the extent to which common signaling mechanisms and regulatory events are operative in the control of proliferation and differentiation in general. And while it would be premature to propose unifying mechanisms to explain the relationship of growth to differentiation, optimistically, the next few years should yield valuable insight into the regulation of this relationship as it is operative during early development and in the maintenance of structural and functional integrity of cells and tissues.

Gary S. Stein
Jane B. Lian

I

Regulation of Cell Proliferation

1

Growth Factors: Their Role in the Control of Cell Proliferation

NANCY E. OLASHAW, JAMES E. OLSON, VLADIMIR DROZDOFF, AND W. J. PLEDGER

Department of Cell Biology
Vanderbilt University School of Medicine
Nashville, Tennessee 37203

Polypeptide growth factors act in a synergistic and sequential manner to promote the proliferation of nontransformed cells in culture. Although the mechanisms by which these factors impart mitogenic information to target cells are incompletely understood, recent studies have defined a series of biochemical and molecular events that occur in response to growth-factor treatment and as cells shift from a quiescent to a proliferative state. To initiate the mitogenic response, growth factors interact with, and consequently activate, specific membrane-bound receptors. Receptor activation, in turn, stimulates the formation of *second messengers,* which transduce the mitogenic signal from the cell membrane to the cell interior. As described below, the receptors for growth factors such as platelet-derived growth factor (PDGF) and epidermal growth factor (EGF) possess an intrinsic ligand-activated tyrosine kinase; accumulating evidence suggests that it is via this activity that these receptors communicate with second messengers. Second messengers participate in a variety of events including, for example, the modification of transactivating factors that, via interaction with DNA response elements, induce the expression of specific genes. Proteins preferentially synthesized as a result of second messenger-mediated gene transcription modulate a host of regulatory processes that lead ultimately to the proliferative response.

Early studies identified the pre-DNA synthetic (G_1) phase of the cell cycle as the primary site of growth-factor action. Using cells arrested in early G_1 (G_0, see below) by mitogen deprivation, numerous investigators characterized events that occurred rapidly in response to growth-factor treatment. While the importance of these early G_1 responses is not to be minimized, data from other studies indicate that growth factor-dependent processes occurring in mid and late G_1 are also essential for proliferation. Thus, the purpose of this chapter is twofold: first, to describe potential mechanisms involved in growth factor-induced receptor activation and second messenger formation and, second, to detail changes in gene expression and other activities that occur throughout the G_1 phase of the cell cycle. Of the numerous growth factors previously characterized, PDGF and EGF have been extensively studied as models for growth-factor action in fibroblast systems. For this reason, and as a comprehensive review of all growth factors is beyond the scope of this chapter, we focus primarily on actions of these factors in fibroblastic growth control

I. GROWTH FACTORS AND RECEPTORS

Polypeptide growth factors form part of a large class of hydrophilic, extracellular signaling molecules, which constitute an important part of the endocrine system. Like classic peptide hormones, they bind to specific receptor proteins on the surface of target cells and regulate a wide variety of cellular functions through activation of several intracellular signals (discussed below). It has been useful to consider growth factors as a distinct class in that, unlike classic hormones, they are also important local mediators for cell regulation. Additionally, individual growth factors may be expressed in a wide variety of cells and can exhibit activity in a number of different target cells and tissues.

Molecular cloning techniques have greatly facilitated the identification of a growing number of growth factors, along with their complementary receptors. Three important themes have become apparent with further characterization of the role of growth factors and receptors in cell regulation. First, sequence and structural comparisons have allowed both growth factors and receptors to be recognized as members of distinct families (see discussion on receptors). Second, strong evidence has been found that, surprisingly, common mechanisms are shared between the different families both for receptor activation and for how the extracellular signal is then subsequently conveyed through activation of intersecting intracellular signaling pathways. This will be discussed in part in this section, in terms of the common structural and functional aspects of the different receptors, and later in terms of the intracellular events following activation. Last, and probably most relevant to the functional effect of growth factors, is the understanding that cell proliferation and physiological responses are not specifically regulated by any one growth factor but are instead under the coordinate control of several factors acting at numerous stages in the growth and development of cells and tissues.

A. Fibroblast Proliferation Is Coordinately Regulated by Multiple Growth Factors

In vitro fibroblast systems have been extremely useful in the development of experimental paradigms for growth-factor action. The isolation of PDGF was prompted by the observation that fibroblasts could proliferate in growth medium containing serum but not platelet-poor plasma, the liquid fraction of unclotted blood. PDGF was identified as the primary factor among several released from platelet secretory granules that enabled fibroblasts to proliferate. This system was important in providing evidence that clearly demonstrated that both the concerted and sequential action of several growth factors was required in order to signal cells to divide.

The cell cycle can be defined as the sequence of events occurring from the completion of mitosis in the parent cell until the completion of the subsequent mitosis in one or both daughter cells.[1] In most cell systems, the cycle is made up of sequential phases consisting of the mitotic or M phase, the presynthetic gap or G_1 phase, the DNA-synthetic or S phase and the postsynthetic or G_2 phase. The majority of cells *in vivo,* however, are not cycling, but remain in a nonproliferating state during most of their life. Similarly, cells in culture may remain viable for an extended period in a growth-arrested or quiescent state referred to as G_0. Nontransformed fibroblastic cells, such as BALB/c 3T3 cells, may be growth-arrested either by growth-factor deprivation or by growth to a confluent density. The differences between noncycling, quiescent cells and those in a proliferating population have been an intriguing area of study, and the reader is referred to several excellent reviews for more detail.[1–3] Quiescent fibroblasts may be stimulated to reenter the cell cycle by exposure to several mitogenic factors. These have been termed competence factors because exposure to these factors alone is both required and sufficient to render the cells *competent* or responsive to additional factors in plasma that govern the further transition through the cell cycle and initiation of DNA synthesis[4] (see below). Studies employing BALB/c 3T3 cells identified PDGF as the primary competence factor present in serum,[5] although several other factors including fibroblast growth factor (FGF), calcium phosphate crystals, and bombesin, have been subsequently identified as additional competence factors in either BALB/c 3T3 cells or other fibroblastic systems.[6–8]

PDGF is an approximately 30-kDa cationic glycoprotein composed of A and B polypeptide chains, which are encoded by two distinct homologous genes.[9] PDGF in its active form exists as either a disulfide-bonded homodimer or heterodimer of its two chains. PDGF, like most other peptide growth factors, is biologically active at nano- and picomolar concentrations, and interacts with its target cells by binding to a cell-surface receptor that exhibits both high affinity and selectivity for its ligand. The PDGF receptor shares several important common characteristics with other growth-factor receptors, which will be discussed in more detail below. Ligand binding results in rapid activation of receptor tyrosine kinase activity and, subsequently, changes in a variety of cell processes, including redistribution of vinculin and actin,[10] formation of inositol phosphates and consequent calcium mobilization (see below), cellular alkalinization,[11] and the induction of a number of early genes, including the cellular protooncogenes c-*fos* and c-*myc.*[12,13]

It is still unclear which events are actually required for the transition from a quiescent to a proliferative state. The use of mutant receptor constructs has shown that receptor kinase activity is crucial for the mitogenic function of the PDGF receptor[14] and, as will be discussed, for other growth-factor receptors as well.[15] The ability of antisense c-*myc* and c-*fos* oligonucleotides to inhibit DNA synthesis suggests that expression of *myc* and *fos* proteins is also required for mitogensis.[16,17] However, it is difficult to determine from these studies whether these proteins are required exclusively in the initial stage of the mitogenic response, or whether they might also function in regulating progression throughout the cell cycle. In BALB/c 3T3 cells, PDGF stimulation is not itself sufficient to promote a complete mitogenic response, which requires the subsequent and continuous exposure of PDGF-stimulated cells to progression factors found in plasma.[18] Importantly, cells exposed first to plasma and then to PDGF do not progress though the cell cycle, implying that not only may cell proliferation be coordinately regulated by multiple growth factors, but also that the regulatory events occur in a definite sequential order.

Progression factors include insulin-like growth factor 1 (IGF-1), a 7-kDa member of the insulin peptide family first identified as a mediator of growth hormone action,[19] and possibly EGF, a 6-kDa polypeptide isolated as an inducer of precocious eyelid opening and tooth eruption in newborn mice.[20] The specific function of these factors clearly depends on both the target-cell type and the context in which a particular cell is exposed to these factors. The effects of specific growth factors in other cell systems are not always so easily distinguished as in the BALB/c 3T3 system. For example, in C3H10T½ fibroblasts, EGF appears able to substitute for both competence and progression factors, and to act alone as a mitogen, albeit at concentrations significantly above those normally found in serum. However, exposure to PDGF increases the sensitivity of these cells to EGF by more than 10-fold.[21] Thus, the synergistic effect of multiple growth factors may indeed be required to achieve an optimal mitogenic response at growth-factor concentrations normally encountered by cells.

The sequential regulation of mitogenic events by multiple factors observed in the 3T3 system is not restricted to fibroblastic cells. The proliferation of T lymphocytes is similarly regulated in discrete steps. In this system, either plant lectins (Con A or phytohemagglutinin), phorbol esters, or antigens play the role of competence factors in mitogenically activating quiescent cells. Exposure to these factors results in similar activation of intracellular responses and the induction of several of the same early genes as in BALB/c 3T3 cells.[22,23] Treatment with these factors alone is not sufficient to induce DNA synthesis, but enables cells to respond to interleukin 2 (IL-2) through up-regulation of the IL-2 receptor[24]; analogously, PDGF has been shown to up-regulate the IGF-1 receptor in 3T3 cells.[25]

The coordinate regulation of cellular events by multiple growth factors is important in a broader scope, apart from proliferation alone. In fact, in almost all cases, the differentiation and clonal expansion of cell populations from small numbers of stem cells depends on the action of multiple growth factors acting at discrete points during the expansion of the resulting cell lineages. In an elegant *in vitro* model, Zezulak and Green[26] demonstrated that growth hormone regulated two steps in the differentiation of an adipogenic fibroblast line. Growth hormone both promoted differentiation of the cells to preadipocytes, which then became responsive to IGF-I, and served to regulate the clonal expansion of these cells by inducing IGF-I synthesis in the differentiated cells. Optimal formation of differentiated hematopoietic colonies from *in vitro* bone marrow stem-cell cultures has been shown to require the synergistic action of IL-1 and at least one other hematopoietic growth factor.[27] *In vivo,* IL-1 was proposed to play a dual role during hematopoiesis by both stimulating the proliferation of quiescent stem cells and indirectly regulating their differentiation by subsequently up-regulating receptors for various hematopoietic factors in the stimulated cells.[28] Evidence that IL-1 can also induce production of hematopoietic growth factors themselves in a variety of stromal cells[29] illustrates how the control of differentiation may require a complex cascade of growth-factor interactions among several different cell types.

Several important roles for growth factors beyond the scope of this review should be mentioned. The chemotactic activities of several growth factors including PDGF,[30] fibroblast growth factor (FGF),[31] and transforming growth factor β (TGFβ)[32] are central to angiogenesis, wound healing, and tissue development. Several growth factors have been demonstrated to modulate the synthesis of extracellular matrix proteins, and their cellular receptors.[33,34] In light of growing evidence that the extracellular environment at least in part directs cellular responsiveness to growth factors, this provides an indirect pathway for growth control. Probably the widest role assumed by the TGFβ family of peptides, however, is their function as potent growth inhibitors, especially in epithelial and immune cells. The reader is referred to several recent comprehensive reviews for information on this important polypeptide family.[35,36]

B. Growth-Factor Receptors

Almost all of the receptors for polypeptide growth factors characterized have been identified as tyrosine kinases, and share a consistent set of structural and functional features. These receptors each contain three distinct structural regions: an extracellular ligand-binding domain, a hydrophobic domain that makes one pass through the cell membrane, and a hydrophilic intracellular domain containing a highly conserved kinase domain, in which resides the receptor tyrosine kinase activity. A more detailed comparison of structural characteristics has allowed several of the receptors to be placed in distinct families. The insulin-receptor family is characterized by a heterotetrameric receptor formed from two α and two β subunits, which are processed from a precursor molecule encoded by a single gene.[37–39] Two nearly identical insulin receptors and the IGF-I receptor share significant homology in the extracellular binding domain, which contains a single cysteine-rich region, and exhibits numerous conserved cysteine residues, glycosylation, and precursor cleavage sites. Members of the PDGF-receptor family possess extracellular ligand-binding regions characterized by multiple immunoglobulin-like domains, a lack of cysteine-rich regions, and a number of conserved cysteine residues and glycosylation sites.[40] A unique feature of this receptor family is that the conserved tyrosine kinase domain is split into two regions around a short, poorly conserved sequence of approximately 100 amino acids.[41] Members of this family include the receptors for PDGF (termed α and β), colony-stimulating factor 1 (CSF-1),[42] and the protein product of the c-*kit* gene,[43] recently identified as the receptor for stem-cell factor.[44] The FGF-receptor family bears some resemblance to the PDGF-receptor family but has a shorter ligand-binding domain.[15] A third receptor family includes the EGF receptor and the HER-2/neu receptor identified in rat cells.[45,46] In contrast to the insulin-receptor family, the extracellular portion of these receptors contains two cysteine-rich regions, which flank the putative ligand-binding site.[47] Several other receptors, including the nerve growth factor (NGF) receptor[48] and the IGF-II/mannose 6-phosphate receptor[49] appear unrelated to the above families and to each other.

The usefulness of these structural comparisons has been validated by evidence that similar receptors may be functionally related as well, in terms of overlapping specificities for binding several related growth-factor ligands. For example, high levels of insulin can activate the IGF-I receptor,[50] and there is good evidence that many of the cellular responses to IGF-II are mediated through the IGF-I receptor.[51] Similarly, the PDGFα receptor can bind both the A and B chains of PDGF.[52] On the other hand, the activation of the EGF receptor by both EGF and TGFα provides a good example of one receptor interacting with two quite divergent peptide ligands.[53]

Evidence from several lines of investigation strongly suggests that a large part of the functional activity of the different receptor domains is an intrinsic property of each domain as a distinct polypeptide unit. This was initially suggested by the finding of serum forms of several receptors that retained their ability for high-affinity binding of their natural ligands.[54] Several transforming oncogenes for a variety of cells have been identified as the truncated forms of receptors, in that they lack an extracellular region and are constitutively activated in the cell (for an excellent recent review, see reference 55). Several studies utilizing chimeric receptor constructs have implied that mechanistic features of the separate domains may be surprisingly conserved between different receptor families. In one of the first examples, the extracellular-binding domain of the insulin receptor was joined to the transmembrane and tyrosine kinase domains of the EGF receptor.[56] The finding that the tyrosine kinase activity of this hybrid receptor became dependent on insulin binding strongly suggested that the extracellular and transmembrane domains of the insulin and EGF receptors share common elements in the way they up-regulate receptor activity after ligand binding.

Studies employing receptor mutants have provided the best evidence for the crucial role of tyrosine kinase activity in growth factor-induced processes. In the PDGF receptor, mutation of the lysine residue serving as the ATP-binding site does not affect ligand binding, but results in the inability of the receptor to induce DNA synthesis as well as most of the early responses associated with PDGF binding, including the stimulation of inositol phosphate turnover, increases in intracellular calcium, actin reorganization, cellular alkalinization, and induction of early-gene transcription.[57,58] The expression of kinase-deficient mutants of other receptor tyrosine kinases, such as in the EGF system,[59] has similarly demonstrated that kinase activity is necessary both for the mitogenic activity of the receptor and for the early cellular events induced by ligand binding to the wild-type receptor.

The molecular mechanisms by which ligand binding to the extracellular domain of receptors results in a change of kinase activity in their intracellular domains remain an elusive problem. Several early studies observed that EGF binding induced a clustering of EGF receptors on the cell surface. Receptor–receptor interaction was initially interpreted as being involved in the internalization of the receptor (see below) because of the localization of clustered receptors into coated pits. Recently, receptor dimerization has been proposed to regulate receptor kinase activity, possibly by affecting the binding affinity of receptor sites. Evidence in support of this interpretation has been reviewed in Schlessinger[60] and a number of alternative explanations are presented by Staros.[61] An analogous model for the

PDGF receptor has been recently proposed, and the reader is referred to several recent reviews.[9,14]

A second major puzzle in growth-factor receptor function concerns the different mechanisms by which ligand binding to growth-factor receptors is regulated. Several common mechanisms, which include changes in both receptor synthesis and the rate of receptor degradation and conformational changes in the receptor that affect the affinity of ligand binding, are shared between the different receptor families. In general, in a process commonly referred to as down-regulation, ligand binding results in a rapid decrease in the number of binding sites for the ligand on the cell surface. As first observed for the insulin receptor,[62] this occurs most often by an increase in the rate of receptor internalization and degradation. Several studies, both with kinase-defective PDGF receptors,[63] suggest that this process does not require the receptor kinase activity, although in contrast, impaired internalization of the EGF receptor has been observed with some kinase-deficient EGF receptors.[59] On the other hand, receptor phosphorylation can play a role regulating ligand binding. Mutant receptor studies have provided direct evidence that the transmodulation or decrease in EGF binding observed after treatment of cells with protein kinase C agonists can be mediated by the phosphorylation of a specific residue (threonine 654) in the intracellular domain of the receptor.[64] Although several studies have proposed that this effect is mediated through a decrease in the number of high-affinity sites, the exact mechanism remains unclear.

II. SIGNAL TRANSDUCTION

An important aspect of growth control concerns the mechanisms responsible for transducing the mitogenic signal from the ligand–receptor complex at the cell surface to the cell interior, in particular to the nucleus where changes in gene expression occur. This transfer of information is mediated by second messengers, generally defined as molecules that are rapidly formed in response to receptor activation and that convey, either directly or indirectly, information from the membrane to the nucleus. One of the best-characterized signal-transduction systems involves the receptor-mediated hydrolysis of the membrane phospholipid, phosphatidylinositol 4,5-bisphosphate (PIP_2), and the consequent formation of two second messengers, inositol 1,4,5-trisphosphate ($InsP_3$), which effects the release of calcium from intracellular stores, and diacylglycerol, which activates the serine–threonine kinase protein kinase C (PKC).[65] Hydrolysis of PIP_2 is mediated by phosphoinositide-specific phospholipase C (PLC), and several isozymes of PLC have been purified and cloned.[66]

PDGF stimulates hydrolysis of PIP_2 and subsequent formation of $InsP_3$ and diacylglycerol in numerous cell lines, including BALB/c 3T3.[67] Accumulation of $InsP_3$ and diacylglycerol is detectable within minutes of addition of PDGF to quiescent cells. EGF induces this response in cells that overexpress EGF receptors either naturally (e.g., A431 adenocarcinoma cells)[68] or as a result of transfection of EGF receptor cDNA[69] but not, for the most part, in nontransformed cells that possess a normal complement of EGF receptors. EGF treatment of quiescent BALB/c 3T3 cells, for example, does not elicit $InsP_3$ formation.[70] Accumulating evidence indicates that PDGF and, in appropriate cell lines, EGF stimulate PIP_2 hydrolysis via receptor-mediated tyrosine phosphorylation of a specific PLC isozyme, PLC-γ. Initial studies showed that ablation of the tyrosine kinase activity of EGF and PDGF receptors abrogated ligand-induced $InsP_3$ formation, and thus suggested that the activity of a component of the phosphoinositide system was modulated by tyrosine phosphorylation.[71,72] Subsequently, several laboratories[73–79] demonstrated tyrosine phosphorylation of PLC-γ in cells exposed to EGF or PDGF (as compared to unstimulated cells in which tyrosine phosphorylation of PLC-γ was undetectable) and of purified PLC-γ incubated with activated EGF or PDGF (β type) receptors. In intact cells, PLC-γ was shown to associate physically with EGF and PDGF receptors in a ligand-dependent manner.[69,76,78] PLC-γ (but not PLC-β or PLC-δ, which apparently are not substrates of receptor tyrosine kinases[76,79]) shares sequence similarity with regions contained within the N-terminal regulatory domain of $pp60^{src}$ and other members of the *src* nonreceptor tyrosine kinase family.[80–82] Mapping of PLC-γ phosphorylated by EGF receptors both *in vivo* and *in vitro* revealed four tyrosine phosphorylation sites; of these, two were located immediately adjacent to the *src*-homologous (SH_2) regions.[83,84]

Tyrosine phosphorylation of PLC-γ induced by EGF and PDGF is rapid, occurring within minutes of addition of hormone to cells, and correlates in both a time- and dose-dependent manner with ligand-induced $InsP_3$ formation.[75,78] All three PDGF isoforms stimulate PLC-γ phosphorylation at tyrosine in BALB/c 3T3 cells; the order of potency observed (PDGF-BB and PDGF-AB greater than PDGF-AA) parallels the capacity of these isoforms to induce $InsP_3$ formation.[85] As described in the preceding section, the receptors for FGF, NGF, insulin, and CSF-1 also possess tyrosine kinase activity; of these, FGF and NGF induce both tyrosine phosphorylation of PLC-γ and $InsP_3$ formation in target cells,[86,87] whereas insulin and CSF-1 do not stimulate either response.[79,88] Collectively, these findings demonstrate a correlation between tyrosine phosphorylation of PLC-γ and activation of PLC-γ. Definitive proof that tyrosine phosphorylation of PLC-γ modulates its activity was recently provided by Carpenter and co-workers,[89] who demonstrated increased activity of tyrosine-phosphorylated PLC-γ in an *in vitro* assay. It is noted that EGF and PDGF also increase the serine phosphosphorylation of PLC-γ,[74–76] as do cyclic adenosine monophosphate (AMP) agonists[90,91]; whether the activity of PLC-γ is affected by serine phosphorylation is, however, unknown at present.

In addition to PDGF and EGF, numerous hormones (e.g., thrombin, vasopressin, bradykinin, bombesin), acting via receptors that do not possess tyrosine kinase activity, also stimulate PIP_2 hydrolysis.[65] A growing body of data suggests that the receptors for these hormones are coupled to PLC by GTP-binding proteins (G proteins).[92] G protein modulation of PLC activity is indicated by studies demonstrating increased $InsP_3$ formation in systems treated with nonhydrolyzable GTP analogs and with the G-protein agonist aluminum fluoride. Furthermore, in some systems, PIP_2 hydrolysis is sensitive to pertussis toxin, a bacterial agent that precludes receptor–G protein interaction. Thus, two potential mechanisms for PLC activation exist—one involving tyrosine phosphorylation of PLC, and a second dependent on PLC–G protein interaction. As PLC-β and PLC-δ do not appear to be substrates of tyrosine kinases,[76,79] it is likely that activation of these isozymes (i.e., isozymes lacking SH regions) is G protein mediated. Participation of G proteins in receptor tyrosine kinase-mediated PIP_2 hydrolysis, however, cannot be excluded. We have found, for example, that pretreatment of BALB/c 3T3 cells with cyclic AMP agonists allows EGF to subsequently stimulate $InsP_3$ formation without inducing tyrosine phosphorylation of PLC-γ.[93] An involvement of a G protein-dependent pathway in this permissive action of cyclic AMP is suggested by data showing that preexposure of cells to cyclic AMP agonists also potentiates aluminum fluoride-induced $InsP_3$ production (N. Olashaw, unpublished data).

As stated above, both $InsP_3$ and diacylglycerol act as second messengers. $InsP_3$ mobilizes calcium from nonmitochondrial stores and thus activates a host of calcium-modulated processes. Diacylglycerol mediates the phosphorylation of target proteins (e.g., EGF receptors,[94] vinculin,[95] c-*abl*[96]) on serine and threonine via activation of PKC. It is noted that several species of PKC, differing in tissue distribution and kinetic properties, have been identified.[97] An obvious question concerns the role of calcium and PKC in growth control, and data addressing this question are contradictory. Numerous growth-promoting agents stimulate PIP_2 hydrolysis in numerous cell lines at concentrations approximating those required for DNA synthesis. In addition, phorbol esters [e.g., 12-*O*-tetradecanoyl-13-phorbol acetate (TPA)], which share structural homology with diacylglycerol and thus act as exogenous activators of PKC, are mitogenic for a number of cell lines, as is diacylglycerol itself.[98–100] In lymphocytes, TPA acts synergistically with calcium ionophores to stimulate DNA synthesis.[101] Furthermore, Swiss 3T3 cells transfected with and thus overexpressing PKC-α have been shown to manifest enhanced growth capacity in low serum-containing medium.[102]

In other experiments, Smith *et al.*[103] showed that microinjection of PLC-β or PLC-γ into serum-deprived NIH 3T3 cells induced DNA synthesis and morphologic transformation. In contrast, in NIH 3T3 cells transfected with PLC-γ cDNA to overexpress PLC-γ, neither PDGF- nor FGF-induced DNA synthesis was affected despite increased PLC-γ tyrosine phosphorylation and $InsP_3$ formation.[104,105] Interestingly, the enhancement of PDGF-stimulated $InsP_3$ production was not accompanied by an increase in the calcium signal, perhaps owing to activation of desensitization mechanisms, and thus may account, at least in part, for the lack of effect of PLC-γ overexpression on mitogenesis.[104] Microinjection of $InsP_3$ into BALB/c 3T3 cells also failed to elicit DNA synthesis; whether calcium mobilization occurred was not examined.[100] On the other hand, microinjection of quiescent NIH 3T3 cells with antibody to PIP_2 blocked PDGF- and bombesin-stimulated DNA synthesis,[106] and a mixture of antibodies to PLC-β and PLC-γ inhibited proliferation when injected into exponentially growing 3T3 cells.[107] While these studies suggest that PIP_2 hydrolysis is necessary for mitogenesis, data by Escobedo and Williams[108] indicate that it is insufficient. These investigators transfected PDGF receptor-minus cells with a construct encoding PDGFβ receptors that lacked a portion of the kinase insert region, and found that treatment of these cells with PDGF effectively elicited $InsP_3$ formation, but only weakly stimulated DNA synthesis. Other studies, however, question the necessity of PIP_2 hydrolysis for proliferation. Hill *et al.*,[109] for example, used the tyrosine kinase inhibitor genistein to show that PDGF efficiently induced DNA synthesis in C3H10T½ mouse fibroblasts in the absence of calcium mobilization and protein kinase C activation. Last, chronic exposure of BALB/c 3T3 cells to TPA, a treatment that down-regulates protein kinase C activity, has been shown to prevent[100] and to have no effect on[110] PDGF-stimulated DNA synthesis.

The disparate results described above presumably reflect, at least in part, the different cell lines, culture conditions, and experimental protocols used. Interpretation of overexpression experiments, for example, is difficult, as negative results do not necessarily imply that the overexpressed factor is unessential, as it may be required but not sufficient. Positive results may reflect the ability of abnormally high levels of the overexpressed factor to obviate events that normally participate in the mitogenic process. Furthermore, growth factors may elicit similar responses via activation of different pathways, and thus, blockade of a particular pathway would not necessarily preclude DNA synthesis. Diacylglycerol, for example, is also generated by hydrolysis of phosphatidylcholine, a process previously shown to be induced by PDGF in Swiss 3T3 cells.[111] Phosphatidylcholine, a more prevalent phospholipid than phosphatidylinositol, thus represents an alternative and more abundant source of diacylglycerol.

In addition to PLC-γ, other proteins also associate with and are phosphorylated by activated PDGFβ receptors. These proteins include type I phosphatidylinositol 3-kinase (PI-3-K),[112,113] the serine–threonine kinase Raf-1,[114,115] the *ras* GTPase activating protein (*ras* GAP),[116,117] and three *src* family members, $pp60^{src}$, $pp59^{fyn}$ and $pp62^{yes}$.[118–120] Interaction of PI-3-K[121] and Raf-1[122] with, and tyrosine phosphorylation of *ras* GAP[123] by, activated EGF receptors have also been reported. Like PLC-γ, PI-3-K[55] and GAP[124] contain SH_2 regions; thus, this common structural element may act as a recognition site for receptor tyrosine kinases. In support, Anderson *et al.*[125] showed that the SH_2 regions of PLC-γ and GAP, expressed as bacterial fusion proteins, efficiently bound EGF and PDGF receptors present in lysates of ligand-treated cells. Two major autophosphorylation sites on the human PDGFβ receptor have been identified—Tyr 751 in the kinase insert region and Tyr

857 in the second kinase domain[126]—and mutation of these sites has shown that binding of SH_2-containing proteins to the PDGF receptor is regulated by receptor autophosphorylation at distinct sites. For example, mutation of Tyr 751 but not Tyr 857 abolished association with PI-3-K activity,[126] whereas maximal GAP binding required tyrosine at both 751 and 857.[127] The mechanism regulating the binding of Raf-1, which does not possess SH_2,[128] to receptors is at present unknown.

PLC-γ, PI-3-K, Raf-1 and *ras* GAP are present in the cytosol of unstimulated cells and are recruited from the cytosol to the membrane in response to growth-factor stimulation.[55] All, when associated with receptors, are substrates of serine kinases as well as receptor tyrosine kinases. As noted above, PLC-γ is activated by tyrosine phosphorylation; the activity of Raf-1 may be regulated by either tyrosine or serine phosphorylation (see below), whereas the effects of phosphorylation on the activity of PI-3-K and *ras* GAP are unclear at present. The potential biological actions of PI-3-K, Raf-1 and *ras* GAP are described below.

PI-3-K phosphorylates the D-3 position of the inositol ring of phosphatidylinositol, phosphatidylinositol 4-phosphate, and PIP_2 to form phosphatidylinositol 3-phosphate, phosphatidylinositol 3,4-bisphosphate and phosphatidylinositol trisphosphate, respectively.[129,130] Increases in all three products have been observed in PDGF-treated cells[130]; none, however, are substrates of PLC, and their functions at present are obscure, although a role of PI-3-K in actin rearrangement has been proposed.[55] As recently described by Carpenter *et al.*,[131] PI-3-K is a heterodimer of 110-kDa and 85-kDa proteins; whether the 110-kDa subunit, like the 85-kDa subunit,[112] associates with activated PDGF receptors is not known at present. An involvement of PI-3-K in growth control is suggested by studies showing that PI-3-K activity and the 85-kDa subunit associate with the middle T/pp60^{src} complex in polyoma-transformed cells, and that polyoma mutants lacking associated PI-3-K activity are not transforming.[112,132,133] In addition, partial deletion of the kinase insert region of the PDGFβ receptor, which impairs PDGF-stimulable DNA synthesis without affecting $InsP_3$ formation and calcium mobilization, has been found to abolish receptor-associated PI-3-K activity.[113] Analogous deletions in the PDGFα receptor produced similar results; however, despite a reduced DNA synthesis response (30% of wild-type receptor), these transfectants, when under the autocrine stimulation of c-*sis*, were capable of sustained proliferation as monitored by colony formation in soft agar and tumorigenicity in nude mice.[134] Thus, for the PDGFα receptor at least, loss of PI-3-K activity does not totally abrogate mitogenic signaling.

Raf-1, a 74-kDa serine–threonine kinase, is the cellular homolog of v-*raf*, the transforming gene product of the murine sarcoma virus 3611. As described by Morrison *et al.*,[114] EGF and PDGF rapidly stimulate phosphorylation of Raf-1 in quiescent BALB/c 3T3 cells. Phosphorylation occurred predominantly on serine and threonine residues; in PDGF-treated cells, tyrosine phosphorylation of Raf-1 was also detectable and, in an *in vitro* assay, was shown to increase Raf-1 activity.[115] Activation of Raf-1 by EGF- and insulin-stimulated serine–threonine phosphorylation has also been observed.[122,135,136] Studies employing transient transfection assays have demonstrated transcriptional activation of the c-*fos* and β-actin promoters by v-*raf*,[137] and of the serum response element of the c-*fos* enhancer by Raf-1 activated by amino-terminal truncation.[138] Activated Raf-1 and v-*raf* have also been shown to stimulate transcription from the AP-1–related promoter, PEA1, in a manner dependent on functional *raf* kinase activity.[139] These findings suggest that v-*raf*/Raf-1 may phosphorylate factors that regulate gene expression. Previous studies have shown that microinjection of activated Raf-1 into quiescent NIH 3T3 cells induces DNA synthesis and transformation,[140] and that expression of Raf-1 antisense RNA inhibits serum-induced 3T3 cell proliferation,[141] and thus a role of Raf-1 in mitogenesis has been proposed.

GAP is a 110-kDa protein that associates with the product of the *ras* protooncogene, $p21^{ras}$, and other low-molecular-weight GTP-binding proteins.[142] GAP interacts with active (i.e., GTP-bound) but not inactive (i.e., GDP-bound) $p21^{ras}$; binding of GAP increases the intrinsic GTPase activity of $p21^{ras}$ and thus converts it to its inactive form.[143] Previous studies have shown that microinjection of quiescent NIH 3T3 cells with antibody to c-*ras* inhibits DNA synthesis induced by PDGF plus EGF,[144] and thus implicates *ras* as an obligatory intermediate in PDGF-stimulated mitogenesis. Consistent with this possibility, Satoh *et al.*[145] found that treatment of Swiss 3T3 cells with PDGF enhanced the formation of GTP-bound p21. Whether PDGF increases p21-GTP production by decreasing GAP activity is not known; however, lipids produced via hydrolysis of PIP_2 (e.g., diacylglycerol and its phosphorylated derivative, phosphatidic acid) have been shown to inhibit GAP activity *in vitro*.[146] In addition to acting as an attenuator of *ras* action, data showing that mutations that abolish the biological activity of oncogenic forms of *ras* also ablate GAP/p21 interaction suggest that GAP may also function as the downstream effector of *ras*.[143] Potential targets of GAP action are, however, unclear at present.

III. GROWTH-RELATED GENE EXPRESSION

The subject of growth-related gene expression is one that have been reviewed often and in more detail than is possible here. This section will summarize and provide more recent information, but for in-depth analysis, the reader is referred to any of several excellent and exhaustive reviews.[1,147–150]

Growth-related genes, for the purposes of this discussion, are those whose expression is induced in a cell cycle-dependent manner. That is, they are genes that are not expressed, or are expressed in a limited manner, in cells that are proliferatively quiescent. When quiescent cells are induced to reenter the cell cycle by addition of appropriate mitogenic stimuli, the products of growth-related genes (mRNAs and/or proteins) accumulate in the cytoplasm at various times as the cells progress through the cell cycle. Note that many cellular responses to mitogenic stimuli are not directly related to cell division. As such, it is reasonable to assume that a number of genes will be induced whose products are not directly related to cell division. Separating directly from indirectly involved gene products will require functional determinations.

Several approaches have been used in the search for growth-related genes. The most widely used techniques have been differential screening and subtractive hybridization, comparing cDNA made from quiescent and actively cycling cells. The use of cell-cycle and temperature-sensitive mutants has been very productive as well. Preferential synthesis of cell cycle-specific proteins has been assessed through use of two-dimensional gel electrophoresis. Of course, many protooncogenes have been found to be expressed in a cell cycle-dependent manner. However, the genetic basis of the proliferative response is a highly complex one; it has been estimated that perhaps 3% of the 10,000 or so mRNAs in a growing cell are not present when the cell is quiescent.[151] Many messages of potential importance are present in quiescent cells but expressed to a greater extent in growing cells. As a result, the list of known genes with growth- or cell cycle-dependent expression is long and will continue to get longer as technological advances improve the limits of detection.

As described above, control of cell cycling is usually exercised during G_1; in the presence of factors such as adequate nutrients and appropriate stimuli, cells will commit to the cycle by initiating DNA synthesis. In the absence of stimulatory factors, cells arrest in G_1 and may enter a state of quiescence termed G_0. It is important to note, however, that some cultured cell lines and some early embryos have cycles that do not include a G_1.[152] In these cells, it is possible that the biochemical and regulatory events of G_1 are being accomplished in other phases of the cycle; i.e., the cell is preparing for the next S phase while mitosis and cytokinesis are being completed.[1] As a result, the gene expression normally associated with G_1 may be occurring while the daughter cells are still in mitosis. Another point worth considering is that reproductive quiescence is an entirely separate physiological state compared to a cell in G_1 or any other portion of the cell cycle. Indeed, several different *depths* or stages of quiescence may exist for a given cell type. As a result, stimulation from quiescence may elicit a different or modified program of gene expression compared to a cell stimulated to continue in the cell cycle.[153]

One cultured cell line that has been studied in great detail is the BALB/c 3T3 line. In these cells, S phase lasts about 7 hr; G_2, about 3 hr; and M phase, about 1 hr. G_1, in 3T3 cells as well as many other cell lines, can be highly variable. In rapidly cycling 3T3 cells, G_1 is about 6 hr long. If the cells are stimulated from *normal* quiescence, G_1 will take about 12 hr. But when some cell types, such as WI-38 cells, are stimulated from a deep quiescence, one which has taken several days to weeks to establish, onset of DNA synthesis may require 18 hr.[153] As discussed above, commitment to DNA synthesis in BALB/c 3T3 cells is a two-stage process. The first stage is competence, a stable state induced by PDGF; the second is progression from competence to S-phase commitment, requiring the presence of IGF-I.[4] If subcritical levels of IGF-I are provided, cells will arrest in mid-G_1, 6 hr before S, at a point termed the V point.[154] V-point arrest also can be achieved by elevating intracellular cAMP or by mitogenically stimulating in amino acid-deficient media.[155,156]

A. Early Growth-Regulated Genes

The addition of growth factors to quiescent cells elicits a cascade of biochemical changes as well as a stepwise program of transcriptional and translational events that culminate in mitosis and cell division.[157] The earliest genetic events involve induction of what has been termed immediate-early or competence genes.[158] These genes do not require protein synthesis for their induction. Prototypical immediate-early genes are c-*myc* and c-*fos*.

The c-*myc* gene is transiently expressed in a number of different cell types. In 3T3 cells, *myc* message and protein are maximally expressed 1–2 hr after stimulation of quiescent cells; levels of both decrease thereafter, but may be detected up to 24 hr later.[159,160] While only faintly detected in quiescent cells, the 2.2 kb transcript increases 20- to 40-fold on stimulation, a result of increased transcription but primarily enhanced posttranscriptional processing.[148] Transcription of *myc* is superinduced by inhibitors of protein synthesis such as cycloheximide; a similar effect is seen in amino acid-deficient medium where protein synthesis is reduced 60–80%.[161] Several mechanisms have been proposed to account for superinduction, among them an inhibition of transcription shutdown and the stabilization of resulting transcripts.[162] The precise function of the *myc* nuclear phosphoprotein is unknown; however, when overexpressed, cells exhibit a reduced requirement for PDGF in the mitogenic response.[148] It should also be noted that in A431 cells, EGF inhibits proliferation but still stimulates *myc* expression.[148]

The induction of c-*fos* actually precedes that of c-*myc; fos* message levels peak 15–30 min and protein levels, 30–60 min, after a mitogenic stimulus.[163] Increased rates of transcription account for most of the induction. Following maximal expression, the *fos* message is rapidly degraded and is not detectable by 2 hr after stimulation.[148,162] As with *myc, fos* message is superinduced in the presence of cycloheximide.[148] The c-*fos* protein is a nuclear phosphoprotein of 55 kDa; it has been shown to be evenly distributed in the nucleus during G_1, S, and G_2. However, during mitotic prophase, *fos* dissociates from the condensed chromosomes and diffuses into the cytoplasm until telophase, when it reassociates with the chromatin of the assembling nucleus.[164] The protein is thought to have a role as a trans-acting transcription factor.[165] c-*fos* functions transcriptionally as a complex with other proteins, which are themselves the products of immediate-early genes. These proteins are actually a family of immediate-early genes. These proteins are actually a family of DNA-binding proteins termed Jun-A, Jun-B, and Jun-D.[166] The synthesis of the messages for the Jun proteins have kinetics similar to those of c-*fos*: they are not expressed in quiescent cells; they are rapidly induced by serum or growth factors; they peak quickly (about 1 hr) and are nearly undetectable by 3 hr.[167] They are also superinduced by serum in the presence of an inhibitor of protein synthesis.[167]

A number of other clones have been isolated that show a pattern of early induction and rapid breakdown. For example, Stiles and co-workers, using differential colony hybridization, isolated 5 cDNA clones that were inducible by PDGF.[168] Two of them, JE and KC, are low-abundance messages in quiescent cells, but show a 15- to 20-fold induction within 1 hr of stimulation. Both are superinducible with cycloheximide, though KC is much more responsive.[168] A third, JB, shows homology to the third exon of c-*fos* and is designated as r-*fos*.[169] Ten additional clones were identified by Lau and Nathans[158] as immediate-early genes. After stimulation of resting 3T3 cells by serum or PDGF, the mRNAs reached peak levels between 40 and 120 min later. All the inductions were the result of rapid transcriptional activation and were superinducible in the presence of cycloheximide. More recently, Bravo and co-workers used differential screening of about 200,000 recombinant phage plaques to identify 82 independent sequences derived from cells stimulated by serum in the presence of cycloheximide.[162] Of 71 clones analyzed further, 7 showed peak induction in 30 min, while 32 more peaked at 1 hr. All the products exhibited some cycloheximide superinducibility. Stability of the messages varied greatly with half-lives ranging from 10 to 15 min to over 4 hr. Obviously, such an array of responsive sequences verifies the complexity of the genetic program in early mitogenesis and is indicative of the effort still required to adequately assess early-response genes.

B. Late Growth-Regulated Genes

Early-response genes can be separated from late genes most easily by a differential response to the presence of protein-synthesis inhibitors. In addition to being induced later in the cell cycle, late-response genes are highly sensitive to the presence of cycloheximide.[150] In some cases, even very low concentrations of cycloheximide with minimal impacts on protein synthesis can inhibit the induction of late gene products. As with early-response genes, the program of late-gene expression through the cell cycle is highly complex, and several attempts have been made to clone late genes.[157,170] Expressed products cover all aspects of cell reproduction and include structural proteins, and glycolytic and macromolecular synthetic enzymes, as well as several oncogenes. In this discussion, we will focus on a few of the more well known examples of late-response genes, while acknowledging that a great deal of additional information exists. A more thorough survey can be found in several of the reviews previously mentioned.

Mid-G_1 sees the increased expression of several structural proteins; an example is actin.[171] Other genes exhibiting a mid-G_1 induction include proliferin (MRP), major excreted protein (MEP; pII), and calcyclin (2A9). Proliferin is a prolactin-related polypeptide that is secreted into the medium after growth-factor stimulation of 3T3 cells.[172] MEP, identified as pII by Pledger *et al.*,[173] also is a secreted protein but with proteinase activity; it is synthesized and secreted in greater quantities by transformed cells.[174] Calcyclin, originally designated as cDNA clone 2A9, has sequence similarity with the S100 calcium-binding protein as well as with a subunit of the major cellular substrate for tyrosine kinase.[175] The mRNA for calcyclin builds up throughout G_1 to a 10-fold induction by 12 hr after stimulation of quiescent 3T3 cells.[161] Some forms of *ras* exhibit a mid-G_1 induction; c-Ki-*ras* induces about sixfold after serum stimulation[171] while c-Ha-*ras* shows an early though weak twofold induction in stimulated BALB/c 3T3 cells.[161]

As cells progress through late G_1 and into S phase, several genes are dramatically induced. Principal among these are genes for core histones and the enzymes for nucleotide and polyamine metabolism. Histones complex with DNA to form chromatin and, obviously, new histones must be made when new DNA is made. As a result, histone mRNA levels, particularly for the core histones that make up the nucleosomes (H2A, H2B, H3, and H4), are much higher in cells in S phase.[176,177] As indicated in Denhardt *et al.*[148], the messages for a number of enzymes involved in nucleotide metabolism and DNA synthesis show an induction at the G_1–S boundary; they list thymidine kinase, dTMP kinase, dCMP deaminase, ribonucleotide reductase, dihydrofolate reductase, thymidylate synthetase, ornithine decarboxylase, DNA polymerases, polynucleotide ligase, DNA topoisomerase I (and II), and DNA methylase.

Finally, during late G_1–S, the levels of mRNA for proliferating cell nuclear antigen (PCNA) show an induction paralleling DNA synthesis.[178] PCNA (previously known as cyclin) is a nuclear protein whose synthesis increases sixfold to sevenfold in response to PDGF, FGF, or serum stimulation of 3T3 cells. The message for PCNA reaches a maximal level 16–18 hr after stimulation and is apparently regulated at the posttranscriptional level.[179] In addition to PCNA, the oncogene c-*myb* is transcribed late in G_1. c-*myb* appears to have a major role in controlling proliferation in immature cells of hematopoietic origin, where its expression is substantial and apparently required for cell division.[180,181] Both the message and the protein have short half-lives[182]; the protein is localized to the nucleus, where it apparently has DNA-binding ability.[183] Expression of c-*myb* in cells other than those from hematopoietic lines has been questioned[150]; however, we have routinely detected a characteristic 3.8-kb transcript[184] in poly(A)$^+$ RNA

Northern blots from 3T3 cells.[161] This transcript appears transiently, beginning at 9 hr after stimulation, and maximally by 12 hr. It does not appear at all, however, in cells arrested at or released from the V point.[161] It has also been shown in 3T3 fibroblasts that, when c-*myb* is constitutively expressed, the cells no longer require IGF-I for progression to DNA synthesis.[185] Additionally, 3T3 cells constitutively expressing both c-*myc* and c-*myb* could grow in serum-free medium with no supplemental growth factors.[185] These findings suggest a critical role for *myb* in the cellular commitment to DNA synthesis.

The role c-*myb* plays may be starting to come into focus. Recent work has demonstrated that, apparently in all eukaryotic cell types, entry into mitosis is regulated by a complex involving the protein kinase designated p34.[186] p34 is the product of the *cdc2* gene in the fission yeast, *Schizosaccharomyces pombe,* and is the equivalent of the CDC28 gene of *Saccharomyces cerevisiae.* Activation of p34 requires dephosphorylation of the protein and its subsequent association with proteins called cyclins (not the same protein as PCNA). Cyclins exhibit a dramatic fluctuation in abundance in synchrony with the cell cycle: they peak at the onset of mitosis and are rapidly degraded as mitosis proceeds.

Interestingly, activation of the p34 kinase also is required for the G_1- to S-phase transition; it appears that activated p34 is involved in the initiation of DNA replication.[187] As with the G_2- to M-phase transition, the limiting step for activation may be the association of p34 with cyclins.[188] Three G_1-specific cyclins have been identified; they are the products of the CLN1, CLN2 and CLN3 genes.[189,190] CLN2 codes for a 62-kDa polypeptide whose abundance peaks in late G_1 and then is rapidly degraded.[191] The abundance of the CLN2 protein correlates with the G_1 accumulation of the CLN2 transcript. Both the CLN1 and CLN2 transcripts increase more than fivefold during late G_1, then decrease in S phase.[191] In addition, the transcripts for a p34 homolog in human T lymphocytes were found to increase coincident with the G_1 to S transition; inhibition with antisense RNA blocked the onset of DNA synthesis.[192] Perhaps more importantly, the induction of the *cdc2* transcripts required the prior induction of c-*myc* and c-*myb.* While the regulation of the p34 kinase is incompletely understood, it is apparent that one form of regulation is at the level of transcription for both the kinase and the cyclins; it remains a possibility that *myc* and/or *myb* may have a role, direct or indirect, in that transcription process.

IV. CONCLUDING REMARKS

Perhaps the major conclusion emerging from this discussion concerns the complexity of the mechanisms regulating the proliferation of cultured cells. As described above, traverse of the G_1 phase of the cell cycle requires the concerted action of multiple growth factors that, via interaction with specific receptors, induce a variety of processes that function in an integrated and ordered manner to initiate and maintain the proliferative response. Although not discussed in this chapter, it is noted that negative as well as positive signals are induced by growth factors and contribute to the generation of a sustained but controlled mitogenic response. PDGF, for example, has been shown to stimulate, albeit weakly, the expression of the gene encoding the growth-inhibitory agent, interferon-β[193]; tumor necrosis factor (TNF) also stimulates the expression of this gene, and treatment of cells with antiserum to interferon-β has been found to potentiate the mitogenic activity of TNF.[194] A small induction of the tumor-suppressor gene p53[195] by PDGF has also observed.[161] As a final point, the mechanisms by which transformation deregulates cell proliferation require consideration. Loss of negative regulatory controls represents a potential mechanism, as does alteration of the synthesis or activity of any of the components involved in the stimulatory response. As mentioned above, constitutive activation of receptors may occur via removal of regulatory domains and, as has been shown in many systems, transformed cells may acquire the capacity to produce increased amounts or more oncogenic forms of growth factors such as PDGF.

REFERENCES

1. Baserga, R. (1985). "The Biology of Cell Regulation." Harvard Univ. Press, Cambridge, Massachusetts.
2. Cross, F., Roberts, J., and Weintraub, H. (1989). *Annu. Rev. Cell. Biol.* **5,** 341–396.
3. Prescott, D. M. (1976). "Reproduction of Eukaryotic Cells." Academic Press, New York.
4. Pledger, W. J., Stiles, C. D., Antoniades, H. N., and Scher, C. D. (1977). *Proc. Natl. Acad. Sci. U.S.A.* **74,** 4481–4485.
5. Antoniades, H. N., and Scher, C. D. (1977). *Proc. Natl. Acad. Sci. U.S.A.* **74,** 1973–1977.
6. Stiles, C. D., Capone, G. T., Scher, C. D., Antoniades, H. N., Van Wyk, J. J., and Pledger, W. J. (1979). *Proc. Natl. Acad. Sci. U.S.A.* **76,** 1279–1283.
7. Mitchell, P. G., Pledger, W. J., and Cheung, H. S. (1989). *J. Biol. Chem.* **264,** 14071–14077.
8. Rozengurt, E., and Sinnett-Smith, J. (1983). *Proc. Natl. Acad. Sci. U.S.A.* **80,** 2936–2940.
9. Heldin, C. H., and Westermark, B. (1990). *Cell Regulation* **1,** 555–566.
10. Herman, B., and Pledger, W. J. (1985). *J. Cell Biol.* **100,** 1031–1040.
11. Lopez-Rivas, A., Stroobant, P., Waterfield, M. D., and Rozengurt, E. (1984). *EMBO J.* **3,** 939–944.
12. Cochran, B. H., Reffel, A. C., and Stiles, C. D. (1983). *Cell* **33,** 939–947.
13. Rollins, B. J., and Stiles, C. D. (1989). *Adv. Cancer Res.* **53,** 1–32.
14. Williams, L. T. (1989). *Science* **243,** 1564–1570.
15. Ullrich, A., and Schlessinger, J. (1990). *Cell* **61,** 203–212.
16. Holt, J. T., Gopal, T. V., Moulton, A. D., and Nienhuis, A. W. (1986). *Proc. Natl. Acad. Sci. U.S.A.* **83,** 4794–4798.
17. Heikkila, R., Schwab, G., Wickstrom, E., Loke, S. L., Pluznik, D. H., Watt, R., and Neckers, L. M. (1987). *Nature* **328,** 445–449.
18. Russell, W. E., Van Wyk, J. J., and Pledger, W. J. (1984). *Proc. Natl. Acad. Sci. U.S.A.* **81,** 2389–2392.

19. Baxter, R. C. (1986). *Adv. Clin. Chem.* **25,** 49–115.
20. Gill, G. N., Bertics, P. J., and Santon, J. B. (1987). *Mol. Cell. Endocrinol.* **51,** 169–186.
21. Wharton, W., Leof, E., Olashaw, N., O'Keefe, E. J., and Pledger, W. J. (1983). *Exp. Cell Res.* **147,** 443–448.
22. Reed, J. C., Alpers, J. D., Nowell, P. C., and Hoover, R. G. (1986). *Proc. Natl. Acad. Sci. U.S.A.* **83,** 3982–3986.
23. Moore, J. P., Todd, J. A., Hesketh, T. R., and Metcalfe, J. C. (1986). *J. Biol. Chem.* **261,** 8158–8162.
24. Stern, J. B., and Smith, K. A. (1986). *Science* **233,** 203–206.
25. Clemmons, D. R., Van Wyk, J. J., and Pledger, W. J. (1980). *Proc. Natl. Acad. Sci. U.S.A.* **77,** 6644–6648.
26. Zezulak, K. M., and Green, H. (1986). *Science* **233,** 551–553.
27. Moore, M. A., Muench, M. O., Warren, D. J., and Laver, J. (1990). *Ciba Found. Symp.* **148,** 43–58; discussion 58–61.
28. Moore, M. A., and Warren, D. J. (1987). *Proc. Natl. Acad. Sci. U.S.A.* **84,** 7134–7138.
29. Vogel, S. N., Douches, S. D., Kaufman, E. N., and Neta, R. (1987). *J. Immunol.* **138,** 2143–2148.
30. Grotendorst, G. R., Chang, T., Seppa, H. E., Kleinman, H. K., and Martin, G. R. (1982). *J. Cell. Physiol.* **113,** 261–266.
31. Presta, M., Moscatelli, D., Joseph-Silverstein, J., and Rifkin, D. B. (1986). *Mol. Cell. Biol.* **6,** 4060–4066.
32. Postlethwaite, A. E., Keski-Oja, J., Moses, H. L., and Kang, A. H. (1987). *J. Exp. Med.* **165,** 251–256.
33. Albelda, S. M., and Buck, C. A. (1990) *FASEB J.* **4,** 2868–2880.
34. Roberts, A. B., Heine, U. I., Flanders, K. C., and Sporn, M. B. (1990). *Ann. N. Y. Acad. Sci.* **580,** 225–232.
35. Barnard, J. A., Lyons, R. M., and Moses, H. L. (1990). *Biochim. Biophys. Acta* **1032,** 79–87.
36. Sporn, M., and Roberts, A. (1990). *Cell Regulation* **1,** 875–882.
37. Ullrich, A., Bell, J. R., Chen, E. Y., Herrara, R., Petruzzelli, L. M., Dull, T. J., Gray, A., Coussens, L., Liao, Y. C., Tsubokawa, M., Mason, A., Seeburg, P. H., Grunfield., C., Rosen, O. M., and Ramachandran, J. (1985). *Nature* **313,** 756–761.
38. Massague, J., and Czech, M. P. (1982). *J. Biol. Chem.* **257,** 5038–5045.
39. Ullrich, A., Gray, A., Tam, A. W., Yang-Feng, T., Tsubokawa, M., Collins, C., Henzel, W., Le Bon, T., Kathuria, S., Chen, E., *et al.* (1986). *EMBO J.* **5,** 2503–2512.
40. Claesson-Welsh, L., Eriksson, A., Moren, A., Severinsson, L., Ek, B., Ostman, A., Betsholtz, C., and Heldin, C. H. (1988). *Mol. Cell. Biol.* **8,** 3476–3486.
41. Yarden, Y., Escobedo, J. A., Kuang, W. J., Yang-Feng, T. L., Daniel, T. O., Tremble, P. M., Chen, E. Y., Ando, M. E., Harkins, R. N., Francke, U., *et al.* (1986). *Nature* **323,** 226–232.
42. Sherr, C. J., Rettenmier, C. W., Sacca, R., Roussel, M. F., Look, A. T., and Stanley, E. R. (1985). *Cell* **41,** 665–676.
43. Qiu, F. H., Ray, P., Brown, K., Barker, P. E., Jhanwar, S., Ruddle, F. H., and Besmer, P. (1988). *EMBO J.* **7,** 1003–1011.
44. Zsebo, K. M., Williams, D. A., Geissler, E. N., Broudy, V. C., Martin, F. H., Atkins, H. L., Hsu, R. Y., Birkett, N. C., Okino, K. H., Murdock, D. C. *et al.* (1990). *Cell* **63,** 213–224.
45. Ullrich, A., Coussens, L., Hayflick, J. S., Dull, T. J., Gray, A., Tam, A. W., Lee, J., Yarden, Y., Libermann, T. A., Schlessinger, J., *et al.* (1984). *Nature* **309,** 418–425.
46. Coussens, L., Yang-Feng, T. L., Liao, Y. C., Chen, E., Gray, A., McGrath, J., Seeburg, P. H., Liberman, T. A., Schlessinger, J., Francke, U., Levinson, A., and Ullrich, A. (1985). *Science* **230** 1132–1139.
47. Lax, I., Burgess, W. H., Bellot, F., Ullrich, A., Schlessinger, J., and Givol, D. (1988). *Mol. Cell. Biol.* **8,** 1831–1834.

48. Johnson, D., Lanahan, A., Buck, C. R., Sehgal, A., Morgan, C., Mercer, E., Bothwell, M., and Chao, M. (1986). *Cell* **47,** 545–554.
49. Morgan, D. O., Edman, J. C., Standring, D. N., Fried, V. A., Smith, M. C., Roth, R. A., and Rutter, W. J. (1987). *Nature* **329,** 301–307.
50. Rechler, M. M., and Nissley, S. P. (1985). *Annu. Rev. Physiol.* **47,** 425–442.
51. Rechler, M. M., and Nissley, S. P. (1990). *Hand. Exp. Pharmacol.* **95,** 263–367.
52. Matsui, T., Heidaran, M., Miki, T., Popescu, N., La Rochelle, W., Kraus, M., Pierce, J., and Aaronson, S. (1989). *Science* **243,** 800–804.
53. Ibbotson, K. J., Twardzik, D. R., D'Souza, S. M., Hargreaves, W. R., Todaro, G. J., and Mundy, G. R. (1985). *Science* **228,** 1007–1009.
54. Weber, W., Gill, G. N., and Spiess, J. (1984). *Science* **224,** 294–297.
55. Cantley, L. C., Auger, K. R., Carpenter, C., Duckworth, B., Graziani, A., Kapeller, R., and Soltoff, S. (1991). *Cell* **64,** 281–302.
56. Riedel, H., Dull, T. J., Schlessinger, J., and Ullrich, A. (1986). *Nature* **324,** 68–70.
57. Escobedo, J. A., Barr, P. J., and Williams, L. T. (1988). *Mol. Cell. Biol.* **8,** 5126–5131.
58. Westermark, B., Siegbahn, A., Heldin, C. H., and Claesson-Welsh, L. (1990). *Proc. Natl. Acad. Sci. U.S.A.* **87,** 128–132.
59. Chen, W. S., Lazar, C. S., Poenie, M., Tsien, R. Y., Gill, G. N., and Rosenfeld, M. G. (1987). *Nature* **328,** 820–823.
60. Schlessinger, J. (1988). *Trends Biochem. Sci.* **13,** 443–447.
61. Staros, J. V., Fanger, B. O., Faulkner, L. A., Palaszewski, P. P., and Russo, M. W. (1989). Mechanism of transmembrane signaling by the epidermal growth factor receptor/kinase. In "Receptor Phosphorylation." V. K. Moudgil, CRC Press, Boca Raton, Florida, pp. 227–242.
62. Green, A., and Olefsky, J. M. (1982). *Proc. Natl. Acad. Sci. U.S.A.* **79,** 427–431.
63. Severinsson, L., Ek, B., Mellstrom, K., Claesson-Welsh, L., and Heldin, C. H. (1990). *Mol. Cell. Biol.* **10,** 801–809.
64. Lin, C. R., Chen, W. S., Lazar, C. S., Carpenter, C. D., Gill, G. N., Evans, R. M., and Rosenfeld, M. G. (1986). *Cell* **44,** 839–848.
65. Abdel-Latif, A. A. (1986). *Pharmacol. Rev.* **38,** 227–272.
66. Rhee, S. G., Suh, P.-G., Ryu, S.-H., and Lee, S. Y. (1989). *Science* **244,** 546–550.
67. Olashaw, N. E., and Pledger, W. J. (1988). *Adv. Second Messenger Phosphoprotein Res.* **22,** 139–173.
68. Sawyer, S. T., and Cohen, S. (1981). *Biochemistry* **20,** 6280–6286.
69. Margolis, B., Bellot, F., Honegger, A. M., Ullrich, A., Schlessinger, J., and Zilberstein, A. (1990). *Mol. Cell. Biol.* **10,** 435–441.
70. Besterman, J. M., Watson, S. P., and Cuatrecasas, P. (1986). *J. Biol. Chem.* **261,** 723–727.
71. Molenaar, W. H., Bierman, A. J., Tilly, B. C., Verlaan, I., Defize, L. H. K., Honegger, A. M., Ullrich, A., and Schlessinger, J. (1988). *EMBO J.* **7,** 707–710.
72. Escobedo, J., Barr, P. J., and Williams, L. T. (1988). *Mol. Cell. Biol.* **8,** 5126–5131.
73. Wahl, M. I., Daniel, T. O., and Carpenter, G. (1988). *Science* **241,** 968–970.
74. Wahl, M. I., Nishibe, S., Suh, P.-G., Rhee, S. G., and Carpenter, G. (1989). *Proc. Natl. Acad. Sci. U.S.A.* **86,** 1568–1572.
75. Wahl, M. I., Olashaw, N. E., Nishibe, S., Rhee, S. G., Pledger, W. J., and Carpenter, G. (1989). *Mol. Cell. Biol.* **9,** 2934–2943.
76. Meisenhelder, J., Suh, P.-G., Rhee, S. G., and Hunter, T. (1989). *Cell* **57,** 1109–1122.
77. Margolis, B., Rhee, S. G., Felder, S., Merivc, M., Lyall, R., Levitzki, A., Zilberstein, A., and Schlessinger, J. (1989). *Cell* **57,** 1101–1107.
78. Kumjian, D., Wahl, M. I., Rhee, S. G., and Daniel, T. O. (1989). *Proc. Natl. Acad. Sci. U.S.A.* **86,** 8232–8236.
79. Nishibe, S., Wahl, M. I., Wedegaertner, P. B., Kim, J. J., Rhee, S. G., and Carpenter, G. (1990). *Proc. Natl. Acad. Sci. U.S.A.* **87,** 424–428.

80. Mayer, B. J. Hamaguchi, M., and Hanafusa, H. (1988). *Nature* **332,** 272–275.
81. Stahl, M. L., Ferenz, C. R., Kelleher, K. L., Kriz, R. W., and Knopf, J. L. (1988). *Nature* **332,** 269–272.
82. Suh, P.-G., Ryu, S. H., Moon, K. H., Suh, H. W., and Rhee, S. G. (1988). *Cell* **54,** 161–169.
83. Wahl, M. I., Nishibe, S., Kim, J. W., Kim, H., Rhee, S. G., and Carpenter, G. (1990). *J. Biol. Chem.* **265,** 3944–3948.
84. Kim, J. W., Sim, S. S., Kim, U.-H., Nishibe, S., Wahl, M. I., Carpenter, G., and Rhee, S. G. (1990). *J. Biol. Chem.* **265,** 3940–3947.
85. Olashaw, N. E., Kusmik, W., Daniel, T. O., and Pledger, W. J. (1991). *J. Biol. Chem.* **266,** 10234–10240.
86. Burgess, W. H., Dionne, C. A., Kaplow, J., Mudd, R., Friesel, R., Zilberstein, A., Schlessinger, J., and Jaye, M. (1990). *Mol. Cell. Biol.* **10,** 4770–4777.
87. Kim, U.-H., Fink, D., Kim, H. S., Park, D. J., Contreras, M. L., Guroff, G., and Rhee, S. G., (1991). *J. Biol. Chem.* **266,** 1359–1362.
88. Downing, J. R., Margolis, B. L., Zilberstein, A., Ashmun, R. A., Ullrich, A., Sherr, C. J., and Schlessinger, J. (1989). *EMBO J.* **8,** 3345–3350.
89. Nishibe, S., Wahl, M. I., Hernandez-Sotomayor, S. M. T., Tonks, N. K., Rhee, S. G., and Carpenter, G. (1990). *Science* **250,** 1253–1256.
90. Kim, U.-H., Kim, J. W., and Rhee, S. G. (1989). *J. Biol. Chem.* **264,** 20167–20170.
91. Olashaw, N. E., Rhee, S. G., and Pledger, W. J. (1990). *Biochem. J.* **272,** 297–303.
92. Harden, T. K. (1990). *Am. Rev. Respir. Dis.* **141,** S119–S122.
93. Olashaw, N. E., and Pledger, W. J. (1988). *J. Biol. Chem.* **262,** 1111–1114.
94. Downward, J., Waterfield, M. D., and Parker, P. J. (1985). *J. Biol. Chem.* **260,** 14538–14546.
95. Werth, D. K., Niedel, J. E., and Pastan, I. (1983). *J. Biol. Chem.* **258,** 11423–11426.
96. Pendergast, A. M., Traugh, J. A., and Witte, O. N. (1987). *Mol. Cell. Biol.* **7,** 4280–4289.
97. Ogita, K., Kikkawa, M. S., Shearman, K., Ase, K., Berry, N., Kishimoto, A., and Nishizuka, Y. (1989). *Adv. Prostaglandin, Thromboxane Leukotriene Res.* **19,** 49–56.
98. Frantz, C. N., Stiles, C. D., and Scher, C. D. (1979). *J. Cell. Physiol.* **100,** 413–424.
99. Rozengurt, E., Rodriguez-Pena, A., Coombs, M., and Sinnett-Smith, J. (1984). *Proc. Natl. Acad. Sci. U.S.A.* **81,** 5748–5752.
100. Suzuki-Sekimori, R., Matuoka, K., Nagai, Y., and Takenawa, T. (1989) *J. Cell. Physiol.* **140,** 432–438.
101. Guy, G. R., Gordon, J., Michell, R. H., and Brown, G. (1985). *Biochem. Biophys. Res. Commun.* **131,** 484–491.
102. Eldar, H., Zisman, Y., Ullrich, A., and Livneh, E. (1990). *J. Biol. Chem.* **265,** 13290–13296.
103. Smith, M. R., Ryu, S.-H., Suh, P.-G., Rhee, S. G., and Hsiang, H.-F. (1989). *Proc. Natl. Acad. Sci. U.S.A.* **86,** 3659–3663.
104. Margolis, B., Zilberstein, A., Franks, C., Felder, S., Kremer, S., Ullrich, A., Rhee, S. G., Skorecki, K., and Schlessinger, J. (1990). *Science* **248,** 607–610.
105. Cuadrado, A., and Molloy, C. J. (1990). *Mol. Cell. Biol.* **10,** 6069–6072.
106. Matuoka, K., Fukami, F., Nakanishi, O., Kawai, S., and Takenawa, T. (1988). *Science* **239,** 640–643.
107. Smith, M.R., Liu, Y.-L., Kim, H., Rhee, S. G., and Kung, H.-F. (1990). *Science* **247,** 1074–1077.
108. Escobedo, J. A., and Williams, L. T. (1988). *Nature* **335,** 85–87.
109. Hill, T. D., Dean, N. M., Mordan, L. J., Lau, A. F., Kanemitsu, M. Y., and Boynton, A. L. (1990). *Science* **248,** 1660–1663.
110. Coughlin, S. R., Lee, W. M. F., Williams, P. W., Giels, G. M., and Williams, L. T. (1985). *Cell* **43,** 243–251.
111. Larrodera, P., Cornet, M. E., Diaz-Meco, M. T., Lopez-Barahona, M., Diaz-Laviada, I., Guddal, P. H., Johansen, T., and Moscat, J. (1990). *Cell* **61,** 1113–1120.

112. Kaplan, D. R., Whitman, M., Schaffhausen, B., Pallas, D. C., White, M., Cantley, L., and Roberts, T. M. (1987). *Cell* **57,** 1021–1029.
113. Coughlin, S. R., Escobedo, J. A., and Williams, L. T. (1989). *Science* **243,** 1191–1194.
114. Morrison, D. K., Kaplan, D. R., Rapp, U., and Roberts, T. M. (1988). *Proc. Natl. Acad. Sci. U.S.A.* (1988). **85,** 8855–8859.
115. Morrison, D. K., Kaplan, D. R., Escobedo, J. A., Rapp, U. R., Roberts, T. M., and Williams, L. T. (1989). *Cell* **58,** 649–657.
116. Molloy, C. J., Bottaro, D. P., Fleming, T. P., Marshall, M. S., Gibbs, J. B., and Aaronson, S. A. (1989). *Nature* **342,** 711–714.
117. Kaplan, D. R., Morrison, D. K., Wong, G., McCormick, F., and Williams, L. T. (1990). *Cell* **61,** 125–133.
118. Ralston, R., and Bishop, J. M. (1985). *Proc. Natl. Acad. Sci. U.S.A.* **82,** 7845–7849.
119. Gould, K. L., and Hunter, T. (1988). *Mol. Cell. Biol.* **8,** 3345–3356.
120. Kypta, R. M., Goldberg, Y., Ulug, E. T., and Courtneidge, S. A. (1990). *Cell* **62,** 481–492.
121. Bjorge, J. D., Chan, T.-O., Antczak, M., Kung, H.-J., and Fujita, D. J. (1990). *Proc. Natl. Acad. Sci. U.S.A.* **87,** 3816–3820.
122. App, H., Hazan, R., Zilberstein, A., Ullrich, A., Schlessinger, J., and Rapp, U. (1991). *Mol. Cell. Biol.* **11,** 913–919.
123. Ellis, C., Moran, M., McCormick, F., and Pawson, T. (1990). *Nature* **343,** 377–380.
124. Vogel, U. S., Dixon, R. A. F., Schaber, M. D., Diehl, R. E., Marshall, M. S., Scolnick, E. M., Sigal, I. S., and Gibbs, J. B. (1988). *Nature* **335,** 90–93.
125. Anderson, D., Koch, C. A., Grey, L., Ellis, C., Moran, M., and Pawson. T. (1990). *Science* **250,** 979–982.
126. Kazlauskas, A., and Cooper, J. A. (1989). *Cell* **58,** 1121–1133.
127. Kazlauskas, A., Ellis, C., Pawson, T., and Cooper, J. A. (1990). *Science* **247,** 1578–1581.
128. Bonner, T. I., Oppermann, H., Seeburg, P., Kerby, S. B., Gunnell, M. A., Young, A. C., and Rapp, U. R. (1986). *Nucleic Acids Res.* **14,** 1009–1015.
129. Whitman, M., Downes, C. P., Keeler, M., Keller, T., and Cantley, L. (1988). *Nature* **332,** 644–646.
130. Auger, K. R., Serunian, L. A., Soltoff, S. P., Libby, P., and Cantley, L. C. (1989). *Cell* **57,** 167–175.
131. Carpenter, C. L., Duckworth, B. C., Auger, K. R., Cohen, B., Schaffhausen, B. S., and Cantley, L. C. (1990). *J. Biol. Chem.* **265,** 19704–19711.
132. Whitman, M., Kaplan, D. R., Schaffhausen, B., Cantley, L., and Roberts, T. M. (1985). *Nature* **315,** 239–242.
133. Courtneidge, S. A., and Heber, A. (1987). *Cell* **50,** 1031–1037.
134. Heidaran, M. A., Pierce, J. H., Lombardi, D., Ruggiero, M., Gutkind, J. S., Matsui, T., and Aaronson, S. A. (1991). *Mol. Cell. Biol.* **11,** 134–142.
135. Blackshear, P. J., Haupt, D. M., App, H., and Rapp, U. R. (1990). *J. Biol. Chem.* **265,** 12131–12134.
136. Kovacina, K. S., Yonezawa, K., Brautigan, D. L., Tonks, N. K., Rapp, U. R., and Roth, R. A. (1990). *J. Biol. Chem.* **265,** 12115–12118.
137. Jamal, S., and Ziff, E. (1990). *Nature,* **344,** 463–466.
138. Kaibuchi, K., Fukumoto, Y., Oku, N., Hori, Y., Yamamoto, K., Toyoshima, K., and Takai, Y. (1989). *J. Biol. Chem.* **264,** 20855–20858.
139. Wasylyk, C., Wasylyk, B., Heidecker, G., Huleilel, M., and Rapp, U. R. (1989). *Mol. Cell. Biol.* **9,** 2247–2250.
140. Smith, M. R., Heidecker, G., Rapp, U. L., and Kung, H.-F. (1990). *Mol. Cell. Biol.* **10,** 3828–3833.
141. Kolch, W., Heidecker, G., Lloyd, P., and Rapp, U. R. (1991). *Nature* **349,** 426–428.

142. Trahey, M., and McCormick, F. (1987). *Science* **238,** 542–545.
143. McCormick, F. (1989). *Cell* **56,** 5–8.
144. Mulcahy, L. S., Smith, M. R., and Stacey, D. W. (1985). *Nature* **313,** 241–243.
145. Satoh, T., Endo, M., Nakafuku, M., Nakamura, S., and Kaziro, Y. (1990). *Proc. Natl. Acad. Sci. U.S.A.* **87,** 5993–5997.
146. Tsai, M.-H., Yu, C.-L., and Stacey, D. W. (1990). *Science* **250,** 982–985.
147. Hochhauser, S. J., Stein, J. L., and Stein, G. S. (1981). *Int. Rev. Cytol.* **71,** 95–243.
148. Denhardt, D. T., Edwards, D. R., and Parfett, C. L. J. (1986). *Biochem. Biophys. Acta* **865,** 83–125.
149. Pardee, A. B. (1989). *Science* **246,** 603–608.
150. Travali, S., Koniecki, J., Petralia, S., and Baserga, R. (1990). *FASEB J.* **4,** 3209–3214.
151. Williams, J. G., and Penman, S. (1975). *Cell* **6,** 197–206.
152. Prescott, D. M. (1987). *Int. Rev. Cytol.* **100,** 93–128.
153. Owen, T. A., Carter, R., Whitman, M. M., Soprano, D. R., and Soprano, K. J. (1990). *J. Cell. Physiol.* **142,** 137–148.
154. Pledger, W. J., Stiles, C. D., Antoniades, H. N., and Scher, C. D. (1978). *Proc. Natl. Acad. Sci. U.S.A.* **75,** 2839–2843.
155. Leof, E. B., Wharton, W. O'Keefe, E., and Pledger, W. J. (1982). *J. Cell. Biochem.* **19,** 93–103.
156. Stiles, C. D., Isberg, R. R., Pledger, W. J., Antoniades, H. N., and Scher, C. D. (1979). *J. Cell. Physiol.* **99,** 395–406.
157. Nikaido, T., Bradley, D. W., and Pardee, A. B. (1991). *Exp. Cell Res.* **192,** 102–109.
158. Lau, L. F., and Nathans, D. (1987). *Proc. Natl. Acad. Sci. U.S.A.* **84,** 1182–1186.
159. Morgan, C. J., and Pledger, W. J. (1989). *J. Cell. Physiol.* **141,** 535–542.
160. Persson, H., Gray, H. E., and Godeau, F. (1986). *Mol. Cell. Biol.* **5,** 2903–2912.
161. Olson, J. E., and Pledger, W. J. (1991). Manuscript in preparation.
162. Almendral, J. M., Sommer, D., Macdonald-Bravo, H., Burckhardt, J., Perera, J., and Bravo, R. (1988). *Mol. Cell. Biol.* **8**(5), 2140–2148.
163. Kruijer, W., Cooper, J. A., Hunter, T., and Verma, I. M. (1984). *Nature* **312,** 711–715.
164. Rahm, M., Hultgardh-Nilsson, A., Jiang, W., Sejersen, T., and Ringertz, N. R. (1990). *J. Cell. Physiol.* **143,** 475–482.
165. Setoyama, C., Frunzio, R., Liau, G., Mudryj, M., and de Crombrugghe, B. (1986). *Proc. Natl. Acad. Sci. U.S.A.* **83,** 3213–3217.
166. Nakabeppu, Y., Ryder, K., and Nathans, D. (1988). *Cell* **55,** 907–915.
167. Ryder, K., and Nathans, D. (1988). *Proc. Natl. Acad. Sci. U.S.A.* **85,** 8464–8467.
168. Hendrickson, S. L., Cochran, B. H., Reffel, A. C., and Stiles, C. D. (1985). In "Mediators in Cell Growth and Differentiation" (R. J. Ford, and A. L. Maizel, eds.), pp. 71–85, Raven Press, New York.
169. Cochran, B. H., Zullo, J. Verma, I. M., and Stiles, C. D. (1984). *Science* **226,** 1080–1082.
170. Zumstein, P., and Stiles, C. D. (1987). *J. Biol. Chem.* **262**(23), 11252–11260.
171. Campisi, J., Gray, H. E., Pardee, A. B., Dean, M., and Sonenshein, G. E. (1984). *Cell* **36,** 241–247.
172. Parfett, C. L. J., Hamilton, R. T., Howell, B. W., Edwards, D. R., Nilsen-Hamilton, M., and Denhardt, D. T. (1985). *Mol. Cell. Biol.* **5,** 3289–3292.
173. Pledger, W. J., Hart, C. A., Locatell, K. L., and Scher, C. D. (1981). *Proc. Natl. Acad. Sci. U.S.A.* **78,** 4481–4485.
174. Scher, C. D., Dick, R. L., Whipple, A. P., and Locatell, K. L. (1983). *Mol. Cell. Biol.* **3,** 70–81.
175. Ferrari, S., Calabretta, B., deRiel, J. K., Battini, R., Ghezzo, F., Lauret, E., Griffin, C., Emanuel, B. S., Gurrieri, F., and Baserga, R. (1987). *J. Biol. Chem.* **262**(17), 8325–8332.
176. Stein, G. S., Stein, J. L., Baumbach, L., Leza, A., Lichtler, A., Marashi, F., Plumb, M., Rickles, R., Sierra, F., and Van Dyke, T. (1982). *Ann. N. Y. Acad. Sci.* **397,** 148–167.
177. Plumb, M., Stein, J. L., and Stein, G. S. (1983). *Nucleic Acids Res.* **11,** 2391–2410.

178. Bravo, R. (1986). *Exp. Cell Res.* **163,** 287–293.
179. Chang, C. D., Ottavio, L., Travali, S., Lipson, K. E., and Baserga, R. (1990). *Mol. Cell. Biol.* **10,** 3289–3296.
180. Gewirtz, A. M., Anfossi, G., Venturelli, D., Valpreda, S., Sims, R., and Calabretta, B. (1989). *Science* **24,** 180–183.
181. Venturelli, D., Mariano, T., Szczylik, C., Valtieri, M., Lange, B., Crist, W., Link, M., and Calabretta, B. (1990). *Cancer Res.* **50,** 7371–7375.
182. Thompson, C. B., Challoner, P. B., Neiman, P. E., and Groudine, M. (1986). *Nature* **319,** 374–380.
183. Bender, T. P., and Kuehl, W. M. (1986). *Proc. Natl. Acad. Sci. U.S.A.* **83,** 3204–3208.
184. Mushinski, J. F., Potter, M., Bauer, S. R., and Reddy, E. P. (1983). *Science* **220,** 795–798.
185. Travali, S., Reiss, K., Ferber, A., Petralia, S., Mercer, W. E., Calabretta, B., and Baserga, R. (1991). *Mol. Cell. Biol.* **11**(2), 731–736.
186. Nurse, P. (1990). *Nature* **344,** 503–508.
187. Blow, J. J., and Nurse, P. (1990). *Cell* **62,** 855–862.
188. D'Urso, G., Marraccino, R. L., Marshak, D. R., and Roberts, J. M. (1990). *Science* **250,** 786–791.
189. Hadwiger, J. A., Wittenberg, C., Richardson, H. E., de Barros Lopes, M., and Reed, S. I. (1989). *Proc. Natl. Acad. Sci. U.S.A.* **86,** 6255–6259.
190. Richardson, H. E., Wittenberg, C., Cross, F., and Reed, S. I. (1989). *Cell* **59,** 1127–1133.
191. Wittenberg, C., Sugimoto, K., and Reed, S. (1990). *Cell* **62,** 225–237.
192. Furukawa, Y., Piwnica-Worms, H., Ernst, T. J., Kanakura, Y., and Griffin, J. D. (1990). *Science* **250,** 805–808.
193. Kohase, M., May, L. T., Tamm, I., Vilcek, J., and Sehgal, P. B. (1987). *Mol. Cell. Biol.* **7,** 273–280.
194. Kohase, M., Hendriksen-DeStefano, D., May, L., Vilcek, J., and Sehgal, P. B. (1986). *Cell* **45,** 659–666.
195. Finlay, C. A., Hinds, P. W., and Levine, A. J. (1989). *Cell* **57,** 1083–1093.

2

Nuclear Protooncogenes and Growth Regulation

BRUNO CALABRETTA AND DONATELLA VENTURELLI

Department of Microbiology and Immunology
and Jefferson Cancer Institute
Thomas Jefferson Medical College
Philadelphia, Pennsylvania 19107

I. INTRODUCTION

During the past few years, numerous homologs of retroviral transforming genes have been isolated and structurally characterized (Bishop, 1983; Weinberg,

MOLECULAR AND CELLULAR APPROACHES TO THE CONTROL OF PROLIFERATION AND DIFFERENTIATION

1985; Varmus, 1987). In most cases, the function of the encoded proteins is unknown, although there are exceptions, notably the protooncogene c-*sis,* which encodes the β-chain of platelet-derived growth factor (PDGF) (Waterfield *et al.,* 1983); c-*fms,* which encodes the macrophage colony stimulating factor (M-CSF) receptor (Sherr *et al.,* 1985); c-*erbB,* which encodes a truncated version of the epidermal growth factor (EGF) receptor (Downward *et al.,* 1986); and c-*kit,* which encodes the receptor for a newly discovered growth factor that stimulates the proliferation of early erythroid progenitors and mast cells (Zsebo *et al.,* 1990; Huang *et al.,* 1990; Anderson *et al.,* 1990). Protooncogenes are broadly subdivided into two major groups: cytoplasmic and nuclear. This distinction is obviously based on the cellular localization of the encoded proteins, but acquired a broader meaning in relationship to the model of tumorigenic conversion of primary embryo fibroblasts that is based on the cooperation between the cytoplasmic oncogene c-*ras* and the nuclear oncogene c-*myc* (Land *et al.* 1983).

The protooncogenes that encode proteins localized in the nucleus appear to participate in the regulation of the proliferation of mammalian cells, although the biochemical pathways involved are largely unknown. One member of this groups, c-*myc,* offers some insights into the way in which studies in the field of molecular oncogenesis have evolved. Interest in this oncogene was initially stimulated by the discovery of its frequent involvement in the chromosomal translocations of Burkitt's lymphomas (Taub *et al.,* 1982; Dalla-Favera *et al.,* 1983) and the less-frequent findings of gene amplification in tumor cell lines (Collins and Groudine, 1982; Dalla-Favera *et al.,* 1982; Little *et al.,* 1983). Subsequently, the observation that c-*myc* expression is induced by PDGF (Kelly *et al.,* 1983) not only led to the powerful hypothesis that two oncogenes, c-*sis* and c-*myc,* work "in concert," as suggested by Leder and co-workers (Kelly *et al.,* 1983), but also, of greater importance, established a seminal link between protooncogene expression and proliferative response. Innumerable reports have since extended that initial finding to additional oncogenes in many other model systems utilized to study the molecular basis of mammalian cell proliferation. The discovery that the nuclear protooncogene c-*jun* encodes the transcriptional activator AP-1 (Bohmann *et al.,* 1987; Angel *et al.,* 1988) had a similar profound impact, because it gave experimental credence to the notion that oncogenes participate in the regulation of normal cell growth (and in the abnormal growth of neoplastic cells) by affecting the transcriptional apparatus.

In summarizing the relevant literature on the role of nuclear protooncogenes in the process of cell-growth regulation, we attempt in this chapter to correlate the kinetics of nuclear protooncogene expression during the proliferative response with their functional requirements during cell proliferation and with their DNA-binding and transactivating activities.

II. EXPRESSION OF NUCLEAR PROTOONCOGENES DURING PROLIFERATION

A. c-*fos*

Induction of c-*fos* expression is among the earliest events following serum or mitogen stimulation of fibroblasts and T lymphocytes (Greenberg and Ziff, 1984; Kruijer *et al.*, 1984; Muller *et al.*, 1984; Reed *et al.*, 1986). The expression of c-*fos* is transient, and c-*fos* mRNA levels return to baseline within minutes. In continually proliferating cells, c-*fos* mRNA is barely detectable but is still inducible (Bravo *et al.*, 1986). Exposure to cycloheximide overinduces c-*fos* expression (Greenberg and Ziff, 1984; Kruijer *et al.*, 1984; Muller *et al.*, 1984; Reed *et al.*, 1986).

B. c-*jun*

Three cellular homologues of the v-*jun* transforming gene have been isolated (c-*jun, jun-B* and *jun-D*). They are closely related based on sequences in the DNA-binding and activation domains; c-*jun* and *jun-B* are readily induced by growth factors and mitogens (Ryder and Nathans, 1988; Ryder *et al.*, 1988; Ryseck *et al.*, 1989), whereas *jun-D* mRNA levels vary little in response to growth-factor stimulation (Ryder *et al.*, 1989).

C. c-*myc*

Accumulation of c-*myc* mRNA has been observed when resting cells such as quiescent fibroblasts and lymphocytes are stimulated with mitogens or growth factors (Kelly *et al.*, 1983; Reed *et al.*, 1985; Kaczmarek *et al.*, 1985). In each cell type, the pattern of c-*myc* mRNA accumulation follows a similar kinetics: there is an early increase, beginning 1 or 2 hr after the proliferative stimulus, and levels of expression are maximal in mid to late G_0–S transition, declining slightly thereafter. In the subsequent cell cycle, c-*myc* mRNA levels are significantly lower than during the initial G_0–S transition and are invariant throughout the phases of the cell cycle of proliferating cells (Thompson *et al.*, 1985). Expression of c-*myc* is not affected by treatment with cycloheximide, indicating that new protein synthesis is not required (Kelly *et al.*, 1983; Reed *et al.*, 1986).

D. c-*myb*

Expression of c-*myb* is restricted to T lymphocytes and is a relatively late event: induction occurs 36–40 hr after phytohemagglutinin (PHA) stimulation of peripheral blood mononuclear cells (PBMC), and maximal expression coincides with entry into S phase (Torelli *et al.,* 1985; Pauza, 1987). Interleukin 2-(IL-2) directly induces c-*myb* mRNA expression in T lymphocytes expressing the IL-2 receptor (Stern and Smith, 1986). In exponentially growing cells, c-*myb* mRNA levels, unlike those of c-*myc,* are higher during S phase than in the other phases of the cell cycle (Thompson *et al.,* 1986). Exposure to cycloheximide abrogates c-*myb* expression, indicating that protein synthesis is required (Reed *et al.,* 1986).

E. c-*ets*

Recent analysis of the expression of two members of the *ets* gene family, *ets*-1 and *ets*-2, during cell proliferation, revealed high *ets*-1 mRNA levels in resting T lymphocytes that declined after induction of T-cell proliferation, but induction of *ets*-2 mRNA levels by a combination of signals that trigger DNA synthesis in T lymphocytes (Bhat *et al.,* 1990). Like c-*myb* expression, *ets*-2 mRNA induction is prevented by pretreatment with cycloheximide (Bhat *et al.,* 1990).

III. PROPERTIES OF NUCLEAR PROTOONCOGENES

A. DNA-Binding Activity

DNA-binding activity of the proteins encoded by nuclear protooncogenes has been demonstrated in many instances. Four nuclear protooncogenes (c-*jun* family members, c-*myb,* c-*ets* family members and c-*myc*) recognize specific nucleotide core sequences.

c-*jun,* which encodes the transcription activator protein AP-1, has been shown to bind to a specific heptameric consensus sequence TGACTCA (Bohmann *et al.,* 1987; Angel *et al.,* 1988). Its DNA-binding activity resides in the carboxy-terminal region of the protein; Jun-B has extensive amino acid sequence similarity to c-*jun* in the region that encodes the DNA-binding domain and, as expected, binds to the same DNA consensus sequence (Nakageppu *et al.,* 1988); Jun-D, the third number of this family, behaves similarly (Nakageppu *et al.,* 1988). The proteins encoded by c-*ets*-1 and c-*ets*-2 genes bind to a 14-base pair sequence from the oncogene-responsive domain of the polyoma enhancer, in which the ACTTCCT appears to be the essential portion of the domain (Wasylyk *et al.,* 1990). The DNA-binding activity also appears to be localized at the carboxy-terminal region of the c-*ets*–encoded protein (Wasylyk *et al.,* 1990).

c-*myb* encodes a protein that binds to a specific core sequence (py AACG/TG) (Biedenkapp *et al.,* 1988). The DNA-binding activity of c-*myb,* unlike that of the c-*jun* and c-*ets* gene families, is localized in the amino-terminal portion of the

TABLE I

Properties of Nuclear Protooncogenes

Oncogene	DNA-binding site	Trans-activation domain
c-*fos*	No specific DNA-binding site	No direct trans-activating activity
c-*jun*	TGAC/GTCA	Negatively charged helical region, glutamine-rich region, proline-rich region
c-*myc*	CACGTG	
c-*myb*	pyAACG/TG	Negatively charged domain adjacent to DNA-binding domain
c-*ets*	ACTTCCT	Apparently localized in the aminoterminal region portion of the protein

protein (Klempnauer and Sippel, 1987). The c-*fos* product has been shown to bind nonspecifically to DNA (Renz *et al.,* 1987); however, when complexed to c-*jun*–encoded proteins, the c-*fos* product has a marked stimulatory effect on their binding to AP-1 sites (Chiu *et al.,* 1988; Halazonetis *et al.,* 1988). It has been known for a long time that the human c-*myc* protein is a DNA-binding protein exhibiting a high nonspecific activity for double-stranded DNA (Persson *et al.,* 1984; Watt *et al.,* 1985). However, very recently, it has been shown that a purified carboxylterminal fragment of human c-*myc* binds *in vitro* in a sequence-specific manner to the sequence CACGTG (Blackwell *et al.,* 1990) (Table I).

B. Transactivating Activity

Among the nuclear protooncogenes, c-*jun* is the best-known transcription factor and is now recognized as identical to the transcription factor AP-1. The activation domain of c-*jun* is not well characterized, although c-*jun*–encoded proteins appear to contain several motifs with potential activation function adjacent to the DNA-binding domain (Mitchell and Tjian, 1989). These motifs include a negatively charged α-helical region and both a glutamine- and proline-rich stretch (Mitchell and Tjian, 1989).

The transactivating properties of c-*myb* are also under intense investigation; c-*myb* up-regulates the expression of reporter genes linked to *myb*-binding sites (Weston and Bishop, 1989; Sakura *et al.,* 1989) and the cellular gene MIM-1, whose expression is promyelocytic-specific, appears to be directly regulated by c-*myb* and contains *myb*-binding sites in the 5′ flanking region (Ness *et al.,* 1989).

More recently it has also been suggested that c-*ets*-1 and c-*ets*-2 transactivate the expression of reporter genes linked to c-*ets* binding sites; the c-*ets* binding domain is contiguous with the AP-1 binding site in the polyoma (Py) enhancer; this association generates a responsive element that is highly stimulated by the concomitant expression of c-*jun* and c-*ets* (Wasylyk *et al.*, 1990).

The activation domain of c-*ets* appears to be localized in the amino-terminal portion of the protein (Wasylyk *et al.*, 1990). Transactivating activity of c-*fos* and c-*myc* is suggested by the observation that both proteins stimulate the transcription of transfected mouse α-1 collagen and human heat-shock protein 70 genes in transient expression functional assays (Setoyama *et al.*, 1986; Kingston *et al.*, 1984; Kaddurah-Daouk *et al.*, 1987); however, the interpretation of those findings is complicated by the lack of evidence that the oncogene products interact directly with the promoters of the genes whose transcription they appear to stimulate. In addition, in the case of c-*fos*, transactivation may have resulted from the formation of the AP-1 complex.

IV. FUNCTIONAL SIGNIFICANCE OF NUCLEAR PROTOONCOGENE EXPRESSION DURING CELL PROLIFERATION

Two main approaches have been used to investigate the functional significance of nuclear protooncogene expression in cellular proliferation: (1) constitutive expression of nuclear protooncogenes introduced into recipient cells by standard techniques of gene transfer; and (2) inhibition of the function of endogenous nuclear protooncogenes in target cells by introduction of constructs containing inducible promoters that express the antisense transcripts or by exposure to synthetic oligodeoxynucleotides complementary to the protooncogene-encoded mRNA.

A. c-*fos*

Holt *et al.* (1986) first analyzed the functional significance of c-*fos* expression in proliferating NIH 3T3 cells by inducing the production of c-*fos* antisense transcripts from a transfected c-*fos* cDNA driven by the steroid-responsive mouse mammary tumor virus (MMTV) promoter; the production of c-*fos* antisense transcripts blocked the utilization of the endogenous c-*fos* mRNA and was associated with decreased colony formation and inhibition of the proliferation of cells in which the c-*fos* antisense transcriptional unit has been integrated. Using a similar approach, Nishikura and Murray (1987) demonstrated that inhibition of c-*fos* mRNA expression in the presence of c-*fos* antisense transcripts induced by exposure to dexamethasone in mouse 3T3 cells, transfected with an antisense c-*fos* gene, fused with the MMTV promoter, prevented G_0–G_1 transition induced by PDGF, but had no effect on 3T3 cells during exponential growth. These conflicting results raise the question of whether c-*fos* is required only at the G_0–G_1 transition or is also essential in continuously cycling cells. The transient expression of c-*fos* during the G_0–G_1 transition is consistent with a role for c-*fos* in that phase of the cell cycle, whereas in continuously cycling cells, residual levels of c-*fos* expression might be needed for a long-term effect.

B. c-*myc*

Armelin *et al.* (1984) provided the first demonstration that the c-*myc* product functions as an intracellular mediator of the mitogenic response induced by PDGF (Armelin *et al.,* 1984). For optimal growth, BALB/c 3T3 fibroblasts require the sequential addition of PDGF and IGF-1; Armelin *et al.* (1984) showed that PDGF can be replaced by c-*myc* in BALB/c 3T3 lines containing an exogenous c-*myc* cDNA driven by the MMTV promoter. In those experiments, a significant number of BALB/c 3T3 cells underwent DNA synthesis in the absence of PDGF, following activation of the MMTV promoter with hydrocortisone (Armelin *et al.,* 1984). Subsequently Kaczmarek *et al.* (1985) showed that direct microinjection of c-*myc* protein into BALB/c 3T3 cells, plated in PDGF-deprived serum, induced DNA synthesis in these cells. Those experiments clearly demonstrated that c-*myc* acts as a *competence* factor for BALB/c 3T3 cells, since it carried out part of the mitogenic response triggered by PDGF.

The functional significance of c-*myc* expression in PHA-stimulated T lymphocytes was analyzed by inhibiting myc protein synthesis by exposure to c-*myc* antisense oligodeoxynucleotides (Heikkila *et al.,* 1987). This treatment resulted in down-regulation of myc protein synthesis and inhibition of G_1–S transition. Early activation events such as expression of IL-2 receptor and transferrin receptor were not affected.

C. c-*myb*

The nuclear protooncogene c-*myb* is preferentially expressed in hematopoietic cells (Westin *et al.,* 1982; Slamon *et al.,* 1984; Kastan *et al.,* 1989), although c-*myb* transcripts are occasionally found in nonhematopoietic tissues such as small-cell lung cancer (Griffin and Baylin, 1985), teratocarcinoma (Janssen *et al.,* 1986), neuroblastoma cell lines (Thiele *et al.,* 1988), primary colon tumors (Torelli *et al.,* 1987) and colon carcinoma cell lines (Alitalo *et al.,* 1984). In mouse cells, the mechanism that regulates the preferential expression of c-*myb* in hematopoietic tissues is likely to involve a transcription block in the first intron; this block was detected in cells that do not express c-*myb* or during the process of terminal differentiation of c-*myb*–expressing cells (Bender *et al,* 1987; Reddy and Reddy, 1989). In normal T lymphocytes, c-*myb* is expressed at the G_1–S transition (Torelli *et al.,* 1985; Stern and Smith, 1986; Thompson *et al.,* 1986; Pauza, 1987). Down-regulation of c-*myb* protein synthesis by exposure to c-*myb* antisense oligodeoxynucleotides prevented entry of mitogen- or antigen-stimulated T lymphocytes into S phase, and was accompanied by down-regulation of expression of genes directly involved in DNA synthesis, such as DNA polymerase-α and histone H3 (Gewirtz *et al.,* 1989; Venturelli *et al.,* 1990). In fibroblasts, c-*myb* can be expressed if the c-*myb* cDNA is under the control of a constitutive promoter, such as SV40 early promoter, and is linked to the SV40 polyadenylation signal (Clarke *et al.,* 1988; Travali *et al.,* 1991). Constitutive expression of such a construct in BALB/c 3T3 cells allows their proliferation in 1% serum or in a serum-free, PDGF-supplemented medium (Travali *et al.,* 1991), whereas the parental cells require both PDGF and IGF-I for growth. Introduction of constitutively expressed c-*myc* and c-*myb* into BALB/c 3T3 cells completely abrogates the requirement for PDGF and IGF-I, and allows the growth of BALB/c 3T3 cells in serum-free medium (Travali *et al.,* 1991).

V. MOLECULAR MECHANISMS OF PROLIFERATION CONTROL BY NUCLEAR PROTOONCOGENES

A useful model for study of the role of nuclear protooncogenes in cell proliferation is BALB/c 3T3 fibroblasts. These cells are tightly growth regulated since, when quiescent, they enter into S phase upon the sequential addition of PDGF and IGF-I. PDGF renders these cells *competent* to respond to *progression factors* like IGF-I, which in turn are responsible for the entry into S phase and subsequent proliferation (Stiles *et al.,* 1979). One approach to the understanding of the genetic basis for mammalian cell proliferation is to identify genes whose expression is affected, at the transcriptional or posttranscription level, by growth factors, and which are able to effect the proliferative response elicited by a growth factor. Accordingly, the functional relevance of protooncogenes in the proliferative response is best demonstrated by their ability to replace a growth factor; c-*myc* protein appears to replace PDGF, whereas the myb protein, if expressed at sufficient levels in fibroblasts, can abrogate the requirement for IGF-I. Although this model is probably simplistic, since multiple genes might be required for entry into S phase, an understanding of the mechanisms by which c-*myc* and c-*myb* participate in the regulation of the competence and progression programs in the cell cycle could serve as useful model for study of the functional relevance of other genes. One would expect c-*myc* to regulate the genetic events that define the state of competence since it belongs to the large family of *competence genes* inducible with PDGF (Cochran *et al.,* 1983; Lau and Nathans, 1987; Almendral *et al.,* 1988); however, it is not known whether *myc* induces the expression or affects the function of the entire repertoire of competence genes or only of a subset of these genes. Much of the uncertainty derives from the lack of sufficient information on the function of c-*myc;* there is no firm evidence that c-*myc* acts as a transcriptional activator. However, the c-*myc* protein sequence includes two structural motifs—*helix–loop–helix* and a *leucine zipper* found in several proteins known to regulate gene activity (Almendral *et al.,* 1988). It appears that the helix–loop–helix motif interacts *in vitro* in a sequence-specific manner to the sequence CACGTG, suggesting that some of the biological functions of the myc protein are accomplished by sequence-specific DNA bindings (Blackwell *et al.,* 1990).

c-*myb* is not normally expressed in fibroblasts; however, constitutively expressed c-*myb* appears to be directly involved in the regulation of the G_1–S transition in these cells (Travali *et al.,* 1991). There is evidence that c-*myb* binds to a specific DNA sequence, whose core motif is pyAACG/TG, and transactivates the expression of reporter genes linked to c-*myb* DNA-binding consensus sequences, although it is not yet known whether the *in vitro* transactivating activity of c-*myb* is relevant for the *in vivo* role of this gene. In the words of Weston and Bishop, only "the identification of cellular genes whose transcription is directly stimulated by the myb proteins would provide unambiguous evidence of the importance of their transactivation functions."

Since c-*myb* plays an important role in the regulation of G_1–S transition, it is reasonable to ask whether the expression of this gene is functionally linked to that of genes encoding proteins of the DNA-synthesizing machinery that are coordinately regulated at the onset of the S phase.

The expression of DNA polymerase-α is up-regulated in a temperature-sensitive (ts) fibroblast cell line constitutively expressing human myb mRNA driven by the SV40 promoter at the nonpermissive temperature, whereas in the parental cell line DNA polymerase-α mRNA levels are readily down-regulated (Venturelli *et al.,* 1990) (Fig. 1). Further, the mRNA levels of two other G_1–S boundary genes, PCNA and histone H3, are still detectable at the nonpermissive temperature in the cell line constitutively expressing c-*myb,* but disappear in the parental line at the nonpermissive temperature (Travali *et al.,* in press). Proof that c-*myb* directly regulates the expression of DNA polymerase-α and other G_1–S transition genes awaits the demonstration of a direct interaction of myb protein with the G_1–S transtion gene promoters; in the meantime, the results discussed establish that there is a functional link between high levels of *myb* expression and that of DNA polymerase-α and other G_1–S boundary genes directly implicated in DNA synthesis.

VI. REGULATION OF NUCLEAR PROTOONCOGENE EXPRESSION

A. Autoregulation

A common characteristic of c-*myc,* c-*myb* and c-*fos* is the rapid rate of degradation of the encoded proteins (Lusher and Eisenmann, 1988; Klempnauer *et al.,*

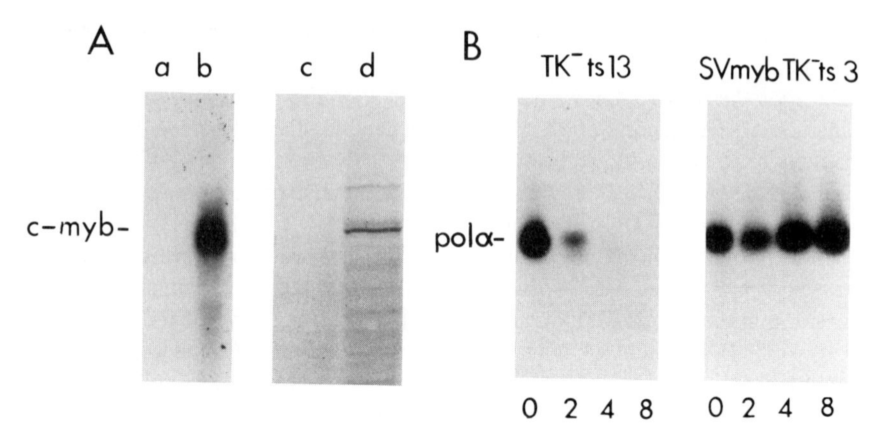

Fig. 1. Expression of c-*myb* and DNA polymerase-α (polα) mRNAs in TK-ts13 cells and human MYB-expressing Tk-ts13 cells at the restrictive temperature of 39.6°C. (A) c-*myb* mRNA and protein levels were determined by Northern and Western blot techniques in Tk-ts13 cells (lanes a and b) and in SVmyb Tk-ts13 (lanes c and d). (B) DNA polymerase-α mRNA levels were determined by reverse transcriptase-polymerase chain reaction technique (Saiki *et al.*, 1988; Rappollee *et al.*, 1989) at each time point (in hr). Total RNA was isolated from 1×10^6 cells, reverse-transcribed and amplified with DNA polymerase-α–specific primers as described (Venturelli *et al.*, 1990). The amplification products were separated through a 2% agarose gel, and transferred to a nitrocellulose filter. The resulting blot was hybridized with a synthetic 50-base oligomer, end-labeled with [α-32]-complementary to the amplified DNA polymerase-α mRNA.

1986; Curran *et al.,* 1984); in contrast, c-*jun*–encoded protein is more stable (Angel *et al.* 1988). Most likely, a key factor in the activity of nuclear protooncogenes is the availability of sufficient amounts of encoded protein at critical points of cell-cycle progression. A sufficient amount of these proteins must be produced to counteract the relative transiency of the stimulatory signal (for instance, the interaction growth factor–receptor), and the probability that there is a narrow window in which the nuclear protooncogene-encoded protein is needed (for instance, c-*myb* might be required only at the G_1–S transition or briefly during the S phase). Accordingly, if nuclear protooncogenes can up-regulate their own expression, the mitogenic signal that they encode would be amplified at the critical points in which protooncogene expression is required.

There is evidence that c-*jun* is positively autoregulated by its product through the interaction with an AP-1 binding site located between nucleotides −72 to −63 of the c-*jun* 5′ flanking region (Angel *et al.*, 1988). The significance of this positive autoregulation during the proliferative response is presently unclear for at least two reasons: (1) the functional significance of c-*jun* expression during mammalian cell proliferation remains unclear; and (2) c-*fos,* which interacts with c-*jun* to form the active AP-1 complex, is known to be negatively autoregulated, likely through the binding of the Jun/AP-1–Fos complex to the AP-1 binding sites contained in the c-*fos* promoter (Sassone-Corsi, *et al.*, 1988); this negative autoregulation of c-*fos* would then antagonize the positive autoregulation of c-*jun* Jun–Fos complex formation. In principle, the Jun–Jun complex, although less active than the Jun–Fos complex, could still interact with AP-1 sites to maintain an autoregulatory loop.

The protooncogene c-*myb* is also positively regulated by its product through the interaction with two closely spaced *myb*-binding sites approximately 600 bp upstream from the cap site (Nicolaides *et al.*, submitted). The functional significance of c-*myb* autoregulation during cellular proliferation is also unclear. A primary mechanism in the regulation of c-*myb* expression involves, at least in mouse cells, a transcriptional block in cells not expressing c-*myb* (Bender *et al.*, 1987; Reddy and Reddy, 1989); the autoregulation of c-*myb* could be part of the tissue-specific mechanisms that allow the generation of c-*myb* transcripts at levels sufficiently high for optimal biological activity, despite the low activity of the c-*myb* promoter.

B. Coordinated Regulation

Studies on the functional significance of nuclear protooncogene expression in proliferating cells suggest that three different nuclear protooncogenes (c-*fos,* c-*myc,* and c-*myb*) regulate similar stages of the proliferative response in some systems but not in others. For instance, inhibition of c-*myc* or c-*myb* protein synthesis by antisense oligodeoxynucleotides in PHA-stimulated PBMC prevents, in each case, G_1–S transition but not early activation events. This observation must be reconciled with the differential regulation of c-*myc* and c-*myb* expression in PHA-stimulated T lymphocytes; induction of c-*myc* expression is an early event during G_0–G_1 transition and does not require protein synthesis, whereas c-*myb* mRNA expression appears at the G_1–S transition and is prevented by treatment with the protein-synthesis inhibitor, cycloheximide. On the other hand, c-*myc* mRNA expression in BALB/c 3T3 fibroblasts is stimulated by PDGF at the G_0–G_1 transition and replaces, in part, the PDGF requirement in these cells, whereas c-*myb* is not expressed at detectable levels in fibroblasts. When constitutively expressed, c-*myb* plays an important role at the G_1–S boundary, since it abrogates the requirement for the progression factor IGF-I. These observations are consistent with at least two apparently conflicting interpretations:

1. in BALB/c 3T3 cells, c-*myc* and c-*myb* (or a c-*myb* equivalent) regulate two distinct stages of the prereplicative phase of the cell cycle (G_0–G_1 and G_1–S transition); these two stages appear to be under different genetic regulation,

since constitutive expression of c-*myc* does not result in entry into S phase of the cell cycle in the absence of progression factors; and

2. in antigen- or mitogen-stimulated human T lymphocytes, c-*myc* and c-*myb* regulate G_1–S progress but not the G_0–G_1 transition.

The expression of c-*myc* is similarly regulated in normal T lymphocytes and in fibroblasts, since it is an early event in both cell types following mitogen or growth-factor stimulation and is independent of protein synthesis. The expression of c-*myb* is a late event during the prereplicative phase of mitogen-stimulated T lymphocytes and requires new protein synthesis; no direct comparison of the regulation of c-*myb* expression in lymphocytes and fibroblasts can be made, since c-*myb* mRNA is not detectable in fibroblasts. However, two *myb*-related genes were recently identified (Nomura *et al.,* 1988), one of which (B-*myb*) is more widely expressed than c-*myb* among different cell types and, in both fibroblasts and T lymphocytes, behaves like a typical G_1–S boundary gene in its increased expression after serum stimulation and its requirement for new protein synthesis (Reiss *et al.,* in press).

The lack of c-*myb* expression in mouse fibroblasts appears to be due to a transcriptional block in the first intron (Bender *et al.,* 1987; Reddy and Reddy, 1989). Thus, it is possible that constitutive expression of c-*myc* in fibroblasts could stimulate c-*myb* transcription; however, the transcription block in the first intron might simply prevent formation of mature myb mRNA transcripts in detectable amounts. This hypothesis would reconcile the observation that exposure of PHA-stimulated T lymphocytes to c-*myc* or c-*myb* antisense oligodeoxynucleotides blocks, in each case, G_1–S transition, but does not affect early T-lymphocyte activation events such as morphological blast transformation and induction of IL-2 receptor expression. It is also possible that c-*myc* and c-*myb* expression are coordinately regulated and functionally linked; indeed, down-regulation of c-*myc* expression after treatment with c-*myc* antisense oligodeoxynucleotides is associated with a decrease in c-*myb* mRNA levels relative to those in c-*myc* sense-treated cultures (Fig. 2). Although inhibition of myc protein synthesis might block T-lymphocyte activation at a stage that precedes induction of c-*myb* mRNA expression, a functional link between c-*myc* and c-*myb* expression would explain the block in G_1–S progress observed in c-*myc* or c-*myb* antisense-treated, PHA-stimulated T lymphocytes.

The hypothesis of a functional link between nuclear protooncogenes can be also invoked to explain the functional significance of c-*fos* expression during the proliferative response; c-*fos* expression is transiently induced at the G_0–G_1 transition in serum-deprived 3T3 fibroblasts. However, c-*fos* antisense expression inhibits growth of exponentially growing 3T3 fibroblasts, suggesting that c-*fos* is functionally linked to nuclear protooncogenes such as c-*myc* and c-*myb* (or a c-*myb* equivalent) that probably exert their regulatory role in cell-cycle progres-

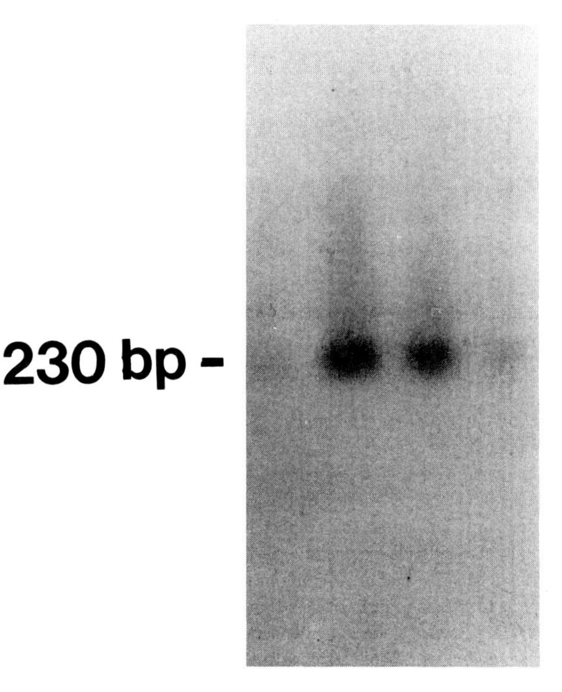

Fig. 2. c-*myb* mRNA expression in PHA-stimulated PBMC, exposed to c-*myc* sense and antisense oligomers. PBMC (2×10^5/ml) were cultured in 24-well plates (COSTAR) in RPMI 1640 supplemented with 10% heat-inactivated fetal bovine serum, 2mM L-glutamine and antibiotics. The cells were preincubated for 6 hr with 40 μg/ml (7 μ*M*) of either c-*myb* sense or c-*myb* antisense oligomer followed by the addition of PHA (10 μg/ml). The oligomers were added to the cultures at time 0 and 24 hr after PHA addition to the final concentration of 10 μg/ml (1.75 μ*M*). Total RNA was extracted from 1.2×10^5 cells for each condition in the presence of 20 μg of *Escherichia coli* ribosomal RNA as described (Rappolee *et al.,* 1988). RNA reverse-transcription, cDNA amplification and hybridization of the amplified products with a *myb* synthetic oligomer corresponding to a 230-nucleotide sequence in the 3′ untranslated region of c-*myb* mRNA (Majello *et al.,* 1986) was performed as described (Venturelli *et al.,* 1990). The sequence of the c-*myc* antisense oligomer is 5′ GAAGCTAACGTGAGGGG and is complementary to 18 nucleotides starting from the second codon of c-*myc* mRNA (Watt *et al.,* 1983). The c-*myb* sense oligomer has the sequence 5′ CCCCTCAACGTTAGCTTC-3′. Lane A, unstimulated PBMC; lane B, PBMC stimulated for 60 hr with PHA (10 μg/ml); lane C, PBMC stimulated for 60 hr with pHA (10 μg/ml) in the presence of c-*myc* sense oligodeoxynucleotides; lane D, PBMC stimulated for 60 hr with PHA (10 μg/ml) in the presence of c-*myc* antisense oligodeoxynucleotides.

sion at different points from those involving c-*fos*. Alternatively, c-*fos* expression might induce long-term genetic changes that allow the proliferative response when the complex and relatively lengthy process of cellular activation is completed (Pardee, 1989).

The best example of coordinated regulation among nuclear protooncogenes is provided by the interaction of members of the *fos* gene family with those of the *jun* gene family (Nakageppu *et al.*, 1988; Chiu *et al.*, 1988; Halazonetis *et al.*, 1988; Rausher *et al.*, 1988; Zerial *et al.*, 1989). The significance of these interactions is at present under intense investigation; the difficulty in understanding the biological relevance of such interactions is attributable, in part, to the lack of sufficient information on the role of each member of the *jun* and *fos* gene families during the proliferative response. Although the regulation of the expression of c-*jun* and c-*fos* gene family members appears to be rather similar, in that they are all growth-related, immediate-early genes (Lau and Nathans, 1987; Almendral *et al.*, 1988), it is likely that they play different roles during the proliferative response. In this regard emerging evidence suggests that Jun-B inhibits the transforming and transactivating activity of c-*jun* (Chiu *et al.*, 1989; Schutte *et al.*, 1989), perhaps by competing with c-*jun* for the formation of c-*jun*–c-*fos* AP-1 complexes needed for transactivation and transformation (Chiu *et al.*, 1988; Schuermann *et al.*, 1989). This raises the possibility, clearly discussed by Minna and coworkers (1989), that the simultaneous expression of different members of the Jun family attenuates the action of c-*jun* by sequestering c-*fos* molecules from the association with c-*jun*. This model also predicts that members of the Jun family can balance the action of c-*jun* and may even act as tumor-suppressor genes in certain circumstances. c-*jun* homodimers bind to the AP-1 binding sites; however, the affinity of this complex is lower than that of the c-*fos*–c-*jun* heterodimer.

A key aspect for understanding the mechanism(s) by which the c-*jun*–c-*fos*–AP-1 complex contributes to the regulation of mammalian cell proliferation is the identification of target genes whose transcriptional activation allows amplification and stabilization of the proliferative response triggered by the transient induction of the c-*fos*–c-*jun* complex. In this regard, an attractive possibility is that the expression of other nuclear protooncogenes is regulated by the binding of the AP-1 complex to AP-1 binding sites. This would determine the formation of a cascade in which the activation of multiple transacting factors transforms a proliferative stimulus into a proliferative response. In our investigation of the functional link between c-*jun* and c-*myb*, we have observed that c-*jun* transactivates the expression of a reporter gene linked to c-*myb* 5′ flanking sequences via an AP-1 binding site at nucleotide position −1331 to −1322 of the human gene (Gualdi *et al.*, submitted). This was observed in fibroblast lines transfected with various segments of the c-*myb* 5′ flanking region linked to the thymidine kinase (TK) cDNA, which served as reporter gene, and a constitutively expressed c-*jun*, cDNA; the expression of the reporter gene was up-regulated in cells constitutively expressing c-*jun* (manuscript in preparation). The significance of these preliminary findings in the context of the proliferative response is still largely speculative. As discussed previously, most nuclear protooncogenes are characterized by the short half-life of the encoded mRNAs and proteins; this feature allows for the rapid extinction of the mitogenic signal encoded by the nuclear protooncogenes, and may prevent the undesired effects resulting from the constitutive transactivation of genes directly involved in DNA synthesis. On the other hand, the transient activation signal cannot be converted into a permanent response unless it is amplified; the existence of functional interactions among nuclear protooncogenes would allow a balanced amplification of the signal. In this scenario, the transient activation of a transactivating factor that operates at a distinct stage of the proliferative response (like the c-*jun* activation at the G_0–G_1 transition) is not lost following the decay of the encoded mRNA and protein, but is transmitted to a second or a third transactivating factor until the actual event of DNA duplication occurs.

VII. SUMMARY

Increasing experimental evidence suggests that nuclear protooncogenes play a key role in the regulation of cell proliferation. The initial observation that the expression of nuclear protooncogenes is regulated by growth factors or by mitogens has provided the basis to determine that the deregulated expression of protooncogenes introduced by transfection in recipient cells alters their growth-factor requirement, and that down-regulation of nuclear protooncogene expression inhibits proliferation or progress throughout specific phases of the cell cycle. Recently, it was shown that some of the proteins encoded by nuclear protooncogenes bind to specific nucleotide sequences and have a transactivating activity; these two features further support the hypothesis that nuclear protooncogenes regulate cell proliferation by activating the expression of other genes that directly or indirectly participate in critical steps of the proliferative response, such as DNA synthesis and cell division. It follows that the identification of target genes regulated by nuclear protooncogenes is a necessary step for a genetic and biochemical analysis of cell proliferation.

ACKNOWLEDGMENT

Supported by NIH grant CA4678. D.V. was supported by National Institutes of Health Training Grant CA 09485. B.C. is a Scholar of the Leukemia Society of America.

REFERENCES

Alitalo, K., Winquist, R., Lin, C. C., de la Chappelle, A., Schwab, M., and Bishop, J. M. (1984). Aberrant expression of an amplified c-*myb* oncogene in two cell lines from a colon carcinoma. *Proc. Natl. Acad. Sci. U.S.A.* **81,** 4534–4538.

Almendral, J. M., Sommer, D., MacDonald-Bravo, H., Burckhardt, J., Perera, J., and Bravo, R. (1988). Complexity of the early genetic response to growth factors in mouse fibroblasts. *Mol. Cell. Biol.* **8,** 2140–2148.

Anderson, D. M., Lyman, S. D., Baird, A., Wignall, J. M., Eisenmann, J., Rauch, C., March, C. J., Boswell, H. S., Gimpal, S. D., Cosman, D., and Williams, D. E. (1990). Molecular cloning of mast cell growth factor, a hematopoietin that is found in both membrane-bound and soluble forms. *Cell* **63,** 235–243.

Angel, P., Allegretto, E. A., Okino, S., Hattori, K., Boyle, W. J., Hunter, T. and Karin, M. (1988). Oncogene *jun* encodes a DNA-binding protein with structural and functional properties of transcriptional factor AP-1. *Nature* **332,** 166–171.

Angel, P., Hattori, K., Smeal, T., and Karin, M. (1988). The *jun* protooncogene is positively autoregulated by its product, Jun/AP-1. *Cell* **55,** 875–885.

Armelin, H. A., Armelin, M. C. S., Kelly, K., Stewart, T., Leder, P., Cochran, B. H., and Stiles, C. D. (1984). Functional role for c-*myc* in mitogenic resonse to platelet-derived growth factor. *Nature* **310,** 655–660.

Bender, T. P., Thompson, C. B., and Kuehl, M. W. (1987). Differential expression of c-*myb* mRNA in murine B-lymphomas by a block of transcription elongation. *Science* **237,** 1473–1476.

Bhat, N. K., Thompson, C. B., Lindsten, T., June, C. H., Fujiwara, S., Koizumi, S., Fisher, R., and Papas, T. S. (1990). Reciprocal expression of human *ets*-1 and *ets*-2 genes during T-cell activation: Regulatory role for the protooncogene *ets*-1. *Proc. Natl. Acad. Sci. U.S.A.* **87,** 3723–3727.

Biedenkapp, H., Borgmeyer, U., Sippel, A. E., and Klempnauer, K-H. (1988). Viral *myb* oncogene encodes a sequence-specific DNA-binding activity. *Nature* **335,** 835–837.

Bishop, J. M. (1983). Cellular oncogenes and retroviruses. *Annu. Rev. Biochem.* **52,** 301–354.

Blackwell, T. K., Kretzner, L., Blackwood, E. M., Eisenmann, R. N., and Weintraub, H. (1990). Sequence-specific DNA binding by the c-*myc* protein. *Science* **250,** 1149–1151.

Bohmann, D., Bos, T. J., Admon, A., Nishimura, T., Vogt, P. K., and Tjian, R. (1987). Human protooncogene c-*jun* encodes a DNA-binding protein with structural and functional properties of transcription factor AP-1. *Science* **238,** 1386–1392.

Bravo, R., Burckhardt, J., Curran, T., and Muller, R. (1986). Expression of c-*fos* in NIH 3T3 cells is very low but inducible throughout the cell cycle. *EMBO J.* **5,** 695–699.

Chiu, R., Angel, P., and Karin, M. (1989). Jun-B differs in its biological properties from, and is a negative regulator of c-*jun. Cell* **59,** 979–986.

Chium, R., Boyle, W. J., Meek, J., Smeal, T., Hunter, T., and Karin, M. (1988). The c-*fos* protein interacts with c-*jun*/AP-1 to stimulate transcription of AP-1–responsive genes. *Cell* **54,** 541–552.

Clarke, M. F., Kukowska-Latallo, J. F., Westin, E., Smith, M., and Prochownik, E. V. (1988). Constitutive expression of a c-*myb* cDNA blocks Friend murine erythroleukemia cell differentiation. *Mol. Cell. Biol.* **8,** 884–892.

Cochran, B. H., Reffel, A. C., and Stiles, C. D. (1983). Molecular cloning of gene sequences regulated by platelet-derived growth factor. *Cell* **33,** 939–947.

Collins, S., and Groudine, M. (1982). Amplification of endogenous *myc*-related DNA sequences in a human myeloid leukemia cell line. *Nature* **298,** 679–681.

Curran, T., Miller, A. D., Zokas, L., and Verma, I. M. (1984). Viral and cellular fos proteins: A comparative analysis. *Cell* **36,** 259–268.

Dalla-Favera, R., Martinotti, S., Gallo, R. C., Erikson, J., and Croce, C. M. (1983). Translocation and rearrangements of the c-*myc* oncogene locus in human undifferentiated B-cell lymphomas. *Science* **219,** 963–967.

Dalla-Favera, R., Wong-Staal, F., and Gallo, R. C. (1982). Oncogene amplification in promyelocytic leukemia cell line HL-60 and primary leukemic cells of the same patient. *Nature* **299,** 61–63.

Downward, J., Yarden, Y., Mayes, E., Sczee, G., Totty, N., Stockwell, P., Ullrich, A., Schlessinger, J., and Waterfield, M. D. (1984). Close similarity of epidermal growth factor receptor and V-*erbB* oncogene protein sequences. *Nature* **307,** 521–527.

Gewirtz, A. M., Anfossi, G., Venturelli, D., Valpreda, D., Sims, R., and Calabretta, B. (1989). G_1–S transition in normal human T lymphocytes requires the nuclear protein encoded by c-*myb*. *Science* **245,** 180–183.

Greenberg, M., and Ziff, E. (1984). Stimulation of 3T3 cells induces transcription of the c-*fos* protooncogene. *Nature* **311,** 433–438.

Griffin, C. A., and Baylin, S. B. (1985). Expression of the c-*myb* oncogene in human small-cell lung carcinoma. *Cancer Res.* **45,** 272–275.

Gualdi, R., et al. (submitted).

Halazonetis, T. D., Georgopoulos, K., Greenberg, M. E., and Leder, P. (1988). c-*jun* dimerizes with itself and with c-*fos,* forming complexes of different DNA-binding affinities. *Cell* **55,** 917–924.

Heikkila, R., Schwab, G., Wickstrom, E., Loke, S. L., Pluznik, D. H., Watt, R., and Neekers L. M. (1987). A c-*myc* antisense oligodeoxynucleotide inhibits entry into s phase but not progress from G_0 to G_1. *Nature* **328,** 445–449.

Holt, J. T., Venkat Gopal, T., Moulton, A. D., and Nienhuis, A. W. (1986). Inducible production of c-*fos* antisense RNA inhibits 3T3 cell proliferation. *Proc. Natl. Acad. Sci. U.S.A.* **83,** 4794–4798.

Huang, E., Nocka, K., Beiler, D. R., Chu, T-Y., Buck, J., Lahm, H-V., Wellner, D., Leder, P., and Besmer, P. (1990). The hematopoietic growth factor KL is encoded by the S1 locus is the ligand of the c-*kit* receptor, the gene product of the BAD FORMAT(@ l)ocus. *Cell* **63,** 225–233.

Janssen, J. W. G., Vernole, P., deBoer, P. A. J., Oosterhuis, J. W., and Collard, J. G. (1986). Sublocalization of c-*myb* to 6q21-q23 by *in situ* hybridization and c-*myb* expression in human teratocarcinoma by 6q rearrangements. **41,** 129–135.

Kaczmarek, L., Calabretta, B., and Baserga, R. (1985). Expression of cell-cycle–dependent genes in phytohemagglutinin-stimulated human lymphocytes. *Proc. Natl. Acad. Sci. U.S.A.* **82,** 5375–5379.

Kaczmarek, L., Hyland, J. K., Watt, R., Rosenberg, M., and Baserga, R. (1985). Microinjected c-*myc* as a competence factor. *Science* **228,** 1313–1315.

Kaddurah-Daouk, R., Greene, J. M., Baldwin, A., Jr., and Kingston, R. (1987). Activation and repression of mammalian gene expression by the c-*myc* protein. *Genes Dev.* **1,** 347–357.

Kastan, M. B., Slamon, D. J., and Civin, C. I. (1989). Expression of protooncogene c-*myb* in normal human hematopoietic cells. *Blood* **73,** 1444–1451.

Kelly, K., Cochran, B. H., Stiles, C. D., and Leder, P. (1983). Cell-specific regulation of the c-*myc* gene by lymphocyte mitogens and platelet-derived growth factor. *Cell* **35,** 603–610.

Kingston, R. E., Baldwin, A. S., and Sharp, P. A. (1984). Regulation of heat-shock protein 70 gene expression by c-*myc*. *Nature* **312,** 280–282.

Klempnauer, K.-H., and Sippel, A. E. (1987). The highly conserved amino-terminal region of the protein encoded by the v-*myb* oncogene functions as DNA-binding domain. *EMBO J.* **6,** 2789–2795.

Klempnauer, K., Symonds, G., Evan, G., and Bishop, J. M. (1986). Identification and characterization of the protein encoded by the human c-*myb* protooncogene. *EMBO J.* **5,** 1903–1911.

Kruijer, W., Cooper, J. A., Hunter, T., and Verma, I. M. (1984). Platelet-derived growth factor induces rapid but transient expression of the c-*fos* gene and protein. *Nature* **312,** 711–716.

Land, H., Parada, L. F., and Weinberg, R. A. (1983). Tumorigenic conversion of primary embryo fibroblasts requires at least two cooperating oncogenes. *Nature* **304,** 602–606.

Lau, L. F., and Nathans, D. (1987). Expression of a set of growth-regulated immediate-early genes in BALB/c 3T3 cells: Coordinate regulation with c-*fos* or c-*myc*. *Proc. Natl. Acad. Sci. U.S.A.* **84,** 1182–1186.

Little, C. D., Nau, M. M., Carney, D. N., Gazdar, A. F., and Minna, J. D. (1983). Amplification and expression of the c-*myc* oncogene in human lung cancer cell lines. *Nature* **306,** 194–196.

Lusher, B., and Eisenman, R. N. (1988). c-*myc* and c-*myb* protein: Degradation effect of metabolic inhibitors and heat shock. *Mol. Cell. Biol.* **8,** 2504–2512.

Majello, B., Kenyon, L. C., and Dalla-Favera, R. (1986). Human c-*myb* protooncogene: Nucleotide sequence of cDNA and organizationof the genomic locus. *Proc. Natl. Acad. Sci. U.S.A.* **83,** 9636–9640.

Mitchell, P. J., and Tjian, R. (1989). Transcriptional regulation in mammalian cells by sequence-specific DNA-binding proteins. *Science* **243,** 371–378.

Muller, R., Bravo, R., Burckhardt, J., and Curran, T. (1984). Induction of c-*fos* gene and protein by growth factors precedes activation of c-*myc*. *Nature* **312,** 711–716.

Murre, C., Schonleber, McCaw, P., and Baltimore, D. (1989). A new DNA-binding and dimerization motif in immunoglobulin enhancer binding, daughterless, MyoD, and myc protein. *Cell.* **56,** 777–783.

Nakageppu, Y., Ryder, K. and Nathans, D. (1988). DNA-binding activities of three murine Jun proteins: Stimulation by Fos. *Cell* **5,** 907–915.

Ness, S. A., Marknell, A., and Graf, T. (1989). The v-*myb* oncogene product binds to and activates the promyelocyte-specific *mim*-1 gene. *Cell* **59,** 1115–1125.

Nicolaides, N., Gualdi, R., Casadevall, C., Manzella, L., and Calabretta, B. (submitted).

Nishikura, K., and Murray, J. N. (1987). Antisense RNA of protooncogene c-*fos* blocks renewed growth of quiescent 3T3 cells. *Mol. Cell. Biol.* **7,** 639–647.

Nomura, N., Takahaski, M., Matsui, M., Ishii, S., Date, T., Sasamato, S., and Ishizaki, R. (1988). Isolation of human cDNA clones of myb-related gene, A-*myb* and B-*myb*. *Nucl. Acids. Res.* **16,** 11075–11089.

Pardee, A. B. (1989). G_1 events and regulation of cell proliferation. *Science* **246,** 603–608.

Pauza, C. D. (1987). Regulation of human T lymphocyte gene expression by interleukin 2: Immediate-response genes include the protooncogene c-*myb*. *Mol. Cell. Biol.* **7,** 342–348.

Persson, H., and Leder, P., (1984). Nuclear localization and DNA-binding properties of a protein expressed by the human c-*myc* oncogene. *Science* **225,** 718–721.

Rappolee, D. A., Mark, D., Benda, M. J., and Werb, Z. (1988). Wound macrophages express TGF-α and other growth factors *in vitro:* Analysis by mRNA phenotyping. *Science* **241,** 708–712.

Rausher, F. J., III, Cohen, D. R., Curran, T., Bos, T. J., Vogt, P. I., Bohmann, D., Tjian, R., and Franza, B. R. (1988). Fos-associated protein p39 is the product of the *jun* protooncogene. *Science* **240,** 1010–1016.

Reddy, C. D., and Reddy, E. P. (1989). Differential binding of nuclear factors to the intron 1 sequences containing the transcriptional pause site correlates with c-*myb* expression. *Proc. Natl. Acad. Sci. U.S.A.* **86,** 7326–7330.

Reed, J. C., Alpers, J. D., Nowell, P. C., and Hoover, R. G. (1986). Sequential expression of protooncogenes during lectin-stimulated mitogenesis of normal human lymphocytes. *Proc. Natl. Acad. Sci. U.S.A.* **83,** 3982–3986.

Reed, J. C., Nowell, P. C., and Hoover, R. G. (1985). Regulation of c-*myc* mRNA levels in normal human lymphocytes by modulators of cell proliferation. *Proc. Natl. Acad. Sci. U.S.A.* **82,** 4221–4224.

Reiss, K., Travali, S., Calabretta, B., and Baserga, R. (1991). Growth-regulated expression of B-*myb* in fibroblasts and hematopoietic cells. *J. Cell. Physiol.* (in press).

Renz, M., Verrier, B., Kurz, C., and Mueller, R.(1987). Chomatin association and DNA-binding properties of the c-*fos* protooncogene product. *Nucleic Acids Res.* **15,** 277–292.

Ryder, K., Lanahan, A., Perez-Albuerve, E., and Nathans D. (1989). Jun-D: A third member of the Jun gene family. *Proc. Natl. Acad. Sci. U.S.A.* **86,** 1500–1503.

Ryder, K., Lau, L. F., and Nathans, D. (1988). A gene activated by growth factors is related to the oncogene v-*jun*. *Proc Natl. Acad. Sci. U.S.A.* **85,** 1487–1491.

Ryder, K., and Nathans D. (1988). Induction of protooncogene c-*jun* by serum growth factors. *Proc. Natl. Acad. Sci. U.S.A.* **85,** 8464–8467.

Ryseck, R.-P., Hirai, S. I., Yaniv, M., and Bravo, R. (1989). Transcriptional activation of c-*jun* during the G_0–G_1 transition in mouse fibroblasts. *Nature* **344,** 535–537.

Saiki, R. K. et al. (198). Primer-directed enzymatic amplification of DNA with a thermostable DNA polymerase. *Science* **239,** 487–491.

Sakura, H., Kaney-Ishi, C., Nagase, T., Nakagoshi, H., Gonda, T. J., and Ishi, S. (1989). Delineation of three functional domains of the transcriptional activator encoded by the c-*myb* protooncogene. *Proc. Natl. Acad. Sci. U.S.A.* **86,**5758–5762.

Sassone-Corsi, P., Sisson, J. C., and Verma, I. M. (1988). Transcriptional autoregulation of the protooncogene *fos. Nature* **334,** 314–319.

Schuermann, M., Neuberg, M., Hunter, J., Jenuween, T., Ryseek, R., Bravo, R., and Muller, R. (1989). The leucine repeat motif in Fos protein mediates complex formation with Jun/AP-1 and is required for transformation. *Cell* **56,** 507–516.

Schutte, J., Viallet, J., Nau, M., Segal, S., Fedorko, J., and Minna, J. (1989). Jun-B inhibits and c-*fos* stimulates the transforming and transactivating activities of c-*jun*. *Cell* **59,** 987–997.

Setoyama, C., Frunzio, R., Liau, G., Mudryo, M., and de Crobrugghe, B. (1986). Transcriptional activation encoded by the v-*fos* gene. *Proc. Natl. Acad. Sci. U.S.A.* **83,** 3213–3217.

Sherr, C. J., Rettenmier, C. W., Sacca, R., Roussel, M. F., Look, A. T., and Stanley, E. R. (1985). The c-*fms* protooncogene product is related to the receptor for the monocyte–phagocyte growth factor, CSF-1. *Cell* **41,** 665–676.

Slamons, D. J., Kernion, J. B., Verma, I. M., and Cline, M. J. (1984). Expression of cellular oncogenes in human malignancies. *Science* **224,** 256–262.

Stern, J. B., and Smith, K. A. (1986). Interleukin-2 induction of T-cell G_1 progression and c-*myb* expression. *Science* **233,** 203–206.

Stiles, C. D., Capone, G. T., Scher, C. D., Antoniades, H. W., Van Wyk, S. J., and Pledger, W. J. (1979). Dual control of cell growth by somatomedins and platelet-derived growth factor. *Proc. Natl. Acad. Sci. U.S.A.* **76,** 1279–1283.

Taub, R., Kirsch, I., Morton, C., Lenoir, G., Swan, D., Tronick, S., Aaronson, S., and Leder, P. (1982). Translocation of the c-*myc* gene into the immunoglobulin heavy chain locus in human Burkitt's lymphoma and murine plasmacytoma cells. *Proc. Natl. Acad. Sci. U.S.A.* **79,** 7837–7841.

Thiele, C. J., Cohen, P. S., and Israel, M. A. (1988). Regulation of c-*myb* expression in human neuroblastoma cells during retinoic acid–induced differentiation. *Mol. Cell. Biol.* **8,** 1677–1683.

Thompson, C. B., Challoner, P. B., Neiman, P. E., and Groudine, M. (1985). Levels of c-*myc* oncogene mRNA are invariant throughout the cell cycle. *Nature (London)* **314,** 363–369.

Thompson, C. B., Challoner, P, B., Neiman, P. E., and Groudine, M. (1986). Expression of the c-*myb* protooncogene during cellular proliferation. *Nature* **319,** 374–380.

Torelli, G., Selleri, L., Donelli, A., Ferrari, S., Emilia, G., Venturelli, D., Moretti, L., and Torelli, U. (1985). Activation of c-*myb* expression by phytohemagglutinin stimulation in normal human T lymphocytes. *Mol. Cell. Biol.* **5,** 2874–2877.

Torelli, G., Venturelli, D., Colo, A., Zanni, C., Selleri, L., Moretti, L., Calabretta, B., Torelli, U. (1987). Expression of c-*myb* protooncogene and other cell cycle–related genes in normal and neoplastic colonic mucosa. *Cancer Res.* **47,** 5266–5269.

Travali, S., Reiss, K., Ferber, A., Petralia, S., Mercer, W. E., Calabretta, B., and Baserga, (1991) R. Constitutively expressed c-*myb* abrogates the requirement for IGF-1 in 3T3 fibroblasts. *Mol. Cell. Biol.* **11,** 731–736.

Travali, S., Ferber, A., Reiss, K., Sell, C., Koniecki, J., Calabretta, B., and Baserga, R. Effect of the myb gene product on the expression of the PCNA gene in fibroblasts. *Oncogene,* (in press).

Varmus, H. (1987). Cellular and viral oncogenes. *In* "The Molecular Basis of Blood Diseases" (P. Stamatoyannpoulos, A. Nienhuis, P. Leder, and A. Majerus eds.), pp. 271–346. Saunders, Philadelphia.

Venturelli, D., Travali, S., and Calabretta, B. (1990). Inhibition of T-cell proliferation by a *myb* antisense oligomer is accompanied by selective down-regulation of DNA polymerase-α expression. *Proc. Natl. Acad. Sci. U.S.A.* **87,** 5963–5967.

Wasylyk, B., Wasylyk, C., Flores, P., Begue, A., Leprince, D., and Stehelin, D. (1990). The c-*ets* protooncogenes encode transcription factors that cooperate with c-*fos* and c-*jun* for transcriptional activation. *Nature* **346,** 191–193.

Waterfield, M. D., Scrace, G. T., Whittle, N., Stroobant, P., Johnson, A., Wasteson, A., Westermark, B., Heldin, C.-H., Huang, J. S., and Deuel, T. (1983). Platelet-derived growth factor is structurally related to the putative transforming protein p28^{sis} of simian sarcoma virus. *Nature* **304,** 35–39.

Watt, R. A., Shatzmann, A. R., and Rosenberg, M. (1985). Expression and characterization of the human c-*myc* DNA-binding protein. *Mol. Cell. Biol.* **5,** 448–456.

Watt, R. A., Stanton, L. W., Marcu, K. B., Gallo, R. C., Croce, C. M., and Rovera, G. (1983). Nucleotide sequence of cloned cDNA of the human c-*myc* oncogene. *Nature* **303,** 725–728.

Weinberg, R. A. (1985). The action of oncogenes in the cytyoplasm and nucleus. *Science* **230,** 770–776.

Westin, E. H., Gallo, R. C., Arya, S. K., Eva, A., Souza, L. M., Baluda, M. A., Aaronson, S. A., and Wong-Staal, F. (1982). Differential expression of the AMV gene in human hematopoietic cells. *Proc. Natl. Acad. Sci. U.S.A.* **79,** 2194–2198.

Weston, K., and Bishop, J. M. (1989). Transcriptional activation by the v-*myb* oncogene and its cellular progenitor c-*myb. Cell* **58,** 85–93.

Zerial, M., Toschi, L., Ryseck, R.-P., Schiermann, M., Muller, R., and Bravo, R. (1989). The product of a novel growth factor–activated gene, *fos* B, interacts with Jun proteins enhancing their DNA-binding activity. *EMBO J.* **8,** 805–813.

Zsebo, K. M., Williams, D. A., Geissler, E. N., Broudy, V. C., Martin, F. H., Atkins, H. L., Hsu, R-Y., Birkett, N. C., Okino, K. H., Murdock, D. C., Jacobsen, F. W., Langley, K. E., Smith, K. A., Takeishi, T., Cahanach, B. M., Galli, S. J., and Suggs, S. V. (1990). Stem-cell factor is encoded at the S1 locus of the mouse and is the ligand for the c-*kit* tyrosine kinase receptor. *Cell* **63,** 213–224.

3

The Control of Mitotic Division

POTU N. RAO

Department of Medical Oncology
The University of Texas M. D. Anderson Cancer Center
Houston, Texas 77030

I. INTRODUCTION

Eukaryotic cells go through a sequence of events during their proliferation cycle, which can be divided into four phases, *viz.*, G_1, the pre-DNA synthesis period; S, the period of DNA replication; G_2, the post-DNA synthesis period; and M, the mitotic period. Lack of nutrients can drive the cells into a lag phase or G_0 phase (Fig. 1). In eukaryotic cells this cell cycle is regulated at three different transition points, from G_0 to G_1, G_1 to S, and G_2 to M. In this chapter we focus our

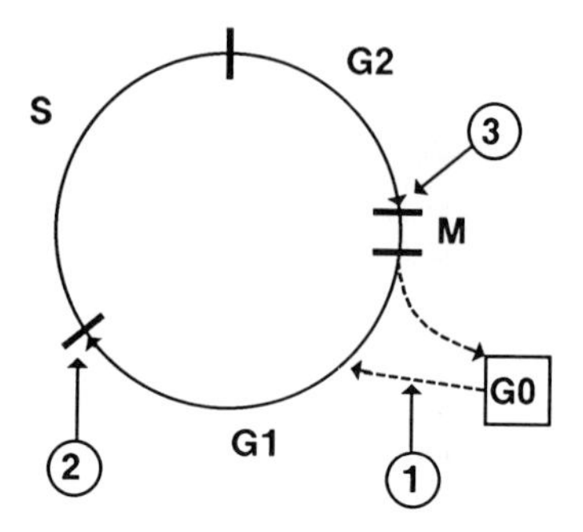

Fig. 1. Schematic presentation of the eukaryotic cell cycle. G_0, lag or noncycling phase; G_1, pre-DNA synthetic period; S, period of DNA synthesis; G_2, post-DNA synthetic period; M, mitosis. Numbered arrows indicate the points of commitment from one phase to the next: 1, G_0 to G_1; 2, G_1 to S phase; and 3, G_2 to M phase transition.

attention exclusively on the mechanisms that regulate the transition from G_2 to M phase.

Although the cytological aspects of mitosis have been known for over a century, only now are we grasping the complexities of its regulation at the molecular level. The past 2 or 3 years particularly have been critical, bringing some significant breakthroughs in this field. In general, four different approaches have been taken in the study of events leading to mitosis. They are (1) the cell biological approach, (2) the genetic approach, (3) the biochemical approach, and (4) the immunochemical approach. Each approach will be discussed in detail in this chapter.

II. THE CELL BIOLOGICAL APPROACH TO THE STUDY OF MITOSIS

A. Mitotic Regulation in Multinucleate Cells

Earlier studies used either naturally occurring (e.g., *Physarum polycephalum*) or experimentally produced (by the use of Sendai virus) multinucleate cells to study the effect of cytoplasmic factors on the initiation of mitosis. In experiments with *Pelomyxia*, Daniels (1951, 1952a,b, 1954, 1958) showed that when two plasmodia in different phases of the life cycle were fused, the subsequent mitosis was synchronous. These studies suggested that substances present in the plasmodium more advanced in the cell cycle, i.e., closer to mitosis, could speed up the initiation of mitosis in the lagging nuclei. The results also indicated that these substances are diffusable, becoming equilibrated throughout the plasmodium in order to bring about synchronous mitosis. Similar results were obtained with *Physarum polycephalum.* Following fusion of microplasmodia, mitotic synchrony was quickly established (Guttes, *et al.*, 1959, 1969). The coalescence of plasmodia of different ages resulted in the formation of abnormal nuclei, which were perhaps the result of premature stimulation of nuclear division (Rusch *et al.*, 1966). The studies of Rusch *et al.* (1966) on *Physarum* in which various sizes of plasmodia from different stages in the cell cycle were fused, indicated that mitotic inducers synthesized throughout interphase, reaching critical levels just before mitosis. These investigators also demonstrated that DNA and RNA synthesis were early steps in the *synthesis of mitotic stimuli* during the G_2 phase. Studies by Commins *et al.* (1965, 1966) showed that the mitotic stimuli were at least in part proteins, since the addition of actidione, a specific inhibitor of protein synthesis, at any stage during interphase up to 20 min before metaphase blocked the initiation of mitosis.

B. Mitotic Factors Induce Premature Chromosome Condensation in Interphase Nuclei

Clear-cut evidence for the presence of mitotic inducers in mitotic cells came from the cell-fusion studies of Rao and Johnson (1970) and Johnson and Rao (1970). The fusion of a mitotic cell with an interphase cell results in chromosome condensation and the dissolution of the nuclear membrane in the interphase nucleus. This phenomenon is called premature chromosome condensation (PCC), whose products are prematurely condensed chromosomes (PCCs) (Johnson and Rao, 1970). The induction of PCC may be regarded as equivalent to the initiation of mitosis in many respects. The inducers of mitosis, which are present in eukaryotic cells during mitosis, can induce PCC in a variety of cell types without exhibiting any species specificity (Johnson *et al.*, 1970). It has become evident that during the induction of PCC, prelabeled proteins from the mitotic cell become associated with the prematurely condensed chromosomes of the unlabeled interphase cell (Rao and Johnson, 1974).

C. Mitotic Factors Induce Meiosis in *Xenopus* Oocytes

Studies on *Xenopus* oocyte maturation proved to be valuable in the study of mitotic factors. Immature oocytes of *X. laevis* are physiologically arrested at the G_2-prophase transition of the first meiotic division. A brief exposure of the immature oocytes to progesterone leads to oocyte maturation, i.e., meiosis, as indicated by germinal vesicle breakdown (GVBD), chromosome condensation, and spindle formation in 6 to 8 hr. Masui and Markert (1971) have shown that the cytoplasm of mature oocytes contains a factor that can induce maturation within 2 to 3 hr after microinjection into fully grown immature oocytes; this maturation is independent of new protein synthesis (Drury and Schroderet-Slatkine, 1975; Wasserman and Masui, 1976). This factor was named maturation-promoting factor or MPF. MPF-like activity is also present in early cleavage-stage amphibian embryos; its peak coincides with the time embryonic nuclei enter mitosis (Wasserman and Smith, 1978).

On the basis of these studies, we wanted to determine whether extracts from mitotic HeLa cells could induce meiosis in immature oocytes after microinjection. The results of our experiments clearly indicated that extracts of mitotic HeLa cells can indeed induce GVBD and chromosome condensation in *Xenopus* oocytes, just as mature oocyte extracts can (Sunkara *et al.*, 1979). Subsequently a number of investigators have shown that extracts of mitotic cells from a wide variety of species including yeast can induce GVBD in *Xenopus* oocytes following microinjection (Nelkin *et al.*, 1980; Kishimoto *et al.*, 1982; Weintraub *et al.*, 1982; Halleck *et al.*, 1984). These results proved for the first time that the factors responsible for the initiation of mitosis and meiosis are similar, if not identical. Therefore, the term MPF now generally stands for M-phase (mitosis or meiosis)–promoting factor (Gerhart *et al.*, 1984). The characterization of MPF and its mechanism of activation will be discussed later.

III. THE GENETIC APPROACH

A. The ts Mutants

Diverse biochemical and biological processes must be coordinated to insure the creation of an exact replica of a cell at the end of mitosis. Mitchison (1973) compared the orderly development of morphogenesis in multicellular organisms to the sequence of events leading to cell division during the cell cycle. The common thread that ties these two phenomena is the fact that they are gene regulated. Temperature-sensitive (ts) mutants have become a valuable tool in probing the role of genes and their products in the regulation of cell cycle in prokaryotic and eukaryotic cells. The ts mutants are conditional lethal mutations, i.e., at permissive low temperature, the cells grow and complete the cell cycle normally, but at nonpermissive high temperature, they grow poorly and become arrested at a specific point in the cell cycle, because a specific gene product required for the execution of that step is either missing or defective and nonfunctional. Horowitz and Leubold (1951) isolated ts mutants of *Escherichia coli* K12 and used them to support the one gene–one enzyme hypothesis. Subsequently ts mutants were isolated in animal viruses, yeast, *Drosophila,* and mammalian cells.

B. The Cell-Division Cycle (cdc) Mutants of Yeast and Mammalian Cells

The pioneering studies of Hartwell and his colleagues, who isolated a number of cell-division cycle (cdc) mutants that are temperature-sensitive in the budding yeast *Saccharomyces cerevisiae*, have paved the way for the genetic analysis of cell-cycle regulation in eukaryotic cells (reviewed by Hartwell *et al.*, 1974). Their studies have helped to formulate a model of the cell cycle as a series of events organized in a dependent temporal sequence, each under the control of a specific gene (Hartwell *et al.*, 1973, 1974). The expression of the gene CDC28 is necessary for the *start* function, i.e., G_1 to S-phase transition in *S. cerevisiae* (Hartwell *et al.*, 1974; Hartwell, 1978). Subsequently Reed *et al.* (1985) showed that CDC28 encodes a 34-kDa protein kinase that bears significant homology to the $cdc2^+$ gene product, an important regulator for G_1 to S and G_2 to M transitions in *Schizosaccharomyces pombe.* Recently Reed and Wittenberg (1990) established that the CDC28 protein kinase in *S. cerevisiae* has a mitotic function as well. The fact that it has taken so long to detect the role of CDC28 in the initiation of mitosis could be attributable either to the residual activity of the protein or to a relatively lower requirement for it in the G_2 to M-phase transition at the nonpermissive temperature (Reed and Wittenberg, 1990).

The first ts mutants of mammalian cells were isolated by Naha (1969) from BSC-1 monkey cells and by Thompson *et al.* (1970) from the L line of mouse cells. A number of other ts mutants were isolated from Syrian hamster, Chinese hamster, and mouse cells (see review by Basilico, 1977). Most of these ts mutants fail to synthesize DNA or protein at the nonpermissive temperature. Some of these mutants are arrested either in G_1 or S phase. An interesting ts mutant of baby hamster kidney (BHK) cells, isolated by Nishimoto *et al.* (1978) and designated ts BN2, enters mitosis prematurely without completing DNA replication at the nonpermissive temperature (39.5°C). As a result the chromosomes of the ts BN2 appear fragmented, resembling the prematurely condensed chromosomes (PCCs) of the S phase cells that were fused to mitotic cells (Nishimoto, 1988). Besides nuclear-membrane breakdown and chromosome condensation, other mitosis-specific events, i.e., phosphorylation of histone H3 and appearance of mitosis-specific antigens, appear when ts BN2 cells are shifted to 39.5°C (Aziro *et al.*, 1983). The appearance of mitosis-specific events at 39.5°C is dependent on new protein synthesis (Nishimoto *et al.*, 1981). Since the ts BN2 mutation behaves as a recessive mutation in somatic cell hybrids (Nishimoto and Basilico, 1978), it is presumed that the mutational defect is in the gene that acts as a negative regulator for the initiation of chromosome condensation and nuclear-membrane breakdown, i.e., mitosis (Ohtsubo *et al.*, 1989). This gene, designated RCC1, has been cloned from HeLa cells by DNA-mediated transfer; it complements the ts BN2 mutation (Kai *et al.*, 1986).

Another example of a gene that acts as a negative regulator of mitosis has been reported in *S. cerevisiae* by Weinert and Hartwell (1988). In general, the cells do not enter mitosis until DNA synthesis is completed. Their exposure to chemical or physical agents that damage DNA also delays the initiation of mitosis. This mitotic delay is due to the time required for the cell to repair the DNA damage. Cell-fusion studies revealed significantly greater chromosome damage when cells were forced to enter mitosis prematurely by fusing them with mitotic cells soon after exposure to clastogens, as compared to those that entered mitosis naturally or those in which chromosome condensation was induced after G_2 delay (Hittelman and Rao, 1974a, b; 1975; Rao and Rao, 1976). Therefore, there appears to be a *surveillance mechanism* to check the integrity of the genome and its duplication or repair before the cell enters mitosis, where the duplicated DNA is segregated between the two daughter cells. In *S. cerevisiae* the gene RAD9 seems to perform this function (Weinert and Hartwell, 1988). The product of RAD9, a protein of 1309 amino acids with a M_r 148,412, may bind to specific sequences on the DNA and repress the synthesis of protein kinases required for the initiation of mitosis, or may induce synthesis of proteins that inhibit the activation of mitotic factors (Schiestl *et al.*, 1989).

C. The Network of Genes That Regulate Mitosis in Fission Yeast

A significant breakthrough has been achieved in recent years with regard to our understanding of the genetic regulation of the G_2–M transition by the use of ts mutants of the fission yeast *S. pombe*. The contributions to this field made by the laboratory of Paul Nurse are noteworthy. They isolated a number of cdc mutants of *S. pombe* that helped to dissect the regulatory network of genes controlling the entry of cells into mitosis (Nurse *et al.*, 1976). This network consists of at least four genes encoding for protein kinase homologs that coordinate their functions for the orderly transition of cells from G_2 to mitosis (Lee and Nurse, 1988). The central gene in this network is $cdc2^+$ which is required for the G_1–S as well as the G_2–M transition in *S. pombe* (Nurse and Bisset, 1981). $cdc2^+$ encodes a 34-kDa protein kinase, whose function is essential for entry of cells into mitosis (Beach *et al.*, 1982; Simanis and Nurse, 1986). A homolog of $cdc2^+$, found in human cells (Draetta *et al.*, 1987), can correct the defect in $cdc2^-$ mutants of *S. pombe* (Lee and Nurse, 1987), indicating that this gene, which is essential for cell division, is highly conserved during evolution.

During the G_2–M transition, the $cdc2^+$ gene is under the influence of both positive and negative regulators of the network (Fig. 2). The gene products of nim 1^+ and wee 1^+ are inhibitory, whereas the product of $cdc25^+$ is an activator (Fantes, 1979; Russel and Nurse, 1986; Russel and Nurse, 1987a, b). These regulatory genes can be mutated or overexpressed in *S. pombe.* Since wee 1^+ is an inhibitor of $cdc2^+$, its deletion advances cells into mitosis before they achieve their normal size, thus leading to the formation of small (wee) daughter cells. The wee phenotype can also be obtained by overexpressing the activators $cdc25^+$ and nim 1^+. The $cdc2^+$ also interacts with two other genes, $suc1^+$ (Hayles *et al.*, 1986a,b) and $cdc13^+$(Booher and Beach, 1987). Immunoprecipitation studies indicate a close physical association between the gene products of $cdc2^+$ and $cdc13^+$ on the one hand (Moreno *et al.*, 1989; Booher *et al.*, 1989; Hagan *et al.*, 1988) and between $cdc2^+$and $suc1^+$ on the other (Brizuela *et al.*, 1987). Cyclin, the gene product of $cdc13^+$, is required both before and during mitosis as shown in Fig. 2. $suc1^+$ is an

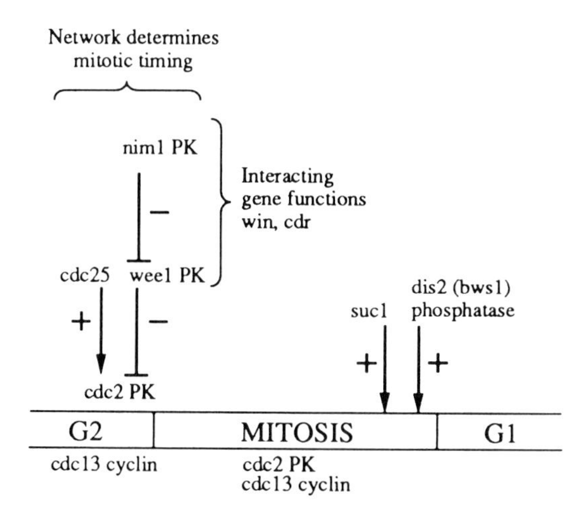

Fig. 2. Regulatory genes in fission yeast. The gene products of cdc2$^+$, cdc25$^+$, wee 1$^+$, and nim 1$^+$ act in a regulatory gene network that determines the cell-cycle timing of mitosis. cdc13$^+$ is required for mitotic onset and together with cdc2$^+$ is also required later during mitosis. The genes suc1$^+$ and dis2$^+$ (allelic with bws$^+$) are probably required toward the end of mitosis. The gene products of cdc2$^+$, wee 1$^+$ and nim 1$^+$ are protein kinases; dis2$^+$, a phosphatase; and cdc13$^+$, a cyclin. (Reprinted with permission from P. Nurse, 1990 and *Nature* **344**: 503–508, copyright © 1990 Macmillan Magazines LTD.)

allele-specific suppressor of cdc2 mutants (Hayles *et al.*, 1986a,b). Overexpression of suc1$^+$ may result in mitotic delay (Hindley *et al.*, 1987; Hayles *et al.*, 1986b), whereas its deletion results in mitotic arrest (Moreno *et al.*, 1989). In *Aspergillus nidulans*, the function of the gene nim A is essential for the entry of cells into mitosis (Osmani *et al.*, 1988). Overexpression of nim A can advance cells into mitosis even when DNA synthesis is inhibited. As is the case with the mitotic regulatory genes in *S. pombe, viz.*, cdc2$^+$, cdc13$^+$, nim 1$^+$, and wee 1$^+$, the nim A gene in *A. nidulans* encodes a potential protein kinase (Osmani *et al.*, 1988). The involvement of a network of genes that encode protein kinase activities suggest the importance of protein phosphorylation in the regulation of the initiation of mitosis in eukaryotic cells. The importance of histone H1 phosphorylation as a mitotic trigger was suggested as early as 1974 by Bradbury *et al.*, (1974a,b) in their studies with *Physarum polycephalum.*

IV. THE BIOCHEMICAL APPROACH

A. Purification of MPF

As indicated earlier, MPF activity is present in both mitotic and meiotic cells. On microinjection into immature oocytes, extracts from M phase cells induce maturation, i.e., GVBD, chromosome condensation, and spindle formation, without exhibiting any species specificity. MPF activity oscillates during the division cycle, with its activity reaching a peak in M phase and precipitously dropping at the end of M phase (Wasserman and Smith, 1978; Sunkara *et al.*, 1979; Doree *et al.*, 1983; Gerhart *et al.*, 1984; Picard *et al.*, 1987). An inactive precursor of MPF present in immature oocytes becomes activated in an autocatalytic manner during oocyte maturation. Autoamplification of MPF has been demonstrated in starfish and amphibian oocytes, wherein microinjection of a small amount of MPF resulted in severalfold increases in MPF activity (Wasserman and Masui, 1975; Doree, 1982; Cyert and Kirschner, 1988). There is a close correlation between MPF activity and protein phosphorylation as a cell enters mitosis (Bradbury *et al.*, 1974a; Davis *et al.*, 1983) or meiosis (Maller *et al.*, 1977; Wu and Gerhart, 1980).

Earlier attempts to purify MPF have met with only limited success (Masui and Markert, 1971; Wasserman and Masui, 1976; Wasserman and Smith, 1978; Drury, 1978; Wu and Gerhart, 1980; Adlakha *et al.*, 1985; Nguyen-Gia *et al.*, 1986). In all these studies MPF activity was measured by injecting a small amount of the crude or partially purified extracts from M-phase cells (i.e., mature oocytes or mitotic cells) into fully grown immature oocytes of *Xenopus* and then scoring the injected oocytes for the frequency of GBVD. The MPF activity could be quantitated in units, where a unit of MPF activity was defined as the amount required to induce GVBD in 50% of the oocytes injected with 40 nl of the extract (Gerhart *et al.*, 1984). Even under the best conditions, no greater than 200-fold purification of MPF has been achieved. The purification of MPF proved to be more difficult than anticipated because it was highly unstable (Adlakha and Rao, 1987).

B. The Basic Components of MPF

1. $p34^{cdc2}$ Protein Kinase

The observation that the introduction of somatic cell nuclei into maturing oocytes leads to condensation of chromosomes of the interphase nucleus (Ziegler and Masui, 1973) has led to the development of an *in vitro* system for monitoring MPF activity in which nuclei from interphase cells are induced to undergo premature chromosome condensation by incubation with cell-free extracts (Lohka and Masui, 1984; Lohka and Maller, 1985; Miake-Lye and Kirschner, 1985; Lohka *et al.*, 1987). Using this MPF-monitoring system, Lohka *et al.* (1988) purified MPF from *Xenopus* oocytes 3000-fold by passing the extracts through a combination of ammonium sulfate precipitation and a battery of six chromatographic procedures. In these purified fractions, the enrichment of two proteins of 45 kDa and 32 kDa correlated with MPF activity. These fractions also exhibited protein kinase activity that phosphorylated *in vitro* the 45-kDa protein as well as other substrates, i.e., histone H1, phosphatase inhibitor 1, and casein (Lohka *et al.*, 1988). The active fractions not only induced chromosome condensation in interphase nuclei *in vitro* but also were capable of inducing germinal-vesicle breakdown and chromosome condensation when injected into immature oocytes of *X. laevis*. The antibody raised against the gene product of cdc2 of *S. pombe* ($p34^{cdc2}$) recognized the 32-kDa protein of *X. laevis*, suggesting homology between these proteins (Gautier *et al.*, 1988). The 13-kDA protein encoded by $suc1^{+}$ ($p13^{suc1}$) in *S. pombe* has been shown to interact with and bind to $p34^{cdc2}$ kinase (Brizuela *et al.*, 1987). Dunphy *et al.* (1988) have shown that the MPF activity of *Xenopus* oocyte extracts is inhibited when they are mixed with $p13^{suc1}$. They also found that a p13 affinity column efficiently retained the *Xenopus* homolog of cdc2 (p32) and another 42-kDa polypeptide. These studies suggest that MPF may consist of a complex of at least two proteins, one of them being $p34^{cdc2}$. Using the phosphorylation of histone H1 as the *in vitro* assay, Labbe *et al.* (1988, 1989) and Arion *et al.* (1988) purified MPF activity from starfish oocytes. In both these studies, levels of a protein of 34 kDa correlated with histone H1 kinase activity and cross-reacted with anti $p34^{cdc2}$ antibody. These studies further confirm that the homologs of $p34^{cdc2}$ of *S. pombe* are also present in both *Xenopus* and starfish and are central to the mitosis-specific histone H1 kinase activity.

2. *Cyclins*

The other component, the 42-kDa protein, that correlated with MPF activity in the purified fractions of *Xenopus* oocyte extracts (Lohka, *et al.*, 1988) is cyclin. Although cyclins are known to be synthesized during the cell cycle and destroyed at the end of mitosis in sea urchin embryos (Evans *et al.*, 1983), their role as regulators of mitosis was not recognized until Swenson *et al.* (1986) showed that injection of mRNA of clam cyclin A into *Xenopus* oocytes induced GVBD and chromosome condensation. These observations were further confirmed when identical results were obtained with mRNA for cyclin from sea urchin eggs (Pines and Hunt, 1987). Minshull *et al.* (1989) have shown that in *Xenopus* oocytes there are mRNAs for two cyclins, which are the major translation products in a cell-free system using the extracts from activated eggs. In this cell-free system, entry into mitosis of the nuclei is blocked when these mRNAs are cleaved with antisense oligonucleotide-directed endogenous RNase H. Nuclei could enter mitosis only when one or the other of the cyclin mRNAs was functional, suggesting that translation of cyclin mRNA is essential for entry into mitosis (Minshull *et al.*, 1989). Murray and Kirschner (1989) developed a cell-free system in which activated *Xenopus* egg extracts undergo multiple cell cycles *in vitro* as monitored by initiation of DNA synthesis (S phase) or mitosis (M phase) in the demembranated sperm nuclei in the presence of an ATP-regenerating system. In this cell-free system, destruction of cyclin mRNA arrests the extracts in interphase, and the addition of exogenous cyclin mRNA drives the extract into M phase. As in clam and sea urchin eggs, cyclin accumulates during the cell cycle in this cell-free system and is destroyed at the end of M phase (Murray and Kirschner, 1989).

Immunoprecipitation of clam oocyte extracts with an anti-$p34^{cdc2}$ polyclonal antibody removed the clam homolog of $p34^{cdc2}$ as well as cyclins A and B (Draetta *et al.*, 1989). In a different experiment, a $p13^{suc1}$-sepharose column known to bind $p34^{cdc2}$ also retained cyclins A and B. These results suggest that the mitosis-specific histone H1 kinase activity, presumed to be the same as MPF (Arion *et al.*, 1988), consists of a complex in which cyclins A and B bind individually, but not jointly, to $p34^{cdc2}$ (Draetta *et al.* 1989).

The universal nature of cyclins as mitotic regulators received even greater acceptance when it was shown that in fission yeast, the cell-cycle gene cdc13 encodes a protein of 56 kDa that exhibits sequence homology with B cyclins (Hagan *et al.*, 1988; Booher and Beach, 1988; Soloman *et al.*, 1988). As in other systems the $p56^{cdc13}$ in *S. pombe* is destroyed at the end of mitosis (Moreno *et al.*, 1989; Booher *et al.*, 1989).

V. THE IMMUNOLOGICAL APPROACH

A. Mitosis-Specific Monoclonal Antibodies

In our laboratory we have taken an immunological approach to the study of mitotic regulation in mammalian cells. Our discovery of the phenomenon of premature chromosome condensation (Johnson and Rao, 1970) made us aware that certain factors in mitotic cells are either absent or inactive in interphase cells. Therefore, we decided to use the extracts of mitotic HeLa cells as the immunogen to produce mouse monoclonal antibodies specific to mitotic cells. As a result we have isolated a panel of 13 mitosis-specific monoclonal antibodies designated MPM-1 to MPM-13. Of this panel, the antibody MPM-2 has been studied extensively because it has some interesting properties. By indirect immunofluorescence MPM-2 stains mitotic cells but not interphase cells without exhibiting any species specificity (Davis *et al.*, 1983). On immunoblots it recognizes a family of phosphopeptides in the extracts of mitotic cells but not in those of interphase cells. Treatment of the extracts with alkaline phosphatase removed the antigenicity to MPM-2. Thus, this antibody recognizes a phosphorylated epitope common to this family of proteins that are phosphorylated as the cell enters mitosis and dephosphorylated as it completes mitosis and enters G_1. These studies have provided further evidence in support of an earlier suggestion by Bradbury *et al.* (1974a,b) that protein phosphorylation could be the trigger for the initiation of mitosis. Furthermore, the relative antigenic reactivity of HeLa cells to MPM-2 during the cell cycle (Davis *et al.*, 1983) shows a perfect correlation with the MPF activity of extracts from HeLa cells collected at different points in the cell cycle (Sunkara *et al.*, 1979) (Fig. 3).

B. Relationship between the Expression of MPM-2 Antigens and the Appearance of MPF Activity in *Xenopus* Oocytes

To determine precisely the role of MPM-2 antigens in the activation of MPF, we monitored the appearance of MPM-2 antigens in relation to the appearance of MPF activity during the progesterone-induced maturation of *Xenopus* oocytes (Kuang *et al.*, 1989). MPM-2 antigens were detected on immunoblots of oocyte extracts made at different times following progesterone stimulation; MPF activity was measured by injecting these extracts into immature oocytes and scoring them for the appearance of a white spot and GVBD (Fig. 4). MPM-2 antigens began to appear about 50 min after the exposure of oocytes to progesterone, when MPF activity was still undetectable. By 150 min, when MPF activity was first detectable, MPM-2 antigens were more numerous and were expressed at much higher levels. No MPM-2 antigens were detectable in the extracts of oocytes that were not stimulated with progesterone (Fig. 4A). These results indicate a general correlation between the appearance of MPM-2 antigens and MPF activity, although expression of some MPM-2 antigens at a low level preceded the detection of MPF activity. Inhibition of protein synthesis in progesterone-stimulated oocytes resulted in the failure of both the MPM-2 antigens and MPF activity to appear (Fig. 4B).

Extracts from M-phase cells that possess MPF activity can induce oocyte maturation by activating the endogenous pool of inactive MPF present in the immature oocyte. MPF can induce oocyte maturation in the absence of new protein synthesis. We have also found that MPM-2 antigens can appear along with MPF activity in MPF-induced oocytes even when protein synthesis is inhibited. However, the expression of MPM-2 antigens was either absent or greatly reduced in cycloheximide-treated oocytes. Thus, the expression of MPM-2 antigens correlates with the presence of MPF activity in progesterone-stimulated oocytes. Since these antigens appear even in the absence of new protein synthesis, we can conclude that the appearance of MPM-2 antigens must be due to phosphorylation of preexisting proteins.

C. MPM-2 Inhibits MPF Activity

The concurrent appearance of MPM-2 antigens and MPF activity in progesterone-stimulated oocytes suggests that protein phosphorylation detected by MPM-2 might be involved in the induction of M phase. Therefore, the effect of

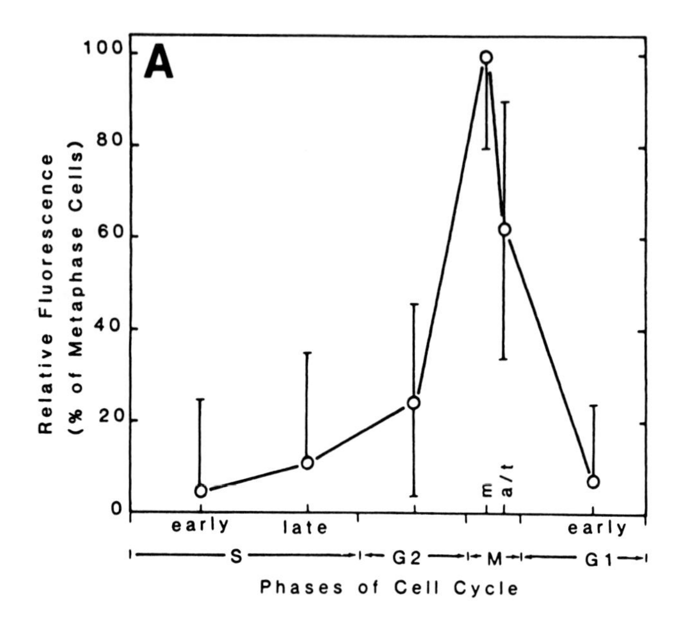
A
Relative Fluorescence (% of Metaphase Cells)
100
80
60
40
20
0
m
a/t
early
late
early
S
G2
M
G1
Phases of Cell Cycle

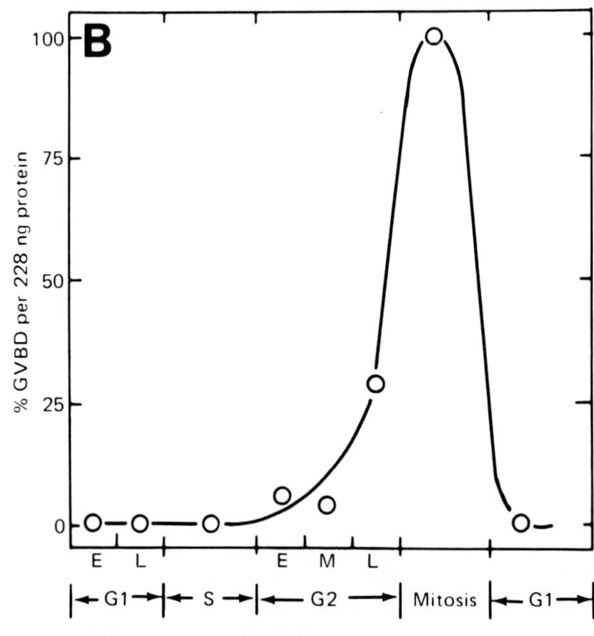
B
% GVBD per 228 ng protein
100
75
50
25
0
E
L
E
M
L
G1
S
G2
Mitosis
G1
Phase of cell cycle

Fig. 3. (A) Relative antigenic reactivity of synchronized HeLa cells to MPM-2 antibody during cell cycle. Cytocentrifuge preparations of synchronized HeLa cells were stained by indirect immunofluorescence with antibody MPM-2. The fluorescence intensity from 25 cells from each population was measured with a Leitz MPV microscope photometer. The fluorescence from blank areas on the slide adjacent to the cells was subtracted from each measurement. The fluorescence intensity from the nonmetaphase populations was normalized to that from 25 metaphase cells included on the same slide. The mean fluorescence intensities were plotted, and the bars represent the standard deviation. Metaphase cells (m) fluoresced more intensely than anaphase–telophase cells (a/t), and mitotic cells fluoresced more intensely than interphase cells (Reprinted with permission from Davis *et al.*, 1983). (B) Meiotic maturation-promoting activity of cell extracts during HeLa cell cycle. Because 228 ng of mitotic protein induced GVBD in 100% of the cases, the percentage activity for other phases of the cell cycle was normalized to that amount of protein. E, early; M, mid-; and L, late (Reprinted with permission from Sunkara *et al.*, 1979.)

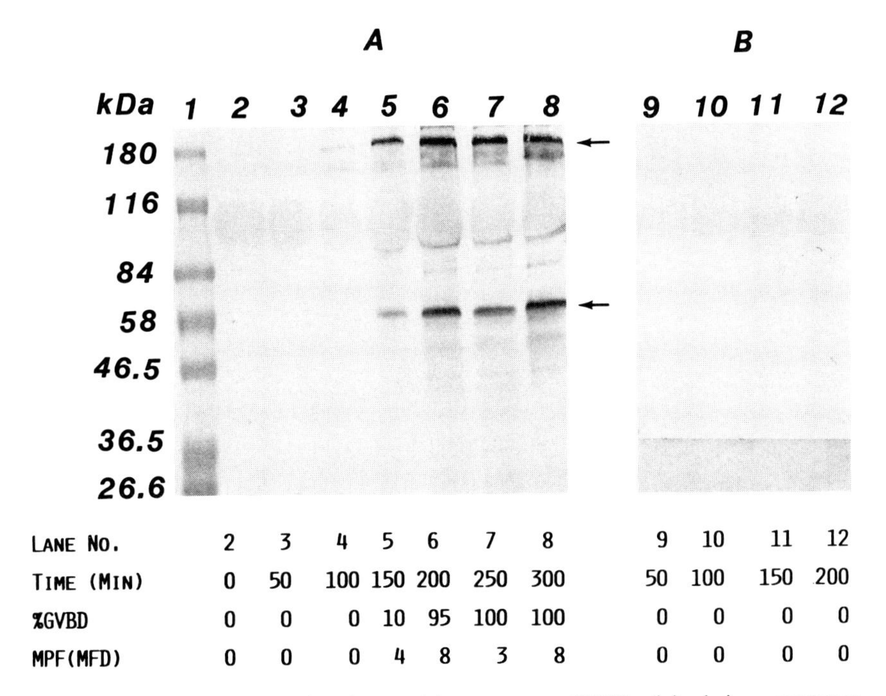

Fig. 4. Expression of MPM-2 antigens and the appearance of MPF activity during progesterone-induced oocyte maturation. Defolliculated *X. laevis* oocytes were incubated in Barth's solution in the absence (A) or presence (B) of cycloheximide (100 μg/ml). Progesterone was added to the medium at time zero. Every 50 min thereafter, oocytes were scored for GVBD, and a batch of oocytes was taken to make extracts. The extracts were analyzed for MPF activity by microinjection into immature oocytes and for the expression of MPM-2 antigens on immunoblots. Proteins from the extracts were separated by $NaDodSO_4$-8% PAGE, electrophoretically transferred to nitrocellulose sheets, and stained with MPM-2. The time of sampling, percentage GVBD, and relative MPF activity (expressed as the maximum fold dilution, MFD) are indicated below each lane. In cycloheximide-treated oocytes, neither MPM-2 antigens nor MPF activity was detectable (B). In oocytes not treated with cyclohex-imide (A), MPM-2 antigens of about 180 kDa appeared as early as 50 min after progesterone stimulation, whereas MPF activity was not detectable until 150 min. A 58-kDa band appeared with the detection of MPF activity. Lane 1 shows protein size standards. (Reprinted with permission from Kuang *et al.*, 1989.)

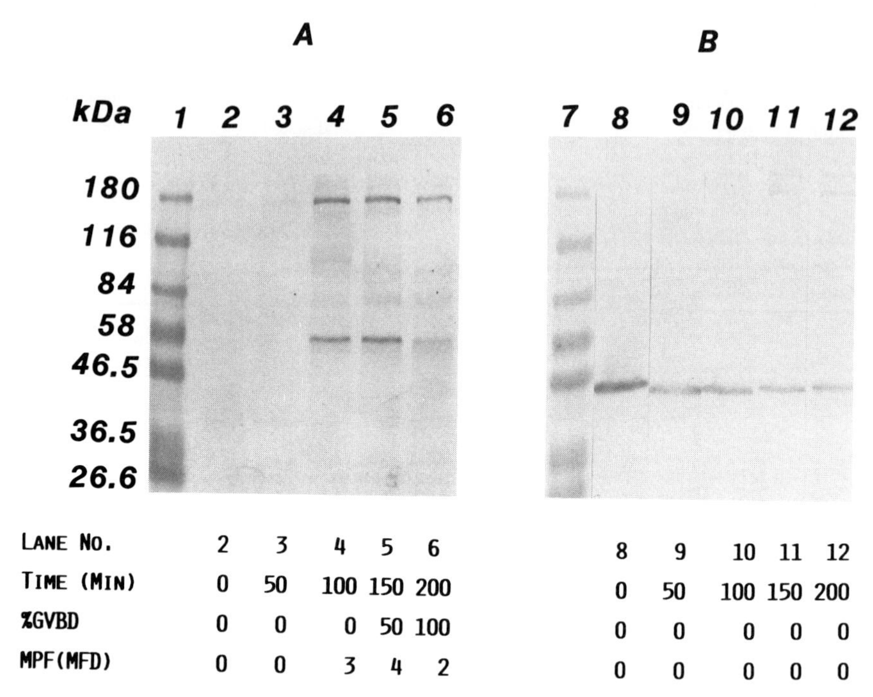

Fig. 5. Effect of MPM-2 on the appearance of MPF activity and MPM-2 antigen expression in progesterone-stimulated oocytes. *Xenopus* oocytes were injected with 70 nl of MPM-2 (20 mg/ml) or TBS and then immediately exposed to progesterone (10 μg/ml). Time-course studies were performed as described earlier. The time of sampling, percentage GVBD, and MPF activity (MFD) are indicated below each lane. (A) Oocytes injected with TBS. (B) Oocytes injected with MPM-2. The dark band in B, lanes 8–12, is the heavy chain of the injected antibody. (Reprinted with permission from Kuang *et al.*, 1989.)

MPM-2 injection on the maturation of oocytes was examined. We observed that 70 nl of MPM-2 at 20 mg/ml significantly blocked progesterone-induced oocyte maturation (Table I) as well as expression of MPM-2 antigens (Fig. 5). Furthermore, we have shown that MPM-2 can neutralize MPF upon mixing and also immunodeplete MPF activity from M-phase cell extracts (Kuang *et al.*, 1989). Immunodepletion of M-phase cell extracts with MPM-2 results not only in the loss of MPF activity (Table II) but also in the loss of MPM-2 antigens from the supernatant (Fig. 6). From these data it is clear that MPM-2 recognizes either MPF itself or a protein that regulates MPF activity, and that the kinase that phosphorylates MPM-2 antigens may be a key component in the regulation of M-phase induction (Kuang *et al.*, 1989). Our latest studies indicate that MPM-2 recognizes an M phase-specific protein kinase in mature oocytes of *Xenopus*. This kinase, named MPM-2 kinase, can phosphorylate histone H1 *in vitro* and is distinctly

TABLE I

MPM-2 Inhibits Progesterone-Induced Maturation of *Xenopus* Oocytes[a]

Experiment	$GVBD_{50}$[b] (hr)	Percentage GVBD	
		MPM-2	MPM-7
1	7.0	10	100
2	6.0	20	100
3	3.0	20	100
4	3.0	30	100
5	4.0	10	100
6	6.0	0	100
7	3.0	30	100
8	3.0	20	100

[a]In all the experiments, 70 nl of the antibody (MPM-2 or MPM-7) at 20 mg/ml was injected into each oocyte. MPM-2–injected oocytes were scored for GVBD when MPM-7–injected oocytes had matured.

[b]Time required for 50% of the control (MPM-7–injected) oocytes to mature after exposure to progesterone. This time varies among oocytes from different frogs. (Reprinted with permission from Kuang *et al.*, 1989.)

different from the p34[cdc2] kinase complex (Kuang *et al.,* 1991). Therefore, the histone H1 kinase activity observed in M-phase cell extracts appears to be owing not only to the cdc2 kinase but also to some MPM-2 antigens. Further studies are necessary to identify the MPM-2 antigens to see whether they correspond to the known proteins encoded by the genetic network regulating the initiation of mitosis.

TABLE II.

Neutralization of MPF Activity by *In Vitro* Mixing of MPM-2 with Mitotic HeLa Cell Extract or Mature Oocyte Extract[a]

Antibody	Mitotic HeLa extract		Mature oocyte extract	
	Undiluted	Diluted	Undiluted	Diluted
MPM-2	0/10[b]	0/10	0/10	0/10
MPM-7	10/10	7/10	10/10	5/10
None (TBS)	10/10	7/10	10/10	3/10

[a]Five μl of the extract (10 mg/ml) was mixed with 5 μl antibody (MPM-7 or MPM-2, 10 mg/ml or TBS) in an Eppendorf tube. The mixture was incubated on ice for 3 hr (HeLa cell extract) or 2 hr (oocyte extract) and then injected, with or without 1:1 dilution with EB, and scored for GVBD 2 hr after injection. (Reprinted with permission from Kuang *et al.*, 1989.)

[b]Oocytes showing GVBD, number/number injected.

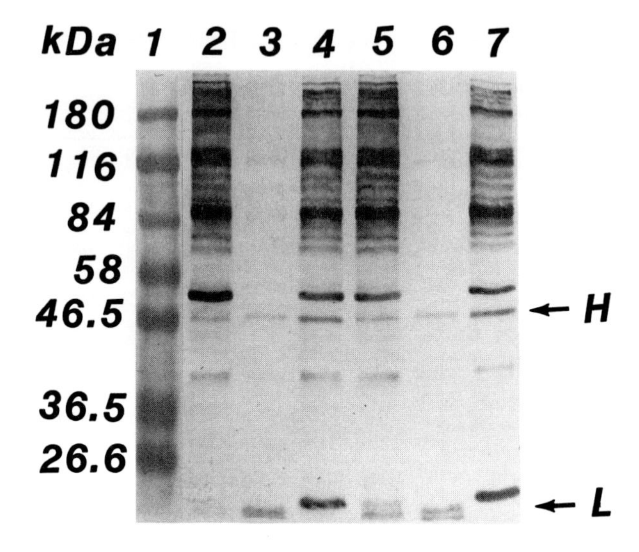

Fig. 6. Immunoblot of mitotic HeLa cell extract (100 μl) was mixed with an equal volume of protein A beads to which antibody was bound. After rotating at 4°C for 3 hr, the beads were pelleted, and the supernatant was assayed for MPF activity. Thirty-three μl of each supernatant was mixed with 3X $NaDodSO_4$/PAGE sample buffer for immunoblot analysis with MPM-2. Lanes: 1, molecular mass markers; 2, nucleolus-specific monoclonal antibody RDA-1; 3, MPM-2; 4, MPM-7; 5, mouse IgG; 6, MPM-2; and 7, MPM-7. Arrows indicate the heavy (H) and light (L) chains of IgG, which are probably due to trace amounts of immunoaffinity beads in the supernatant. (Reprinted with permission from Kuang *et al.*, 1989.)

VI. THE CURRENT MODEL FOR MITOTIC REGULATION

The convergence of these approaches has helped us to draw a composite picture of mitotic control in eukaryotic cells. The initial cell biological studies established the presence of MPF in mitotic and meiotic cells. These studies also pointed to the fact that protein phosphorylation is critical for the initiation of mitosis (Bradbury *et al.*, 1974a,b). Studies with mitosis-specific monoclonal antibodies have confirmed the importance of protein phosphorylation for the initiation of mitosis and dephosphorylation for the end of mitosis (Davis *et al.*, 1983). However, our knowledge of mitotic regulation at the molecular level advanced even more rapidly during the last 2 or 3 years, primarily owing to convergence of biochemical and genetic approaches to this problem. For the first time, MPF was purified to a very high degree from *Xenopus* and starfish oocytes (Lohka *et al.*, 1988; Labbe *et al.*, 1989). This highly purified MPF contained two polypeptides of 32 kDa and 45 kDa. The 32-kDa protein is the *Xenopus* homolog of $p34^{cdc2}$ of *S. pombe.* The 45-kDa protein was found to be cyclin encoded by the gene $cdc13^{+}$ in *S. pombe.*

The homologs of these genes are also found in other systems. These two proteins form a complex that triggers the initiation of mitosis. Although $p34^{cdc2}$ is present throughout the cell cycle at about the same levels, the protein kinase activity fluctuates in a cyclical manner, reaching a peak at mitosis and suddenly decreasing to the lowest levels at the end of mitosis (Bradbury *et al.*, 1974a,b; Labbe *et al.*, 1988, 1989; Draetta and Beach, 1988; Gautier *et al.*, Moreno *et al.*, 1989; Booher *et al.*, 1989). The protein kinase activity is correlated with the dephosphorylation of specific phosphothreonine and phosphotyrosine residues in $p34^{cdc2}$ (Morla *et al.*, 1989; Gould and Nurse, 1989; Draetta *et al.*, 1988; Dunphy and Newport, 1989). In light of these facts a *universal model* for mitotic regulation has been proposed by Paul Nurse (1990), and has been reproduced in Fig. 7. The essential features of this model may be summarized as follows.

1. The 34^{cdc2} protein kinase plays a central role in the initiation and maintenance of M phase, as it is the catalytic subunit of the MPF complex. The activation of $p34^{cdc2}$ kinase is associated with dephosphorylation of specific threonine and tyrosine residues in this polypeptide.
2. At the G_2 to M transition, $p34^{cdc2}$ forms a complex with cyclin, the gene product of $cdc13^+$ in *S. pombe*, leading to the activation of its histone H1 kinase activity. Cyclins are also known to be present in other species.
3. In *S. pombe* two other genes, i.e., $cdc25^+$ and $suc1^+$, have a regulatory role in the initiation of mitosis. The gene product of $cdc25^+$ functions as an activator of $p34^{cdc2}$ either by activating a phosphatase or inhibiting a kinase that acts on $p34^{cdc2}$ (Gould and Nurse, 1989). The accumulation of the gene product of $cdc25^+$, a phosphoprotein ($p80^{cdc25}$), to a critical level might determine the onset of mitosis in *S. pombe* (Moreno *et al.*, 1990). In contrast the role of $suc1^+$ is not clearly understood. However, it seems to act as a negative regulator of mitosis since its overexpression can result in mitotic delay in *S. pombe* (Hayles *et al.*, 1986a,b; Hindley *et al.*, 1987). On the other hand, $suc1^+$ might have a positive role in the completion of mitosis, since its total deletion can result in mitotic arrest (Moreno *et al.*, 1989).
4. The $p34^{cdc2}$ kinase complex may phosphorylate certain proteins that are essential for the execution of specific events during mitosis such as chromosome condensation, nuclear-membrane breakdown, spindle formation, and cytoskeletal rearrangement. For example, histone H1 and lamins, which are involved in chromosome condensation and nuclear envelope dissolution, respectively, could be the substrates for $p34^{cdc2}$ kinase.
5. The completion of mitosis is associated with the breakdown of the $p34^{cdc2}$ and cyclin complex. Cyclin is degraded and $p34^{cdc2}$ is dephosphorylated at the threonine and tyrosine residues. A phosphatase may be activated to dephosphorylate the proteins phosphorylated at the beginning of mitosis. It has been suggested that cyclin destruction is induced by $p34^{cdc2}$ kinase

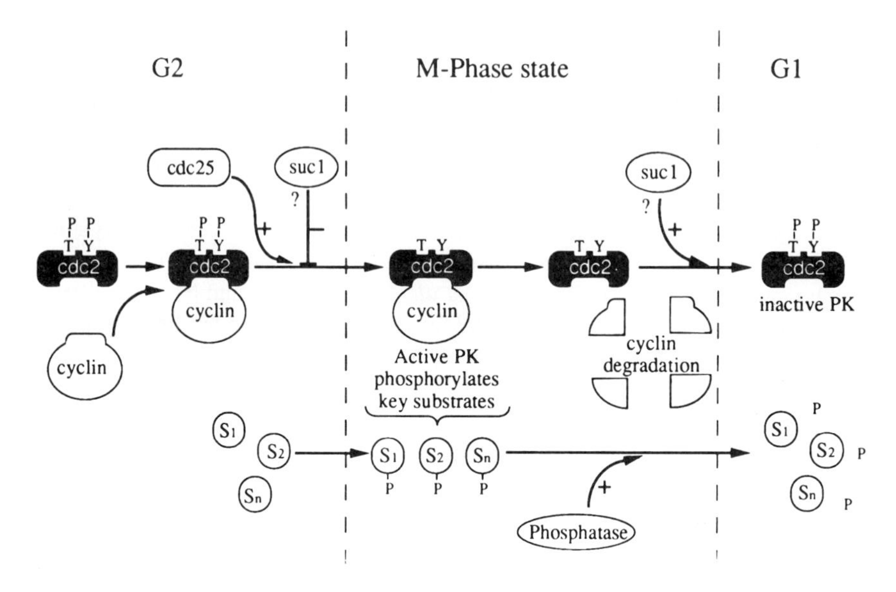

Fig. 7. Universal M-phase control mechanism. The precise order and timing of some of these events such as the relationship between $p34^{cdc2}$ phosphorylation changes and cyclin, $cdc25^{+}$, and $suc1^{+}$ function are unknown at present. The key substrates could include H1 histone, $p60^{src}$, lamins, centrosomal proteins, and other proteins that need to be displaced from chromatin to allow chromosome condensation. (Reprinted with permission from P. Nurse, 1990 and *Nature* **344:** 503–508. Copyrignt © 1990 Macmillan Magazines LTD.)

itself, thereby completing a negative feedback loop to end mitosis (Murray *et al.*, 1989; Felix *et al.*, 1989). The direct evidence that $p34^{cdc2}$ kinase can indeed trigger cyclin proteolysis was provided recently by Felix *et al.* (1990).

REFERENCES

Adlakha, R. C., and Rao, P. N. (1987). Regulation of mitosis by nonhistone protein factors in mammalian cells. *In* "Molecular Regulation of Nuclear Events in Mitosis and Meiosis" (R. A. Schlegel, M. S. Halleck, and P. N. Rao, eds.), pp. 179–226. Academic press, New York.

Adlakha, R. C., Wright, D. A., Sahasrabuddhe, C. G., Davis, F. M., Prashad, N., Bigo, H., and Rao, P. N. (1985). Partial purification and characterization of mitotic factors from HeLa cells. *Exp. Cell Res.* **160**, 471–482.

Arion, D., Meijer, L., Brizuela, L., and Beach, D. (1988). cdc2 is a component of the M phase–specific histone H1 kinase: Evidence for identity with MPF. *Cell* **55**, 371–378.

Aziro, K., Nishimoto, T., and Takahashi, T. (1983). Histone H1 and H3 phosphorylation during premature chromosome condensation in a temperature-sensitive mutant (ts BN2) of baby hamster kidney cells. *J. Biol. Chem.* **258**, 4534–4538.

Basilico, C. (1977). Temperature-sensitive mutations in animal cells. *Adv. Cancer Res.* **24**, 223–266.

Beach, D., Durkacz, B., and Nurse, P. (1982). Functionally homologous cell cycle–control genes in budding and fission yeast. *Nature* **300**, 706–709.
Booher, R. N., Alfa, C. E., Hyams, J. S., and Beach, D. H. (1989). The fission yeast cdc2/cd13/suc1 protein kinase: Regulation of catalytic activity and nuclear localization. *Cell* **58**, 485–497.
Booher, R., and Beach, D. (1987). Interaction between cdc13$^+$ in the control of mitosis in fission yeast: Dissociation of the G_1 and G_2 roles of the cdc2$^+$ protein kinase. *EMBO J.* **6**, 3441–3447.
Booher, R., and Beach, D. (1988). Involvement of cdc13$^+$ in mitotic control in *Schizosaccharomyces pombe*: Possible interaction of the gene product with microtubules. *EMBO J.* **7**, 2321–2327.
Bradbury, E. M., Inglis, R., and Matthews, H. R. (1974a). Control of cell division by very lysine-rich histone (F1) phosphorylation. *Nature* **249**, 257–261.
Bradbury, E. M., Inglis, R., Matthews, H. R., and Langan, T. (1974b). Molecular basis of control of mitotic cell division in eukaryotes. *Nature* **249**, 553–556.
Brizuela, L., Draetta, G., and Beach, D. (1987). p13^{suc1} acts in fission yeast cell-division cycle as a component of the p34^{cdc2} protein kinase. *EMBO J.* **6**, 3507–3514.
Cummins, J. E., Blomquist, J. C., and Rusch, H. P. (1966). Anaphase delay after inhibition of protein synthesis between late prophase and prometaphase. *Science* **154**, 1343–1344.
Cummins, J. E., Brewer, E. N., and Rusch, H. P. (1965). The effect of actidione on mitosis in the slime mold *Physarum polycephalum*. *J. Cell Biol.* **25**, 337–341.
Cyert, M. S., and Kirschener, M. W. (1988). Regulation of MPF activity *in vitro*. *Cell* **53**, 185–195.
Daniels, E. W. (1951). Studies on the effect of X-irradiation upon *Pelomyxa carolinensis* with special reference to nuclear division and plasmogamy. *J. Exp. Zool.* **117**, 189–209. *Cell* **53**, 185–195.
Daniels, E. W. (1952a). Some effects on cell division in *Pelomyxa carolinensis* following X-irradiation, treatment with bis (6-chloroethyl)-methylamine and experimental plasmogamy (fusion). *J. Exp. Zool.* **120**, 509–523.
Daniels, E. W. (1952b). Cell division in the giant amoeba *Pleomyxa carolinensis* following X-irradiation. *I. J. Exp. Zool.* **120**, 525–545.
Daniels, E. W. (1954). Cell division in the giant amoeba *Pelomyxa carolinensis* following X-irradiation. II. *J. Exp. Zool.* **127**, 427–462.
Daniels, E. W. (1955). X-irradiation of the giant amoeba *Pelomyxa carolinensis I. J. Exp. Zool.* **130**, 183–197.
Daniels, E. W. (1958). X-irradiation of the giant amoeba *Pelomyxa carolinensis: II. J. Exp. Zool.* **137**, 425–442.
Davis, F. M., Tsao, T.-Y., Fowler, S. K., and Rao, P. N. (1983). Monoclonal antibodies to mitotic cells. *Proc. Natl. Acad. Sci. U.S.A.* **80**, 2926–2930.
Doree, M. (1982). Protein synthesis is not involved in initiation or amplification of the maturation promoting factor (MPF) in starfish oocytes. *Exp. Cell Res.* **139**, 126–133.
Doree, M., Peaucellier, G., and Picard, A. (1983). Activity of the maturation-promoting factor and the extent of protein phosphorylation oscillate simultaneously during meiotic maturation of star fish oocytes. *Dev. Biol.* **99**, 489–501.
Draetta, G., and Beach, D. (1988). Activation of cdc2 protein kinase during mitosis in human cells: Cell cycle–dependent phosphorylation and subunit rearrangement. *Cell* **54**, 17–26.
Draetta, G., Brizuela, L., Potashkin, J., and Beach, D. (1987). Identification of p34 and p13 human homologs of the cell-cycle regulators of fission yeast encoded by cdc2$^+$ and suc1$^+$. *Cell* **50**, 319–325.
Draetta, G., Luca, F., Westendorf, J., Brizuela, L., Ruderman, J., and Beach, D. (1989). cdc2 protein kinase is complexed with both cyclin A and B: Evidence for proteolytic inactivation of MPF. *Cell* **56**, 829–838.
Draetta, G., Piwnica-Worms, H., Morrison, D., Druker, B., Roberts, T., and Beach, D. (1988). Human cdc2 protein kinase is a major cell cycle–regulated tyrosine kinase substrate. *Nature* **336**, 738–743.

Drury, K. C. (1978). Method for the preparation of active maturation-promoting factor (MPF) from *in vitro* matured oocytes of *Xenopus laevis. Differentiation* **10,** 181–186.

Drury, K. C., and Schroderet-Slatkine, S. (1975). Effects of cycloheximide on the "autocatalytic" nature of the maturation promoting factor (MPF) in oocytes of *Xenopus laevis. Cell* **4**, 269–274.

Dunphy, W. G., Brizuela, L., Beach, D., and Newport, J. (1988). The *Xenopus* cdc2 protein is a component of MPF, a cytoplasmic regulator of mitosis. *Cell* **54**, 423–431.

Dunphy, W. G., and Newport, J. W. (1989). Fission yeast p13 blocks mitotic activation and tyrosine dephosphorylation of the *Xenopus* cdc2 protein kinase. *Cell* **58**, 181–191.

Evans, T., Rosenthal, E. T., Youngblom, J., Distel, D., and Hunt, T. (1983). Cyclin: A protein specified by maternal mRNA in sea urchin eggs that is destroyed at each cleavage division. *Cell* **33**, 389–396.

Fantes, P. (1979). Epistatic gene interactions in the control of division in fission yeast. *Nature* **279**, 428–430.

Felix, M. A., Labbe, J. C., Doree, M., Hunt, T., and Karsenti, E. (1990). Triggering of cyclin degradation in interphase extracts of amphibian eggs by cdc2 kinase. *Nature* **346**, 379–382.

Felix, M. A., Pines, J., Hunt, T., and Karsenti, E. (1989). A postribosomal supernatant from activated *Xenopus* eggs that displays posttranslationally regulated oscillation of its $cdc2^{+}$ mitotic kinase activity. *EMBO J.* **8**, 3059–3069.

Gautier, J., Matsukawa, T., Nurse, P., and Maller, J. (1989). Dephosphorylation and activation of *Xenopus* $p34^{cdc2}$ protein kinase during the cell cycle. *Nature* **339**, 626–629.

Gautier, J., Norbury, C., Lohka, M., Nurse, P., and Maller, J. (1988). Purified maturation-promoting factor contains the product of a *Xenopus* homolog of the fission yeast cell-cycle gene $cdc2^{+}$. *Cell* **54**, 433–439.

Gerhart, J., Wu, M., and Kirschner, M. (1984). Cell cycle dynamics of an M phase–specific cytoplasmic factor in *Xenopus laevis* oocytes and eggs. *J. Cell Biol.* **98**, 1247–1255.

Gould, K. L., and Nurse, P. (1989). Tyrosine phosphorylation of the fission yeast $cdc2^{+}$ protein kinase regulates entry into mitosis. *Nature* **342**, 39–45.

Guttes, E., Devi, V. R., and Guttes, S. (1969). Synchronization of mitosis in *Physarum polycephalum* by coalescence of postmitotic and premitotic plasmodial fragments. *Experientia* **25**, 615–616.

Guttes, E., Guttes, S., and Rusch, H. P. (1959). Synchronization of mitoses by the fusion of the plasmodia of *Physarum polycephalum. Fed. Proc.* **18**, 479.

Hagan, I., Hayles, J., and Nurse, P. (1988). Cloning and sequencing of the cyclin-related $cdc13^{+}$ gene and a cytological study of its role in fission yeast mitosis. *J. Cell Sci.* **91**, 587–595.

Halleck, M. S., Reed, J. A., Lumley-Sapanski, K., and Schlegel, R. A. (1984). Injected mitotic extracts induce condensation of interphase chromatin. *Exp. Cell Res.* **153**, 561–569.

Hartwell, L. H. (1978). Cell division from a genetic perspective. *J. Cell Biol.* **77**, 627–637.

Hartwell, L. H., Culotti, J., Pringle, J. R., and Reid, B. J. (1974). Genetic control of the cell-division cycle in yeast: A model. *Science* **183**, 46–51.

Hartwell, L. H., Mortimer, R. K., Culotti, J., and Culotti, M. (1973). Genetic control of the cell-division cycle in yeast: V. Genetic analysis of cdc mutants. *Genetics* **74**, 267–286.

Hayles, J., Aves, S., and Nurse, P. (1986a) $Suc1^{+}$ is an essential gene involved in both the cell cycle and growth in fission yeast. *EMBO J.* **5**, 3373–3379.

Hayles, J., Beach, D. Durkacz, B., and Nurse, P. (1986b). The fission yeast cell-cycle control gene cdc2: Isolation of a sequence of suc1 that suppresses cdc2 mutant function. *Mol. Gen. Genet.* **202**, 291–293.

Hindley, J., Phear, G., Stein, M., and Beach, D. (1987). $Suc1^{+}$ encodes a predicted 13-kilodalton protein that is essential for cell viability and is directly involved in the division cycle of *Schizosaccharomyces pombe. Mol. Cell. Biol.* **7**, 504–511.

Hittelman, W. N., and Rao, P. N. (1974a). Premature chromosome condensation. I. Visualization of X-ray-induced chromosome damage in interphase cells. *Mutat. Res.* **23**, 251–258.

Hittelman, W. N., and Rao, P. N. (1974b). Premature chromosome condensation. II. The nature of chromosome gaps produced by alkylating agents and ultraviolet light. *Mutat. Res.* **23**, 259–266.

Hittelman, W. N., and Rao, P. N. (1974c). Bleomycin-induced damage in prematurely condensed chromosomes and its relationship to cell-cycle progression in CHO cells. *Cancer Res.* **34**, 3433–3439.

Hittelman, W. N., and Rao, P. N. (1975). The nature of adriamycin-induced cytotoxicity in Chinese hamster cells as revealed by premature chromosome condensation. *Cancer Res.* **35,** 3027–3035.

Hortowitz, N. H., and Leupold, V. (1951). Some recent studies bearing on the one gene–one enzyme hypothesis. *Cold Spring Harb. Symp. Quant. Biol.* **16**, 65–74.

Johnson, R. T., and Rao, P. N. (1970). Mammalian cell fusion: Induction of premature chromosome condensation in interphase nuclei. *Nature (London)* **226**, 717–722.

Johnson, R. T., Rao, P. N., and Hughes, S. D. (1970). Mammalian cell fusion. III. A HeLa cell inducer of premature chromosome condensation active in cells from a variety of animal species. *J. Cell. Physiol.* **76**, 151–158.

Kai, R., Ohtsubo, M., Sekiguchi, M., and Nishimoto, T. (1986). Molecular cloning of a human gene that regulates chromosome condensation and is essential for cell proliferation. *Mol. Cell. Biol.* **6**, 2027–2032.

Kishimoto, T., Kuriyama, R., Kondo, H., and Kanatani, H. (1982). Generality of the action of various maturation-promoting factors. *Exp. Cell Res.* **137**, 121–126.

Kuang, J., Penkala, J. E., Wright, D. A., Saunders, G. F., and Rao, P. N. (1991). A novel M phase-specific H1 kinase recognized by the mitosis-specific monoclonal antibody MPM-2. *Dev. Biol.* **144,** 54–64.

Kuang, J., Zhao, J-Y., Wright, D. A., Saunders, G. F., and Rao, P. N. (1989). Mitosis-specific monoclonal antibody MPM-2 inhibits *Xenopus* oocyte maturation and depletes maturation-promoting activity. *Proc. Natl. Acad. Sci. U.S.A.* **86**, 4982–4986.

Labbe, J. C., Lee, M. G., Nurse, P., Picard, A., and Doree, M. (1988). Activation at M phase of a protein kinase encoded by a starfish homologue of the cell cycle–control gene $cdc2^+$. *Nature* **335**, 251–254.

Labbe, J. C., Picard, A., Peaucellier, G., Cavadore, J. C., Nurse, P., and Doree, M. (1989). Purification of MPF from starfish: Identification as the H1 histone kinase $p34^{cdc2}$ and a possible mechanism for its periodic activation. *Cell* **57**, 253–263.

Lee, M. G., and Nurse, P. (1987). Complementation used to clone a human homologue of the fission yeast cell cycle–control gene cdc2. *Nature* **327**, 31–35.

Lee, M., and Nurse, P. (1988). Cell cycle–control genes in fission yeast and mammalian cells. *Trends Genet.* **4**, 287–290.

Lohka, M. J., Hayes, M. K., and Maller, J. L. (1988). Purification of maturation-promoting factor, an intracellular regulator of early mitotic events. *Proc. Natl. Acad. Sci. U.S.A.* **85**, 3009–3013.

Lohka, M. J., Kyes, J., and Maller, J. L. (1987). Metaphase protein phosphorylation in *Xenopus laevis* eggs. *Mol. Cell. Biol.* **7**, 760–768.

Lohka, M. J., and Maller, J. L. (1985). Induction of nuclear-envelope breakdown, chromosome condensation, and spindle formation. *J. Cell Biol.* **101**, 518–523.

Lohka, M. J., and Masui, Y. (1984). Effects of Ca^{2+} ions on the formation of metaphase chromosomes and sperm pronuclei in cell-free preparations from inactivated *Rana pipiens* eggs. *Dev. Biol.* **103**, 434–442.

Maller, J. W., Wu, M., and Gerhart, J. C. (1977). Changes in protein phosphorylation accompanying maturation of *Xenopus laevis* oocytes. *Dev. Biol.* **58**, 295–312.

Masui, Y., and Markert, C. L. (1971). Cytoplasmic control of nuclear behavior during meiotic maturation of frog oocytes. *J. Exp. Zool.* **177**, 129–146.

Miake-Lye, R., and Kirschner, M. W. (1985). Induction of early mitotic events in a cell-free system. *Cell* **41**, 165–175.

Minshull, J., Blow, J. J., and Hunt, T. (1989). Translation of cyclin mRNA is necessary for extracts of activated *Xenopus* eggs to enter mitosis. *Cell* **56**, 947–956.

Mitchison, J. M. (1973). Differentiation in the cell cycle. *In* "The Cell Cycle in Development and Differentiation" M. Balls, and F.S. Billet, eds.), pp. 1–11. Cambridge University Press, London.

Moreno, S., Hayles, J., and Nurse, P. (1989). Regulation of $p34^{cdc2}$ protein kinase. *Cell* **58**, 361–372.

Moreno, S., Nurse, P., and Russel, P. (1990). Regulation of mitosis by cyclic accumulation of $p80^{cdc25}$mitotic inducer in fission yeast. *Nature* **344**, 549–552.

Morla, A. O., Draetta, G., Beach, D., and Wang, J. Y. L. (1989). Reversible tyrosine phosphorylation of cdc2: Dephosphorylation accompanies activation during entry into mitosis. *Cell* **58**, 193–203.

Murray, A. W., and Kirschner, M. W. (1989). Cyclin synthesis drives the early embryonic cell cycle. *Nature* **339**, 275–280.

Murray, A. W., Solomon, M. J., and Kirschner, M. W. (1989). The role of cyclin synthesis and degradation in the control of maturation-promoting factor activity. *Nature* **339**, 280–286.

Naha, P. M. (1969). Temperature-sensitive conditional mutants of monkey kidney cells. *Nature (London)* **223**, 1380–1381.

Nelkin, B., Nicholas, C., and Vogelstein, B. (1980). Protein factor(s) from mitotic CHO cells induce meiotic maturation in *Xenopus laevis* oocytes. *FEBS Lett.* **109**, 233–238.

Nguyen-Gia, P., Bomsel, M., Labrousse, J. P., and Weintraub, H. (1986). Partial purification of the maturation promoting factor MPF from unfertilized eggs of *Xenopus laevis. Eur. J. Biochem.* **161,** 771–777.

Nishimoto, T. (1988). The 'BN2' gene, a regulator for the onset of chromosome condensation. *Bioessays* **9**, 121–124.

Nishimoto, T., and Basilico, C. (1978). Analysis of a method for selecting temperature-sensitive mutants of BHK cells. *Somat. Cell Genet.* **4**, 323–340.

Nishimoto, T., Eilen, E., and Basilico, C. (1978). Premature chromosome condensation in a ts DNA^- mutant of BHK cells. *Cell* **15**, 475–483.

Nishimoto, T., Ishida, R., Ajiro, K., Yamamoto, S., and Takahashi, T. (1981). The synthesis of protein(s) for chromosome condensation may be regulated by a post-transcriptional mechanism. *J. Cell. Physiol.* **109,** 299–308.

Nurse, P. (1990). Universal control mechanism regulating onset of M phase. *Nature* **344**, 503–508.

Nurse, P., and Bisset, Y. (1981). Gene required in G_1 for commitment to cell cycle and in G_2 for control of mitosis in fission yeast. *Nature* **292**, 558–560.

Nurse, P., Thuriaux, P., and Nasmyth, K. (1976). Genetic control of cell-division cycle in the fission yeast *Schizosaccharomyces pombe. Mol. Gen. Genet.* **146**, 167–178.

Ohtsubo, M., Okazaki, H., and Nishimoto, T. (1989). The RCC1 protein, a regulator for the onset of chromosome condensation locates in the nucleus and binds to DNA. *J. Cell Biol.* **109**, 1389–1397.

Osmani, S., Pu, R. T., and Morris, N. R. (1988). Mitotic induction and maintenance by overexpression of a G_2-specific gene that encodes a potential protein kinase. *Cell* **53**, 237–244.

Picard, A., Labbe, J. C., Peaucellier, G., LeBouffant, F., Le Peauch, C. J., and Doree, M. (1987). Changes in the activity of the maturation-promoting factor are correlated with those of a major cyclic AMP and calcium-independent protein kinase during the first mitotic cell cycles in the early starfish embryo. *Dev. Growth Diff.* **29**, 93–103.

Pines, J., and Hunt, T. (1987). Molecular cloning and characterization of the mRNA for cyclin from sea urchin eggs. *EMBO J.* **6**, 2987–2995.

Rao, P. N., and Johnson, R. T. (1970). Mammalian cell fusion: Studies on the regulation of DNA synthesis and mitosis. *Nature (London)* **225**, 159–164.

Rao, P. N., and Johnson, R. T. (1974). Regulation of cell cycle in hybrid cells. *In* "Control of Proliferation in Animal Cells" (Vol. 1, pp. 785–800). *Cold Spring Harbor Conferences on Cell Proliferation*, Cold Spring Harbor, New York.

Rao, A. P., and Rao, P. N. (1976). The cause of G_2 arrest in CHO cells treated with anticancer drugs. *J. Natl. Cancer Inst.* **57**, 1139–1143.

Reed, S. I., Hadwiger, J. C., and Lorincz, A. T. (1985). Protein kinase activity associated with the product of the yeast cell-division cycle CDC28. *Proc. Natl. Acad. Sci. U.S.A.* **82**, 4055–4059.

Reed, S. I., and Wittenberg, C. (1990). Mitotic role for the cdc28 protein kinase of *Saccharomyces cerevisiae. Proc. Natl. Acad. Sci. U.S.A.* **87**, 5697–5701.

Rusch, H. P., Sachenmaier, W., Behrens, K., and Gruter, V. (1966). Synchronization of mitosis by the fusion of the plasmodia of *Physarum polycephalum. J. Cell Biol.* **31**, 204–209.

Russel, P., and Nurse, P. (1986). cdc25$^+$ functions as an inducer in the mitotic control of fission yeast. *Cell* **45**, 145–153.

Russel, P., and Nurse, P. (1987a). Negative regulation of mitosis by wee 1$^+$, a gene encoding a protein kinase homolog. *Cell* **49**, 559–567.

Russel, P., and Nurse, P. (1987b). The mitotic inducer nim 1$^+$ functions in a regulatory network of protein kinase homologs controlling the initiation of mitosis. *Cell* **49**, 569–576.

Schiestl, R. H., Reynolds, P., Prakash, S., and Prakash, L. (1989). Cloning and sequence analysis the *Saccharomyces cerevisiae* RAD9 gene and further evidence that its product is required for cell-cycle arrest induced by DNA damage. *Mol. Cell. Biol.* **9**, 1882–1896.

Simanis, V., and Nurse, P. (1986). The cell cycle–control gene cdc2$^+$ of fission yeast encodes a protein kinase potentially regulated by phosphorylation. *Cell* **45**, 261–268.

Solomon, M., Booher, R., Kirschner, M., and Beach, D. (1988). Cyclin in fission yeast. *Cell* **54**, 738–739.

Sunkara, P. S., Wright, D. A., and Rao, P. N. (1979). Mitotic factors from mammalian cells induce germinal vesicle breakdown and chromosome condensation in amphibian oocytes. *Proc. Natl. Acad. Sci. U.S.A.* **76**, 2799–2802.

Swenson, K. I., Farrell, K. M., and Ruderman, J. V. (1986). The clam embryo protein cyclin A induces entry into M phase and the resumption of meiosis in *Xenopus* oocytes. *Cell* **47**, 861–870.

Thompson, L. H., Mankovitz, R., Baker, R. M., Wright, J. A., Till, J. E., Siminovitch, L., and Whitmore, G. F. (1970). Isolation of temeprature-sensitive mutants of L-cells. *Proc. Natl. Acad. Sci. U.S.A.* **66**, 377–384.

Wasserman, W. J., and Masui, Y. (1975). Effects of cycloheximide on a cytoplasmic factor initiating meiotic maturation in *Xenopus* oocytes. *Exp. Cell Res.* **91**, 381–388.

Wasserman, W. J., and Masui, Y. (1976). A cytoplasmic factor promoting oocyte maturation: Its extraction and preliminary characterization. *Science* **191**, 1266–1268.

Wasserman, W. J., and Smith, L. D. (1978). The cyclic behavior of a cytoplasmic factor controlling nuclear-membrane breakdown. *J. Cell Biol.* **78**, R15–R22.

Weinert, T. A., and Hartwell, L. H. (1988). The RAD9 gene controls the cell-cycle response to DNA damage in *Saccharomyces cerevisiae. Science* **241**, 317–322.

Weintraub, H., Buscaglia, M., Ferrez, M., Weiller, S., Boulet, A., Fabre, F., and Baulieu, E. E. (1982). "MPF" activity in *Saccharomyces cerevisiae. C.R. Seances Acad. Sci. Ser.3.* **295**, 787–790.

Wu, M., and Gerhart, J. C. (1980). Partial purification and characterization of the maturation-promoting factor from eggs of *Xenopus laevis. Dev. Biol.* **79**, 465–477.

Yamashita, K., Davis, F. M., Rao, P. N., Sekiguchi, M., and Nishimoto, T. (1985). Phosphorylation of nonhistone proteins during premature chromosome condensation in a temperature-sensitive mutant, ts BN_2. *Cell Struct. Funct.* **10**, 259–270.

Ziegler, D., and Masui, Y. (1973). Control of chromosome behavior in amphibian oocytes. I. The activity of maturing oocytes inducing chromosome condensation in transplanted brain nuclei. *Dev. Biol.* **35**, 283–292.

4

Cell Cycle and Cell-Growth Control

KENNETH J. SOPRANO AND STEPHEN C. COSENZA

Department of Microbiology and Immunology
Temple University School of Medicine
Philadelphia, Pennsylvania 19140

I. INTRODUCTION: A HISTORICAL OVERVIEW OF THE CELL CYCLE

The mammalian cell cycle has been the focus of a large number of studies at both the genetic and, more recently, the molecular level. Until recently, these

MOLECULAR AND CELLULAR APPROACHES TO THE CONTROL OF PROLIFERATION AND DIFFERENTIATION

studies have largely been descriptive in nature, but have revealed the complexity of the cell cycle. The cell cycle can be described as the period between the formation of a daughter cell, by the division of a mother cell, and the subsequent time at which the cell divides to form two more daughter cells (Mitchison, 1971). This period was historically divided into two parts called interphase and mitosis. With the advent of radiographic and cytophotometric techniques, interphase was then further divided into four phases. The pioneering studies of Howard and Pelc in 1951 were the first to use radiographic techniques to determine when DNA was being actively synthesized. They employed ^{32}P to label a population of rapidly dividing plant root tips to determine that the bulk of the DNA of eukaryotic cells was replicated during interphase and not during mitosis (Howard and Pelc, 1951, 1953). These studies, along with additional microscopic observations, provided a means to categorize a population of cells into those that are in the act of division (mitotic cells), those that are synthesizing DNA, and those that are not involved with either of these processes. These same investigators then went on to introduce the concept of the deterministic cell cycle. They divided the cycle into four stages: (1) the presynthetic gap or gap 1 (G_1), the interval between mitosis and the onset of DNA synthesis; (2) DNA synthesis (S phase), the time of bulk incorporation of radiolabeled DNA precursors and subsequent DNA replication; (3) the post-synthetic gap or gap 2 (G_2), the interval between the end of S phase and the initiation of mitosis; and (4) mitosis, the period during which the chromosomes condense, and cytokinesis is completed (Howard and Pelc, 1953).

The cell cycle described above is the accepted model for those cells that are actively dividing. However, it soon became obvious that in a whole adult animal, not all cells were actively dividing at all times. Thus, a further modification of this model was introduced by Cowdry in 1950. He divided the cells and tissues of the whole adult animal in three categories based on their ability to proliferate:

1. *Vegetative intermitotics*, tissue that contained either rapidly growing or continuously growing cells, based on the fact that mitotic cells were frequently found by microscopic examination. These cells included the epithelial lining of the intestine, the hematopoietic cells that form the bone marrow, and the basal layer of the epidermis.
2. *Reverting postmitotics*, a category of cells in which mitotic figures were rarely seen under normal circumstances unless the cells were treated with an appropriate stimulus. Two examples are liver cells, which begin to regenerate upon wounding, and mature T and B cells, which begin to proliferate after sensitization.
3. *Fixed postmitotics*, terminally differentiated cells such as neurons, polymorphonuclear cells, and erythrocytes, in which mitotics were never found.

With the addition of these nondividing cells, the diagram of the mammalian cell cycle had to be modified to take into account the fact that some cells continually

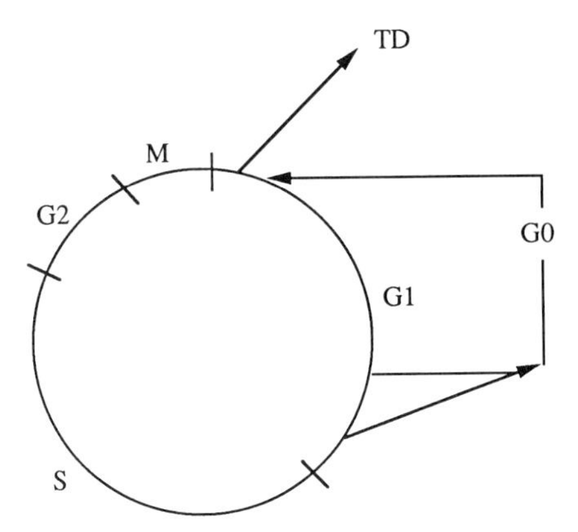

Fig. 1. Model of the mammalian cell cycle. During active cell growth, cells pass through G_1 (the prereplicative phase), S (DNA synthesis), G_2 (the postsynthetic period), and M (mitosis). During G_1, if culture conditions become dense and or growth factors become limiting, cells may exit the cell cycle and arrest in mid G_1 and enter G_0 (quiescence). Quiescent cells can reenter the cell cycle. When they do, they progress through G_1 and then enter S. Some cells have the capacity to leave the cell cycle during G_1 and terminally differentiate (TD). These cells will never reenter the cell cycle.

divide, others leave the cycle but retain the ability to reenter it, and still others leave and never proliferate again (reviewed in Pardee, 1982; Baserga, 1985; Wier and Scott, 1986). Thus this modified model (shown in Fig. 1) includes actively proliferating cells (vegetative intermitotics), which cycle through G_1, S, G_2, and M; (fixed postmitotics), nondividing cells or terminally differentiated (TD cells); and the nondividing or quiescent (G_0) cells (reverting postmitotic cells).

It should be noted that this diagram also takes into account more recent studies that provide evidence for the role of G_1 in regulating traversal into, out of, and through the cycle. It is on these more recent studies that this chapter about cell cycle and cell-growth control will focus. We shall begin by describing the features that distinguish G_0 and G_1 as distinct stages of the cell cycle. Next, we shall summarize a number of important events that occur during G_1, and describe experiments that show that modification of these events can cause nongrowing cells to reenter the cell cycle and proliferate, and cause growing cells to leave the cell cycle and either temporarily or permanently cease proliferating.

II. G_1: THE PREREPLICATIVE PHASE OF THE CELL CYCLE

A large proportion of the evidence suggesting that the prereplicative phase (G_1) of the cell cycle regulates mammalian cell proliferation is based on the finding that cell reproduction can be blocked by causing the cell to leave the cycle during G_1.

Conversely, proliferation can be maintained by preventing the exit from G_1 (Prescott, 1987). The nature and amount of external growth factors (Carpenter and Cohen, 1979; Stiles *et al.*, 1979; Barnes and Sato, 1980; Rozengurt, 1983) and nutrients can influence cells to either continue to divide or to exit the cycle in G_1 and enter into either a quiescent (G_0) or a differentiated (TD) state. It should be noted that very few *normal* cells (i.e., nontumor or nontransformed cells) are blocked at other stages of the cell cycle, suggesting that once the cell progresses through G_1, it is committed to enter S phase and eventually divide (Pardee *et al.*, 1978; Wharton *et al.*, 1982; Baserga, 1985; Prescott, 1987). Conversely, transformed cells frequently do not undergo growth arrest, even under suboptimal growth conditions. As a consequence, they continue to proliferate, utilize all of the available nutrients and growth factors, and then eventually die (Burstin *et al.*, 1974; Cholon and Studzinski, 1974; Pardee and James, 1975; Bartholomew *et al.*, 1976).

The length of G_1 is the most variable of the four phases of the cell cycle. The variability originates from differences in the length of G_1 from cell to cell within a population, as well as from differences in the duration of G_1 exhibited by the same cell from one cycle to the next. The consequence of this G_1 variability is that the average generation time of mammalian cells can differ markedly depending on the nutrient makeup of its environment (Tobey *et al.*, 1987), the time the cell has been in quiescence (Augenlicht and Baserga, 1974), and in the case of normal fibroblasts, the number of population doublings (Cristofalo and Sharf, 1973). Therefore, different rates of cell proliferation can be achieved by changing the average length of G_1. This strongly implies that there are control points or switches that regulate G_1 progression (Prescott, 1987). It logically follows that events regulating G_1 progression also regulate the cell cycle as well as cell growth. The ability to modulate the length of G_1 has enabled investigators to provide important information about the nature and timing of both the growth regularity as well as the growth-regulated events that occur as cells progress through G_1 toward DNA synthesis (Cosenza *et al.*, 1988).

III. G_0–G_1: DIVIDING VERSUS NONDIVIDING STATES

One approach that has been very fruitful in understanding the problem of cell cycle and cell-growth control has been to understand the nature and specific characteristics of (1) a cell when it is dividing compared to when it is not dividing and (2) the events that regulate the transition between these two states.

One of the first studies to suggest that the nondividing state was in fact a distinct phase of the cell cycle was done in mice by Quastler and Sherman (1959). Using injections in mice of ^{3}H-thymidine to monitor cell proliferation, they described two different compartments in the lining of the intestinal epithelium of mice: one in which the cells were dividing, and one in which the cells had ceased division. They proposed that actively proliferating cells must decide to continue to divide or cease proliferating after each round of mitosis. The controversial question at that time was whether there was indeed a distinct phase out of the cycle (G_0) with its own biochemical and functional characteristics, or were the nondividing cells simply cells with a very long G_1. In the 30 years following this work, a considerable amount of kinetic, biochemical, and most recently, molecular evidence has accumulated to indicate that G_0 is in fact a distinct stage of the cell cycle.

A. Kinetic Evidence Suggesting That G_0 Is a Distinct Phase

1. In Vitro Kinetic Evidence

The term G_0 was introduced in 1963 by Lajtha to designate the position of a reverting postmitotic cell, which exhibited the ability to leave a state of quiescence and reenter the proliferative cycle upon application of the appropriate stimulus. Culture of fibroblast cells, *in vitro,* can result in a cessation of growth when external conditions become suboptimal for growth. Examples of such conditions include serum depletion (Holley and Kiernan, 1968; Brooks, 1976; Temin, 1971), high cell density (Todaro and Green, 1963; Todaro *et al.*, 1965), nutrient deficiency (Ley and Tobey, 1970), and high cyclic adenosine monophosphate (cAMP) levels (Rozengurt, 1986). It was shown in these original studies, and has been confirmed in many others, that as the cell-growth rate decreased, DNA synthesis and cell division ceased. Concurrently, RNA and protein synthesis also decreased (Pardee *et al.*, 1978). These changes could be totally reversed if the cells were replated at lower densities or if the medium was replaced with fresh growth factors and nutrients. All of these studies showed that when the quiescent cells were restimulated to divide, they first synthesized DNA before proceeding to divide. This suggested that the cells had left the cycle at some point after mitosis, but before DNA synthesis. Subsequent studies showed that G_0 cells contain a 2n amount of DNA (Prescott, 1987), consistent with the idea that such cells leave the cycle during the G_1 phase.

Regardless of whether the cells are made quiescent by serum deprivation or by growth to high density, stimulation by addition of fresh serum or purified growth factors results in a synchronous reentry of the cells into S phase after a lag time (Todaro *et al.*, 1965; Temin, 1971; Brooks, 1976). This lag time varies from cell type to cell type. However, as a result of this lag, the prereplicative phase (or G_1) following stimulation of quiescent cells is *longer* than the G_1 phase following mitosis.

2. In Vivo Kinetic Evidence

There are also *in vivo* kinetic data supporting the fact that G_0 is a distinct phase of the cell cycle. Although the adult rat liver is an extremely metabolically active organ, fewer than 0.2% of the hepatocytes become labeled when a single injection of ^{3}H-thymidine is made into the rat (Baserga, 1985). However, if a part of the liver is surgically removed (70%), the remaining cells enter S phase and divide, ultimately restoring the liver to approximately its previous size (Bucher, 1963). Kinetic studies following partial hepatectomy (Grisham, 1962; Fabrikant, 1968) found that the lag period for reentry of these reverting postmitotic liver cells into S phase is approximately 18 hr. Therefore, one can consider liver cells as an *in vivo* counterpart to the quiescent (or G_0) cells. The literature describes a number of other *in vivo* models of G_0 cell populations including cells from kidney (Baserga *et al.*, 1968), salivary glands (Barka, 1965), and skin (Block *et al.*, 1963).

3. The Concept of Prolonged Growth Arrest

Augenlicht and Baserga in 1974, Rossini *et al.* in 1976, and more recently Owen *et al.* (1987) and Cosenza *et al.* (1988) have shown that WI-38 cells, a normal human diploid fibroblast cell line, can be held in a state of density-dependent growth arrest for prolonged periods. When such cells are treated with fresh serum or growth factors, they reenter the cell cycle and proceed to synthesize DNA and divide. However, the time required to do so is proportionately lengthened according to the length of time the cells are growth arrested. This is illustrated in Fig. 2

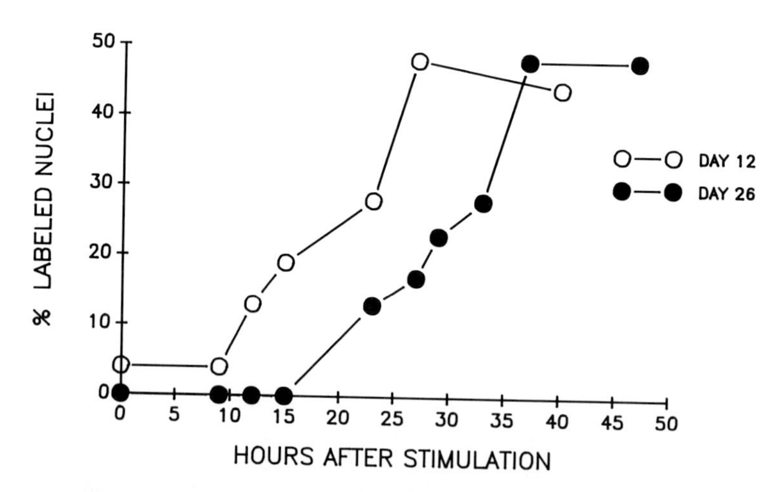

Fig. 2. Kinetics of entry into DNA synthesis following prolonged quiescence. WI-38 cells plated on glass coverslips at 2×10^4 cells/cm^2 were stimulated by replacement of medium with fresh medium containing 10% fetal bovine serum and 0.5 μCI/ml ^{3}H-thymidine either 12 days (open circles) or 26 days (closed circles) after plating, as described in Owen *et al.* (1987) and Cosenza *et al.* (1988). Cells were fixed in cold absolute methanol at the indicated time points and prepared for autoradiography. The number of labeled nuclei from at least 1000 cells was determined per time point. Each data point represents the mean value from two independent stimulation experiments.

for WI-38 cells growth arrested for either 12 or 26 days. The prolongation of the prereplicative phase is consistent with the possibility that there are different levels of quiescence (Owen *et al.*, 1989a). Recent biochemical and molecular evidence confirms this (see below). Similar findings have been obtained with concanavalin A-stimulated lymphocytes (Gunther *et al.*, 1974), a mouse fibroblast cell line designated 10T ½ (Miska and Bosmann, 1980) and Swiss 3T3 cells (O'Farrell and Yanez, 1989).

B. Nonkinetic Evidence That G_0 Exists As a Distinct Phase

1. Biochemical Evidence

The kinetic studies mentioned above as well as the studies about prolonged quiescence provide circumstantial evidence in support of the idea that quiescent cells enter a distinct phase of the cell cycle. However, it can still be argued [and has been by Burns and Tannock (1970), Smith and Martin (1973) and Shields and Smith (1977)] that these cells have never *left* the cycle but merely have a very long G_1 owing to a low probability of transition that regulates the entry of a cell into S phase. The ultimate proof that G_0 is a distinct phase of the cell cycle is found in a number of studies that provide biochemical evidence that quiescent cells are different from cycling cells (Sander and Pardee, 1972; Rovera and Baserga, 1973; Baserga, 1976, 1985; Owen *et al.*, 1989a).

The following are just a few of the observed differences reported between G_0 and G_1 cells (reviewed by Baserga, 1985).

1. It was shown that rRNA and tRNA synthesis both increase upon stimulation of quiescent cells with serum (Prescott, 1976). The ribosome content of stationary-phase hamster fibroblasts was only 70% of that found in G_1 cells (Becker *et al.*, 1971);
2. In line with the differences shown for ribosome content, protein synthesis was also shown to be increased twofold to threefold after stimulation of G_0 cells (Hassell and Engelhardt, 1976; Mostafapour and Green, 1975);
3. DNA-binding proteins specific for stationary-phase cells have been reported (Fox and Pardee, 1971) as well as differences specific for stationary-phase L-mouse fibroblasts in gel electrophoretic patterns of nonhistone chromosomal proteins (Becker and Stanners, 1972);
4. Certain variants of histones are synthesized in G_0 but not G_1 cells (Wu *et al.*, 1982); and
5. G_1 cells have a higher number of S1-sensitive sites than do G_0 cells (Collins *et al.*, 1982).
6. Finally, it should be mentioned that there are significant differences in the abilities of G_0 membranes to transport specific molecules. They have been found

to have a reduced ability to take up phosphate, uridine, hexoses, and some amino acids. When the G_0 cell is stimulated, there is a change in the cell's ability to transport specific molecules within minutes of the stimulus (Pardee *et al.*, 1978). In fact, the rate at which these small molecules are transported across the membrane has been proposed as a primary regulator for the transition from G_0 to G_1 (Lee *et al.*, 1987).

The above examples suggest that, on the whole, there is less overall metabolic activity in a stationary cell. However, on stimulation with the appropriate signal, there is an increase in transcription of RNA, followed by an increase in the synthesis of new molecules necessary for the transition out of G_0 and progression through G_1 into S phase (Johnson *et al.*, 1974).

It should be noted that recent studies utilizing the prolonged quiescence model described above have shown that the quiescent state is not a state of overall gradual deterioration, but rather a metabolically active state. Owen *et al.* (1989a) examined a number of biochemical and molecular parameters in WI-38 cells density arrested for various lengths of time. They found that density-dependent growth arrest of WI-38 cells occurs as a two-stage process. Cells density arrested for 7 to 10 days entered a state referred to as early G_0. In early G_0, the cells stopped bulk DNA synthesis but had not yet undergone any dramatic biochemical changes with respect to RNA or protein content, when compared to actively proliferating cells. In contrast, as the cells remain in a state of density arrest for 10 to 20 days, they leave early G_0 and enter into late G_0. Late G_0 cells exhibited a number of the biochemical changes that have already been discussed. The most interesting finding was that while levels of many of the proteins and gene products commonly associated with G_1 progression declined during late G_0, the levels of other proteins such as fibronectin and collagen increased. This work confirms the idea that G_0 cells are clearly physiologically different from G_1 cells, but that the state of quiescence is not a state of gradual deterioration. Quiescent cells are still metabolically active, and exhibit both decreases and increases in a number of specific biochemical processes. Moreover, even when a quiescence-associated change in a biochemical process does occur, it happens in a two-step fashion.

2. *Molecular Evidence*

Molecular evidence that G_0 is a distinct phase was first provided in cell-fusion studies. Jonak and Baserga (1979) showed that molecules not present in G_0 cells are necessary for entry into S phase (Baserga, 1985). Pereira-Smith *et al.* (1985) also showed that G_0 cells contained an inhibitory substance such that fusion of cytoplasts prepared from quiescent cells to cycling cells led to an inhibition of DNA synthesis. The amount of inhibition was increased relative to the length of time the cells had been held in quiescence before cell fusion. These studies indicate that specific biochemical events must occur in order for a cell to exit quiescence, progress through G_1, and enter DNA synthesis.

More recently, the techniques of differential cDNA analysis, Northern blot hybridization, and RNAse protection have identified a large number of transcripts either not present in G_0 cells or present in dramatically reduced amounts but that become expressed at various times during G_1. These genes include c-*fos*, c-*myc*, c-*jun*, p53, ornithine decarboxylase, to name just a few. In addition, several other studies have identified transcripts present only in growth-arrested cells and not in actively proliferating cells. More details about all of these growth-associated genes will be provided later in this chapter and in several others in this volume.

Having established that G_0 is in fact a distinct stage of the cell cycle, and in light of the large amount of evidence suggesting that the decision to continue to proliferate or to arrest growth occurs as a result of events during G_1, we will now focus our attention on the many biochemical and molecular events that occur when a G_0 cell has been stimulated to reenter G_1 and eventually divide.

IV. CHANGES ASSOCIATED WITH THE G_0–G_1 TRANSITION AND G_1 PROGRESSION

Many changes have been shown to occur after quiescent cells have been stimulated to divide following treatment with the appropriate growth factors. However, while a large number of biochemical events associated with the prereplicative phase have been described, it is necessary to distinguish those events that *regulate* mammalian cell proliferation from those that are growth regulated. Baserga (1985) has classified these biochemical events into three categories:

1. Events necessary and specific for cell-cycle progression that are absent or markedly reduced in noncycling cells, but that respond directly to environmental signals without previous expression of other cell-cycle genes. These would include the SV40 large-T antigen (Mueller *et al.*, 1978; Baserga *et al.*, 1982; Soprano *et al.*, 1983) and the cdc2 gene product of *Saccharomyces cerevisiae* (Beach *et al.*, 1982);
2. Events necessary and specific for cell cycle progression that do not directly respond to environmental signals, but instead respond to other intracellular signals. These would include enzymes such as DNA polymerase and ribonucleotide reductase, which are essential for proper DNA synthesis, but do not regulate the entry into S phase; and
3. Events that are housekeeping functions required for normal cell cycle progression, but can be found in nonproliferating cells. These are required for biochemical reactions including maintenance of divalent cation level, transport functions, and generation of ATP and RNA polymerase II activity.

The components listed in categories two and three play a crucial role in cell viability and growth. As a result, the cell will cease to divide in their absence, but if these gene products are up-regulated, they will have little or no effect on cell-cycle progression.

A. G_1-Specific Transcription

Studies by Williams and Penman in 1975, utilizing hybridization in solution, compared the complexity, relative abundance classes, and homology of cytoplasmic poly(A)$^+$ RNA in quiescent and proliferating 3T6 cells. They were able to show that the two populations of cells had both high- and low-complexity RNA classes with only slight differences in the observed levels. More important, they were able to show that at least 90% of the mRNAs found in proliferating cells were also found in nonproliferating cells. However, they calculated that at least 1400 high-complexity and 400 low-complexity RNA species were different in proliferating cells compared to quiescent cells. Each of these could code for a protein unique to actively growing cells.

Evidence for the necessity of transcription of unique copy genes during the transition out of G_0 and progression through G_1 comes from work with a cell-cycle mutant designated tsAF8. TsAF8 cells were isolated from the Syrian hamster BHK-21 cell line by Meiss and Basilico in 1972. This G_1 mutant was shown to be arrested during mid-G_1 several hours before S phase (Burstin *et al.*, 1974). In 1975, Rossini *et al.*, (1976), showed that levels of RNA polymerase II were dramatically reduced at the nonpermissive temperature (npt) in both isolated nuclei and cell extracts as measured by ^{3}H-α-amanitine binding (Rossini *et al.*, 1980). These data suggested that the mutation exhibited by tsAF8 was associated with the gene coding for RNA polymerase II. Direct proof came when the ts mutation could be complemented by the introduction of the gene coding for RNA polymerase II (Ingles and Shales, 1982). In addition to these studies, Waechter *et al.*, (1984) have shown that microinjection of purified RNA polymerase II protein into tsAF8 cells allowed them to enter S phase even at the nonpermissive temperature. These studies, as well as the numerous cDNA clones corresponding to mRNA molecules synthesized upon serum stimulation, suggest that there is a need for the synthesis of unique copy genes for the transition out of G_0 and progression through G_1 into S phase.

B. G_1-Specific Translation

The requirement for protein synthesis during G_1 was first shown by Terasima and Tolmach in 1966. They reported that treatment of L5 cells with puromycin during G_1 delayed the entry into S phase by the amount of time in which protein synthesis was inhibited. A more detailed study using cyclohexamide (CHM)-treated mitotically synchronized Chinese hamster ovary (CHO) cells showed that when mitotic cells were treated with CHM for 1 to 5 hr, the cells were delayed in the initiation of DNA synthesis by the length of the CHM treatment. If the cells were treated after division, the delay was greater than the duration of CHM treatment. The delay was shown to be roughly proportional to the duration of the CHM treatment. Treatment of the cells during S phase stopped DNA synthesis; however, after removal of the CHM, cells immediately resumed DNA synthesis (Schneiderman *et al.*, 1971). Similar results were obtained in a system utilizing isoproterenol-stimulated mouse salivary glands. The addition of CHM at various times before or after stimulation showed that the synthesis of labile proteins was required during the prereplicative phase for entry into DNA synthesis (Saski *et al.*, 1969; Novi and Baserga, 1971).

V. NUTRIENT-RELATED CONTROL POINTS WITHIN G_1

A. The Restriction Point

As a cell is cycling, it must make a decision either to commit itself to DNA synthesis or to exit the cell cycle and enter into a state of quiescence. The earliest work using serum deprivation suggested that when serum is withdrawn from chicken cells, they enter the stationary phase at some point in mid-G_1 (Temin, 1971). Pardee proposed in 1974 (Pardee, 1974) that when cells have reached quiescence by any physiological means, they exit and enter at the same point during G_1. This point was referred to as the *switching* point. The point at which this switch was made was referred to as the R point or restriction point. Since the lag period between the time of stimulation and reentry into DNA synthesis was the same, it was concluded at the time that they all exited and reentered at the same position or restriction point. However, the exact time during G_1 at which the R point exists is not so clearly defined as first believed. Depending on the nature of the arresting conditions and the method used to measure reentry into DNA synthesis, the time before S phase can vary between 2 and 3 hours (Pardee, 1974; Rubin and Steiner, 1975; Yen and Pardee, 1978; Campisi *et al.*, 1982). Still other studies place it as close as 12 to 40 min before S phase (Blair *et al.*, 1981). Interestingly, this control point seems to be lost or relaxed in several of the tumor-cell lines tested, since transformed cells required close to 90% inhibition of total protein synthesis before inhibition of DNA synthesis was observed (Pardee, 1974; Yen and Pardee, 1978; Medrano and Pardee, 1980). This is consistent with the fact described previously that many transformed cell lines will continue to proliferate under poor growth conditions until they eventually cease dividing and stop at random points in the cell cycle.

Studies on the nature of the R point suggested that it is regulated by a labile protein or complex of proteins, which form a replicase complex near the start of DNA synthesis. Studies on the stability of the R protein or protein complex concluded that the half-life was 2.5 hr in nontransformed cells but was longer in their transformed counterparts (Campisi *et al.*, 1982). Although the synthesis of this putative protein(s) was localized at the beginning of G_1 (Schneiderman *et al.*, 1971; Roscow *et al.*, 1979) and a 68-kDa protein with the required characteristics of an R protein was identified by 2D gel electrophoresis (Croy and Pardee, 1983), attempts to clone the R point gene have thus far been unsuccessful.

B. The V and W Control Points

In a model system employing density-arrested BALB/c 3T3 cells, Pledger, Stiles, and their colleagues have proposed two additional control points for the G_0 to G_1 transition following stimulation with serum. These control points have been termed V and W and represent two points that cells must traverse if they are to successfully progress toward S phase and undergo DNA synthesis. Based on a series of kinetic experiments using two chief components of serum, platelet derived growth factor (PDGF) and platelet poor plasma (PPP), these investigators were able to show that both components are necessary for entrance into S phase from density-dependent quiescence, while neither would suffice alone (Pledger *et al.*, 1977). In a more-detailed study (Pledger *et al.*, 1978), these investigators showed that the events initiated by PDGF and PPP occurred in a temporal fashion. They found that PDGF treatment made the cells *competent* to respond to further growth factors contained in PPP. The competent cells always entered S phase 12 hr after they were transferred to PPP. Further experiments utilizing sequential addition and withdrawal of PPP to PDGF-treated cells enabled them to locate the two control points. The V point was located midway through the G_0–G_1 phase, past which the cells could not progress without PPP. The second point, W, was similar to Pardee's R point in that it was located just before the start of DNA synthesis, when cells required no additional factors (i.e., they were committed to enter S phase). This indicated that PDGF and PPP were responsible for different activities. Treatment of quiescent BALB/c 3T3 cells with PDGF induced a competent state so that they were now ready to respond to progression factors found in PPP, resulting in the entry into S phase. A more detailed discussion on the competence and progression system as well as the molecular events and the factors involved in each component will be discussed in another chapter in this volume.

The possible role of epidermal growth factor (EGF) as a competence factor in human diploid fibroblasts was also suggested in a recent report using the long-term quiescent WI-38 cell-model system described previously. When cultures of WI-38 cells reach high cell densities, they cease to proliferate and enter a state of quiescence. They can remain in this state for extended periods; however, as mentioned previously, the longer the cells remain growth arrested, the longer the time required to reenter S phase following growth stimulation (Augenlicht and Baserga, 1974; Rossini and Baserga, 1978; Owen *et al.*, 1987; Cosenza *et al.*, 1988). Using defined medium supplemented with EGF, IGF-I, and dexamethasone, Owen *et al.* (1989b) addressed the question of whether the long-term cells have different growth factor requirements. They found that both the long- and short-term quiescent cells had the same qualitative and quantitative growth-factor requirements. The defined medium stimulated both populations of cells to enter S phase to the same extent and with the same kinetics as did serum. However, the long-term quiescent cells displayed a difference in the time during which either serum or individual growth factors were required to be present. Owen *et al.*, (1989b) found that the long-term quiescent cells required 14 hr in the presence of EGF before they were able to respond to IGF-I or insulin, whereas the short-term quiescent cells could respond almost immediately. Therefore it seems that, analogous to the competence-progression system of 3T3 cells, WI-38 cells probably use EGF as a competence factor, which allows them to respond to progression factors such as IGF-I or insulin and enter DNA synthesis. A more detailed discussion on the competence progression system as well as the molecular events and the factors involved in each component will be discussed in another chapter in this volume.

C. A Control Point within G_0

Further studies using the prolonged-quiescence WI-38 model system have led to the identification of yet another, previously unknown control point. Cosenza *et al.* (1988) set out to map, at a molecular level, the location of the prolongation of the prereplicative phase exhibited by WI-38 cells stimulated after long-term growth arrest. They analyzed the changes in the pattern of expression of a number of representative early G_1 (c-*fos*, c-*myc*, c-Ha-*ras*, ODC, and vimentin), mid-G_1 (p53, calcyclin) and late G_1–S (TK and histone H3) growth-associated genes. As shown in Fig. 3, they found that the time of induction and/or maximal accumulation of all of the transcripts analyzed except c-*fos*, c-*myc* and c-Ha-*ras* was delayed by a period nearly equal to the length of the prolongation of the prereplicative phase. Moreover, once early G_1 growth-associated gene expression occurred, the pattern of expression and the time required to progress through G_1 and enter S was the same in all cell groups regardless of how long they were growth arrested. They concluded that the prolongation occurs at a point closely following stimulation and suggested that certain as-yet-undefined molecular events must

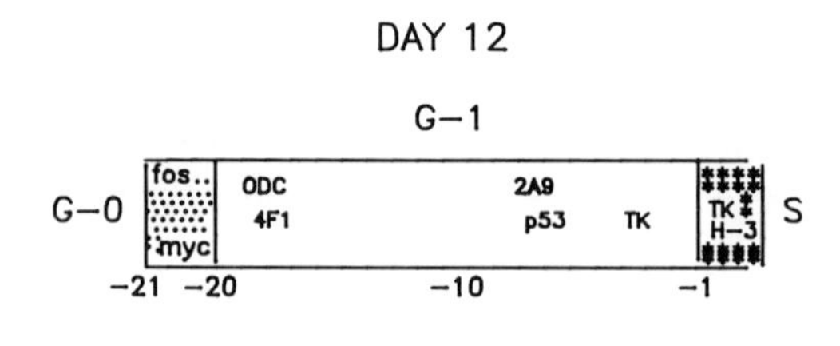

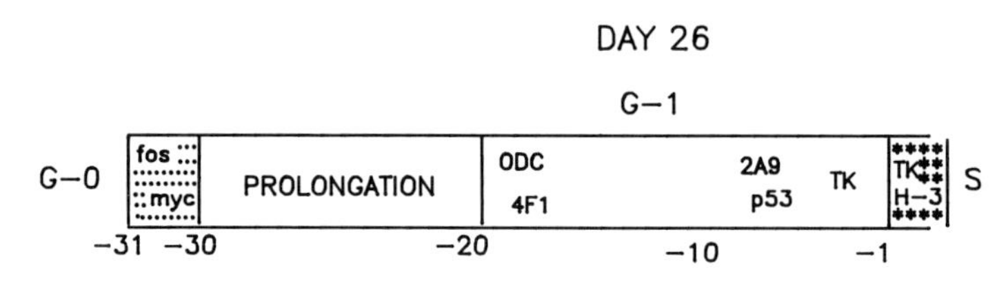

Fig. 3. Diagramatic representation of the time of maximal expression of growth-associated genes in WI-38 cells stimulated following prolonged quiescence. Total cytoplasmic RNA was isolated from 12-day to 26-day growth-arrested WI-38 cells at various times after stimulation as described in Owen *et al.* (1987) and Cosenza *et al.* (1988). The time of maximal expression determined for each gene is plotted relative to the time of entry into S.

occur in cells stimulated after prolonged periods before early G_1 genes become transcribed. However, once these events occur, there do not appear to be any additional delay points along the path to S. It is clear from their data that, at least in this model system, growth-associated gene expression is temporally and coordinately regulated with respect to time of entry into DNA synthesis. These experiments are among the first to use analysis of the expression of G_0, G_1, and S phase-specific genes as molecular markers in a model system in which the length of the G_1 phase can be modulated. As a result, they do provide clear molecular evidence to support the hypothesis first proposed by Hartwell and supported, at least on a theoretical basis, by many others (see Baserga 1976, 1985; Pardee, 1982, 1987; Prescott 1987; and Hartwell and Weinert 1989 for reviews) that there is a defined order to the molecular events that occur as a cell progresses out of quiescence through G_1 into S.

The existence of such a program and its delay in long-term, growth-arrested cells taken together with previously described data showing that changes occurring during prolonged quiescence happen in a two-step rather than a gradual fashion, suggested that a controlling event or *metabolic switch* must exist and be turned on in quiescent cells before this program can be activated. In cells growth arrested for short periods (*early* G_0 cells), this switch is turned on at the time of stimulation, and thus the cell does not become blocked at this control point. However, in cells growth arrested for extended periods (*late*

G_0 cells), the cells are blocked at this control point and cannot pass through it until certain other events occur, which are necessary for activation of the metabolic switch. While the nature of these events or this switch remains to be determined, one can perhaps obtain a clue from the growth-factor studies described in the previous section of this article. Since both short- and long-term quiescent cells can respond to EGF, but long-term cells cannot respond immediately to insulin or IGF-I, it is possible that passage through the early G_0 to late G_0 control point involves EGF-dependent synthesis or activation of IGF-I receptors, binding proteins and/or intracellular mediators, which have degraded during the prolonged time in quiescence. Once these *factors* reach the critical level necessary for the cell to respond to IGF-I, the cell progresses through G_1 in precisely the same way and time as cells stimulated after short periods of growth arrest.

It should be noted that there is additional evidence to support this hypothesis. Studies by Ferrari *et al.* (1988) and Owen *et al.* (1989a) showed that WI-38 cells growth arrested for up to 20 days could be restimulated with only PPP and did not exhibit a prolongation of the prereplicative phase. In contrast, WI-38 cells density arrested for more than 20 days required full serum and did exhibit the characteristic prolongation of the prereplicative phase. Thus a clear change in growth responsiveness occurs during prolonged growth arrest.

D. The G_1pm and G_1ps Control Points

Zetterberg and Larsson (1985) made a detailed kinetic analysis of the point at which cells were induced to enter G_0. They analyzed individual cells located at different positions between mitosis and S phase by time-lapse cinematographic analysis of cells after exposure to brief periods of serum deprivation or cycloheximide-induced inhibition of protein synthesis. Their studies showed that cells younger than 3.5 hr were extremely sensitive to the presence of serum. If serum was removed for longer than 1 hr or up to 8 hr (which was considered prolonged serum starvation), the cells would exhibit a delay in the onset of mitosis equal to the actual time in serum-free medium plus an additional 8 hr. However, by 4 hr after mitosis, a rapid change occurred, such that more than 90% of the cells were serum independent. They proposed that G_1 can therefore be divided into one serum-dependent postmitotic phase (G_1pm), which has a relatively constant duration of 3 to 4 hr, followed by a serum-independent pre-DNA-synthetic phase (G_1ps). The G_1ps phase does not have a constant range, since some cells started DNA synthesis immediately on leaving G_1pm, while others remained in G_1ps for 10 hr or more. This phase would therefore account for the variability of G_1 and would be similar to a point after Pardee's R point (Pardee, 1974).

They went on to test the ability of purified EGF, insulin, or PDGF to counteract the intermitotic delay of G_1pm cells caused by brief serum deprivation. These experiments were performed by adding each individual growth factor to the serum-free medium. The supplemented medium was then removed when serum containing medium was added back to the cultures. The results showed that all three growth factors could prevent the intermitotic delay if the period of serum deprivation was only 2 hr; however, if the cells underwent prolong deprivation (8 hr), only PDGF could prevent the intermitotic delay. Therefore, while EGF and insulin could delay entrance into G_0, only PDGF induced all the necessary genetic requirements to allow the cells to become committed to enter S phase. It should be mentioned that the problem with these experiments is that cells were placed back into complete medium after the various treatments. Under these experimental conditions, one cannot determine whether PDGF alone was all that was required for cells to progress into S phase, or whether G_1pm represents a competence state that occurs during the first 3.5 hr after mitosis, followed by G_1ps, which is primarily responsive to progression factors found in PPP. With this in mind, it is apparent that Swiss 3T3 cells (Larrson *et al.*, 1989) sense the environmental conditions within the first 3.5 hr after mitosis (G_1pm) and make the decision to divide or exit into G_0.

E. Exposure to Serum during G_1 of the Previous Cell Cycle Controls Subsequent Entry into S phase

As already mentioned, entry into S phase is regulated by events initiated by the action of growth factors. We have discussed some of the possible models for control with respect to the decision of a cell to proliferate or to cease dividing and enter into a state of quiescence. This decision is dependent not only on the environment in which a daughter cell enters but also on the environment in which the mother cell has just exited. This *memory* can be discussed in relationship to the effect of serum or serum deprivation on entry into DNA synthesis of daughter cells.

Scher *et al.* (1979), studying the function of PDGF during the growth of BALB/c 3T3 cells, were able to show that not only does PDGF cause density-dependent cells to become competent to divide (Pledger *et al.*, 1977), but it also has a second function. It prevents cells from exiting the cell cycle and entering into G_0. These studies differed from the reports described above, in that the role of each of the two main components of serum, PDGF and PPP, was examined individually. Interestingly, their results show that PDGF exerts its effect in the previous cycle. In these studies, density-arrested BALB/c 3T3 cells were stimulated with PDGF followed by 5% PPP in the presence of methotrexate. This caused the cells to arrest in S phase. The S-phase cultures were then released and treated with PDGF for 2-hr periods and then placed back into PPP. Scher *et al.* (1979) found that any 2-hr treatment within 0 to 14 hr following the release from methotrexate (when cells were in S, G_2, or M) enabled 60–70% of the daughter cells to enter the next round of DNA synthesis without further addition of PDGF. If the daughter cells had progressed through approximately 7–9 hr of G_1 in 5% PPP without a previous 2-hr treatment of PDGF during S, or G_2 of the previous cycle, only 10% of the cells could enter into DNA synthesis. These results indicated that in order for a cell to enter S phase during the next generation, PDGF was required for at least a 2-hr period during or after its current S phase.

The effects of serum deprivation on the initiating of DNA synthesis in second generations of RAT 3Y1 cells were examined in a study published by A. Okuda and G. Kimura (1982, 1984). They found that cells that had traversed S phase and G_2 in the absence of serum required additional time to enter S phase in the second generation. The cells traversed S phase, G_2, and divided in the absence of serum. To determine the effect of serum deprivation during this time, they isolated mitotic cells, replated them in the presence of serum, and followed their entry into DNA synthesis by autoradiography. The length of the delay was approximately equal to the time of serum deprivation. In cells that traversed S and G_2 in the absence of serum (6 hr), the delay in the second generation was approximately 6 hr. In this report, 3 hr of serum deprivation at either the beginning or end of S phase resulted in an approximate 3-hr delay. They also reported that in this system, a 4-hr serum deprivation following mitosis resulted in a 4-hr delay in the cells' entry into S phase. It should be noted that these data are somewhat contradictory to the data from Zetterberg and Larsson (1985) and Larsson *et al.* (1989), who reported a much longer delay in the intermitotic times of serum-deprived G_1pm cells. In Okuda and Kimura's second study (1984), cells were grown in different combinations of growth factors in order to investigate the effects that individual growth factors had in the previous generation. For these experiments, they blocked the cells at S phase by aphidicolin treatment. After release, the cells were treated with various combinations of EGF, insulin, transferrin (TRS), serum, or Dulbecco's minimum essential medium (DMEM) for 6 hr. Mitotic cells were then collected and replated in DMEM *plus serum*. The results showed that only the cells that were treated with EGF, insulin, and TRS, exhibited a G_1 similar to cells exposed to serum. When individual growth factors were used during S phase and G_2, the time of entry into S phase of the daughter cells was the same as when no serum was present during the previous S and G_2 phases.

One can conclude from all the studies described in this section, as well as in the previous section, that events leading to DNA synthesis are influenced by events that occur in both previous and present generations, and that the potential for a daughter cell to respond to growth factors in a subsequent cycle is determined by events that occur in the previous cycle. The ultimate response by a cell to growth factors is the eventual entry into DNA synthesis followed by its division into two daughter cells. The molecular events occurring along this path have been the topic of many reports. The molecular signaling pathway that occurs from the time that growth factors bind to their respective receptors, present on the cell membrane, to their activation of certain genes required for cell proliferation, will be discussed in another chapter of this volume.

VI. MOLECULAR CONTROL POINTS WITHIN G_1

Normal cell proliferation depends on environmental signals, which can both stimulate and inhibit cell growth. The cell responds to growth factors through secondary signaling systems that result in a number of genetic responses (induction of gene sequences) within the cell. One of the most profound differences between G_0 and G_1 cells is the rapid induction of new mRNA species, which occurs within minutes after stimulation of quiescent cells (Pardee and James, 1975; Baserga, 1985; Prescott, 1987). The previously described evidence that RNA polymerase II-directed transcripts were necessary for the entry into S phase (Rossini *et al.*, 1980; Ingles and Shales, 1982), further suggests that new transcripts are required for progression through G_1. Furthermore, the fact that competence has a *memory*, in that the competent state can be transferred to quiescent cells, also suggests that synthesis of regulatory molecules is a functional part of the competence process (Smith and Stiles, 1981; Olashaw and Pledger, 1985). Finally, the finding that oncogenic retroviruses contain modified cellular genes that upon constitutive and/or overexpression, result in their transforming phenotype provides strong evidence that a group of genes has the ability to regulate normal and abnormal mammalian cell growth (Muller and Verma, 1984). A number of other chapters in this volume will discuss in detail the identification and functional analysis of growth-associated genes including the protooncogenes. However, with respect to growth control, it is important to consider the patterns of expression exhibited by growth-associated genes and whether these patterns are important in regulating cell-cycle progression.

A. Expression of Growth-Associated Genes following Serum Stimulation of Quiescent Cells

With the estimation that 3% of the mRNA species in logarithmically growing mouse fibroblasts are absent in quiescent cells (Williams and Penman, 1975), many laboratories (Cochran *et al.*, 1983; Linzer and Nathans, 1983; Matrisian *et al.*, 1985) in the early 1980s set out to identify newly synthesized mRNA species after serum stimulation of quiescent cells or mRNAs made in mutant cells conditionally blocked in mid G_1 (Lee *et al.*, 1983; Hirschhorn *et al.*, 1984). The finding that at least two protooncogenes, c-*fos* (Greenberg and Ziff, 1984; Kruijer *et al.*, 1984; Muller *et al.*, 1984; Bravo *et al.*, 1986) and c-*myc* (Campisi *et al.*, 1982, 1984; Kelly *et al.*, 1983), were expressed transiently at high levels immediately following stimulation of fibroblasts with growth factors, supported the notion that a set of genes must be under the control of growth factors. The concept that growth-factor stimulation results in the enhanced expression of specific cellular genes is now well established.

The original studies used cDNA cloning techniques, with modifications designed to amplify mRNA levels. They resulted in the isolation of hundreds of newly synthesized molecules now termed immediate-response genes. The important characteristics of all immediate-response genes is that they are superinduced by serum in the absence of protein synthesis (Cochran *et al.*, 1983; Henrickson and Scher, 1983; Cochran *et al.*, 1984; Muller *et al.*, 1984). This suggests that they are direct responders to serum factors, mostly to PDGF in the mouse G_0–G_1 system, and therefore do not require *de novo* protein synthesis.

Stiles and co-workers were able to isolate several cDNA clones that were superinduced by purified PDGF in quiescent mouse BALB/c 3T3 cells (Cochran *et al.*, 1983). Two of the clones, KC and JE, have been extensively studied. Both KC and JE have been used by many investigators as molecular markers for the mitogenic response to growth factors, antimitogenic agents, and other biological response modifiers. In general, both are induced within 1 hr following stimulation, much like c-*fos* and c-*myc*, and were thought to play a role in competence. However, unlike c-*fos* and c-*myc*, JE is not an intracellular mediator of growth-factor action, but rather has been shown to encode a small secretory glycoprotein, which is homologous to several cytokines (Rollins *et al.*, 1987, 1988).

The laboratory of Dr. Dan Nathans has identified some of the most interesting immediate-early genes. Their interest in identifying immediate-response genes came from the observation that during tumor-virus replication, many of the viral genes expressed during the early stages of an infection are regulatory in nature. They speculated that, like the viral system, many of the immediate-response genes would also be regulatory genes. Nathans' group initially cloned several cDNA clones by differential screening between actively growing BALB/c 3T3 and serum-deprived cells (Linzer and Nathans, 1983). One clone, designated 28H6, was found to be maximally expressed between 12 and 18 hr after stimulation, and was subsequently termed proliferin, since it had 40% homology to prolactin (Linzer and Nathans, 1984). It has recently been shown to be identical to mitogen-regulated protein (MRP) of Swiss 3T3 cells (Parfett *et al.*, 1985). In subsequent studies, they utilized the observation that cyclohexamide amplifies direct-responding messages, while at the same time, it inhibits secondary-responding messages that require protein synthesis (Lau and Nathans, 1985). They isolated five cDNA clones whose expression was very similar to those of c-*fos* and c-*myc*. The clone designated 3CH96 turned out to have a high amount of homology to the viral gene v-*jun* (Ryder and Nathans, 1988). It has now been shown that there is a family of molecules encoding genes with the transactivating functions of AP-1 (Bochman *et al.*, 1987). There are at least three murine members (Ryder *et al.*, 1989) and two human members (Mattei *et al.*, 1990). A second clone, 3CH77, now termed nur-77, has recently been shown to be homologous to members of the thyroid hormone nuclear-receptor superfamily (Hazel *et al.*, 1988). This gene is induced as early as is c-*fos* after quiescent cells are stimulated to proliferate (Lau and Nathans, 1985; Hazel *et al.*, 1988).

A different approach was utilized by Baserga's laboratory for the identification of genes expressed during the transition from G_0 to G_1 (Hirschorn *et al.*, 1984). These investigators utilized a temperature-sensitive cell line, ts13. When incubated at the nonpermissive temperature, ts13 cells are known to be blocked 6 hr before S phase (Rossini and Baserga, 1978). This allowed them to divide the clones into early and late genes. Three of the clones (2A9, 2F1, and 4F1) have been extensively characterized. The clone designated as 2A9, subsequently named calcyclin, has strong homology to the S-100 protein and other calcium-binding proteins (Calabretta *et al.*, 1986). It also contains a domain highly homologous to the so-called p11 or p10 subunit of a protein complex, which is the major substrate for viral tyrosine–specific kinase. Since there is a large body of evidence for the role of both calcium and tyrosine kinases in proliferation, it is possible, although not yet proven, that this gene may fit as a link between growth factor–receptor interaction and nuclear signaling (Baserga *et al.*, 1986). The second clone, which was designated as 4F1, has subsequently been shown to be vimentin, a cytoskeleton protein (Ferrari *et al.*, 1986). Both of these clones have also been utilized by investigators as molecular markers for proliferation (Calabretta *et al.*, 1985; Rittling *et al.*, 1985, 1986; Cosenza *et al.*, 1988).

A recent report from Bravo's laboratory (Almendral *et al.*, 1988) stresses the importance of the induction of new genes following stimulation of quiescent cells. They were able to show that the stimulation of proliferation of quiescent cells by growth factors is accompanied by a complex genetic program. By differential screening, they were able to identify 82 independent genes that were found to be induced immediately following serum stimulation. All the messages were shown to have very short half-lives, much like that of c-*fos*. The regulation of the stability of these genes is all the same, i.e., cyclohexamide stabilizes the transcripts, suggesting that, like c-*fos* and other immediate-early genes, there may be a common mechanism controlling the shutoff of the genes and the half-lives of the mRNAs. The authors proposed that genes whose expression is very transient may not be essential during the cell cycle, but only for G_0–G_1 transition, while others such as c-*myc*, whose half-life is slightly longer, may be essential for G_1 progression into S phase. The need for a certain number of these gene products for particular stage-specific functions has been investigated and will be reviewed later in this chapter.

B. Mapping of G_1 Based on the Time of Maximal Expression of Growth-Associated Genes

The role of growth-associated genes during mammalian growth has been widely studied using a variety of tissue-culture systems. The best characterized of these has been the study of gene expression following stimulation of quiescent cells. As described in the previous section, expression of many genes is found to be maximal very soon after stimulation; however, the time of maximal expression of many others occurs at other times during G_1 progression. Based on the temporal pattern of gene expression, the transit from G_0–G_1 to S can be molecularly divided into discrete periods: early, mid, and late G_1 and S phase-specific (Baserga, 1985; Rittling *et al.*, 1986, Denhardt *et al.*, 1987; Toscani *et al.*, 1988; Cosenza *et al.*, 1988). The correlation of the time of maximal accumulation of steady-state levels of message to the pattern of protein accumulation is good, but not 100% (Baserga, 1985). This is also not to say that the time of maximal accumulation of message is also the time that the particular gene activity is needed for proliferation, but the system does allow mapping a few landmarks known to occur during G_0–G_1 progression. It should be noted that there is considerable overlap between the various groups, and there may be some cell-to-cell variation in the time at which some of these genes are expressed.

The immediate-response genes cloned by the differential-screening technique make up the majority of genes that can be classified as early genes. Examples of this class of molecules already described above are c-*fos* (Greenberg and Ziff, 1983; Cochran *et al.*, 1983), c-*myc* (Greenberg and Ziff, 1984, Kelly *et al.*, 1983), JE (Cochran *et al.*, 1983), *nur*-77 (Lau and Nathans, 1985), *jun*-A (Lau and Nathans, 1985; Ryder and Nathans, 1988), and vimentin (Hirschhorn *et al.*, 1984; Rittling *et al.*, 1986; Cosenza *et al.*, 1988). Representative genes induced during mid to late G_1 are ornithine decarboxylase (ODC) (Campisi *et al.*, 1984; Kahana and Nathans, 1984), p53 (Reich and Levine, 1984) and calcyclin (Hirschhorn *et al.*, 1984; Rittling *et al.*, 1986; Cosenza *et al.*, 1988). The last group of genes are those induced at the G_1–S border and during S, and include (but are not restricted to) the gene products necessary for DNA synthesis. Examples include histones (H3 and H4) (Owen *et al.*, 1987; Seiler-Tuyns and Birnsteil, 1981), proliferating cell nuclear antigen (PCNA) (Bravo *et al.*, 1987), and thymidine kinase (TK) (Johnson *et al.*, 1982; Coppock and Pardee, 1985; Rittling *et al.*, 1986; Cosenza *et al.*, 1988).

The study of cell cycle-regulated genes during mammalian cell proliferation is important, since growth-associated genes are clearly important for the regulation of processes involved in cell growth. However, for practical reasons, the majority of these studies have been carried out using tissue-culture conditions that study the transition from G_0–G_1 and entry into S phase in fibroblasts and peripheral blood lymphocytes. Since the observed patterns and/or roles that these genes presumably play in this transition may be very different or nonexistent during active proliferation, it is also necessary to define these patterns of gene expression in cells that have not entered G_1 from G_0.

C. Growth-Associated Gene Expression during Active Proliferation

As stated previously, events that occur during G_1 regulate cell growth. This includes not only the transition from quiescence to active growth, but also the maintenance of growth in cells that are already cycling. Thus we shall now review what is known about the events that occur during G_1 in actively proliferating cells, in order to compare them with molecular events known to occur after stimulation of quiescent cells. Since activation of many genes studied in the G_0–G_1 transition systems seems to be dependent on environmental signals used for the stimulation process, the activation and role of the same genes may be entirely different in nongrowth-arrested, actively proliferating cells. The majority of the previous studies analyzing growth-associated gene expression during active growth have concentrated on only a few of the known genes that are cell-cycle dependent in the G_0–G_1 transition systems. Most of these studies employed drugs or counter-flow elutriation methods in order to synchronize their cultures.

Since it has been well established that c-*myc* is important for cell growth (Muller and Verma, 1984), this gene has been the focus of a few studies involving synchronous cultures (Hann *et al.*, 1985; Thompson *et al.* 1985; Griep and Westphal, 1988; Neckers *et al.*, 1988; Prochownik *et al.*, 1988). In order to study the pattern of expression and protein accumulation of c-*myc* during each stage of the cell cycle, Thompson *et al.* (1985) and Hann *et al.* (1985) examined c-*myc*expression in cell populations collected by counter flow elutriation. Both studies concluded that message and protein levels were invariant during active proliferation. In the same studies, they were able to show cell cycle-dependent expression of thymidine kinase and histone. Data from a more recent study using M_1 cells arrested at different stages of the cell cycle suggested that c-*myc* levels may be cell cycle-dependent during active proliferation (Neckers *et al.*, 1988). These investigators arrested the cells during early G_1 by a density-dependent mechanism previously shown to synchronize the cells 6 hr before S phase. A second population of cells was synchronized at the G_1–S border by aphidicolin treatment (Tsuda, 1987). A comparison of c-*myc* and transferrin-receptor expression in both of these populations of cells to steady-state levels found in actively proliferating populations, found that the levels of both of these messages were not constant throughout G_1. C-*myc* was expressed in cells arrested in early G_1 at levels similar to those found in an actively growing sample. However, low levels were detected in cells arrested at the G_1–S border. It was concluded that c-*myc* expression is higher during early G_1 while cells are actively proliferating. In addition, the temporal pattern of expression observed for c-*myc* steady-state levels of mRNA was also observed at the protein level. The discrepancy between the work of Neckers and that reported by Thompson most likely reflects the differences in the method of obtaining *synchronous* populations. This is confirmed in very recent work by Cosenza *et al.* (1991) in which growth-associated gene expression was analyzed at frequent intervals in a highly synchronized population of cells. These investigators also found that c-*myc* was expressed at high levels in early G_1 and then reduced levels in late G_1 and S (see below and see Fig. 4).

A number of isolated studies have investigated the expression and activity of various genes in various cell lines. These include Hsp70 (Milarski and Morimoto, 1986), DNA polymerase alpha (Wahl *et al.*, 1988), primase p49 (Tseng *et al.*, 1989), thymidine kinase (Sherley and Kelly, 1988), and cyclin (Pines and Hunter, 1989). Milarski and Morimoto (1986) were able to show, using mitotically synchronized HeLa cells obtained by mitotic shake-off, that HSP70 was cell-cycle regulated during active proliferation. They reported, on both the RNA and protein levels, barely detectable levels observed during G_1, followed by an induction as the cells entered S phase. Wahl *et al.* (1988) analyzed the expression of DNA polymerase alpha using samples obtained from elutriation as well as RNA samples from Milarski's laboratory. They were able to show, by both methods, that there was cell-cycle regulation of histone H3. However, they observed no fluctuations in the levels of DNA polymerase alpha in samples from the elutriation method, with only a slight increase at S phase followed by a 2.8-fold decline in G_2 in the mitotically synchronized samples. Sherley and Kelly (1988) obtained cells in G_1 by elutriation, and unlike the previous examples, replated and harvested cells at various times after progression. They then analyzed the expression and activity of thymidine kinase, and analyzed histone expression. Unlike the studies by Thompson *et al.* (1985), they did not observe an induction of thymidine kinase RNA during S phase but did see an increase in enzymatic activity. On the other hand, they were able to show an induction of histone message.

Combining all the data from the various reports cited, while keeping in mind the tremendous experimental variations, it appears that genes shown to be cell-cycle regulated in the G_0–G_1 transition systems may or may not show the same or any cell-cycle regulation during active proliferation. The question of cell-cycle regulation during active proliferation is still unanswered, and an exhaustive survey of many growth-associated genes employing mitotically synchronized cells should provide the answer. This type of study facilitates the mapping of G_1 in a manner similar to that described in the G_0–G_1 system (Denhardt *et al.*, 1987).

Such an extensive study has recently been reported by Cosenza *et al.* (1991). These investigators were able to isolate, by a noninductive, drug-free system, a population of highly synchronized Swiss 3T3 cells within mitosis (> 90%) in numbers sufficient to determine the pattern of expression of 10 representative growth-associated genes at frequent intervals. Their results, summarized in Fig. 4 and Table I, show that

1. growth-associated gene expression is not constant in Swiss 3T3 cells traversing G_1 after exiting mitosis. The steady state mRNA levels of these genes become elevated and reduced at very specific times following mitosis, just as they do after quiescent cells are stimulated to proliferate;
2. the expression of four representative immediate-early genes (c-*fos*, c-*jun*, c-*myc* and JE), previously shown to be associated with the G_0–G_1 transition, is also associated with the M–G_1 transition; and
3. while the pattern of expression of some growth-associated genes was iden-

tical to that described following exit from G_0 (i.e., c-*fos*, JE, and c-*myc* were expressed in early G_1, while ODC and p53 were expressed in mid G_1), the pattern of expression of other early G_1 genes was completely different (i.e., c-*jun* and *nur*-77 were expressed throughout G_1, while vimentin and calyclin were expressed in late G_1 and S).

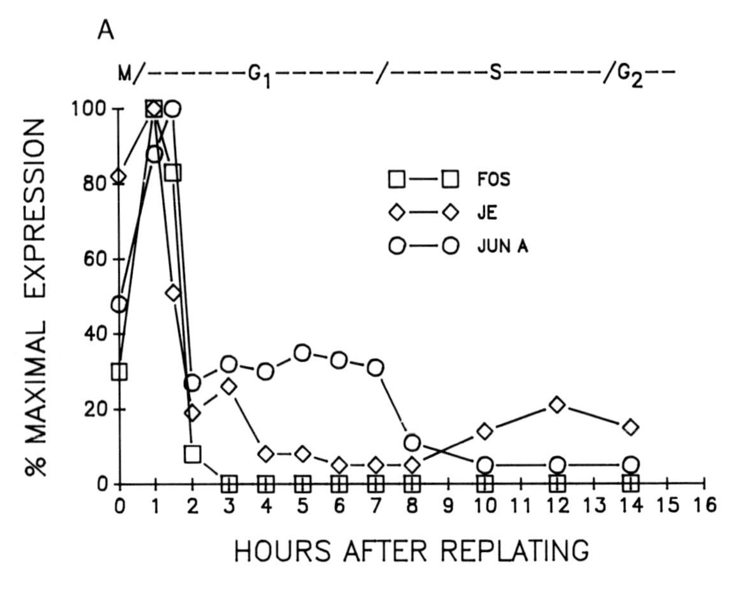

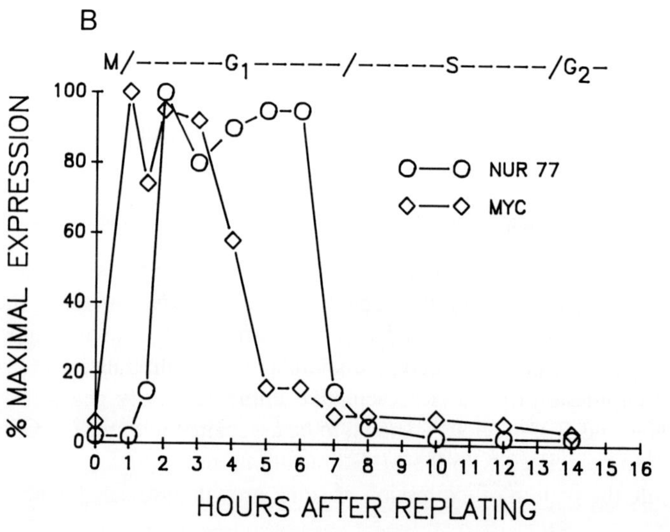

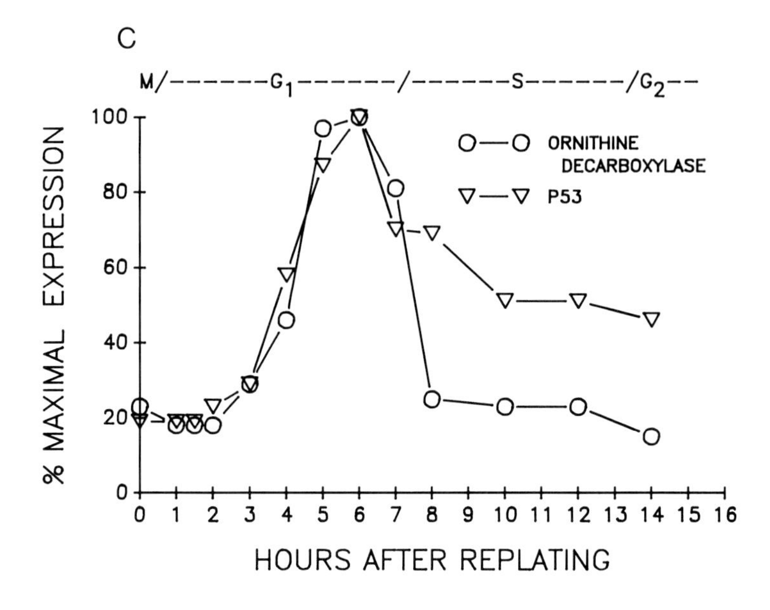

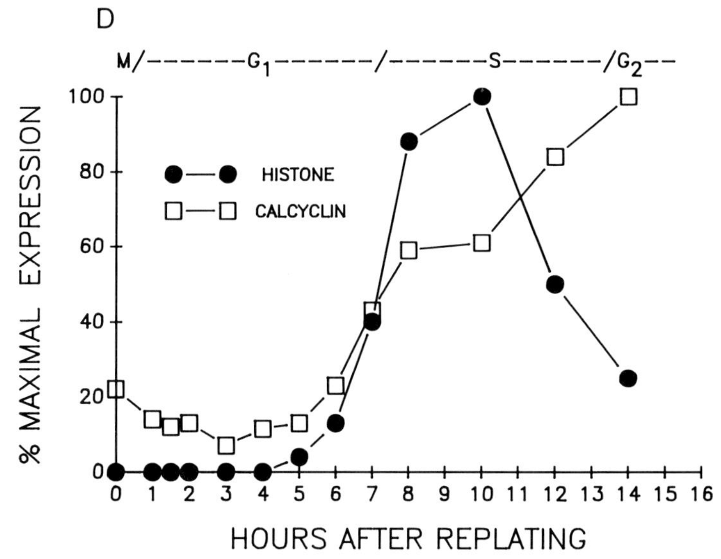

Fig. 4. Expression of growth-associated genes after replating mitotically synchronized swiss 3T3 cells. Mitotically synchronized Swiss 3T3 cells were isolated by the large-scale mitotic shake-off method and immediately replated into prewarmed conditioned medium. At the indicated time points, cells were harvested, and total cytoplasmic RNA was isolated. Steady-state levels of a variety of representative growth-associated transcripts were determined by either Northern blot or RNase protection. Autoradiograph signals were quantitated using a scanning densitometer, and the relative integrated density was determined and expressed relative to the highest sample. (A)c-*fos*, JE, jun A (c-*jun*); (B) c-*myc*, *nur*-77; (C) ODC, p53; (D) histone, calcyclin.

TABLE I

Summary of Growth-Associated Gene Expression during Active Proliferation of Swiss 3T3 Cells

	Time of maximal expression[a]	
Gene	G_0–G_1	M–G_1
c-fos	Early G_1	Early G_1
JE	Early G_1	Early G_1
jun A	Early G_1	Early G_1
c-myc	Early G_1	Early–Mid G_1
nur-77	Early G_1	Early–Late G_1
ODC	Mid-G_1	Mid-G_1
p53	Mid-G_1	Mid-G_1
Vimentin	Early G_1	Late G_1–S
Calcyclin	Late G_1–S	S–G_2
Histone	S	S

[a]The times of maximal expression shown in the G_0–G_1 system are representative times derived from published reports utilizing murine cell lines. These times may vary from cell line to cell line and from species to species. The times of maximal expression indicated in the M–G_1 system were taken from Cosenza *et al.* (1991).

Studies such as these permit the construction of a temporal *molecular map* of G_1. While time of expression does not always dictate the time when a gene product functions, a comparison of the temporal patterns of expression can be used to distinguish events likely to be specific for processes involved in G_1 progression from those that represent processes involved in the exit from M or G_0. Thus, genes such as c-*fos*, JE, c-*myc*, ODC, and p53, which exhibited the same temporal patterns of expression in both model systems, most likely mediate functions common to all cells entering and progressing through G_1. Conversely, genes such as *nur*-77, vimentin, and calcyclin, which exhibited different patterns of expression, most likely mediate functions with at least different temporal requirements, if not different functions entirely during G_1. The nature of these differences depends on the origin of the G_1 cell. Obviously the precise functions and why they are apparently different still remain to be determined. Suffice it to say at this point that growth-associated gene expression is highly regulated during G_1, regardless of whether the cell enters G_1 from mitosis or quiescence. The importance of this regulation for initiation and maintenance of cell proliferation can be determined only by altering these observed patterns.

VII. MODULATION OF CELL GROWTH BY MODULATION OF GROWTH-ASSOCIATED GENE EXPRESSION

A. Inappropriate Expression

Since the first reports showing that the transforming genes encoded by retroviruses had cellular counterparts, numerous investigations into the functions that these genes play during both normal and abnormal cell proliferation (i.e., transformation) have been undertaken. Many of these studies have demonstrated that introduction of cellular oncogenes into cells can lead to profound physiological consequences including changes in cell growth, immortalization, reduction in growth-factor requirements, and changes in gene expression (Kelly *et al.*, 1983). Many of the original oncogene studies described experiments which suggested that cooperation between oncogenes is required for transformation of normal cells into tumor cells (Bishop 1983, 1985, 1987; Parada *et al.*, 1982; Land *et al.*, 1983a, 1983b, 1983c; Ruley, 1983; Jenkins *et al.*, 1984; Eliyahu *et al.*, 1984). An example of cooperation between two cellular oncogenes can be seen from studies involving v-*myc* and v-Ha-*ras*. When normal cells, such as rat embryo fibroblasts, were transfected with either of these genes alone, there was incomplete transformation. However, if the genes were transfected concomitantly, the cells were able to exhibit fully transformed phenotypes (Land *et al.*, 1983a,b,c). These results, in combination with those of many other experiments, suggested that oncogenes act in distinct and complementary ways on the cellular phenotype. By using the above experimental system of cooperation, investigators were able to replace the functions of *myc* or *ras* with other genes in order to test their roles in transformation. From such studies, it was possible to place genes into two functional classes (Land *et al.*, 1983a,b,c) depending on the ability of each to replace *myc* or *ras* in the cotransfection transformation assay. The first class includes the three *ras* oncogenes (Barbacid, 1987) and polyomavirus middle T, since each could replace *ras* in the cotransfection assay. The second class encompasses N-*myc*, p53, polyomavirus large-T antigen, the N-terminal proximal portion of SV40 large-T, and adenovirus Ela oncogenes, since each of these could replace c-*myc* in the cotransfection assays (Land *et al.*, 1983a,b,c). It was subsequently shown that very high expression of c-*myc* alone could produce transformed phenotypes (Land *et al.*, 1986; Lee *et al.*, 1985). Utilizing special culture conditions, high levels of activated *ras* could also transform without cooperation of class two oncogenes (Spandidos and Wilkie, 1984; Lee *et al.*, 1985).

There are now a large number of examples in the literature in which inappropriate expression of growth-associated genes results in changes in growth-factor responsiveness as well as modulation of endogenous genes. At least three reports show that high levels of c-*myc* can replace PDGF and make 3T3 cells grow in PPP. Armelin *et al.* (1984) were able to show that cell lines that contained high constitutive levels of c-*myc* required only plasma plus dexamethasone to proliferate. If c-*myc* was microinjected into quiescent cells, they could be stimulated to enter S phase with plasma alone (Kaczmarek *et al.*, 1985). The levels of endogenous c-*myc* were also shown to regulate growth-factor requirements. When the level of c-*myc* was elevated by cyclohexamide treatment, the requirement for PDGF was lost (Kaczmarek *et al.*, 1986). In a slightly different system, WI-38 cells arrested for fewer than 15 days contained high levels of c-*myc* mRNA and could be stimulated to enter S phase with PPP alone. Conversely, cells growth arrested for longer periods had lower levels of c-*myc* mRNA and other growth-associated gene transcripts. Full serum was required to stimulate these cells to enter S phase (Ferrari *et al.* 1988). In a recent report, it was shown that increased levels of normal human c-*myc*, transduced into primary mouse fibroblast as a minigene, altered the duration of G_1 in both serum-stimulated and exponentially growing cells (Karn *et al.*, 1989). There was a rough correlation with the levels of c-*myc* proteins and the rate at which the cells entered S phase. They showed that the length of G_1 could vary by as much as 30%. Control cells had an average G_1 of 6.5 hr, whereas the average G_1 of the recombinant cell lines was shortened to 4.6 hr. There was no observable effect on the other stages of the cell cycle. Thus, as stated earlier, there is considerable evidence that overexpression or inappropriate expression of one or more growth-associated genes can alter growth factor requirements and G_1 progression.

B. Down-Regulation of Expression

Another approach used to study the role of growth-associated gene expression in control of cell proliferation has been to inhibit the production of a growth-associated gene product and determine the effect on cellular proliferation and/or differentiation. Three major techniques are commonly employed for such studies. Microinjection of specific antibodies has been used by several investigators as an approach to study c-*fos* (Riabowol *et al.*, 1988), p53 (Mercer *et al.*, 1982, 1984a), a variety of viral transforming proteins (Antman and Livingston, 1980; Floros *et al.*, 1981; Mercer *et al.*, 1983), and RNA transcription enzymes such as RNA polymerase I (Mercer *et al.*, 1984b). The major problem with this approach is the limited number of cells that can be analyzed. A second approach utilizes antisense molecules to selectively suppress gene expression. The earliest antisense experiments relied on the use of expression vectors, which produced antisense molecules by transcription from inducible promoters such as the LTR from MMTV or from constitutively active promoters such as the SV40 enhancer region (see Green *et al.*, 1986 for review). This technique has been used to produce stable recombinant cell lines producing antisense c-*myc* (Lachman *et al.*, 1986; Griep and Westphal, 1988; Prochownik *et al.*, 1988), c-*fos* (Nishikura and Murray, 1987; Mercola *et al.*, 1987; Holt *et al.*, 1986; Edwards *et al.*, 1988), and thymidine kinase (Kim and Wold, 1985). Some of the difficulties with this technique include controlling gene expression, since most of the inducible promoters are leaky; positional effects after integration, which could influence transcription; and the low transfection efficiency of certain cell lines. A recent modification of this technique utilizes synthetic antisense oligodeoxynucleotides. The antisense oligomers, 18–20 bases in length, are added directly to the medium. Thus there is very little manipulation of the cultured cells (see Marcus-Sekura (1988); and Stein and Cohen (1988) for reviews). This system has been utilized successfully to reduce the levels of a number of virally encoded genes (Smith *et al.*, 1986), c-*myc* (Heikkila *et al.*, 1987; Wickstrom *et al.*, 1988), PCNA (Jaskulski *et al.*, 1988), c-*myb* (Gewirtz *et al.*, 1988, 1989), c-*abl* (Caracciolo *et al.*, 1989), and bFGF (Becker *et al.*, 1989). Among the difficulties with this system is the fact that the amount required for total inhibition of protein synthesis may vary from gene to gene, and the secondary structure of the antisense molecule may influence its ability to hybridize to the complementary message (Wickstrom *et al.*, 1988). The antisense may also be broken down by nucleases in the medium; however, this has not been an experimental problem in most systems. Moreover, the stability of the oligomer can be increased by using analogs that possess modifications at the phosphate–oxygen bond by replacement with a number of different functional groups. These changes effectively increase the stability, but they also decrease the ease of synthesis and in some cases, decrease permeability (Stein and Cohen, 1988; Marcus-Sekura, 1988). The studies involving inhibition of c-*myc* and c-*fos* are the most interesting with respect to cell-cycle control.

Antisense studies focusing on the role of c-*myc* during cellular proliferation and differentiation have shown that in all cases, a decrease in the endogenous levels of c-*myc* resulted in a reduction in the growth of the target cells. The first report describing the use of antisense molecules specific for c-*myc* was from Heikkila *et al.* in 1987. This group used exogenously added antisense oligodeoxynucleotides complementary to human c-*myc* to study the cell-cycle progression of normal human peripheral blood lymphocytes (PBLs) following PHA stimulation. They found that preincubation for 4 hr with 30 μM of antisense specific for c-*myc* resulted in a 90% reduction in the level of endogenous myc protein. They found that this amount of antisense inhibited mitogen-stimulated lymphocyte proliferation by at least 75%. It is interesting to note that although entry into S phase was inhibited, the induction of mid and late G_1-specific molecular markers subsequent to the appearance of c-*myc* was not inhibited. This suggested that the cells were able to progress through a large part of G_1 and stop at a point very close to S phase. In a more recent report, investigators using expression vectors constitutively expressing antisense c-*myc* studied the proliferation and growth of F-MEL cells (Prochownik *et al.*, 1988). They found that these transformed cells that expressed high levels of c-*myc* antisense transcripts could grow as well as the control cells in medium containing 10% fetal calf serum, but were significantly growth inhibited when grown in medium supplemented with 5 or 2% fetal calf serum. These investigators were also able to show that clones expressing high levels of antisense transcripts did differentiate faster in the presence of DMSO, suggesting that high levels of c-*myc* protein inhibit the entrance into the differentiation pathway, owing to its growth-promoting activity. This is an agreement with previous reports showing that F-MEL cells transfected with and constitutively expressing exogenous c-*myc* transcripts failed to differentiate. Therefore the above results clearly indicate that inhibiting the expression of c-*myc* will result in decreased cellular proliferation which, at least for these particular cells, increases the potential for entering the differentiation pathway (Griep and Westphal, 1988).

It should be noted, however, that recent studies by Whitman *et al.* (1990) have indicated that modulation of expression of c-*myc* in and of itself does not induce differentiation. These investigators showed that in F9 cell aggregates induced by retinoic acid to differentiate into visceral endoderm containing embryoid bodies, c-*myc* expression was reduced within minutes after addition of retinoic acid, the differentiating agent. However, the reduction in c-*myc* levels was the result of inhibition of growth by aggregation, since F9 cells allowed to aggregate in the absence of retinoic acid (and which therefore did not differentiate) also showed a reduction in c-*myc* levels concomittant with a reduction in histone, a molecular marker for growth. They also observed that later during the course of differentiation, when growth resumes, levels of c-*myc* increased in parallel. Therefore, it would appear that reduction in the levels of c-*myc*, at least in this cell system, represent an effect of the differentiation process and not the cause.

A function for c-*fos* in normal growth was first suggested by the rapid and tremendous increase in the transcription of c-*fos* within minutes after treatment of quiescent cells with a variety of mitogenic compounds (Greenberg and Ziff, 1984; Kruijer *et al.*, 1984). On the other hand, its association with cell growth has also been somewhat confused by reports showing that increased expression of c-*fos* accompanies the cessation of growth and induction of differentiation in HL60 cells (Gonda and Metcalf, 1984), Swiss 3T3 cells (Toscani *et al.*, 1988), F9 cells (Muller and Verma, 1984) and PC12 cells (Curran *et al.*, 1985).

In light of these results, antisense studies involving inducible vectors containing c-*fos* antisense sequences were undertaken by a variety of laboratories. The first paper employing such a vector came in 1986 by Holt *et al.* (1986). These investigators showed that induction of the antisense transcript by dexamethasone treatment resulted in the inhibition of growth of three Swiss 3T3 clones harboring the antisense hybrid vectors. The doubling times for all the control cells were roughly 24 hr in the absence of DEX; however, in the presence of DEX, the antisense clones exhibited cell-doubling times that ranged from 48 to more than 72 hours. These data suggested that c-*fos* was indeed necessary for normal growth. A subsequent paper studied not only the effect of antisense c-*fos* on active growth, but also on the G_0–G_1 transition in NIH 3T3 cells (Nishikura and Murray, 1987). It was shown that there was only a small inhibitory effect on active growth due to the presence of antisense c-*fos*, and this was observed after only 2 days growth in the presence of dexamethasone. However, they were able to show that induction of c-*fos* antisense RNA inhibited the stimulation of quiescent cells to proliferate. Another recent report used microinjection of antibodies specific for c-*fos* to address the question of the role of c-*fos* during cell growth (Riabowol *et al.*, 1988). This study showed that microinjection of c-*fos* antibodies blocked the entry of serum-stimulated quiescent rat embryo fibroblasts when injected within 6 hr following serum stimulation. They also showed that if the antibodies were injected into actively proliferating cells, a large number of cells were prevented from entering S phase. Studies using synchronized cells suggested that c-*fos* activity may be required in early G_1 during active proliferation. These studies provide evidence that c-*fos*, like c-*myc*, is necessary for the G_0–G_1 transition as well as for continual proliferation.

Most recently, as a logical followup to their observations concerning the induction of c-*fos* and c-*jun* in cells entering G_1 from mitosis, Cosenza and Soprano have completed a study to determine the role of immediate-early gene expression on active cell growth and, in particular, the M–G_1 transition. They used the antisense oligomer approach to examine whether the observed expression of c-*fos* and c-*jun* during the first 2 hr after exit from mitosis (described previously) was required for G_1 progression and entry into S. As shown in Fig. 5, they found that treatment of actively proliferating Swiss 3T3 cells with either anti–c-*fos* or anti–c-*jun* oligodeoxynucleotides inhibited both DNA synthesis and cell division. Moreover, Fig. 6 shows that just a 2-hr treatment of mitotic cells with 10μ*M* anti–c-*fos* or anti–c-*jun* would block progression through G_1 and entry into S. Interestingly, addition of anti–c-*fos* at 2 hr after completion of mitosis (after the peak of c-*fos* mRNA induction) had no effect on entry into S, whereas addition of anti–c-*jun* did. These results suggested that expression of both immediate-early genes during the M–G_1 transition was required for subsequent progression through G_1 and entry into S. However, once the cells had passed through the first 2 hr of G_1, they no longer needed c-*fos* but still needed c-*jun*. The role of c-*jun* later in G_1 remains to be determined. However, it is interesting that Carter *et al.* (1991) have reported that in cells stimulated to enter G_1 from quiescence, c-*jun* is induced both at the G_0–G_1 border and at late G_1–S. It is possible that the ability of anti–c-*jun* to block entry into S, when added in mid G_1, is the result of inhibiting this as-yet-undefined late G_1 function of c-*jun*.

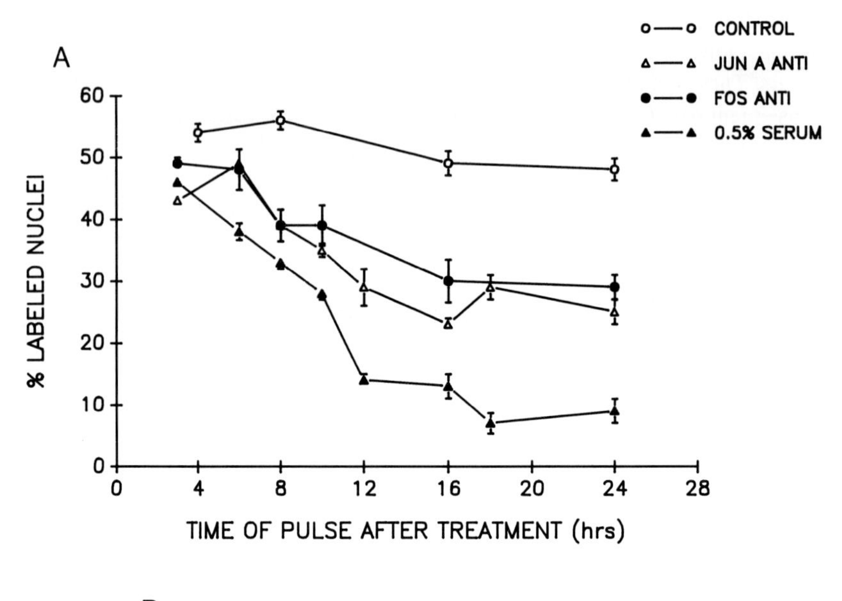

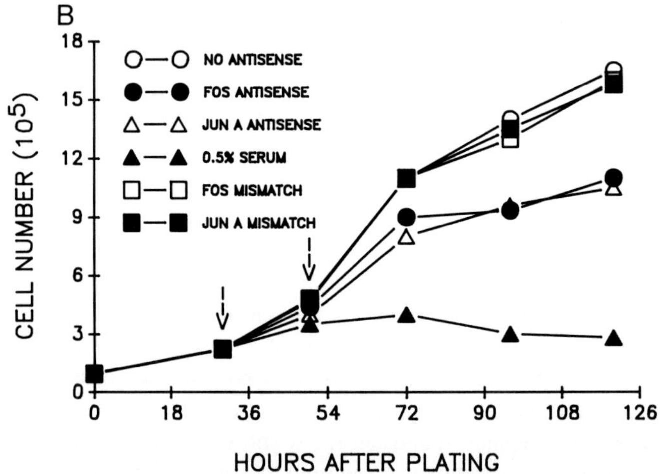

Fig. 5. Effect of antisense c-*fos* and c-*jun* oligodeoxynucleotide treatment on actively proliferating Swiss 3T3 cells. Two groups of actively growing Swiss 3T3 cells were treated with 10μM anti–c-*fos* or anti–c-*jun* oligodeoxynucleotides. One group (shown in panel A), which was grown on coverslips, was incubated with ^{3}H-thymidine and analyzed at the indicated time points for entry into DNA synthesis. The second group (shown in panel B) was assayed for ability to divide by determination of cell number. Controls for these experiments included incubation with no oligodeoxynucleotides or incubation with mismatch oligodeoxynucleotides in which only 2 bases were altered. The 0.5% serum control represented in the maximal level of inhibition that could be expected.

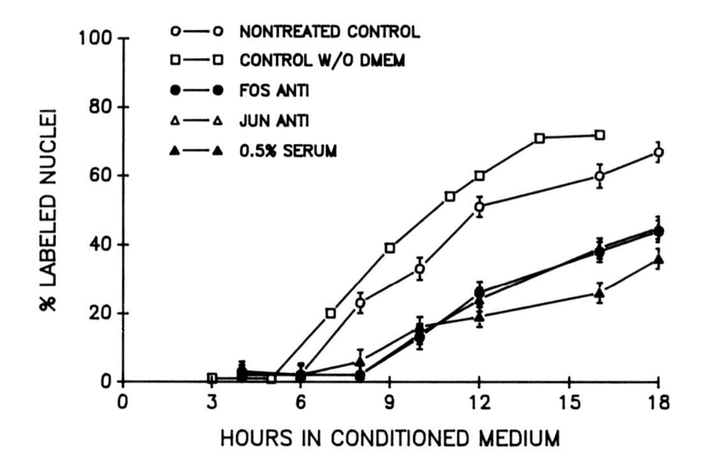

Fig. 6. Effects of antisense c-*fos* and antisense c-*jun* on the entry into S of mitotically synchronized Swiss 3T3 cells. Swiss 3T3 cells were isolated by large-scale mitotic shake-off, replated and treated for 2 hr with serum-free medium containing 10μ*M* of antisense oligodeoxynucleotides. At the indicated times, three coverslips per treatment were harvested, fixed, and prepared for autoradiography. The mean percentage labeled nuclei from three independent coverslips per time point along with the standard error bars were plotted. Symbols: (closed circles) c-*fos* antisense; (open triangle) jun A antisense; (open circles) sense-treated controls. In addition to the sense-treated controls, two additional controls were used. One group of cells was not incubated for 2 hr in serum-free medium. These cells (open squares) entered S 2 hr earlier but with the same kinetics and efficiency. A second group of cells was treated with 0.5% fetal bovine serum (closed triangles). Since it is known that serum is required to progress through G_1, this was considered a control for the maximal amount of inhibition that could be expected by treatment with the antisense of oligodeoxynucleotides.

VIII. CONCLUSION

We have reviewed cell cycle and cell-growth control at the kinetic, biochemical, and molecular level. A tremendous amount of progress has been made over the past 40 years since the first description of the cell cycle by Howard and Pelc in 1951. We can now induce nongrowing cells to proliferate and make proliferating cells quiescent by simply modulating the expression of a single gene or group of genes. Thus, we do know at least some of the events that regulate the cell cycle and cell growth. The task of the next 40 years will be to determine the mechanism by which these regulatory events function.

ACKNOWLEDGMENTS

This work was supported by a grant from the W.W. Smith Charitable Trust.

REFERENCES

Almendral, J. M., Sommer, D., Macdonald-Bravo, H., Buckhardt, J., Pereara, J., and Bravo, R. (1988). Complexity of the early genetic response to growth factors in mouse fibroblasts. *Mol. Cell. Biol.* **8**, 2140–2148.

Antman, K. H., and Livingston, D. M. (1980). Intracellular neutralization of SV40 tumor antigens following microinjection of specific antibodies. *Cell* **19**, 627–635.

Armelin, H. A., Armelin, M. C. S., Kelly, K., Stewart, T., Leder, P., Cochran, B. H., and Stiles, C.D. (1984). Functional role for c-*myc* in mitogenic response to platelet-derived growth factor. *Nature* **310**, 655–660.

Augenlicht, L. H., and Baserga, R. (1974). Changes in the G_0 state of WI-38 fibroblasts at different times after confluence. *Exp. Cell Res.* **89**, 255–262.

Barbacid, M. (1987). Ras genes. *Annu. Rev. Biochem.* **56**, 779–827.

Barka, T. (1965). Induced cell proliferation: The effect of isoproterenol. *Exp. Cell Res.* **37**, 662–679.

Barnes, D., and Sato, G. (1980). Serum-free cell culture: A unifying approach. *Cell* **22**, 649–655.

Bartholomew, J. C., Yokota, H., and Ross, P. (1976). Effect of serum on the growth of $BALB_c$ 3T3 mouse fibroblasts an SV40-transformed derivative. *J. Cell. Physiol.* **88**, 277–286.

Baserga, R., Estensen, R. D., and Petersen, R. O. (1968). Cell proliferation in mouse kidney after a single injection of folic acid. *Lab. Invest.* **19**, 92–96.

Baserga, R. (1976). "Multiplication and Division in Mammalian Cells." Marcel Dekker, New York.

Baserga, R. (1985). "The Biology of Cell Reproduction." Harvard University Press, Cambridge, Massachusetts.

Baserga, R., Waechter, D. E., Soprano, K. J., and Galanti, N. (1982). Molecular biology of cell division. *Ann. N. Y. Acad. Sci.* **397**, 110–120.

Baserga, R., Battini, R., Calabretta, B., Kaczmarek, L., Ferrari, S., Rittling, S., and Gibson, C. (1986). Cell-cycle genes and growth regulation. *In* "Biological Regulation of Cell Proliferation" (R. Baserga, P. Foa, D. Metcalf, and E. E. Polli, eds.), pp. 13–23. Raven Press, New York.

Beach, D., Durkacz, and Nurse, P. (1982). Functionally homologous cell cycle–control genes in budding and fission yeast. *Nature* **300**, 706–709.

Becker, H., Stanners, C. P., and Kudlow, J. E. (1971). Control of macromolecular synthesis in proliferating and resting Syrian hamster cells in monolayer culture. *J. Cell. Physiol.* **77**, 43–50.

Becker, H., and Stanners, C. P. (1972). Control of macromolecular synthesis in proliferating and resting hamster cells in monolayer culture III. Electrophoretic patterns of newly synthesized proteins in synchronized proliferating cells and resting cells. *J. Cell. Physiol.* **80**, 51–61.

Becker, D., Meier, C. B., and Herlyn, M. (1989). Proliferation of human malignant melanomas is inhibited by antisense oligodeoxynucleotides targeted against basic fibroblast growth factor. *EMBO J.* **8**, 3685–3691.

Bishop, J. M. (1983). Cellular oncogenes and retroviruses. *Annu. Rev. Biochem.* **52**, 301–354.

Bishop, J. M. (1985) Viral oncogenes. *Cell* **42**, 23–38.

Bishop, J. M. (1987) The molecular biology of cancer. *Science* **235**, 305–311.

Blair, D. G., Oskarsson, M., Wood, T. G., McClements, W. L., Fischinger, P. J., and Vande Woude, G. F. (1981). Activation of the transforming potential of a normal cell sequence: A molecular model for oncogenesis. *Science* **212**, 941–943.

Block, P., Seiter, I., and Oehlert, W. (1963) Autoradiographic studies of the initial cellular response to injury. *Exp. Cell Res.* **30**, 311–321.

Bochman, D., Bos, T., Admon, A., Nishimura, T., Vogt, P., and Tijan, R. (1987). Human protooncogene c-*jun* encodes a DNA-binding protein with structural and functional properties of transcription factor AP-1. *Science* **238**, 1386–1392.

Bravo, R., Burckhardt, J., Curran, T., Muller, R. (1986). Expression of c-*fos* in NIH 3T3 cells is very low but inducible throughout the cell cycle. *EMBO J.* **5**, 695–700.

Bravo, R., Frank, R., Blundell, P. A., MacDonald-Bravo, H. (1987). Cyclin-PCNA is the auxiliary protein of DNA polymerase delta. *Nature* **326**, 515–517.

Brooks, R. F. (1976). Regulation of the fibroblast cell cycle by serum. *Nature* **260**, 248–250.

Bucher, N. L. R. (1963). Regeneration of mammalian liver. *Int. Rev. Cytol.* **15**, 245–300.

Burns, F. J., and Tannock, I. F. (1970). On the existence of a G_0-phase in the cell cycle. *Cell Tissue Kinet.* **33**, 321–334.

Burstin, S. J., Meiss, H. K., and Basilico, C. (1974) A temperature-sensitive cell-cycle mutant of the BHK cell line. *J. Cell Physiol.* **84**, 397–408.

Calabretta, B., Kaczmarek, L., Mars, W., Ochoa, D., Gibson, C., Hirschhorn, R., and Baserga, R. (1985). Cell-cycle-specific genes differentially expressed in human leukemias. *Proc. Natl. Acad. Sci. U.S.A.* **82**, 4463–4467.

Calabretta, B., Battini, R., Kaczmarek, L., DeReil, J. K., and Baserga, R. (1986). Molecular cloning of the cDNA for a growth factor–inducible gene with strong homology to S-100, a calcium-binding protein. *J. Biol. Chem.* **261**, 12628–12632.

Campisi, J., Medrano, E. E., Morreo, G., and Pardee, A. B. (1982). Restriction-point control of cell growth by a labile protein: Evidence for increased stability in transformed cells. *Proc. Natl. Acad. Sci. U.S.A.* **79**, 436–440.

Campisi, J., Gray, H. E., Pardee, A. B. Dean, M., and Sonenshein, G. E. (1984) Cell-cycle control of c-*myc* but not c-*ras* is lost following chemical transformation. *Cell* **36**, 241–247.

Caracciolo, D., Valtieri, M., Venturelli, D., Peschile, C., Gewirtz, A. M., and Calabretta, B. (1989). Lineage-specific requirement of c-*alb* function in normal hematopoiesis. *Science* **245**, 1107–1110.

Carpenter, G., and Cohen, S. (1979). Epidermal growth factor. *Annu. Rev. Biochem.* **48**, 193–216.

Carter, R., Cosenza, S. C., Pena, A., Lipson, K., Soprano, D. R., and Soprano, K. J. (1991). A potential role for c-*jun* in cell cycle progression through late G_1 and S. *Oncogene* **6**, 229–235.

Cholon, J. J., and Studzinski, G. P. (1974). Effect of aminonucleoside on serum stimulation of nonhistone nuclear protein and DNA synthesis in normal and SV40-transformed human fibroblasts. *Cancer Res.* **34**, 588–593.

Cochran, B. H., Reffel, A. C., and Stiles, C. D. (1983). Molecular cloning of gene sequences regulated by platelet-derived growth factor. *Cell* **33**, 939–947.

Cochran, B. H., Zullo, J., Verma, I., and Stiles, C. D. (1984). Expression of the c-*fos* gene and of a *fos*-related gene is stimulated by platelet-derived growth factor. *Science* **226**, 1080–1082.

Collins, J. M., Glock, M. S., and Chu, A. K. (1982). Nuclease S1-sensitive sites in parental deoxyribonucleic acid of cold and temperature-sensitive mammalian cells. *Biochemistry* **21**, 3414–3419.

Coppock, D. L., and Pardee, A. B. (1985). Regulation of thymidine kinase activity in the cell cycle by a labile protein. *J. Cell. Physiol.* **124**, 269–274.

Coppola, J. A., and Cole, M. D. (1986). Constitutive c-*myc* oncogene expression blocks mouse erythroleukaemia cell differentiation but not commitment. *Nature* **320**, 760–763.

Cosenza, S. C., Owen, T. A., Soprano, D. R., and Soprano, K. J. (1988). Evidence that the time of entry into S is determined by events occurring in early G_1. *J. Biol. Chem.* **263**, 12751–12758.

Cosenza, S. C., Carter, R., Pena, A., Donigan, A., Borrelli, M., Soprano, D. R., and Soprano, K. J. (1991), Growth-associated gene expression is not constant in cells traversing G_1 after exiting mitosis. *J. Cell. Physiol.* (in press).

Cowdry, E. V. (1950). "Textbook of Histology." Lea & Febiger, Philadelphia, Pennsylvania.

Cristofalo, V. J., and Sharf, B. B. (1973). Cellular senescence and DNA synthesis. *Exp. Cell Res.* **76**, 419–427.

Croy, R. G., and Pardee, A. B. (1983). Enhanced synthesis and stabilization of M_r 68,000 protein in transformed BALB/c 3T3 cells: Candidate for restriction-point control of cell growth. *Proc. Natl. Acad. Sci. U.S.A.* **80**, 4699–4703.

Curran, T., Van Beveren, C., Ling, N., and Verma, I. M. (1985). Viral and cellular *fos* proteins are complexed with a 39,000 dalton cellular protein. *Mol. Cell. Biol.* **5**, 167–172.

Denhardt, D. T., Edwards, D. R., and Parfett, C. L. J. (1987). Gene expression during the mammalian cell cycle. *Biochim. Biophys. Acta* **865**, 83–125.

Edwards, S. A., Rundell, A. Y. K., Adamson, E. D. (1988). Expression of c-*fos* antisense RNA inhibits the differentiation of F9 cells to parietal endoderm. *Dev. Biol.* **129,** 91–101.

Eliyahu, D., Ras, A., Gruss, P., Givol, D., and Oren, M. (1984). Participation of p53 cellular tumour antigen in transformation of normal embryonic cell line. *Nature* **312**, 646–649.

Fabrikant, J. I. (1968). The kinetics of cellular proliferation in regenerating liver. *J. Cell Biol.* **36**, 551–565.

Ferrari, S., Battini, R., Kaczmarek, L., Calabretta, B., DeRiel, J. K., Philiponis, V., Wei, J. F., and Baserga, R. (1986). Coding sequence and growth regulation of the human vimentin gene. *Mol. Cell. Biol.* **6**, 3614–3620.

Ferrari, S., Calabretta, B., Battini, R., Cosenza, S. C., Owen, T. A., Soprano, K. J., and Baserga, R. (1988). Expression of c-*myc* and induction of DNA synthesis by platelet-poor plasma in human diploid fibroblasts. *Exp. Cell Res.* **174**, 25–33.

Floros, J., Jonak, G., Galanti, N., and Baserga, R. (1981). Induction of cell DNA replication in G_1-specific ts mutants by microinjection of recombinant SV40 DNA. *Exp. Cell Res.* **132,** 215–223.

Fox, T. O., and Pardee, A. B. (1971). Proteins made in the mammalian cell cycle. *J. Biol. Chem.* **246**, 6159–6169.

Gewirtz, A. M., and Calabretta, B. (1988). C-myb antisense oligodeoxynucleotides inhibits normal human hematopoiesis *in vitro. Science* **242**, 1303–1306.

Gewirtz, A. M., Anfossi, G., Venturelli, D., Valpreda, S., Sims, R., and Calabretta, B. (1989). G_1–S transition in normal human T lymphocytes requires the nuclear protein encoded by c-*myb*. *Science* **245**, 180–183.

Gonda, T. J., and Metcalf, D. (1984). Expression of *myb, myc*, and *fos* protooncogenes during the differentiation of murine myeloid leukemia. *Nature* **310**, 249–251.

Green, P., Pines, O., and Inouye, M. (1986). The role of antisense RNA in gene regulation. *Annu. Rev. Biochem.* **55**, 569–597.

Greenberg, M. E., and Ziff, E. B. (1984). Stimulation of 3T3 cells induces transcription of the c-*fos* protooncogene. *Nature* **311**, 433–438.

Griep, A. E., and Westphal, H. (1988) Antisense *myc* sequences induce differentiation of F9 cells. *Proc. Natl. Acad. Sci. U.S.A.* **85**, 6806–6810.

Grisham, J. W. (1962). A morphologic study of deoxyribonucleic acid synthesis and cell proliferation in regenerating rat liver: Autoradiography with thymidine-H^3. *Cancer Res.* **22**, 842–849.

Gunther, G. R., Wang, J. L., and Edelman, G. M. (1974). The kinetics of cellular commitment during stimulation of lymphocytes by lectins. *J. Cell. Biol.* **62**, 366–377.

Hann, S. R., Thompson, C. B., and Eisenman, R. N. (1985). C-*myc* oncogene protein synthesis is independent of the cell cycle in human and avian cells. *Nature* **314**, 366–369.

Hartwell, L. H., and Weinert, T. A. (1989). Checkpoints: Controls that ensure the order of cell-cycle events. *Science* **246**, 629–634.

Hassell, J. A., and Engelhardt, D. L. (1976). The regulation of protein synthesis in animal cells by serum factors. *Biochem.* **15**, 1375–1381.

Hazel, T. G., Nathans, D., and Lau, L. (1988) A gene inducible by serum growth factors encodes a member of the steroid and thyroid hormone receptor superfamily. *Proc. Natl. Acad. Sci. U.S.A.* **85**, 8444–8448.

Heikkila, R., Schwab, G., Wickstrom, E., Lokes, S. L., Pluznik, D. H., Watt, R., and Neckers, L. M. (1987). A c-*myc* antisense oligodeoxynucleotide inhibits entry into S phase but not progress from G_0 to G_1. *Nature* **328**, 445–449.

Hendrickson, S. L., and Scher, C. D. (1983). Platelet-derived growth factor–modulated translatable mRNAs. *Mol. Cell. Biol.* **3**, 1478–1487.

Hirschhorn, R. R., Aller, P., Yuan, Z. A., Gibson, C. W., and Baserga, R. (1984). Cell-cycle-specific CDNAs from mammalian cells temperature sensitive for growth. *Proc. Natl. Acad. Sci. U.S.A.* **81**, 6004–6008.

Holley, R. W., and Kiernan, J. A. (1968). Contact inhibition of cell division in 3T3 cells. *Proc. Natl. Acad. Sci. U.S.A.* **60**, 300–304.

Holt, J. T., Gopal, T. V., Moulton, A. D., and Nienhuis, A. W. (1986). Inducible production of c-*fos* antisense RNA inhibits 3T3 cell proliferation. *Proc. Natl. Acad. Sci. U.S.A.* **83**, 4794–4798.

Howard, A., and Pelc, S. R. (1951). Nuclear incorporation of ^{32}p as demonstrated by autoradiographs. *Exp. Cell Res.* **2**, 178–187.

Howard, A., and Pelc, S. R. (1953). Synthesis of deoxyribonucleic acid in normal and irradiated cells and its relation to chromosome breakage. *Heredity* **6**, 261–273.

Ingles, C. J., and Shales, M. (1982). DNA-mediated transfer of an RNA polymerase II gene: Reversion of the temperature-sensitive hamster cell-cycle mutant tsAF8 by mammalian DNA. *Mol. Cell. Biol.* **2**, 666–673.

Jaskulski, D., DeRiel, J. K., Mercer, E. W., Calabretta, B., and Baserga, R. (1988). Inhibition of cellular proliferation by antisense oligodeoxynucleotides to PCNA cyclin. *Science* **240**, 1544–1546.

Jenkins, J. R., Rudge, K., and Currie, G. A. (1984). Cellular immortilization by a cDNA clone encoding the transformation-associated phosphoprotein p53. *Nature* **312**, 651–653.

Johnson, L. F., Abelson, H. T., Green, H., and Penman, S. (1974). Changes in RNA in relation to growth of the fibroblast. I. Amounts of mRNA, rRNA and tRNA in resting and growing cells. *Cell* **1**, 95–100.

Johnson, L. F., Rao, L. G., and Muench, A. G. (1982). Regulation of thymidine kinase enzyme level on serum-stimulated mouse 3T6 fibroblasts. *Exp. Cell Res.* **138**, 79–85.

Jonak, G., and Baserga, R. (1979). Cytoplasmic regulation of two G_1-specific temperature-sensitive functions. *Cell* **18**:259–269.

Kaczmarek, L., Hyland, J., Watt, R., Rosenberg, M., and Baserga, R. (1985) Microinjected c-*myc* as a competence factor. *Science* **228**, 1313–1315.

Kaczmarek, L., Surmacz, E., and Baserga, R. (1986). Cyclohexamide and puromycin can substitute for PDGF in inducing cellular DNA synthesis in quiescent 3T3 cells. *Biol. Intl. Rpts.* **10**, 455–463.

Kahana, C., and Nathans, D. (1984). Isolation of cloned cDNA encoding mammalian ornithine decarboxylase. *Proc. Natl. Acad. Sci. U.S.A.* **81**, 3645–3649.

Karn, J., Watson, J. V., Lowe, A. D., Green, S. M., and Vedeckis, W. (1989). Regulation of cell-cycle duration by c-*myc* levels. *Oncogene* **4**, 659–663.

Kelly, K., Cochran, B. H., Stiles, C. D., and Leder, P. (1983). Cell-specific regulation of c-*myc* by lymphocyte mitogens and platelet-derived growth factor. *Cell* **35**, 603–610.

Kim, S. K., and Wold, B. J. (1985). Stable reduction of thymidine kinase activity in cells expressing high levels of antisense RNA. *Cell* **42**, 129–138.

Kruijer, W., Cooper, J. A., Hunter, T., and Verma, I. M. (1984). Platelet-derived growth factor induces rapid but transient expression of the c-*fos* gene and protein. *Nature* **312**, 711–716.

Lachman, H. M., Cheng, G., and Skoultchi, A. I. (1986). Transfection of mouse erythroleukemia cells with *myc* sequences changes the rate of induced commitment to differentiation. *Proc. Natl. Acad. Sci. U.S.A.* **83**, 6480–6484.

Lajtha, L. G. (1963). On the concept of the cell cycle. *J. Cell. Comp. Physiol.* **62**, 143–149.

Land, H., Chen, A. C., Morgenstern, J. P., Parada, L. F., and Weinberg, R. A. (1986). Behavior of *myc* and *ras* oncogenes in transformation of rat embryo fibroblasts. *Mol. and Cell. Biol.* **6**, 1917–1925.

Land, H., Parada, L. F., and Weinberg, R. A. (1983c). Tumorigenic conversion of primary embryo fibroblasts requires at least two cooperating oncogenes. *Nature* **304**, 596–602.

Land, H., Parada, L. F., and Weinberg, R. A. (1983b). Cellular oncogenes and multistep carcinogenesis. *Science* **222**, 771–778.

Land, H., Parada, L. F., and Weinberg, R. A. (1983a). Cellular and viral oncogenes cooperate to achieve tumorigenic conversion of rat embryo fibroblasts. *Cancer Cells* **2**, 473–480.

Larsson, O., Latham, C., Zickert, P., and Zetterberg, A. (1989). Cell-cycle regulation of human diploid fibroblasts: Possible mechanisms of platelet-derived growth factor. *J. Cell. Physiol.* **139**, 477–483.

Lau, L. F., and Nathans, D. (1985). Identification of a set of genes expressed during the G_0/G_1 transition of cultured mouse cells. *EMBO J.* **4**, 3145–3151.

Lee, A. S., Delegeane, A. M., Baker, V., and Chow, P. C. (1983). Transcriptional regulation of two genes specifically induced by glucose starvation in a hamster mutant fibroblast cell line. *J. Biol. Chem.* **250**, 597–603.

Lee, W. P., Mitchell, P., and Tjian, R. (1987). Purified transcription factor AP-1 interacts with TPA-inducible enhancer elements. *Cell* **49**, 741–752.

Lee, W. M., Schwab, M., Westaway, D., and Varmus, H. E. (1985). Augmented expression of normal c-*myc* is sufficient for cotransformation of rat embryo cells with a mutant *ras* gene. *Mol. Cell. Biol.* **5**, 3345–3356.

Ley, K. D., and Tobey, R. A. (1970). Regulation of initiation of DNA synthesis in Chinese hamster cells. *J. Cell Biol.* **47**, 453–459.

Linzer, D. I. H., and Nathans, D. (1983). Growth-related changes in specific mRNAs of cultured mouse cells. *Proc. Natl. Acad. Sci. U.S.A.* **80**, 4271–4275.

Linzer, D. I. H., and Nathans, D. (1984). Nucleotide sequence of a growth-related mRNA encoding a member of the prolactin-growth hormone family. *Proc. Natl. Acad. Sci. U.S.A.* **81**, 4255–4259.

Marcus-Sekura, C. J. (1988). Techniques for using antisense oligodeoxynucleotides to study gene expression. *Anal. Biochem.* **172**, 280–295.

Matrisian, L. M., Glaichenhaus, N., Gesnel, M.-C., and Breathnach, R. (1985). Epidermal growth factor and oncogenes induce the transcription of the same cellular mRNA in rat fibroblasts. *EMBO J.* **4**, 1435–1440.

Mattei, M. G., Simon-Chazottes, D., Hirai, S.-I., Ryseck, R.-P., Galcheva-Gargova, Z., Guenet, J.-L., Mattei, J. F., Bravo, R., and Yaniv, M. (1990). Chromosomal localization of the three members of the *jun* protooncogene family in mouse and man. *Oncogene* **5**, 151–156.

Medrano, E. E., and Pardee, A. B. (1980). Prevalent deficiency in tumor cells of cyclohexamide-induced cycle arrest. *Proc. Natl. Acad. Sci. U.S.A.* **77**, 4123–4126.

Meiss, H. K., and Basilico, C. (1972). Temperature-sensitive mutants of BHK-21 cells. *Nature New Biol.* **239**, 66–68.

Mercola, D., Rundell, A., Westwick, J., and Edwards, S. A. (1987). Antisense RNA to the c-*fos* gene: Restoration of density-dependent growth arrest in a transformed cell line. *Biochem. Biophys. Res. Commun.* **147**, 288–294.

Mercer, W. E., Avignolo, C., and Baserga, R. (1984a). The role of the p53 protein in cell proliferation as studied by the microinjection of monoclonal antibodies. *Mol. Cell. Biol.* **4**, 276–281.

Mercer, W. E., Nelson, D., Deleo, A. B., Old, L. J., and Baserga, R. (1982). Microinjection of monoclonal antibody of protein p53–inhibited serum-induced DNA synthesis in 3T3 cells. *Proc. Natl. Acad. Sci. U.S.A.* **79**, 6309–6312.

Mercer, W. E., Nelson, D., Hyland, J. K., Croce, C. M., and Renato, R. (1983). Inhibition of SV40-induced cellular DNA synthesis by microinjection of monoclonal antibodies. *Virology* **127**, 149–158.

Mercer, W. E., Avignolo, C., Galanti, N., Rose, K. M., Hyland, J. K., Jacob, S. T., and Baserga, R. (1984b). Cellular DNA replication is independent from the synthesis or accumulation of ribosomal RNA. *Exp. Cell Res.* **150**, 118–130.

Milarski, K. L., and Morimoto, R. I. (1986). Expression of human HSP70 during the synthetic phase of the cell cycle. *Proc. Natl. Acad. Sci. U.S.A.* **83**, 9517–9521.

Miska, D., and Bosmann, H. B. (1980). Existence of an upper-limit to elongation of the prereplicative period in confluent cultures of C3H/10T1/2 cells. *Biochem. Biophys. Res. Commun.* **93**, 1140–1145.

Mitchison, J. M. (1971). "The Biology of the Cell Cycle." Cambridge, Cambridge University Press.

Mostafapour, M. K., and Green H. (1975). Effects of withdrawal of a serum stimulus on the protein synthesis machinery of cultured fibroblasts. *J. Cell. Physiol.* **86**, 313–320.

Mueller, C., Graessmann, A., and Graessmann, M. (1978). Mapping of early SV40-specific functions by microinjection of different early viral DNA fragments. *Cell* **15**, 579–585.

Muller, R., and Verma, I. M. (1984). Expression of cellular oncogenes. *Curr. Top. Micro. Immonol.* **112**, 73–115.

Muller, R., Bravo, R., Burckhardt, J., and Curran, T., (1984). Induction of c-*fos* gene and protein by growth factors precedes activation of c-*myc*. *Nature* **312**, 716–720.

Neckers, L. M., Tsuda, H., Weiss, E., and Pluznik, D. H. (1988). Differential expression of c-*myc* and the transferrin receptor in G_1-synchronized M1 myeloid leukemia cells. *J. Cell. Physiol.* **135**, 339–344.

Nishikura, K., and Murray, J. M. (1987). Antisense RNA of protooncogene c-*fos* blocks renewed growth of quiescent 3T3 cells. *Mol. Cell. Biol.* **7**, 639–649.

Novi, A. M., and Baserga, R. (1971). Association of hypertrophy and DNA synthesis in mouse salivary glands after chronic administration of isoproterenol. *Am. J. Pathol.* **62**, 295–308.

O'Farrell, M., and Yanez, S. (1989). Variation in the length of the lag phase following serum stimulation of mouse 3T3 cells. *Cell Biol. Int. Rep.* **13**, 453–462.

Okuda, A., and Kimura, G. (1982) Effects of serum deprivation on the initiation of DNA synthesis in the second generation in rat 3Y1 cells. *J. Cell. Physiol.* **110**, 267–270.

Okuda, A., and Kimura, G. (1984). Control in previous and present generations of preparation for entry into S phase and the relationship to resting state in 3Y1 rat fibroblastic cells. *Exp. Cell Res.* **155**, 24–32.

Olashaw, N. E., and Pledger, W. J. (1985). Mechanisms initiating cellular proliferation. *In* "Mediators in Cell Growth and Differentiation" (R. J. Ford, and A. L. Maizel, eds.), pp. 31–44. Raven Press, New York.

Owen, T. A., Carter, R., Whitman, M. W., Soprano, D. R., and Soprano, K. J. (1989a). Evidence that density-dependent growth arrest is a two-stage process. *J. Cell. Physiol.* **142**:137–148.

Owen, T. A., Cosenza, S. C., Soprano, D. R., and Soprano, K. J. (1987). Time of c-*fos* and c-*myc* expression in human diploid fibroblasts stimulated to proliferate after prolonged quiescence. *J. Biol. Chem.* **262**, 15111–15117.

Owen, T. A., Soprano, D. R., and Soprano, K. J. (1989b). Analysis of the growth-factor requirements of WI-38 cells after extended periods of density-dependent growth arrest. *J. Cell Physiol.* **139**, 424–431.

Parada, L. F., Tabin, C. J., Shih, C., and Weinberg, R. A. (1982). Human EJ bladder carcinoma oncogene is homologue of Harvey sarcoma virus *ras* gene. *Nature* **297**, 474–478.

Pardee, A. B. (1974). A restriction point for control of normal cell proliferation. *Proc. Natl. Acad. Sci. U.S.A.* **71**, 1286–1290.

Pardee, A. B. (1982). Molecular mechanisms of the control of cell growth in cancer. *In* "NATO Advanced Study Institute Series" (C. N. Y. Nicolini, ed), Vol. 38, pp. 673–714.

Pardee, A. B. (1987). The yang and yin of cell proliferation: An overview. *J. Cell. Physiol. (Suppl.)* **5**, 107–110.

Pardee, A. B., Dubrow, R., Hamlin, J. L., and Kletzien, R. F. (1978). Animal cell cycle. *Annu. Rev. Biochem.* **47**, 715–750.

Pardee, A. B., and James, L. J. (1975). Directive killing of transformed baby hamster kidney (BHK) cells. *Proc. Natl. Acad. Sci. U.S.A.* **72** 4994–4998.

Parfett, C. L. J., Hamilton, R. T., Howell, B. W., Edwards, D. R., Nilsen-Hamilton, M., and Denhardt, D. T. (1985). Characterization of a cDNA clone encoding murine mitogen-regulated protein: Regulation of mRNA levels in mortal and immortal cell lines. *Mol. Cell. Biol.* **5**, 3289–3292.

Pereira-Smith, O. M., Fischer, S. F., and Smith, J. R. (1985). Senescent and quiescent cell inhibitors of DNA synthesis. *Exp. Cell Res.* **160**, 297–306.

Pines, J., and Hunter, T. (1989). Isolation of a human cyclin cDNA: Evidence for cyclin mRNA and protein regulation in the cell cycle and for interaction with $p34^{cdc2}$. *Cell* **58**, 833–846.

Pledger, W. J., Stiles, C. D., Antoniades, H. N., Scher, C. D. (1977). Induction of DNA synthesis in BALB/c 3T3 cells by serum components: Reevaluation of the commitment process. *Proc. Natl. Acad. Sci. U.S.A.* **74**, 4481–4485.

Pledger, W. J., Stiles, C. D., Antoniades, H. N., Scher, C. D. (1978). An ordered sequence of events is required before BALB/c 3T3 cells become committed to DNA synthesis. *Proc. Natl. Acad. Sci. U.S.A.* **75**, 2839–2843.

Prescott, D. M. (1976). "Reproduction of Eukaryotic Cells." Academic Press, New York.

Prescott, D. M. (1987). Cell reproduction. *Int. Rev. Cytol.* **100**, 93–128.

Prochownik, E. V., Kukowska, J., and Rodgers, C. (1988). C-*myc* antisense transcripts accelerate differentiation and inhibit G_1 progression in murine erythroleukemia cells. *Mol. Cell. Biol.* **8**, 3683–3695.

Quastler, H., and Sherman, F. G. (1959). Cell-population kinetics in the intestinal epithelium of the mouse. *Exp. Cell Res.* **17**, 420–438.

Reich, N. C., and Levine, A. J. (1984). Growth regulation of a cellular tumor antigen p53 in non-transformed cells. *Nature* **308**, 199–201.

Riabowol, K. T., Vosatka, R. J., Ziff, E. B., Lamb, N. J., and Feramisco, J. R. (1988). Microinjection of *fos*-specific antibodies blocks DNA synthesis in fibroblast cells. *Mol. Cell. Biol.* **8**, 1670–1676.

Rittling, S. R., Brooks, K. M., Cristofalo, V. J., and Baserga, R. (1986). Expression of cell cycle–specific genes in young and senescent WI-38 fibroblasts. *Proc. Natl. Acad. Sci. U.S.A.* **83**, 3316–3320.

Rittling, S. R., Gibson, C. W., Ferrari, S., and Baserga, R. (1985). The effect of cyclohexamide on the expression of cell cycle–dependent genes. *Biochem. Biophys. Res. Commun.* **132**, 327–335.

Rollins, B. J., Morrison, E. D., and Stiles, C. D. (1987). A cell-cycle constraint on the regulation of gene expression by platelet-derived growth factor. *Science* **238**, 1269–1271.

Rollins, B. J., Morrison, E. D., and Stiles, C. D. (1988). Cloning and expression of JE, a gene inducible by platelet-derived growth factor and whose product has cytokine-like properties. *Proc. Natl. Acad. Sci. U.S.A.* **85**, 3738–3742.

Rollins, B. J., Stier, P., Ernst, T., and Wong, G. (1989) The human homolog of the JE gene encodes a monocyte secretory protein. *Mol. Cell. Biol.* **9**, 4687–4695.

Rossini, M., Lin, J. C., and Baserga, R. (1976). Effects of prolonged quiescence on nuclei and chromatin of WI-38 fibroblasts. *J. Cell. Physiol.* **88**, 1–12.

Rossini, M., and Baserga, R. (1978). RNA synthesis in a cell cycle–specific temperature-sensitive mutant from a hamster cell line. *Biochemistry* **17**, 858–863.

Rossini, M., Baserga, S., Huang, C. H., Ingles, C. J., and Baserga, R. (1980). Changes in RNA polymerase II in a cell cycle–specific temperature-sensitive mutant of hamster cells. *J. Cell. Physiol.* **103**, 97–103.

Roscow, P. W., Riddle, G. H., and Pardee, A. B. (1979). Synthesis of labile, serum-dependent protein in early G_1 controls animal cell growth. *Proc. Natl. Acad. Sci. U.S.A.* **76**, 4446–4450.

Rovera, G., and Baserga, R. (1973). Effect of nutritional changes on chromatin template activity and nonhistone chromosomal protein synthesis in WI-38 and 3T6 cells. *Exp. Cell Res.* **78**, 118–126.

Rozengurt, E. (1983). Growth factors, cell proliferation and cancer: An overview. *Mol. Biol. Med.* **1**, 169–181.

Rozengurt, E. (1986). Early signals in mitogenic response. *Science* **234**, 161–166.

Rubin, H., and Steiner, R. (1975) Reversible alterations in the mitotic cycle of chick embryo cells in various states of growth regulation. *J. Cell. Physiol.* **85**, 261–270.

Ruley, H. E. (1983). Adenovirus early region 1A enables viral and cellular transforming genes to transform primary cells in culture. *Nature* **304**, 602–606.

Ryder, K., and Nathans, D. (1988). Induction of protooncogene c-*jun* by serum growth factors. *Proc. Natl. Acad. Sci. U.S.A.* **85**, 8464–8467.

Ryder, K., Lanahan, A., Perez-Albuerne, E., and Nathans, D. (1989) Jun-D: A third member of the *jun* gene family. *Proc. Natl. Acad. Sci. U.S.A.* **86**, 1500–1503.

Sander, G., and Pardee, A. B. (1972). Transport changes in synchronously growing CHO and L cells. *J. Cell. Physiol.* **80**, 267–272.

Sasaki, T., Litwack, G., and Baserga, R. (1969). Protein synthesis in the early prereplicative phase of isoproterenol-stimulated synthesis of deoxyribonucleic acid. *J. Biol. Chem.* **244**, 4831–4837.

Scher, C. D., Stone, M. E., and Stiles, C. D. (1979). Platelet-derived growth factor prevents G_0 arrest. *Nature* **281**, 390–392.

Schneiderman, M. H., Dewey, W. C., and Highfield, D. P. (1971). Inhibition of DNA synthesis in synchronized Chinese hamster cells treated in G_1 with cyclohexamide. *Exp. Cell Res.* **67**, 147, 155.

Schneider, C., King, R., and Philipson, L. (1988). Genes specifically expressed at growth arrest of mammalian cells. *Cell* **54**,787–793.

Seiler-Tuyns, A., and Birnsteil, M. L. (1981). Structure and expression in L-cells of a cloned H4 histone gene of the mouse. *J. Mol. Biol.* **151**, 607–625.

Sherley, J. L., and Kelly, T. (1988). Regulation of human thymidine kinase during the cell cycle. *J. Biol. Chem.* **263**, 8350–8358.

Shields, R., and Smith, J. A. (1977). Cells regulate their proliferation through alterations in transition probability. *J. Cell. Physiol.* **87,** 345–356.

Smith, C., Aurelian, L., Reddy, M., Miller, P., and Ts'o, P. (1986). Antiviral effect of an oligo(nucleoside methyphosphonate) complementary to the splice junction of herpes simplex virus type 1 immediate-early pre-mRNAs 4 and 5. *Proc. Natl. Acad. Sci. U.S.A.* **83**, 2787–2791.

Smith, J. A., and Martin, L. (1973). Do cells cycle? *Proc. Natl. Acad. Sci. U.S.A.* **70**, 1263–1267.

Smith, J. C., and Stiles, C. D. (1981). Cytoplasmic transfer of the mitogenic to platelet-derived growth factor. *Proc. Natl. Acad. Sci. U.S.A.* **78**, 4363–4367.

Soprano, K. J., Galanti, N., Jonak, G. J., McKercher, S., Pipas, J. M., Peden, K. W. C., and Basrega, R. (1983). Mutational analysis of SV40-T antigen: Stimulation of cellular DNA synthesis and activation of ribosomal RNA genes by mutants with deletions in the T antigen. *Mol. Cell. Biol.* **3**, 214–219.

Spandidos, D. A., and Wilkie, N. M. (1984). Malignant transformation of early-passage rodent cells by a single mutant human oncogene. *Nature* **310**, 475–496.

Stein, C. A., and Cohen, J. S. (1988). Oligodeoxynucleotides as inhibitors of gene expression: A review. *Cancer Res.* **48**, 2659–2668.

Stiles, C. D., Cappone, G. T., Scher, C. D., Antoniades, H. N., Van Wyk, J. J., and Pledger, W. J. (1979). Dual control of cell growth by somatomedins and platelet-derived growth factor. *Proc. Natl. Acad. Sci. U.S.A.* **76**, 1279–1283.

Temin, H. M. (1971). Stimulation by serum of multiplication of stationary chicken cells. *J. Cell. Physiol.* **77**, 161–170.

Terasima, T., and Tolmach, L. J. (1966). Synthesis of G_1 protein proceeding DNA synthesis in cultured mammalian cells. *Exp. Cell Res.* **44**, 669–671.

Thompson, C. B., Challoner, P. B., Neiman, P. E., and Groudine, M. (1985). Levels of c-*myc* oncogene mRNA are invariant throughout the cell cycle. *Nature* **314**, 363–366.

Tobey, R. A., Anderson, E. C., and Peterson, D. F. (1967). The effect of thymidine incorporation on the duration of G_1 in Chinese hamster cells. *J. Cell. Biol.* **35**, 53–59.

Todaro, G. J., and Green, H. (1963). Quantitative studies on the growth of mouse embryo cells in culture and their development into established lines. *J. Cell. Biol.* **17**, 299–313.

Todaro, G. J., Lazar, G. K., and Green, H. (1965). The initiation of cell division in a contact-inhibited mammalian cell line. *J. Cell. Comp. Physiol.* **66**, 325–334.

Toscani, A., Soprano, D. R., and Soprano, K. J. (1988). Molecular analysis of sodium butyrate–induced growth arrest. *Oncogene Res.* **3**, 223–238.

Tseng, B. Y., Prussak, C. E., and Almazan, M. T. (1989). Primase p49 mRNA expression is serum stimulated but does not vary with the cell cycle. *Mol. Cell. Biol.* **9**, 1940–1945.

Tsuda, H., Neckers, L. M., and Pluznick, D. H. (1987). Enhanced c-*fos* expression in differentiated monomyelocyte cells is associated with differentiation and not with the position of the differentiated cells in the cell cycle. *Exp. Hematol.* **15**, 700–703.

Waechter, D. E., Avignolo, C., Freund, E., Riggenbach, C. M., Mercer, W. E., Mcguire, P. M., and Baserga, R. (1984). Microinjection of RNA polyerase II corrects the temperature-sensitive defect of tsAF8 cells. *Mol. Cell. Biochem.* **60**, 77–82.

Wahl, A. F., Geis, A. M., Spain, B. H., Wong, S. W., Korn, D., Wang, T. S. F. (1988). Gene expression of human DNA polymerase alpha during cell proliferation and the cell cycle. *Mol. Cell. Biol.* **8**, 5016–5025.

Wier, M. L., and Scott, R. E. (1986). Regulation of the terminal event in cellular differentiation: Biological mechanisms of the loss of proliferative potential. *J. Cell Biol.* **102**, 1955–1964.

Wharton, W., Gillespie, G. Y., Russell, S. W., and Pledger, W. J. (1982). Mitogenic activity elaborated by macrophage-like cell lines acts as competence factor(s) for BALB/c 3T3 cells. *J. Cell. Physiol.* **110**, 93–100.

Whitman, M., Shen, Y. Soprano, D. R., and Soprano, K. J. (1990). Molecular analysis of early growth-associated events during the differentiation of F9 cells into embryoid bodies. *Cancer Res.* **50**:3193–3198.

Wickstrom, E. L., Bacon, T. A., Gonzalez, A., Freeman, D. L., Lyman, G. M., and Wickstrom, E. (1988). Human promelocytic leukemia HL-60 cell proliferation and c-myc protein synthesis are inhibited by an antisense pentadecadeoxynucleotide targeted against c-*myc* mRNA. *Proc. Natl. Acad. Sci. U.S.A.* **85**, 1028–1032.

Williams, J. G., and Penman, S. (1975). Messenger RNA sequences in growing and resting fibroblasts. *Cell* **6**, 197–206.

Yen, A., and Pardee, A. B. (1978). Arrested states produced by isoleucine deprivation and their relationship to the low serum produced arrested state in Swiss 3T3 cells. *Exp. Cell Res.* **114**, 389–394.

Zettereberg, A., and Larsson, O. (1985). Kinetic analysis of regulatory events in G_1 leading to proliferation or quiescence of Swiss 3T3 cells. *Proc. Natl. Acad. Sci. U.S.A.* **82**, 5365–5369.

5

Regulation of Gene Expression by Serum Growth Factors

GREGG T. WILLIAMS, ANDREW S. ABLER, AND LESTER F. LAU

Department of Genetics
The University of Illinois College of Medicine
Chicago, Illinois 60612

I. INTRODUCTION

Polypeptide growth factors regulate such cellular processes as proliferation and differentiation. Studies on growth factors and their specific cell-surface receptors (reviewed by Sporn and Roberts, 1990) have recently converged with studies on oncogenes (reviewed by Reddy *et al.,* 1988) and the cell cycle (reviewed by Baserga, 1985). A number of oncogene products have been found to be mutated forms of growth factors, growth-factor receptors, and proteins implicated in mediating the transduction of signals propagated by growth factor–receptor interactions. These findings reinforce the notion that the biochemical consequences of growth-factor actions are key events in the control of cell proliferation. It is now clear that an important activity of growth factors is to trigger the activation of a complex genetic program. Moreover, a number of genes activated by growth factors also have oncogenic potential, pointing to the importance of these genes in growth control. Characterization of how growth factor-induced gene expression regulates proliferation has thus emerged as an important link in the understanding of the control of cell growth.

Recent studies identified a large number of genes that are rapidly activated by growth factors. The characterization of these genes has already yielded a wealth of new information on potential regulators of call proliferation. In this review we shall address the mechanisms by which growth factors regulate gene expression, the identity of the genes activated, and how the expression of these genes might mediate the growth response.

Several related reviews have been published recently (Bravo, 1989; Rollins and Stiles, 1989b; Schönthal, 1990; Lau and Nathans, 1991). These reviews emphasize different aspects of growth factor-regulated gene expression and are complementary to this discussion.

II. GROWTH FACTOR-REGULATED GENE EXPRESSION AND GROWTH CONTROL

A. Experimental Systems

The development of mammalian cell lines was vital for the study of animal cell growth (reviewed by Baserga, 1985). Murine 3T3 fibroblasts, which are contact inhibited, have proven to be especially useful for these studies (Todaro and Green, 1963). Like most animal cells, 3T3 fibroblasts require serum growth factors for proliferation. When deprived of serum, these cells cease to divide and enter a quiescent state known as G_0. They can be stimulated to reenter the cell cycle synchronously at the G_1 phase by the addition of high concentrations of serum (Todaro *et al.*, 1965; Holley and Kiernan, 1968; Dulbecco, 1970; Temin, 1971). Progression toward S phase appears to be governed by regulatory events during G_1 (Croy and Pardee, 1983; Zetterberg and Larsson, 1985); passing these points of control commits the cells to replicate their DNA and divide. Among the multitude of growth factors present in serum, platelet-derived growth factor (PDGF) is the predominant mitogenic agent (Rutherford and Ross, 1976; Antoniades *et al.*, 1979; Heldin *et al.*, 1979). In BALB/c 3T3 cells, PDGF acts in conjunction with insulin-like growth factors and epidermal growth factor (EGF) or TGF-α to induce cell proliferation (reviewed by Stiles, 1983).

The transition from quiescence to a proliferative state is not unique for cultured 3T3 fibroblasts. A similar transition is observed in other experimental systems, for example, activated T lymphocytes and hepatocytes in regenerating liver, both representative cell populations of the living animal. T lymphocytes normally reside in a G_0-like state until their proliferation is triggered by the binding of antigen and lymphokine to cell-surface receptors (reviewed by Altman *et al.*, 1990). The adult liver is another example of a population of quiescent cells that retains proliferative potential. Following partial hepatectomy or chemically induced liver damage, the remaining hepatocytes proliferate until the organ regenerates to approximately its original size (reviewed by van Lancker, 1989).

The transition from quiescence to a growing state appears to be similar for fibroblasts, T cells, and hepatocytes. Many biochemical events that occur in mitogen-stimulated fibroblasts also occur in activated T cells (Weiss and Imboden, 1987), and many of the genes rapidly activated in mitogen-stimulated fibroblasts and T cells are also activated in regenerating liver (Nathans *et al.*, 1988; Sobczak *et al.*, 1989; Bours *et al.*, 1990). Moreover, the time courses of these events relative to DNA synthesis are similar in these systems. While most of the data we discuss in this review are derived from work in cultured fibroblasts, investigations of the genetic program activated during the G_0–G_1transition in T lymphocytes (Zipfel *et al.*, 1989) and in regenerating liver (Mohn *et al.*, 1991) are ongoing. Given the similarities among these systems, results obtained from any one cell type are likely to have general signficance.

B. Identification of Growth Factor-Regulated Genes

Studies directed toward understanding the mechanism for serum or growth-factor stimulation of quiescent cells established that RNA and protein syntheses are required for the onset of DNA replication (reviewed by Rollins and Stiles, 1989b; Lau and Nathans, 1991). These results suggested the possibility that serum may stimulate cell proliferation by inducing the expression of specific genes not expressed in G_0. Support for this interpretation came from solution hybridization studies, which indicated that 3% of the mRNA population isolated from exponentially growing cells is absent in quiescent cells (Williams and Penman, 1975). Furthermore, the appearance and disappearance of specific polypeptides at characteristic times during G_1 have been demonstrated by one- and two-dimensional gel electrophoresis of proteins from serum-stimulated fibroblasts (reviewed by Rollins and Stiles, 1989b; Lau and Nathans, 1991), indicating that there is an orderly and sequential expression of different genes upon mitogenic stimulation.

To describe the temporally regulated gene expression, we define *immediate-early genes* as those that are rapidly activated by mitogenic growth factors without requiring *de novo* protein synthesis; *delayed-early genes* for those expressed later but before DNA synthesis, and *late genes* for those expressed at or after the onset of DNA synthesis. By analogy to the developmental program of DNA viruses, we expect that among the immediate-early genes are those encoding regulators of the genes expressed later, and the latter may encode functions that are directly required for the duplication of cellular components.

The induction of specific mRNA species in response to serum stimulation was shown by two experimental strategies. In the first approach, known cDNA clones were used to probe RNA isolated from quiescent and serum-stimulated cells. While this strategy is limited by the availability of known cDNA clones, a number of genes, including protooncogenes, were shown to be induced by serum stimulation (Kelly *et al.,* 1983; Greenberg and Ziff, 1984; Thompson *et al.,* 1986). In the second approach, cDNA libraries prepared from poly(A)$^+$ RNA isolated from cells stimulated with serum or growth factors were differentially screened with ^{32}P-labeled cDNA derived from either quiescent or stimulated cell mRNA (Cochran *et al.,* 1983; Linzer and Nathans, 1983; Hirschhorn *et al.,* 1984; Edwards *et al.,* 1985). This approach can be enhanced by stimulating cells in the presence of a protein-synthesis inhibitor such as cycloheximide (Lau and Nathans, 1985). Under these conditions, the new RNAs that appear upon stimulation are limited to those transcribed in the absence of *de novo* protein synthesis; and thus are the products of preexisting transcriptional machinery. Moreover, cycloheximide superinduces some immediate-early mRNAs (Cochran *et al.,* 1983; Kelly *et al.,* 1983), thus enriching the cells for these mRNAs. Using differential hybridization screening of cDNA libraries derived from cells enriched in immediate-early mRNAs, various laboratories have isolated a large number of cDNAs derived from serum or growth factor-stimulated mRNAs, including those derived from mRNAs that are normally present in low abundance (Lau and Nathans, 1985; Lau and Nathans, 1987; Sukhatme *et al.,* 1987; Almendral *et al.,* 1988). This second strategy clearly demonstrated that the genetic program activated by growth factors is a complex and multifaceted one.

C. Expression of Immediate-Early Genes

Immediate early genes are defined by three criteria:

1. their transcripts, which are at low or undetectable levels in quiescent cells, rapidly accumulate to detectable levels within 1 hr of growth-factor stimulation;
2. their expression is independent of *de novo* protein synthesis; and
3. their transcriptional activation is mediated by growth-factor stimulation.

Over 60 immediate-early genes have been identified in 3T3 fibroblasts according to these criteria (Lau and Nathans, 1985; Lau and Nathans, 1987; Sukhatme *et al.,* 1987; Almendral *et al.,* 1988; Lim *et al.,* 1989). Immediate-early genes can be subdivided into three classes on the basis of the kinetics of their mRNA accumulation (see Table I). Transcripts of group I genes are detectable within 5 min of growth-factor stimulation and accumulate to peak levels within 30 to 60 min. Transcription of these genes is quickly repressed following activation, and their mRNAs have short half-lives (approximately 10–30 min) (Lau and Nathans, 1987; Almendral *et al.,* 1988). Therefore, their mRNAs accumulate transiently and decay to undetectable levels within 2 hr after stimulation. Many immediate-early genes encoding transcription factors fall into this category. mRNAs of group II genes accumulate somewhat more slowly, reaching peak levels about 2 hr after stimulation. These mRNAs also have somewhat longer half-lives than those of group I RNAs. The accumulation of these mRNAs is therefore more prolonged, and they decay to undetectable levels 4 to 8 hr after stimulation. Transcripts for group III genes often accumulate as quickly as those of group I genes, but they continue to be present at high levels for many hours. This pattern of expression is partly explained by the relatively long half-lives of many group III RNAs, particularly those encoding structural proteins (Almendral *et al.,* 1988; Lau and Nathans, 1991). Transcription of these genes also tends to be more persistent, and they are not efficiently repressed following the initial activation (O'Brien *et al.,*1990).

Virtually all immediate-early mRNAs examined to date are superinduced when cells are stimulated in the presence of a protein-synthesis inhibitor such as cycloheximide, which exerts two effects. First, it blocks the transcriptional repression that normally occurs following the initial activation, thus leading to prolonged transcription of these genes (Greenberg *et al.,* 1986; Lau and Nathans, 1987). Second, it stabilizes the mRNAs from rapid degradation (Lau and Nathans, 1987; Almendral *et al.,* 1988). These findings may be interpreted to suggest that transcriptional repression of immediate-early genes requires the synthesis of a repressor molecule. Furthermore, degradation of immediate-early mRNAs may require the synthesis of a nuclease or translation of the mRNAs (see Section IV).

TABLE I.

Proteins Encoded by Immediate-Early Genes[a]

Proteins	Putative function	Kinetics of expression
Nuclear proteins		
c-Fos	Transcription factor	Group I
FosB	Transcription factor	Group I
Fra-1	Transcription factor	Group II
Fra-2	Transcription factor	Group II
c-Jun	Transcription factor	Group I
Jun B	Transcription factor	Group I
Rel	Transcription factor	Group II
NF-κB (p50)	Transcription factor	
Myc	Transcription factor	Group II
Zif268 (Krox24/Egr-1/NGFI-A/TIS8)	Transcription factor	Group I
Krox20 (Egr-2)	Transcription factor	Group I
Nur77 (NGFI-B/N10/NAK1/TIS1)	Transcription factor	Group I
Nup475	Transcription factor	Group I
SRF($p67^{SRF}$)	Transcription factor	Group II
Secreted Proteins		
KC	Cytokine	Group II
JE	Cytokine	Group II
PC4/TIS7	Cytokine	
Cyr61	Cell–cell communication	Group III
PAI	Blood clotting	
Integral membrane proteins		
Tissue factor	Blood clotting	Group II
Glucose transporter	Glucose transport	Group II
Other		
Pip92	Unknown	Group I
p27	Unknown	Group III
Structural Proteins		
β-Actin		Group III
Tropomyosin		Group III
Fibronectin		Group III
Fibronectin receptor		Group III

[a]See Section V.

While serum was used to stimulate cells in a number of differential screenings, the majority of the genes identified are also activated by purified PDGF or fibroblast growth factor (FGF) and thus fit the criteria for immediate-early genes (Lau and Nathans, 1987; Almendral *et al.,* 1988). Many of them are also inducible by agents that activate protein kinase C (PKC) (Charles *et al.,* 1990). Conversely, genes isolated on the basis of their inducibility by phorbol esters, which activate PKC, have proven to be inducible by serum growth factors (Lim *et al.,* 1989). The kinetics of immediate-early gene activation induced by purified growth factors or phorbol esters are similar to those induced by serum.

Although identified for their expression during the G_0–G_1 transition, the immediate-early genes are regulated differently in rapidly growing cells. Where examined, these genes are expressed at a constant level throughout the cell cycle in logarithmically growing cells (Thompson *et al.,* 1985; Bravo *et al.,* 1986; O'Brien *et al.,* 1990). In addition, a number of genes known to be induced at the G_1–S border in cells stimulated to proliferate from a quiescent state show little change in their expression across the cell cycle (reviewed by Lau and Nathans, 1991). Aside from the histone genes (Heintz, 1989), there is currently little information on changes in gene transcription in cycling mammalian cells.

D. Genes Down-Regulated by Growth Factors

Concurrent with the transcriptional activation of immediate early genes by growth factors is the down-regulation of a number of other genes. These genes, which are expressed at a moderate to high levels in quiescent cells, are not expressed or are expressed at low levels in stimulated cells. Several such growth arrest-specific (*gas*) genes have been identified via cDNA cloning (Schneider *et al.,* 1988; Fornace *et al.,* 1989). The expression of these genes increases as cells grow to confluence or when cells are shifted to medium containing low serum (Schneider *et al.,* 1988; Manfioletti *et al.,* 1990). One of these genes, *gas*3, is regulated on the postranscriptional level. Upon addition of serum, the *gas*3 mRNA becomes unstable, whereas its gene transcription remains relatively unchanged (Manfioletti *et al.,* 1990). While some quiescent cell-specfic proteins are being characterized (Bedard *et al.,* 1987; Bedard *et al.,* 1989; Ching and Wang, 1990; Manfioletti *et al.,* 1990), current information on quiescent cell-specific genes is limited. Further analysis is necessary to formulate generalizations on their regulation and hypothesis on their biological functions.

III. SIGNAL-TRANSDUCTION PATHWAYS

The availability of purified growth factors made possible the characterization of early biochemical events that accompany growth-factor stimulation of quiescent cells. These studies showed that the binding of a growth factor to its cognate cell-surface receptor induces a number of biochemical responses (reviewed by Rozengurt, 1986; Williams, 1989). Many of these responses precede immediate early gene expression and therefore may function in signal transduction from the plasma membrane to the nucleus. These potential signals include ion fluxes, membrane-derived second messengers, and direct activation of protein phosphorylation cascades (Nishizuka, 1986; Rozengurt, 1986; Berridge, 1987; Ralph *et al.*, 1990). We shall consider the possible role of each of these biochemical responses as potential signals in the activation of immediate-early gene expression.

A. Ion Fluxes

Activation of the amiloride-sensitive Na^+/H^+ antiporter is among the most rapid changes induced by mitogens. Although this response may be required for DNA replication (L'Allemain *et al.*, 1984; Pouysségur *et al.*, 1984; Moolenaar, 1986; Grinstein *et al.*, 1989; Vairo *et al.* 1990), activation of Na^+/H^+ antiport appears to play a minimal role in the induction of immediate early genes. Studies of human fibroblasts showed no correlation of Na^+/H^+ antiport activity with the level of expression of the immediate-early gene *zif268/egr-1* (see Section V,A), and pharmacological agents that directly inhibit the Na^+/H^+ antiporter did not affect induction of this gene by serum (Jamieson *et al.*, 1989). In murine T lymphocytes (Moore *et al.*, 1986; Grinstein *et al.*, 1988) and bone marrow macrophages (Vairo *et al.*, 1990), activation of Na^+/H^+ antiport was unnecessary for the expression of c-*fos* and c-*myc*. Results from these heterologous systems indicate that activation of Na^+/H^+ antiport and subsequent changes in the ionic milieu of the cytoplasm are unlikely to function as important signals that activate immediate-early gene expression.

In addition to activation of Na^+/H^+ antiport, many growth factors stimulate an increase in intracellular calcium either through its release from intracellular stores or its entry through voltage-gated calcium channels (Berridge, 1987). The release of calcium from intracellular stores is stimulated by the growth factor-dependent generation of inositol 1,4,5,-trisphosphate (IP_3), a product derived from phospholipase C-catalyzed hydrolysis of phosphatidylinositol 4,5-bisphosphate (PIP_2) (Berridge, 1987). Changes in the levels of intracellular calcium affect a number of cellular processes as diverse as protein phosphorylation, gene expression, and cell motility (Rasmussen and Means, 1989). Many of these responses are mediated through the calcium-binding protein calmodulin (CaM) and CaM-dependent protein kinases (Means and Rasmussen, 1988; Rasmussen and Means, 1989). Although quiescent fibroblasts mobilize calcium in response to serum (Rozengurt, 1986), PDGF (Lopez-Rivas *et al.*, 1987), and EGF (Cutry *et al.*, 1989), the elevation of intracellular calcium alone cannot mimic the effect of these growth factors on immediate-early gene expression. Studies in which calcium inophores were used to raise intracellular calcium levels in fibroblasts showed that the immediate-early genes c-*fos* (Mehment *et al.*, 1990), *cyr61* (O'Brien *et al.*, 1990), *pip92* (Charles *et al.*, 1990) and *zif268/egr-1* (Jamieson *et al.*, 1989) were poorly induced. Moreover, buffering intracellular calcium to constant levels in rat-1 fibroblasts had no effect on induction of c-*fos* and c-*jun* by the peptide endothelin (Rodland *et al.,* 1990). In contrast, elevation of intracellular calcium was reported to induce the expression of c-*fos* and c-*myc* in C3H 10T1/2 cells (Cutry *et al.,* 1989) and lymphocytes (Moore *et al.,* 1986). One explanation for these apparently contradictory results comes from a series of studies examining the effect of *cyclic adenosine monophosphate* (cAMP) on calcium-dependent signal transduction. These studies demonstrated that in quiescent fibroblasts, the elevation of calcium in the presence of high cAMP activates immediate-early gene expression (Ran *et al.,* 1986; Mehment *et al.,* 1990). A similar mechanism involving the second messengers Ca^{2+} and cAMP was proposed to explain the induction of c-*fos* by membrane depolarization in PC12 cells, a rat pheochromocytoma cell line (Sheng *et al.,* 1990; Sheng and Greenberg, 1990).

B. Membrane-Derived Second Messengers

Growth factor–receptor interactions cause profound changes in phospholipid metabolism that function in signal transduction. Both PDGF (Williams, 1989; Larrodera *et al.,* 1990; Pessin *et al.,* 1990) and EGF (Berridge, 1987; Pessin *et al.,* 1990) have been shown to stimulate diacylglycerol (DAG) production. At least two sources of DAG have been identified during stimulation of quiescent fibroblasts with PDGF. The initial burst of DAG production appears to arise from the hydrolysis of PIP_2 (Larrodera *et al.,* 1990; Pessin *et al.,* 1990) and possibly phosphatidylcholine (PC) (Larrodera *et al.,* 1990; Pessin *et al.,* 1990), whereas PC hydrolysis seems to be critical for sustained DAG production (Larrodera *et al.,* 1990; Pessin *et al.,* 1990). The hydrolysis of PIP_2 also generates another second messenger, IP_3 (Berridge, 1987), which mobilizes intracellular calcium as described above.

The production of DAG and the mobilization of calcium from intracellular stores results in the activation of PKC (Nishizuka, 1986; Berridge, 1987). A great deal of evidence implicates PKC in the regulation of gene expression. Much of this information was obtained from the study of tumor-promoting phorbol esters, which bind to and directly activate PKC (Nishizuka, 1986). The treatment of cells with phorbol esters activates many of the same immediate-early genes induced by growth factors (Lim *et al.,* 1989). The genetic response to phorbol esters is clearly mediated by PKC, as down-regulation of this protein kinase blocks induction of immediate-early genes by the phorbol ester TPA (12-O-tetradecanoylphorbol-13-acetate) (Lim *et al.,* 1989; Charles *et al.,* 1990; O'Brien *et al.,* 1990). Although the exact mechanism by which PKC activates gene expression is unknown, the hypothesis that PKC phosphorylates and thereby directly activates transcription factors is consistent with recent evidence showing that a fraction of PKC may be localized in the nucleus (Cambier *et al.,* 1987; Thomas *et al.,* 1988). Phosphorylation studies have shown that a number of purified transcription factors are substrates for PKC *in vitro* (Yamamoto *et al.,* 1988; Gonzalez *et al.,* 1989; Jones and Jones, 1989; Malviya *et al.,* 1990), but, at present, there is no evidence demonstrating that PKC alters the activity of any transcription factor *in vivo.*

C. Protein Phosphorylation Cascades

While the activation of PKC results in signal transduction to the nucleus, growth factors such as PDGF and EGF can induce gene activation under conditions where cellular PKC levels have been down-regulated by chronic treatment with phorbol esters (Lim *et al.,* 1989; Charles *et al.,* 1990; O'Brien *et al.,* 1990). This observation indicates that growth factors utilize additional signal-transduction pathways to activate transcription. One possible pathway may involve a protein phosphorylation cascade initiated at the cell surface. The basis for this hypothesis is the observation that many growth-factor receptors possess an intrinsic ligand-dependent protein tyrosine kinase activity (Ullrich and Schlessinger, 1990). Unlike the majority of protein kinases in cells, which are serine–threonine kinases, these receptors catalyze the phosphorylation of tyrosine residues. The growth factor-dependent activation of a receptor protein tyrosine kinase suggests a mechanism for signal transmission in which the receptor kinase activates effector proteins by direct phosphorylation. However, growth factors appear to regulate cellular activity through altering the phosphorylation state of many proteins on serine and threonine residues (Edelman *et al.,* 1987; Ahn *et al.,* 1990; Ahn and Krebs, 1990; Ralph *et al.,* 1990; Woodgett, 1990). The regulation of serine and threonine phosphorylation by activation of a tyrosine kinase suggests that existence of protein kinases that are activated by phosphorylation on tyrosine residues. If this hypothesis is correct, growth-factor binding would result in activation of a protein kinase cascade.

The importance of receptor tyrosine kinase activity in signal transduction has been established through studies of kinase-deficient receptors. Kinase-deficient mutants of various growth-factor receptors were constructed by site-directed mutagenesis of cloned receptor cDNAs. The biological activity of a given mutant receptor was then analyzed after transfection of the mutagenized cDNA into a cell line expressing low endogenous levels of that receptor. Using this approach, receptors for PDGF (Escobeco *et al.,* 1988; Williams, 1989), EGF (Ullrich and Schlessinger, 1990), and insulin (Chou *et al.,* 1987; Ebina *et al.* 1987; Ullrich and Schlessinger, 1990) were shown to require receptor-kinase activity for full signal transmission. Moreover, a tyrosine kinase minus mutant of the EGF receptor was unable to activate expression of the immediate-early genes c-*fos* and c-*myc,* linking receptor tyrosine kinase activity to gene expression (Honegger *et al.,* 1987).

Identification of the substrates for receptor tyrosine kinases is the first step toward elucidating the mechanisms of signal transduction from the plasma membrane to the nucleus. The development of high-affinity antibodies to phosphotyrosine, specific growth-factor receptors, and other proteins involved in signal transduction has enabled investigators to define a profile of proteins that are phosphorylated on tyrosine residues in response to PDGF. The majority of proteins identified thus far appear to function in the generation of membrane-derived second messengers. These proteins include phospholipase C-γ (PLC-γ) (Meisenhelder *et al.*, 1989), phosphatidylinositol-3-kinase (PI-3 kinase) (Coughlin *et al.*, 1989), and the GTPase-activating protein for Ras (GAP) (Molloy *et al.*,1989; Ellis *et al.*, 1990; Kalauskas *et al.*, 1990). In addition, the tyrosine kinase c-Src (Kypta *et al.*, 1990), and the serine threonine protein kinase Raf-1 (Morrison *et al.*, 1988; Morrison *et al.*, 1989) have been reported to associate with the activated PDGF receptor.

As mentioned above, PLC-γ, PI-3 kinase, and GAP are probably involved in generating membrane-derived second messengers. PI-3 kinase catalyzes the phosphorylation of phosphatidylinositol on the D-3 position of the inositol ring (Whitman *et al.*, 1988). PI-3 kinase is also a substrate for the EGF receptor (Bjorge *et al.*, 1990; Endemann *et al.*, 1990), the insulin receptor (Ruderman *et al.*, 1990), and the CSF-1 receptor (Shurtleff *et al.*, 1990). The role of this enzyme in signal transduction is currently unknown. However, a mutant PDGF receptor, lacking the structural domain that associates with the PI-3 kinase (Coughlin *et al.*, 1989), is deficient in mitogenic signal transduction without inhibiting c-*fos* induction by PDGF (Severinsson *et al.*, 1990). This receptor mutation also prevents actin reorganization in response to PDGF. Perhaps the PI-3 kinase functions in signaling cytoskeletal changes (Majerus *et al.*, 1990).

PLC-γ catalyzes the hydrolysis of PIP_2 to yield the second messengers DAG and IP_3, as discussed above. Tyrosine phosphorylation of PLC-γ may directly activate this enzyme or facilitate its association with other proteins, for example G proteins (Fain *et al.*, 1988; Cockcroft and Bar-Sagi, 1990). A function for tyrosine phosphorylation of PLC-γ is supported by the observation that only growth-factor tyrosine kinase receptors that stimulate PIP_2 hydrolysis recognize PLC-γ as a substrate (Nishibe *et al.*, 1990a). Moreover, a recent study reports that phosphorylation of PLC-γ by EGF receptor tyrosine kinase both *in vivo* and *in vitro* increases PLC-γ activity about threefold (Nishibe *et al.*, 1990b). PLC-γ activity was reduced to a basal level after treatment with a protein tyrosine phosphate phosphatase (Nishibe *et al.*, 1990b). These data constitute the first evidence that PLC-γ activity is increased by tyrosine phosphorylation. Activation of PLC-γ can potentially induce immediate-early gene expression indirectly through DAG production and activation of PKC.

The function for tyrosine phosphorylation of GAP is unknown. GAP activates the GTPase activity of the protooncogene product Ras (Hall, 1990; Parsons, 1990).

Two lines of evidence indicate that Ras proteins play a critical role in mitogenic signal transduction. Microinjection of anti-Ras antibody blocks DNA synthesis after serum stimulation of quiescent NIH 3T3 cells (Mulcahy *et al.,* 1985), and a dominant negative mutation in the c-Ha-*ras* gene that changes a serine residue (Ser 17) to an asparagine residue inhibits growth factor stimulation of cell proliferation (Cai *et al.,* 1990). However, expression of this c-Ha-*ras* mutant in a stable cell line inhibited cell proliferation without blocking c-*fos,* c-*jun,* and c-*myc* induction (Cai *et al.,* 1990). Studies using microinjection of neutralizing antibodies against PKC, or of a synthetic peptide that inhibits PKC, suggest that Ras function is dependent on PKC activity (Gauthire-Rouviere *et al.,* 1990). Perhaps a complex or the association of growth-factor receptor, GAP, Ras and PLC-γ leads to PIP_2 hydrolysis, PKC activation, and signal transmission to the nucleus. This model, however, does not account for the PKC-independent pathways that activate immediate-early genes.

Evidence that the PDGF receptor stimulates the phosphorylation of the protooncogene product Raf-1, a cytoplasmic serine–threonine protein kinase (Morrison *et al.,* 1988; Morrison *et al.,* 1989; Bjorge *et al.,*1990), raised the exciting possibility that Raf-1 functioned to couple growth-factor receptor tyrosine protein kinase activity to changes in protein serine–threonine phosphorylation. This idea was supported by the report that tyrosine phosphorylation of Raf-1 increased Raf-1 protein kinase activity (Morrison *et al.,* 1989; Morrison *et al.,* 1988). Transfection studies have demonstrated that Raf-1 can induce expression of a reporter gene under the control of the c-*fos* enhancer independent of PKC or PKA activity. This result suggests that Raf-1 could regulate immediate-early gene expression through the serum response factor, $p67^{SRF}$ (see Section IV,A) (Kaibuchi *et al.,* 1989). Moreover, additional transient transfection studies with v-*raf,* the transforming counterpart of c-*raf* (Rapp *et al.,* 1988), showed that v-*raf* can transactivate reporter genes driven by promoters derived from c-*fos* and β-actin genes but not from the c-*myc* gene (Jamal and Ziff, 1990). A more detailed analysis of the c-*fos* promoter identified multiple elements that are activated by v-*raf.* These include the dyad symmetry element (see section IV,A), an octanucleotide direct repeat, and a region spanning nucleotides −225 to −99 (Jamal and Ziff, 1990). The results of these experiments suggest that Raf-1 activity may be required for induction of a subset of immediate-early genes.

Although the activation of Raf-1 protein kinase activity by growth-factor receptor–catalyzed tyrosine phosphorylation is an attractive hypothesis, the original observation has become controversial. A recent study was unable to detect tyrosine phosphorylation of Raf-1 in response to either CSF-1 or PDGF (Baccarini *et al.,* 1990). Other studies showed that insulin stimulated Raf-1 protein kinase activity without tyrosine phosphorylation (Blackshear *et al.,* 1990; Kovacina *et al.,* 1990). Since growth factor–receptor protein tyrosine kinases differ in their substrate specificities (Nishibe *et al.,* 1990a), CSF-1, PDGF, and insulin could activate Raf-1 protein kinase by different mechanisms. However, the conflicting data on PDGF were obtained with similar methods, indicating further investigation will be necessary to resolve the controversy surrounding PDGF-dependent activation of Raf-1.

Casein kinase II (CK-II) is another serine–threonine kinase implicated in the regulation of immediate-early gene expression. Serum, EGF, insulin, and insulin-like growth factor I (Ralph *et al.,* 1990) have been shown to stimulate CK-II activity. CK-II, which is located in both the cytoplasm and nucleus (Ralph *et al.,* 1990), catalyzes phosphorylation of a number of nuclear proteins *in vitro,* including Myc (Lüscher *et al.,* 1989), Myb (Lüscher *et al.,* 1990), SV40 large-T antigen (Grasser *et al.* 1988; Krebs *et al.,* 1988), the p53 tumor-suppression protein (Meek *et al.,* 1990), and $p67^{SRF}$ (serum response factor) (Manak *et al.,* 1990). $p67^{SRF}$ is a transcription factor that binds to the serum response element (SRE) found in the c-*fos* promoter (see Section IV, A), and functions in serum induction of this immediate-early gene. $p67^{SRF}$ contains one CK-II phosphorylation site; when phosphorylated, its DNA-binding activity is enhanced (Manak *et al.,* 1990), suggesting a mechanism whereby its activity is regulated by CK-II phosphorylation. At present there are no data that correlate phosphorylation of the CK-II site of $p67^{SRF}$ with serum induction of an immediate-early gene, but the phosphorylation of this site was detected in HeLa cells (Manak *et al.,* 1990), suggesting a physiological function for its modification. The generation and study of mutants in the CK-II site of SRF should be informative.

While the existence of a protein kinase cascade is unproven, experiments with protein kinase and phosphatase inhibitors suggest that immediate-early gene expression is regulated by reversible protein phosphorylation. Okadaic acid, an inhibitor of protein phosphatase 1 and 2A (Cohen *et al.,* 1990), induced the expression of the c-*fos* gene in synoviocytes (Kim *et al.,* 1990). In contrast, the protein kinase inhibitor 2-aminopurine (Zinn *et al.,* 1988; Mahadevan *et al.,* 1989) blocked induction of c-*fos* and c-*myc* in response to mitogenic stimulation in human and murine fibroblasts (Mahadevan *et al.,* 1989). The action of 2-aminopurine has been correlated with a decrease in the phosphorylation state of two nuclear proteins (Mahadevan *et al.,* 1989).

Perhaps the best current example of how protein phosphorylation can effect gene expression comes from the study of cyclic AMP–responsive genes. The activation of these genes appears to be mediated through the interaction of a specific promoter element and transcription factor (Ziff, 1990). Elegant experiments have demonstrated that phosphorylation of this transcription factor, the cyclic AMP response element binding protein (CREB), on a specific serine residue was required for cyclic AMP-dependent activation of gene expression (Gonzalez and Montminy, 1989). This serine residue is phosphorylated by the cyclic AMP–dependent protein kinase, PKA (Gonzalez and Montminy, 1989). Therefore, agents that elevate cyclic AMP can induce expression of cyclic AMP-responsive genes through PKA-dependent phosphorylation of CREB. In addition, recent data suggest that the unphosphorylated form of CREB could act as a transcriptional repressor (Berkowitz and Gilman, 1990), raising the possibility that cyclic AMP antagonists could block gene expression by causing dephosphorylation of CREB or preventing CREB's phosphorylation.

IV. TRANSCRIPTIONAL AND POSTTRANSCRIPTIONAL CONTROL

Immediate-early genes are transcriptionally silent in the quiescent cell. On growth factor stimulation, transcription of these genes is rapidly induced through a mechanism that does not require *de novo* protein synthesis. In many cases, this activation is soon followed by efficient transcriptional repression. What are the mechanisms of rapid transcriptional activation and subsequent repression? Are all immediate-early genes activated by the same mechanism? Is immediate-early gene expression regulated posttranscriptionally? We shall address these questions in this section.

A. Transcriptional Activation

The key to understanding the connection between the signals propagated as a result of growth factor–receptor interactions and gene activation lies in the analysis of immediate-early gene promoters and the protein factors that interact with them. The most detailed analysis of this kind has been carried out for the c-*fos* protooncogene. Thus, studies on c-*fos* will serve as a paradigm for most of our discussion, although other regulatory mechanisms are now beginning to emerge as more immediate-early gene promoters are characterized.

Transcriptional activation of c-*fos* follows kinetics that are characteristic of group I immediate-early genes (see Section I,C). Analysis of the human c-*fos* promoter identified a 20-bp sequence element centered at –310 relative to the transcription start site (Treisman, 1985). Designated the SRE (serum response element) or the DSE (dyad symmetry element), this sequence element consists of the core sequence CC(A/T)$_6$GG (CArG box) flanked by palindromic arms. Oligonucleotides containing the SRE confer serum responsiveness to both minimal c-*fos* and heterologous promoters, although the CArG box core sequence is sufficient to impart serum inducibility (Treisman, 1986; Greenberg *et al.*, 1987; Mohun *et al.*, 1987; Gilman, 1988; Sheng *et al.*, 1988; Leung and Miyamoto, 1989; Siegfried and Ziff, 1989). SRE is required for induction by PDGF, EGF, insulin, nerve growth factor (NGF), TPA, and UV light (Fisch *et al.*, 1987; Buscher *et al.*, 1988; Gilman, 1988; Sheng *et al.*, 1988; Stumpo *et al.*, 1988; Siegfried and Ziff, 1989), but not required for induction by cAMP or calcium-dependent signals (Buscher *et al.*, 1988; Gilman, 1988; Sheng *et al.*, 1988; Fisch *et al.*, 1989a).

The SRE is a specific binding site for a number of protein factors present in a variety of cell types (Gilman *et al.,* 1986; Prywes and Roeder, 1986; Treisman, 1986; Greenberg *et al.,* 1987). Mutations in SRE oligonucleotides that abolish binding to these protein factors *in vitro* also prevent serum induction when fused upstream of a heterologous promoter of a transfected gene (Fisch *et al.,* 1987; Greenberg *et al.,* 1987; Gilman, 1988). A 67-kDa SRE-binding activity purified from HeLa cells has been referred to as SRF (serum response factor) (Treisman, 1987) or $p67^{SRF}$ (Schröter *et al.,* 1987). $p67^{SRF}$ is a ubiquitous nuclear transcription factor, which is glycosylated (Schröter *et al.,* 1990) and phosphorylated (Prywes *et al.,* 1988; Manak *et al.,* 1990), and binds to the SRE as a dimer (Norman *et al.,* 1988; Schröter *et al.,* 1990). Dimerization and DNA-binding functions map within a core domain between amino acids 93 and 222 (Norman *et al.,* 1988; Schröter *et al.,* 1990) (see Section V,A,6). Another protein purified from HeLa cells designated f-EBP (c-*fos* enhancer-binding protein) displays SRE-binding activity similar to that of $p67^{SRF}$(Prywes and Roder, 1987); however, its 62,000 molecular weight suggests that it may be a different protein.

An additional component, a 62-kDa protein designated $p62^{TCF}$ (ternary complex factor), forms a ternary complex with $p67^{SRF}$ and the SRE (Shaw *et al.,* 1989a). $p62^{TCF}$ does not bind to the SRE alone but requires the presence of $p67^{SRF}$ for binding. Together, the $p67^{SRF}$ dimer and $p62^{TCF}$ have a 50-fold greater affinity for the SRE than the $p67^{SRF}$ dimer alone (Shaw *et al.,* 1989a; Schröter *et al.* 1990). Ternary complex formation requires sequences both within and 5′ of the SRE. Mutations that abolish ternary complex formation *in vitro* severely impair the serum response of the c-*fos* promoter in a transfection assay (Shaw *et al.,* 1989a).

Another SRE-binding activity that is distinct from the p62 discussed above is attributable to another 62-kDa polypeptide that has been characterized in extracts from human H9 T lymphoblasts (Ryan *et al.,* 1989). The relationship between this 62-kDa factor and f-EBP is unclear (Ryan *et al.,* 1989), although they appear to bind differently to SRE (Prywes and Roeder, 1987; Ryan *et al.,* 1989). In addition, the muscle actin promoter factors 1 and 2 (MAPF1 and MAPF2) (Walsh and Schimmel, 1987; Walsh, 1989) purified from primary embryonic chicken cells bind to the actin promoter CArG boxes and to the c-*fos* SRE (Walsh, 1989). Based on similar contacts with the SRE observed in dimethyl sulfate interference assays, the 62-kDa factor from H9 cells appears to be related to MAPF1 (Ryan *et al.,* 1989).

A simple model would predict that mitogenic signals transduced to the nucleus activate the binding of positively acting transcription factors to SRE-like elements associated with target genes, thereby activating transcription. However, contrary to this idea, numerous studies using nuclear extracts have shown that the ability of nuclear factors to bind SRE is independent of serum or growth-factor stimulation (Gilman *et al.,* 1986; Greenberg *et al.,* 1987; Hayes *et al.,* 1987; Mohun *et al.,* 1987; Treisman, 1987; Norman *et al.,* 1988; Sheng *et al.,* 1988). Furthermore, these *in vitro* studies have been confirmed by dimethyl sulfate genomic footprinting (Herrera *et al.,* 1989). In human A431 cells, contacts consistent with the binding of $p67^{SRF}$, $p62^{TCF}$, and at least one other protein binding immediately 3′of $p67^{SRF}$, are present before induction by EGF and are unchanged during gene activation and subsequent repression (Herrera *et al.,* 1989). Oddly, *in vitro* studies using nuclear extracts from A431 cells detected an EGF-inducible SRE-binding activity (Prywes and Roeder, 1986). The significance of this apparent discrepancy remains unclear. The sequence immediately 3′ of the c-*fos* SRE resembles an AP-1 site (see Section V,A,1), although no protein has been reported to interact with this sequence *in vitro* when $p67^{SRF}$ is bound to the SRE (Shaw *et al.,* 1989b).

Transcription-factor binding to the SRE independent of transcriptional activity is consistent with several regulatory models. Polypeptide exchange on the SRE could be occurring, although it would have to be extremely rapid and result in identical contacts before and after exchange. Alternatively, and more likely, regulation may be mediated by posttranslational modifications of the preexisting complex or exchange of factors interacting with this complex or both.

The interactions of $p67^{SRF}$ with SRE-like sequence elements appear to mediate the activation of several other immediate early genes, including *actin,* (Minty and Kedes, 1986; Mohun *et al.,* 1987), *zif268* (Changelian *et al.,* 1989; Christy and Nathans, 1989b; Janssen-Timmen *et al.,* 1989), *krox20* (Chavrier *et al.,* 1988; Chavrier *et al.,* 1989), and *cyr61* (Latinkic *et al.,* 1991). These genes contain CArG box sequences but not the extended dyad symmetry found in the c-*fos* SRE. Four such elements are found upstream of the mouse *zif268* (Changelian *et al.,* 1989; Christy and Nathans, 1989b; Janssen-Timmen *et al.,* 1989) and human cardiac *actin* (Minty and Kedes, 1986) genes; one copy is found upstream of the *cyr61* gene (Latinkic *et al.,* 1991); and two copies are found upstream or in the first intron of the *krox20* gene (Chavrier *et al.,* 1989; Chavrier *et al.,* 1988). These CArG boxes conferred serum responsiveness to heterologous promoters individually (Chavrier *et al.,* 1989; Christy and Nathans, 1989b; Janssen-Timmen *et al.,* 1989), and multiple CArG elements provided greater inducibility than did a single element (Christy and Nathans, 1989b). Furthermore, these CArG boxes can bind to $p67^{SRF}$. However, it is not known whether the same multiprotein complex that forms upon the c-*fos* SRE can also form upon these CArG boxes.

In addition to proteins that bind SRE, a PDGF-inducible DNA binding activity called SIF (*sis*/PDGF-inducible factor) interacts with a sequence element located 352–335 bp upstream of the c-*fos* mRNA initiation site (Hayes *et al.,* 1987). The DNA-binding activity of SIF is rapidly induced when quiescent cells are treated with PDGF even in the presence of cycloheximide. This DNA-binding activity is poorly induced by serum, and serum induction of c-*fos* does not require the SIF binding site. The precise role of SIF in immediate-early gene activation is not yet known.

Several group I immediate-early genes whose kinetics of transcriptional activation are virtually identical to those of c-*fos* do not contain SRE or CArG box sequences in their promoters. These genes include c-*jun* (Hattori *et al.,* 1988), *junB*(Schutte *et al.,* 1989). *nur77* (Watson and Milbrandt, 1989), and *3CH96* (Lau and Nathans, 1987) (J.K. Yoon and L.F.L., unpublished). No $p67^{SRF}$ interactions were detected in the regulatory regions of *nur77* or *3CH96* (J.K.Y., unpublished). Detailed c-*jun* promoter analysis (Angel *et al.,* 1988) implicates AP-1 elements as likely participants in serum induction. The upstream sequences of *junB* do not appear to contain SRE or AP-1 elements (Schutte *et al.,* 1989), indicating further regulatory diversity.

The role of AP-1 elements in transcriptional activation by growth factors has been examined in the context of the c-*fos* promoter. A sequence highly homologous to the AP-1 consensus binding site is located directly adjacent to the SRE at −298 to −287 of human c-*fos* gene. AP-1 is a heterodimeric complex comprising products of the Fos and Jun gene families (see Section V,A,1). Examination of the function of this element has led to contradictory data that appear to be a function of the assay system. Initial studies using transient transfection assays showed that eliminating the AP-1 site from the c-*fos* promoter had no effect on induction by either EGF or TPA (Fisch *et al.,* 1987). Surprisingly, when these same promoter constructs were stably integrated, the data revealed that the AP-1 site could mediate induction by both EGF and TPA about as effectively as the SRE (Fisch *et al.,* 1989b). Although these conflicting conclusions are disturbing, they could be explained by a requisite factor that was limiting in the transient assays but not in the cell lines (Fisch *et al,* 1989b). It is clear that under certain experimental conditions, an AP-1 site can mediate a growth-factor response with kinetics similar to an SRE-mediated response (Fisch *et al.,* 1989b). However, since the proteins that bind AP-1 elements are themselves encoded by immediate-early genes and are present at very low levels in the quiescent cell, it is unclear how they could mediate an immediate-early response. Perhaps the constitutively expressed AP-1 family member, *junD* (Hirai *et al.,* 1989; Ryder *et al.,* 1989), plays a primary role in immediate-early gene activation (Hirai *et al.,* 1989).

Several group II immediate early genes, such as c-*myc* (Asselin *et al.,* 1989) and *JE* (Rollins *et al.,* 1988), do not display *fos*-like kinetics of activation. Presumably, these genes are regulated by alternative mechanisms. As expected, they do not have SRE sequences in their 5′ flanking regions. Currently little information is available on the mechanism of serum activation of group II genes.

B. Transcriptional Repression

The rapid activation of many immediate-early genes is followed by transcriptional repression, rendering their transcription rates to low or undetectable levels within 60 to 120 min after stimulation. In contrast to immediate-early gene activation, repression requires *de novo* protein synthesis (Greenberg *et al.*, 1986; Lau and Nathans, 1987), suggesting that a newly synthesized or labile protein acts as a transcriptional repressor. Likely candidates for such repressor molecules are immediate-early gene-encoded proteins. In addition, immediate-early genes are transcriptionally silent in the quiescent cell, suggesting that they are repressed in the quiescent state. In this section we shall summarize the work directed at understanding the two types of repression: one that occurs to repress transcription following activation due to growth factor stimulation and one that maintains the absence of transcription in the quiescent state.

1. Transcriptional Repression following Stimulation

Many significant discoveries related to immediate early gene expression and function have been obtained as a result of studies on autoregulation. *In vivo* competition studies using transfected promoter fragments indicated that c-*fos* expression may be negatively regulated through a sequence upstream of −225 (Sassone-Corsi and Verma, 1987). Initial studies to examine this regulation determined that the c-Fos protein negatively regulates its own gene promoter (Sassone-Corsi *et al.*, 1988; Schönthal *et al.*, 1988). Two regions in the c-*fos* promoter, from −323 to −276 containing the SRE and AP-1 elements and from −277 to −223, were mapped to mediate the repression (Sassone-Corsi *et al.*, 1988). Concurrent studies speculated that the AP-1 element is required for the down-regulation event (Wilson and Treisman, 1988b; Schönthal *et al.*, 1989). Such speculation was based heavily on the emerging knowledge that Fos and Jun heterodimers bind AP-1 sites.

Several recent studies however, have determined that the SRE is the primary target of c-*fos* repression (Konig *et al.*, 1989; Lucibello *et al.*, 1989; Subramaniam *et al.*, 1989; Gius *et al.*, 1990; Rivera *et al.*, 1990). Two of these studies provided convincing data by examining the transcription rates of stably integrated promoter constructs in which an SRE was fused to an otherwise nonresponsive heterologous promoter (Konig *et al.*, 1989; Rivera *et al.*, 1990). These constructs confirmed that the SRE is sufficient to direct rapid transcriptional induction and repression. Detailed mutagenesis of the c-*fos* SRE and analysis of *zif268/egr-1* CArG boxes indicate that the extended dyad symmetry present in the c-*fos* SRE is not critical for repression (Gius *et al.*, 1990; Rivera *et al.*, 1990).

Initial studies mapping the regulatory domains required for repression determined that c-Fos carboxyl-terminal (C-terminal) replacement mutants failed to shut off c-*fos* transcription. These mutations were intriguing in that they were shown to act in a dominant fashion to interfere with the down-regulation of a cotransfected wild type c-*fos* gene (Wilson and Treisman, 1988b). Other studies reported that transfection of a c-Jun expression plasmid could repress c-*fos* transcription (Konig *et al.*, 1989; Lucibello *et al.*, 1989; Schönthal *et al.*, 1989). Moreover, transfection of Fos and Jun together could significantly enhance this repression (Konig *et al.,* 1989; Lucibello *et al.,* 1989; Schönthal *et al.,* 1989). Concurrent mapping studies determined that the leucine-zipper region of the Fos protein was required for repression, although the parental Fos construct used for mutagenesis in these studies did not contain the 64 amino acids in the C-terminus (Lucibello *et al.,* 1989).

Other studies on Fos-mediated repression examined its ability to down-regulate the *zif268/egr-1* gene, which contains multiple CArG boxes in its promoter. Although these studies confirm that the CArG box, or the core of SRE, is the target for repression, they differ from previous studies regarding the role of c-Jun (Gius *et al.,* 1990). c-Jun was found to have no effect on Fos-mediated down-regulation, nor was it capable of mediating down-regulation alone. However, mutant Fos that was unable to either dimerize with c-Jun or to bind AP-1 sites was capable of mediating repression (Gius *et al.,* 1990). Mapping studies determined that the critical residues for repression were located within the C-terminal 27 amino acids (Gius *et al.,* 1990), consistent with previous studies (Wilson and Treisman, 1988b). Furthermore, transfer of 180 Fos C-terminal amino acids to Jun conferred upon it the ability to repress *zif268/egr-1* expression. The differences between these studies on *egr-1* repression and those that suggested that c-Jun can synergize with c-Fos in down-regulating c-*fos* promoter activity (Konig *et al.,* 1989; Lucibello *et al.,* 1989; Schönthal *et al.,* 1989) may be attributable to sequence differences between the c-*fos* SRE and the *egr-1* CArG elements or to the different promoter contexts in which these elements were examined. It is possible that the oncogenic potential of v-Fos is related to its inability to act as a negative regulator of transcription due to the loss of C-terminal sequences (Wilson and Treisman, 1988b; Gius *et al.,* 1990).

Recent studies have begun to address the mechanisms of Fos autorepression. Mutations of the serine residues between amino acids 362 and 364 of Fos result in hypophosphorylated c-Fos protein that was unable to repress c-*fos* transcription following stimulation, but had no effect on its transactivation function (Ofir *et al.,*1990). This apparent requirement for phosphorylation was relieved by mutation of these serine residues to aspartic and glutamic acid residues, thereby conferring a negative charge. It is still unclear what the precise role of c-Fos is in repression. Perhaps it functions to disrupt the interactions of positively acting factors at the SRE. However, no interactions between c-Fos and the SRE or the SRE/p67^{SRF} complex have been detected (Gius *et al.,* 1990; Rivera *et al.,* 1990). Alternatively, c-Fos may repress transcription indirectly by enhancing the activity or expression of a repressor of c-*fos* transcription (Rivera and Greenberg, 1990).

Examination of the down-regulation of other immediate-early genes will likely lead to the discovery of alternative repression mechanisms. Immediate-early genes such as c-*jun, junB,* and *nur77* do not have SRE or CArG boxes sequences in their promoters and are likely to be repressed by proteins other than Fos. Possible candidates for alternative repressors are the Zif268 and Krox20 (see

Section V,A,4,a). These immediate-early genes encode DNA-binding proteins that interact with the regulatory regions of several immediate-early genes (Chavrier *et al.*, 1990; Christy and Nathans, 1989a; Lemaire *et al.*, 1990). In addition, the *cyr61* gene, whose serum resposiveness is mediated by a CArG box, is repressed with very different kinetics from that of c-*fos,* suggesting a different mechanism (O'Brien *et al.*, 1990). Thus, as with activation, the repression of immediate-early gene transcription appears to be complex and may be mediated by multiple mechanisms.

2. *Negative Regulation of Basal Transcription*

The absence of transcription of c-*fos* in the quiescent cell is due to at least two effects: a nearly complete repression of transcriptional initiation and a block of transcription elongation. The element essential for the elongation block in c-*fos* has been localized to 21 bp from the end of the first exon (Lamb *et al.*, 1990). Coinjection of this element, designated FIRE (*fos* intragenic regulatory element), with the SRE sequence induced c-*fos* expression in quiescent cells, whereas injection of SRE sequence alone did not (Lamb *et al.*, 1990). Although transcription elongation blocks are found in other protooncogenes such as *myc* and *myb* (Spencer and Groudine, 1990), whether this is a common feature of regulation of immediate-early genes is not known.

Examination of the role of upstream sequences showed that basal levels of c-*fos* transcription were elevated threefold in constructs lacking the AP-1 element adjacent to the SRE (−295) (Lucibello *et al.*, 1989; Konig *et al.*, 1989). C-Fos or c-Jun overexpression led to a reduction in c-*fos* basal activity in the wild-type promoter, while depriving the cells of c-Fos or c-Jun using antisense RNAs or transfected binding sites enhances c-*fos* promoter activity (Schönthal *et al.*, 1988; Konig *et al.*, 1989; Schönthal *et al.*, 1989). These data suggest that basal transcription of c-*fos* is negatively regulated by c-Fos or c-Jun or by a Fos–Jun-dependent protein that is targeted to the AP-1 site at −295.

This control of basal expression operates only if the SRE and AP-1 element are placed adjacent to each other. Insertion mutations between the SRE and the AP-1 element result in constitutive overexpression of c-*fos* (Shaw *et al.*, 1989b). In addition, mutations that prevent $p67^{SRF}$ binding result in increased expression when transfected into HeLa cells (Leung and Miyamoto, 1989; Shaw *et al.*, 1989b). This result suggests that $p67^{SRF}$ is not only required for serum induction, but also for the repression before induction. We note that when SRE mutations that abolish the *in vitro* binding of $p67^{SRF}$ are examined in NIH 3T3 cells, these mutations caused no change in the basal levels of the transfected gene (Greenberg *et al.*, 1987; Treisman, 1987; Gilman, 1988; Rivera *et al.*, 1990). This difference between the results from HeLa cells versus those from NIH 3T3 cells may be explained by the fact that some of the NIH 3T3 studies used constructs that did not contain AP-1 elements adjacent to the SRE (Treisman, 1987; Gilman, 1988). Alternatively, results from HeLa cells may be explained by the prevalence of a positively acting trancription factor that can interact at a site adjacent to the SRE when $p67^{SRF}$ binding is abolished (Rivera *et al.,* 1990).

C. Posttranscriptional Regulation

1. mRNA Degradation

The degradation of immediate-early mRNAs appears to follow two major themes. Those immediate-early mRNAs encoding cytoskeletal or other structural proteins tend to have long half-lives of 2 to 10 hr, perhaps reflecting a need for the synthesis of their encoded proteins throughout the proliferative cycle. These include mRNAs encoding actin, fibronectin, fibronectin receptor, and α-tropomyosin (Bravo, 1989; Lau and Nathans, 1991). In contrast, many immediate-early mRNAs encoding regulatory proteins have much shorter half-lives of about 10 to 15 min (Lau and Nathans, 1991).

Mechanistically, the stabilization of immediate-early mRNAs by cycloheximide suggests that a newly synthesized or labile nuclease (or nuclease regulator) may be required for mRNA degradation. Such a factor has been implicated in c-*myc* mRNA degradation (Brewer and Ross, 1989). Alternatively, cycloheximide stabilization may indicate the coupling of mRNA degradation to the process of translation (Lau and Nathans, 1987; Wilson and Treisman, 1988a), similar to the degradation of tubulin mRNA (Gay *et al.,* 1989). It is possible that both a newly synthesized or labile component and translation participate in rapid mRNA degradation.

Several sequence motifs appear to be important for rapid mRNA degradation. Repeats of the sequence AUUUA, identified in virtually all short-lived, immediate-early mRNAs (Shaw and Kamem, 1986), contribute substantially to cytoplasmic transcript instability (Jones and Cole, 1987; Kabnick and Housman, 1988; Wilson and Treisman, 1988a; Shyu *et al.,* 1989). In addition, sequences within the 5′-untranslated (Jones and Cole, 1987; Kabnick and Housman, 1988; Shyu *et al.,* 1989) and coding regions (Kabnick and Housman, 1988; Shyu *et al.,* 1989) influence message instability. Recent studies examining c-*fos* mRNA degradation suggest that the poly(A) tail can protect the mRNA from degradation; removal of the tail occurs before the transcribed portion of the message is degraded (Shyu *et al.,* 1991). Two determinants mediate deadenylation: one in the protein-coding region and the other an AU-rich element in the 3′ untranslated region containing AUUUA pentanucleotides (Kabnick and Housman, 1988; Shyu *et al.,* 1989).

Recently, a protein that binds specifically to the sequence AUUUA has been identified (Malter, 1989). Such a protein may expose the poly(A) tail to ribonuclease digestion (Berstein *et al.*, 1989; Shyu *et al.*, 1991). In addition, an *in vitro* extract that can rapidly degrade c-*myc* and histone mRNAs but not globin mRNA has been described (Brewer and Ross, 1988; Berstein *et al.*, 1989). Further fractionation of this extract will likely provide insight into the mechanism of immediate-early mRNA degradation.

2. *Posttranslational Modification*

Posttranslational modification may serve as an additional mode of regulation on the activity of immediate-early proteins. Many immediate-early transcription factors, including members of the Fos and Jun families (Müller *et al.*, 1987; Cohen *et al.*, 1989), Zif268 (NGFI-A)(Day *et al.*, 1990), SRF (Manak *et al.*, 1990), and Nur77 (NGFI-B)(Fahrner *et al.*, 1990; Hazel *et al.*, 1991), are postranslationally modified by phosphorylation. The role of phosphorylation in modulating the activity of transcription factors is well documented (Gonzalez and Montminy, 1989; Mylin *et al.*, 1989; Raychaudhuri *et al.*, 1989 Jackson *et al.*, 1990; Tanaka and Herr, 1990; Lüscher *et al.*, 1990). Furthermore, serine phosphorylation at the C-terminus of c-Fos is required for its transrepression function (Ofir *et al.*, 1990). Although the functional significance of phosphorylation in regulating the activities of immediate-early transcription factors has not been extensively studied to date, a possible function of such regulation may be in specifying the activities of immediate-early proteins induced by agents that cause distinct biological responses (See Section VII).

V. PROTEINS ENCODED BY IMMEDIATE-EARLY GENES

Characterization of the proteins encoded by immediate-early genes is crucial toward understanding the genetic program activated by growth factors. The hypothesis that immediate-early proteins mediate the downstream cellular responses to growth factors has been supported by the discovery that a number of them are known or likely transcription factors, oncoproteins, and cytokines. Moreover, expression of antisense RNAs in some cases results in inhibition of cell proliferation (Holt *et al.*, 1986; Nishikura and Murray, 1987; Heikkila *et al.*, 1987; Holt *et al.*, 1988). To date, more than 25 immediate-early gene products can be related to known proteins on the basis of sequence comparisons and/or biochemical properties (See Table I). Some of these proteins have been extensively reviewed elsewhere, and thus our discussion here will be brief.

A. Nuclear Proteins

1. Fos and Jun Family Members

Considerable excitement has surrounded the *fos* and *jun* protooncogenes, particularly since the discovery that they encode proteins that can form heterodimers and constitute the major components in the transcription factor AP-1 (reviewed by Distel and Spiegelman, 1990; Vogt and Bos, 1990). C-*fos* is the cellular counterpart of the transforming genes of the FBJ and FBR murine osteosarcoma viruses (reviewed by Curran, 1988) and is one of the first protooncogenes found to be rapidly activated by growth factors (Greenberg and Ziff, 1984; Müller *et al.*, 1984). More recent studies showed that other Fos-related proteins are encoded by immediate early genes as well, including Fos B (Zerial *et al.*, 1989), Fra-1 (Fos-related antigen 1) (Cohen and Curran, 1988), and Fra-2 (Nishina *et al.*, 1990). V-*jun* is the transforming oncogene of avian sarcoma virus 17 (reviewed by Vogt and Bos, 1990). Both c-*jun* and a related gene, *jun B,* are immediate early genes (Lamph *et al.*, 1988; Ryder *et al.*, 1988; Ryder and Nathans, 1988; Ryseck *et al.*,1988; Quantin and Breathnach, 1988). Another member of the *jun* family, *jun D,* is expressed constitutively and is not inducible by growth factors (Hirai *et al.*,1989; Ryder *et al.*, 1989).

Each of the Jun proteins can either form homodimers, or form heterodimers with another Jun protein or any of the Fos proteins through their leucine-zipper regions (Landschulz *et al.*, 1989). The resulting homodimers or heterodimers can bind to AP-1 sites (consensus ATGACTCAT) and the cAMP response elements (CRE) (consensus ATGACGTCAT) (reviewed by Distel and Spiegelman, 1990; Vogt and Bos, 1990). Although it is clear that the Jun and Fos proteins can regulate transcription, the precise difference in target gene specificity or activity of each Fos–Jun heterodimer or Jun–Jun homodimer is not well understood. The large number of possible dimers underscores the possibility for a wide range of regulatory activities. Furthermore, recent work demonstrated that Fos and Jun can independently interact with the glucocorticoid receptor and alter its activity (Diamond *et al.*, 1990; Jonat *et al.*, 1990; Lucibello *et al.*, 1990; Schüle *et al.*, 1990; Yang-Yen *et al.*, 1990), adding further complexity to the potential roles of Fos and Jun proteins.

2. Rel and NF-kB

The protooncogene *rel* (reviewed by Rice and Gilden, 1988) is an immediate early gene activated by serum or TPA (Bull *et al.*, 1989), and its product can function in regulating transcription (Bull *et al.*, 1990). Rel is homologous, but not identical, to the DNA-binding subunit (p50) of NF-κB (Ghosh *et al.*, 1990;

Kiernan *et al.*, 1990), a transcription factor that binds the κB site in the immunoglobulin enhancer (Lenardo and Baltimore, 1989). NF-κB plays a significant role in the inducible expression of a large number of genes, including cytokines, cytokine receptors, major histocompatibility antigens, and genes in a number of viruses. In its active form, the p50 subunit of NF-κB is cytosolic and is complexed to an inhibitory polypeptide, IκB, which upon induction is phosphorylated. The phosphorylated IκB dissociates from p50, allowing it to translocate to the nucleus and function in transcriptional regulation (Ghos and Baltimore, 1990). The homology of Rel to p50 suggests that Rel might also act as an inducible transcription factor. Recently, the gene for p50 itself has been found to be an immediate-early gene in mitogen-stimulated T lymphocytes (Bours *et al.*, 1990).

3. *Myc*

The c-*myc* protooncogene (reviewed by Erisman and Astrin, 1988) is among the first genes to be recognized as an immediate-early gene (Kelly *et al.*, 1983). Overexpression of c-*myc* or microinjection of c-Myc protein into 3T3 cells partially alleviates the cell's requirement for exogenous growth factors for DNA synthesis (Kaczmarek *et al.*, 1985). Also, *myc* antisense oligonucleotides inhibit G_1 cells from progressing into S phase (Heikkila *et al.*, 1987; Holt *et al.*, 1988), indicating that Myc function is important for cell growth.

A region of Myc required for transformation bears significant sequence similarity to the helix–loop–helix family of transcription regulators including the myogenic differentiation factor Myo D, immunoglobulin kappa-chain enhancer-binding proteins, and Drosophila gene products that play roles in cell type and sex determination (Murre *et al.*, 1989). Recently it was found that a purified C-terminal fragment of c-Myc binds specifically to the sequence CACGTG *in vitro*(Blackwell *et al.*, 1990). Current evidence indicates that Myc can activate transcription (Cole, 1990; Kato *et al.*, 1990), and may function in regulating both transcriptional regulation and DNA replication (Collum and Alt, 1990).

4. *Nuclear Proteins Containing Zinc Fingers*

a. Zif268 (Krox24, Egr-1, NGF-IA, TIS8). This zinc-finger protein–encoding cDNA and its homologs have been isolated from various cDNA libraries, including serum-stimulated fibroblasts (Christy *et al.*, 1988; Lemaire *et al.*, 1988; Sukhatme *et al.*, 1988), NGF-stimulated PC12 cells (Milbrandt, 1987), mitogen-activated lymphocytes (Wright *et al.*, 1990), and TPA-stimulated fibroblasts (Varnum *et al.*, 1989a). The protein has been variously named Zif268/Egr-1/Krox24/NGFI-A/TIS8 (Milbrandt, 1987; Christy *et al.*, 1988; Lemaire *et al.*, 1988; Sukhatme *et al.*, 1988; Varnum *et al.*, 1989a) Zif268 is a nuclear phosphoprotein of 533 amino acids containing three tandemly repeated zinc-finger sequences related to those of the transcription factor Sp1 (Christy *et al.*, 1988; Day *et al.*, 1990). It binds specifically to a G/C-rich sequence (GCGG(or T)GGGCG), which can be found in its own promoter and the promoters of other genes, including immediate-early genes (Christy and Nathans, 1989a; Cao *et al.*, 1990; Lemaire *et al.*, 1990). In a transfection assay, Zif268 can function as a transcription activator (Lemaire *et al.*, 1990). There is a group of zinc-finger proteins related to Zif268 that can bind DNA with similar specificity. One of these is the Wilm's tumor locus zinc-finger protein, which appear to act as a tumor suppressor (Rauscher *et al.*, 1990). Another is Krox20 described below.

b. Krox20 (Egr-2). An immediate early gene related to *zif268* has been named *krox20* in mouse (Chavrier *et al.*, 1989) and *egr-2* in human (Joseph *et al.*, 1988). The *krox20* gene encodes a protein of 470 amino acids with three zinc-finger repeats. Although the protein sequences of Krox20 and Zif268 are largely different, their zinc-finger sequences are virtually identical. Krox20 can function as a transcription activator, and binds to the same DNA sequence as Zif268 (Chavrier *et al.*, 1990). *Krox20* is expressed in a segment-specific manner in the developing nervous system; current data suggest that it may play a role in hindbrain segmentation (Wilkinson *et al.*, 1989).

c. Nup475. A gene activated by serum growth factors in 3T3 fibroblasts was found to encode a 319-amino acid protein, Nup475, that contains two tandemly repeated cysteine- and histidine-containing zinc-finger sequences (DuBois *et al.*, 1990). The zinc finger occurs with the spacing $CX_8CX_5CX_3H$, which differs from any previously defined zinc-finger proteins. Consistent with the hypothesis that it may bind DNA in a zinc-dependent manner, the Nup475 protein has been localized to the nucleus (DuBois *et al.*, 1990) and bacterially produced Nup475 binds zinc. However, the specific DNA sequence to which Nup475 binds has not yet been identified.

Part of the nup475 cDNA is similar to a cDNA called TIS11, which was isolated from a cDNA library derived from phorbol ester-induced mRNAs (Varnum *et al.*, 1989b). Compared to nup475, the TIS11 cDNA has a different 5′-untranslated region sequence, and a single base deletion changes its protein reading frame through much of the coding-region sequence (DuBois *et al.*, 1990). Since currently there is no reported biochemical analysis of the putative TIS11-encoded protein, it is not clear how TIS11 is related to Nup475.

5. Steroid Receptor

An immediate-early gene encoding another class of zinc-finger proteins has been identified as *nur77/N10/TIS1* in mouse (Hazel *et al*, 1988; Ryseck *et al.*,

1989b; Varnum *et al.*, 1989a), *NGFI-B* in rat (Milbrandt, 1988), and *NAK1* in human (Nakai *et al.*, 1990). The nur77 mRNA is undetectable in resting mouse 3T3 cells, but accumulates rapidly and transiently upon stimulation by serum, PDGF, FGF, or TPA with *fos*-like kinetics (Hazel *et al.*, 1988; Ryseck *et al.*, 1989b). *nur77* encodes a 601-amino acid protein that shares significant sequence similarity with members of the steroid–thyroid hormone receptor superfamily, which are ligand-dependent transcription factors (reviewed by Evans, 1988; Beato, 1989). Characterization of the Nur77 protein shows that it is synthesized rapidly and transiently following stimulation, has a short half-life of 30 to 40 min, and is found in the nucleus (Fahrner *et al.*, 1990; Hazel *et al.*, 1991). It is posttranslationally modified (Fahrner *et al.*, 1990; Hazel *et al.*, 1991), primarily by phosphorylation on serine residues (Hazel *et al.*, 1991). Recent work in our laboratory has shown that Nur77 can function as a transcriptional activator (Davis *et al.*, 1991).

6. Serum-Response Factor

The $p67^{SRF}$ serum-response factor (see Section IV) appears to play a role in the transcriptional activation of c-*fos* and a number of other immediate-early genes. The molecular cloning of $p67^{SRF}$ made possible the characterization of its pattern of expression and the discovery that it is encoded by an immediate-early gene (Norman *et al.*, 1988). $p67^{SRF}$ is a phosphoprotein of 508 amino acids, and most likely binds DNA as a dimer (Norman *et al.*, 1988). Its DNA-binding domain is homologous to that of the yeast transcription factor MCM1/PRTF/GRM, which regulates the expression of genes involved in mating and is required for the maintenance of minichromosomes (Hayes *et al.*, 1988; Jarvis *et al.*, 1989; Passmore *et al.*, 1989).

B. Secreted Proteins

1. KC and JE

The *KC* and *JE* genes were first identified as PDGF-inducible genes in mouse fibroblasts (Cochran *et al.*, 1983). Both encode secreted proteins that may function as growth-regulatory factors (Rollins *et al.*, 1988; Oquendo *et al.*, 1989). The human homolog of KC was isolated as a melanoma growth factor (Richmond *et al.*, 1988), and the human homolog of JE is a monocyte chemoattractant (Rollins *et al.*, 1989a). Both KC and JE are members of a superfamily of inducible proteins that may function as mediators of the inflammatory response (reviewed by Stoeckle and Barker, 1990).

2. Cyr61 (CEF-10)

cyr61 was identified as an immediate-early gene inducible by serum growth factors in mouse fibroblasts (Lau and Nathans, 1985), and is expressed during liver regeneration (Nathans *et al.,* 1988). Its chicken homolog, called CEF-10, is inducible by v-*src* (Simmons *et al.,* 1989). *cyr61* encodes a 370-amino acid polypeptide with 38 cysteine residues that contains an N-terminal secretory signal sequence (O'Brien *et al.,* 1990). While Cyr61 is secreted, the secreted protein is not found in the culture medium but is associated with the extracellular matrix and the cell surface (Yang and Lau, 1991). Once associated with the ECM, Cyr61 has a half-life of greater than 24 hr, whereas intracellular and cell surface-associated Cyr61 is short-lived. Cyr61 also binds heparin with high affinity. These observations suggest biochemical similarities between Cyr61 and the heparin-binding growth factor family of proteins (Burges and Maciag, 1989) and the protooncogene product Int-1 (Wnt-1)(Bradley and Brown, 1990), and are consistent with the hypothesis that Cry61 may play role in cell–cell communication.

3. PC4 (TIS7)

A cDNA isolated from NGF-stimulated PC12 cells named PC4 (Tirone and Shooter, 1989). encodes a secreted protein with sequence similarity to the rat γ-interferon. The mouse homolog of this gene, TIS7, was isolated as a TPA-inducible gene in fibroblasts (Varnum *et al.,* 1989c). An identical or highly related cDNA was isolated from Newcastle disease virus-infected murine C243 cells (Skup *et al.,* 1982). Despite the sequence similarity with interferon, attempts to demonstrate an interferon-like activity in conditioned media of cells expressing PC4 have been unsuccessful (Tirone and Shooter, 1989).

4. Plasminogen Activator Inhibitor

The plasminogen activator 1 inhibitor (PAI) (Ny *et al.,* 1986) was identified as an immediate-early gene by differential screening of a serum-stimulated cell cDNA library (Almendral *et al.,* 1988; Bravo, 1989). PAI blocks the conversion of plasminogen to plasmin by inhibiting plasminogen activator. Plasmin is a serine protease that cleaves fibrin, and by blocking plasmin formation, PAI inhibits proteolysis of fibrin clots during wound healing. The activation of PAI and tissue factor (see below) may be related to the process of wound healing induced by platelet factors.

C. Strucutral Proteins

A number of cytoskeletal and extracellular-matrix proteins are encoded by immediate-early genes, including β-actin (Elder *et al.*, 1988), α-tropomysin, fibronectin, and α and β subunits of the fibronectin receptor, and p27, a putative actin-associated protein (Almendral *et al.*, 1989; Ryseck *et al.*, 1989a). One distinctive feature of these structural protein genes is that their expression is far more prolonged than that of most other immediate-early genes (Lau and Nathans, 1991). Their mRNAs accumulate for many hours, and typically have half-lives that are longer than 2 hr (Almendral *et al.*, 1988).

D. Integral Membrane Proteins

1. Mouse Tissue Factor

A cDNA identified by differential screening encodes a 294-amino acid protein homologous to the human tissue factor (Hartzell *et al.*, 1989), which initiates the extrinsic pathway of blood coagulation (Nemerson, 1988). Genomic blotting analysis indicates that this clone is the mouse homolog of tissue factor (mTF). Consistent with this interpretation, mTF was found to be a glycosylated integral membrane protein (Hartzell *et al.*, 1989). The induction of the mTF gene by serum growth factors such as PDGF may be related to processes such as wound healing, although mTF may potentially play a role in cell movement during growth (Hartzell *et al.*, 1989).

2. Glucose Transporter

The molecular cloning of the glucose transporter cDNA made possible the characterization of its gene expression and the finding that it is encoded by an immediate-early gene (Hiraki *et al.*, 1988). Its mRNA accumulates rapidly following stimulation by serum or purified PDGF, FGF, and EGF, consistent with the observation that glucose transporter activity increases after mitogenic stimulation (Hiraki *et al.*, 1988).

E. Other Immediate Early Proteins

The proposed functions and biochemical analysis of the immediate-early gene–encoded proteins described above are largely based upon sequence comparisons with proteins of known functions. The immediate-early genes encoding potential transcription factors may regulate the downstream genomic responses to growth factors, including the activation of delayed-early genes. The large number of potential regulatory proteins identified thus far support the contention that other immediate-early genes, many of which are still uncharacterized, may encode regulatory molecules as well.

Although the immediate-early proteins identified thus far include such regulatory molecules as nuclear transcription factors and probable cytokines, few that may potentially play a regulatory role in the cytoplasm have been identified. One might expect that upon growth-factor stimulation, a number of cytoplasmic regulators would be induced. These proteins may regulate signal transduction, protein synthesis, and the activities of key biosynthetic and metabolic enzymes (Pardee *et al.*, 1978). A candidate for such a regulatory molecule is Pip92, an immediate-early protein that is localized in the cytoplasm (Charles *et al.*, 1990). The coregulation of *pip92* with other genes such as c-*fos* and the short half-lives of the pip92 mRNA and protein (5–10 min for both) are consistent with the possibility that it may be a cytoplasmic regulatory protein.

The apparent absence of protein kinases among the immediate-early proteins identified to date is particularly striking. Although a number of protein kinase activities are induced by mitogenic signals that clearly play a role in the growth response, these protein kinases appear to be regulated by posttranslational modification rather than activation of their genes (Lau and Nathans, 1991).

VI. DELAYED-EARLY GENE EXPRESSION AND CELL CYCLE PROGRESSION

After the initial wave of immediate-early gene expression, a second group of genes are expressed approximately 3–8 hr after mitogen stimulation, before the onset of DNA synthesis. Many of these *delayed-early genes* require *de novo* protein synthesis for their expression, suggesting that they are regulated by immediate-early transcription factors. However, at present there are few known examples of delayed-early genes that are direct targets of specific immediate-early transcription factors. Although a number of delayed-early genes have been identified (reviewed by Denhardt *et al.*, 1986b; Lau and Nathans, 1991), undoubtedly many others are yet to be described. Identification of delayed-early genes and the immediate-early transcription factors that regulate them is therefore an important task.

By analogy to the viral developmental program, we anticipate that many delayed-early genes may encode the machineries necessary for the biosynthesis of macromolecules and DNA replication. Consistent with this expectation, the delayed-early genes identified thus far include those encoding biosynthetic enzymes such as ornithine decarboxylase (Katz and Kahana, 1987) and asparagine synthetase (Greco *et al.*, 1989). Also among this group is ADP/ATP translocase

(Battini *et al.*, 1987), whose function relates to the storage of cellular energy. Other delayed early genes include structural proteins such as vimentin (Ferrari *et al.*, 1986) and calcium-binding proteins (Calabretta *et al.*, 1986; Jackson-Grusby *et al.*, 1987).

A group of secreted serum proteases are encoded by delayed-early genes, including cathespin L (Denhardt *et al.*, 1986a), a protease with specificity for extracellular matrix proteins, and stromelysin (Matrisian, 1990), a metalloproteinase. The expression of these proteases may be related to cell motility during growth. Other delayed-early proteins have been suggested to be regulators of growth. One such protein, proliferin (Linzer *et al.*, 1985), is homologous to members of the prolactin–growth hormone family. However, attempts to define its effect on cell growth have been unsuccessful to date. Another protein, a 68-kDa labile protein that is synthesized 2 hr before S phase, was proposed to regulate the G_1 to S transition (Croy and Pardee, 1983). At present the identity and function of this protein remain unclear.

Several delayed early proteins have oncogenic potential. Ras (Lu *et al.*, 1989; Barbacid, 1990), for example, may play a role in signal transduction. Other proteins such as Myb (Lüscher *et al.*, 1990; Shen-Ong, 1990) and p53 (Fields and Jang, 1990; Levine and Momand, 1990; Raycroft *et al.*, 1990) can function as transcription regulators. These and perhaps other delayed-early transcription factors may regulate the expression of late genes that are directly required for the G_1 to S transition.

VII. ACTIVATION OF IMMEDIATE-EARLY GENES BY OTHER SIGNALING AGENTS

The activation of immediate-early genes is not unique to serum growth factors. We have already commented on the similarities in immediate-early gene expression among fibroblasts, T lymphocytes, and hepatocytes in regenerating liver. Other extracellular ligands and signaling agents such as growth inhibitors, differentiation factors, classical hormones, and neurotransmitters also rapidly induce specific changes in gene expression. Although the immediate-early genes described in this review were largely first characterized in mitogen-stimulated fibroblasts, many of the same genes, particularly those encoding nuclear proteins, are also activated by various other agents that cause disparate responses in different cell types (Herschman, 1989). For example, a number of immediate-early genes are induced in the central nervous system after peripheral nerve stimulation or after chemically induced convulsions or electrically induced seizures (reviewed by Sheng *et al.*, 1988). c-*fos* is also induced in the suprachiasmatic nucleus after stimulation by light (Kornhauser *et al.*, 1990). How can stimulation that evokes widely different biological responses induce the same genes? These seemingly paradoxical observations suggest that the immediate-early proteins, particularly nuclear regulators, may act as signal-transduction molecules rather than effectors of the biological responses. According to this hypothesis, the target genes of immediate-early transcription factors may be specific to the inducing signal and/or the recipient cell type.

A cell system that has been studied in testing this hypothesis is the rat phenochromocytoma cell line PC12, which can be induced by various agents to initiate different cellular responses (reviewed by Greene and Tischler, 1982). Upon treatment with NGF, these cells initially undergo proliferation but later differentiate into sympathetic neuron-like cells. In contrast, EGF stimulates proliferation but not differentiation. These cells also have excitable membranes that can be depolarized by neurotransmitters or elevated levels of external KCl. All three types of stimulations induce the transcription of a number of immediate-early genes encoding transcription factors, including c-*fos*, c-*jun*, *jun B*, *nur77*, and *zif268* (Bartel *et al.*, 1989). However, there are quantitative differences in their induction by different agents. For example, NGF and EGF activate *zif268* and *jun B* to a higher level than membrane depolarization, whereas c-*fos* and *nur77* are activated to a greater extent by membrane depolarization. In light of the possibility that different levels of Fos and Jun proteins may form different dimers that have unique activities, the quantitative differences in immediate-early gene activation may partly explain the distinct downstream effects of growth factors and membrane-depolarizing agents in PC12 cells.

Another possible mechanism of regulating immediate-early transcription factors is posttranslational modification. In PC12 cells, NGF, membrane depolarization, and phorbol esters are known to activate distinct sets of serine–threonine kinases. This results in the differential phosphorylation of the rate-limiting enzyme in catecholamine biosynthesis, tyrosine hydroxylase, thus altering its activity in different ways (Haycock, 1990). Therefore differential phosphorylation is known to play a regulatory role in PC12 cells. Moreover, the role of phosphorylation in modulating the activity of transcription factors is also well documented (Gonzalez and Montminy, 1989; Lüscher *et al.*, 1990). Among the immediate-early transcription factors, both Fos and Jun are phosphorylated and differentially modified when induced by different agents (Curran and Morgan, 1986; Müller *et al.*, 1987). In addition, phosphopeptide analysis of Nur77 has shown that Nur77 is differentially phosphorylated upon induction by growth factors or membrane-depolarizing agents (Hazel *et al.*, 1991). Differential phosphorylation may thus be another way in which the activity or target-gene specificity of immediate-early transcription factors can be modulated, thereby contributing to the distinct downstream responses specific to the stimulating agent. A direct test of this possibility will require analysis of the target-gene specificities of immediate-early transcription factors induced under different conditions of stimulation.

Although the similarities among the genetic programs induced in different systems are striking, cell-type differences do exist. For example, among the immediate-early genes inducible in fibroblasts, *cyr61* and *TIS10* are not induced under any condition tested in PC12 cells, whereas *3CH134* is constitutively expressed (Bartel *et al.*, 1989; Herschman, 1989). These observations suggest that certain genes inducible in one cell type may be repressed in another, possibly owing to differences in chromosomal structure or the availability of necessary factors for their activation (Herschman, 1989).

VIII. SUMMARY AND CONCLUSION

In this review, we have emphasized the complexity of the genomic response to polypeptide growth factors. In a number of cell systems, growth factors rapidly activate a large number of immediate-early genes that encode a diverse group of regulatory molecules. Much is yet to be learned about the pathways of signal transduction from the plasma membrane to the nucleus where transcriptional activation occurs. In this regard, we expect significant progress from the biochemical analysis of the transcription factors that induce immediate-early genes. An understanding of how these preexisting factors are activated to induce transcription will help to establish the link between the membrane signaling events and the nucleus.

Characterization of the proteins encoded by immediate-early genes has already uncovered a number of previously unknown regulatory molecules. Of particular interest are the nuclear proteins, many of which have been shown to regulate transcription in transfection assays. The future challenge is to identify the targets of these transcription regulators, thus delineating the next steps in the genetic program for growth. Given the number of immediate-early transcription factors already identified, the complexity of delayed-early genes under their control is likely to be substantial. It is possible that the large number of genes activated in fact comprise several overlapping programs, of which only a part is directly required for proliferation. Other functions of growth factor-regulated genes may be related to specific processes in the organism, for example, differentiation, organogenesis, and wound healing.

Dozens of immediate-early genes are yet to be analyzed, and many others may remain undetected. Judging from the apparent abundance of regulatory molecules among the immediate-early genes analyzed thus far, characterization of other immediate-early genes currently unknown is likely to yield further insight into the mechanisms by which growth factors regulate cell proliferation.

ACKNOWLEDGMENTS

Research in our laboratory is supported by grants from the National Institutes of Health (CA46565 and CA52220). G.T.W. is a National Cancer Center Postdoctoral Fellow. L.F.L. is the recipient of an American Cancer Society Junior Faculty Award and is a Pew Scholar in the Biomedical Sciences.

REFERENCES

Ahn, N. G., and Krebs, E. G. (1990). Evidence for an epidermal growth factor–stimulated protein kinase cascade in Swiss 3T3 cells. *J. Biol. Chem.* **265,** 11495–11501.

Ahn, N. G., Weiel, J. E., Chan, C. P., and Krebs, E. G. (1990). Identification of multiple epidermal growth factor–stimulated protein serine/threonine kinases from Swiss 3T3 cells. *J. Biol. Chem.* **265,** 11487–11494.

Almendral, J. M., Sommer, D., MacDonald Bravo, H., Burckhardt, J., Perera, J., and Bravo, R. (1988). Complexity of the early genetic response to growth factors in mouse fibroblasts. *Mol. Cell. Biol.* **8,** 2140–2148.

Almendral, J. M., Santaren, J. F., Perera, J., Zerial, M., and Bravo, R. (1989). Expression, cloning, and cDNA sequence of a fibroblast serum-regulated gene encoding a putative actin-associated protein (p27). *Exp. Cell Res.* **181,** 518–530.

Altman, A., Mustelin, T., and Coggeshall, K. M. (1990). T-lymphocyte activation: A biological model of signal transduction. *Crit. Rev. Immunol.* **10,** 347–391.

Angel, P., Hattori, K., Smeal, T., and Karin, M. (1988). The *jun* protooncogene is positively autoregulated by its product, Jun/AP-1. *Cell* **55,** 875–885.

Antoniades, H. N., Scher, C. D., and Stiles, C. D. (1979). Purification of human platelet-derived growth factor. *Proc. Natl. Acad. Sci. U.S.A.* **76,** 1809–1813.

Asselin, C., Nepveu, A., and Marcu, K. B. (1989). Molecular requirements for transcriptional initiation of the murine c-*myc* oncogene. *Oncogene* **4,** 549–558.

Baccarini, M., Sabatini, D. M., App, H., Rapp, U. R., and Stanley, E. R. (1990). Colony-stimulating factor (CSF-1) stimulates temperature-dependent phosphorylation and activation of the RAF-1 protooncogene product. *EMBO J.* **9,** 3649–3657.

Barbacid, M. (1990). *ras* oncogenes: Their role in neoplasia. *Eur. J. Clin. Invest.* **20,** 225–235.

Bartel, D. P., Sheng, M., Lau, L. F., and Greenberg, M. E. (1989). Growth factors and membrane depolarization activate distinct programs of early response gene expression: dissociation of *fos* and *jun* induction. *Genes Dev.* **3,** 301–313.

Baserga, R. (1985). "The Biology of Cell Reproduction." Harvard University Press, Cambridge, Massachusetts.

Battini, R., Ferrari, S., Kaczmarek, L., Calabretta, B., Chen, S. T., and Baserga, R. (1987). Molecular cloning of a cDNA for a human ADP/ATP carrier which is growth regulated. *J. Biol. Chem.* **262,** 4355–4359.

Beato, M. (1989). Gene regulation by steroid hormones. *Cell* **56,** 335–344.

Bedard, P. A., Balk, S. D., Gunther, H. S., Morisi, A., and Erikson, R. L. (1987). Repression of quiescence-specific polypeptides in chicken heart mesenchymal cells transformed by Rous sarcoma virus. *Mol. Cell. Biol.* **7,** 1450–1458.

Bedard, P. A., Yannoni, Y., Simmons, D. L., and Erikson, R. L. (1989). Rapid repression of quiescence-specific gene expression by epidermal growth factor, insulin, and pp60$^{v\text{-}src}$. *Mol. Cell. Biol.* **9,** 1371–1375.

Berkowitz, L. A., and Gilman, M. Z. (1990). Two distinct forms of active transcription factor CREB (cAMP response element binding protein). *Proc. Natl. Acad. Sic. U.S.A.* **87,** 5258–5262.

Berridge, M. J. (1987). Inositol trisphosphate and diacylglycerol: Two interacting second messengers. *Annu. Rev. Biochem.* **56,** 159–193.

Berstein, P., Peltz, S. W., and Ross, J. (1989). The poly(A)-binding protein complex is a major determinant of mRNA stability *in vitro. Mol. Cell. Biol.* **9,** 659–670.

Bjorge, J. D., Chan, T. -O., Antczak, M., Kung, H. -J., and Fujita, D. J. (1990). Activated type I phosphatidylinositol kinase is associated with the epidermal growth factor (EGF) receptor following EGF stimulation. *Proc. Natl. Acad. Sci. U.S.A.* **87,** 3816–3820.

Blackshear, P. J., Haupt, D. M., App, H., and Rapp, U. R. (1990). Insulin activates the Raf-1 protein kinase. *J. Biol. Chem.* **265,** 12131–12134.

Blackwell, T. K., Kretzner, L., Blackwood, E. M., Eisenman, R. N., and Weintraub, H. (1990). Sequence-specific DNA binding by the c-Myc protein. *Science* **250,** 1149–1151.

Bours, V., Villalobos, J., Burd, P. R., Kelly, K., and Siebenlist, U. (1990). Cloning of a mitogen-inducible gene encoding a κB DNA-binding protein with homology to the *rel* oncogene and to cell-cycle motifs. *Nature* **348,** 76–80.

Bradley, R. S., and Brown, A. M. C. (1990). The protooncogene *int*-1 encodes a secreted protein associated with the extracellular matrix. *EMBO J.* **9,** 1569–1575.

Bravo, R., Burckhardt, J., Curran, T., and Müller, R. (1986). Expression of c-*fos* in NIH 3T3 cells is very low but inducible throughout the cell cycle. *EMBO J.* **5,** 695–700.

Bravo, R. (1989). Growth factor–inducible genes in fibroblasts. *In "Growth Factors, Differentiation Factors, and Cytokines"* (A. Habenicht, ed.), pp. 324–343. Springer-Verlag, Heidelberg, FRG.

Brewer, G., and Ross, J. (1988). Poly(A) shortening and degradation of the 3′ A+U-rich sequences of human c-*myc* mRNA in a cell-free system. *Mol. Cell. Biol.* **8,** 1697–1708.

Brewer, G., and Ross, J. (1989). Regulation of c-*myc* mRNA stability *in vitro* by a labile destabilizer with an essential nucleic acid component. *Mol. Cell. Biol.* **9,** 1996–2006.

Bull, P., Hunter, T., and Verma, I. M. (1989). Transcriptional induction of the murine c-*rel* gene with serum and phorbol-12-myristate-13-acetate in fibroblasts. *Mol. Cell. Biol.* **9,** 5239–5243.

Bull, P., Morely, K. L., Hoekstra, M. F., Hunter, T., and Verma, I. M. (1990). The mouse c-*rel* protein has an N-terminal regulatory domain and a C-terminal transcriptional transactivation domain. *Mol. Cell. Biol.* **10,** 5473–5485.

Burgess, W. H., and Maciag, T. (1989). The heparin-binding (fibroblast) growth factor of proteins. *Annu. Rev. Biochem.* **58,** 575–606.

Buscher, M., Rahmsdorf, H. J., Litfin, M., Karin, M., and Herrlich, P. (1988). Activation of the c-*fos* gene by UV and phorbol ester: Different signal-transduction pathways converge to the same enhancer element. *Oncogene* **3,** 301–311.

Cai, H., Szeberenyi, J., and Cooper, G. M. (1990). Effect of a dominant inhibitory Ha-*ras* mutation on mitogen signal transduction in NIH 3T3 cells. *Mol. Cell. Biol.* **10,** 5314–5323.

Calabretta, B., Battini, R., Kaczmarek, L., De Riel, J. K., and Baserga, R. (1986). Molecular cloning of the cDNA for a growth factor–inducible gene with strong homology to S-100, a calcium-binding protein. *J. Biol. Chem.* **261,** 12628–12632.

Cambier, J. C., Newell, M. K., Justement, L. B., McGuire, J. C., Leach, K. L., and Chen, Z. Z. (1987). Ia binding ligands and cAMP stimulate nuclear translocation of PKC in B lymphocytes. *Nature* **327,** 629–632.

Cao, X. M., Koski, R. A., Gashler, A., McKiernan, M., Morris, C. F., Gaffney, R., Hay, R. V., and Sukhatme, V. P. (1990). Identification and characterization of the *Egr-1* gene product, a DNA-binding zinc-finger protein induced by differentiation and growth signals. *Mol. Cell. Biol.* **10,** 1931–1939.

Changelian, P. S., Feng, P., King, T. C., and Milbrandt, J. (1989). Structure of the *NGFI-A* gene and detection of upstream sequences responsible for its transcriptional induction by nerve growth factor. *Proc. Natl. Acad. Sci. U.S.A.* **86,** 377–381.

Charles, C. H., Simske, J. S., O'Brien, T. P., and Lau, L. F. (1990). Pip92: a short-lived, growth

factor–inducible protein in BALB/c 3T3 and PC12 cells. *Mol. Cell. Biol.* **10,** 6769–6774.
Chavrier, P., Zerial, M., Lemaire, P., Almendral, J., Bravo, R., and Charnay, P. (1988). A gene encoding a protein with zinc fingers is activated during G_0–G_1 transition in cultured cells. *EMBO J.* **7,** 29–35.
Chavrier, P., Janssen-Timmen, U., Mattéi, M. G., Zerial, M., Bravo, R., and Charnay, P. (1989). Structure, chromosome location, and expression of the mouse zinc-finger gene *krox-20:* Multiple gene products and coregulation with the protooncogene c-*fos. Mol. Cell. Biol.* **9,** 787–797.
Chavrier, P., Vesque, C., Galliot, B., Vigneron, M., Dolle, P., Duboule, D., and Charnay, P. (1990). The segment-specific gene *Krox-20* encodes a transcription factor with binding sites in the promoter region of the Hox-1.4 gene. *EMBO J.* **9,** 1209–1218.
Ching, G., and Wang, E. (1990). Characterization of two populations of statin and the relationship of their syntheses to the state of cell proliferation. *J. Cell Biol.* **110,** 255–261.
Chiu, R., Angel, P., and Karin, M. (1989). Jun B differs in its biological properties from, and is a negative regulator of, c-Jun. *Cell* **59,** 979–986.
Chou, C. K., Dull, T. J., Russell, D. S., Gherzi, R., Lebwohl, D., Ullrich, A., and Rosen, O. M. (1987). Human insulin receptors mutated at the ATP-binding site lack protein tyrosine kinase activity and fail to mediate postreceptor effects of insulin. *J. Biol. Chem.* **262,** 1842–1847.
Christy, B., and Nathans, D. (1989a). DNA-binding site of the growth factor–inducible protein Zif268. *Proc. Natl. Acad. Sci. U.S.A.* **86,** 8737–8741.
Christy, B., and Nathans, D. (1989b). Functional serum response elements upstream of the growth factor–inducible gene *zif268. Mol. Cell. Biol.* **9,** 4889–4895.
Christy, B. A., Lau, L. F., and Nathans, D. (1988). A gene activated in mouse 3T3 cells by serum growth factors encodes a protein with "zinc finger" sequences. *Proc. Natl. Acad. Sci. U.S.A.* **85,** 7857–7861.
Cochran, B. H., Reffel, A. C., and Stiles, C. D. (1983). Molecular cloning of gene sequences regulated by platelet-derived growth factor. *Cell* **33,** 939–947.
Cockcroft, S., and Bar-Sagi, D. (1990). Effect of H-*ras* proteins on the activity of polyphosphoinositide phospholipase C in HL60 membranes. *Cell. Signalling* **2,** 227–234.
Cohen, D. R., and Curran, T. (1988). *fra-1:* a serum-inducible, cellular immediate-early gene that encodes a *Fos*-related antigen. *Mol. Cell. Biol.* **8,** 2063–2069.
Cohen, D. R., Ferreira, P. C., Gentz, R., Franza, B. R., Jr., and Curran, T. (1989). The product of a *fos*-related gene, *fra*-1, binds cooperatively to the AP-1 site with Jun: Transcription factor AP-1 is comprised of multiple protein complexes. *Genes Dev.* **3,** 173–184.
Cohen, P., Holmes, C. F. B., and Tsukitani, Y. (1990). Okadaic acid: A new probe for the study of cellular regulation. *Trends Biochem. Sci.* **15,** 98–102.
Cole, M. D. (1990). The *myb* and *myc* nuclear onocgenes as transcriptional activators. *Curr. Opin. Cell Biol.* **2,** 502–508.
Collum, R. G., and Alt, F. W. (1990). Are Myc proteins transcription factors? *Cancer Cells* **2,** 69.
Coughlin, S. R., Escobedo, J. A., and Williams, L. T. (1989). Role of phosphatidylinositol kinase in PDGF-receptor signal transduction. *Science* **243,** 1191–1194.
Croy, R. G., and Pardee, A. B. (1983). Enhanced synthesis and stabilization of M_r 68,000 protein in transformed BALB/c 3T3 cells: Candidate for restriction-point control of cell growth. *Proc. Natl. Acad. Sci U.S.A.* **80,** 4699–4703.
Curran, T. (1988). The *fos* oncogene. *In "The Oncogene Handbook"* (E. P. Reddy, A. M. Skalka and T. Curran eds.), pp. 307–325. Elsevier, Amsterdam.
Curran, T., and Morgan, J. I. (1986). Barium modulates c-*fos* expression and posttranslational modification. *Proc. Natl. Acad. Sci. U.S.A.* **83,** 8521–8524.
Cutry, A. F., Kinniburgh, A. J., Krabak, M. J., Hui, S. -W., and Wenner, C. E. (1989). Induction of c-*fos* and c-*myc* protooncogene expression by epidermal growth factor and transforming growth factor alpha is calcium-independent. *J. Biol. Chem.* **264,** 19700–19705.

Davis, I.J., Hazel, T.G., and Lau, L.F. (1991). Transcriptional activation by Nur77, a growth factor-inducible member of the steroid hormone receptor superfamily. *Mol. Endocriniology*(in press).

Day, M. L., Fahrner, T. J., Aykent, S., and Milbrandt, J. (1990). The zinc-finger protein NGFI-A exists in both nuclear and cytoplasmic forms in nerve growth factor–stimulated PC12 cells. *J Biol. Chem.* **265,** 15253–15260.

Denhardt, D. T., Hamilton, R. T., Parfett, C. L., Edwards, D. R., St. Pierre, R., Waterhouse, P., and Nilsen Hamilton, M. (1986a). Close relationship of the major excreted protein of transformed murine fibroblasts to thiol-dependent cathepsins. *Cancer Res.* **46,** 4590–4593.

Denhardt, D. T., Edwards, D. R., and Parfett, C. L. (1986b). Gene expression during the mammalian cell cycle. *Biochim. Biophys. Acta* **865,** 83–125.

Diamond, M. I., Miner, J. N., Yoshinaga, S. K., and Yamamoto, K. R. (1990). Transcription factor interactions: Selectors of positive or negative regulation from a single DNA element. *Science* **249,** 1266–1272.

Distel, R. J., and Spiegelman, B. M. (1990). Protooncogene c-*fos* as a transcription factor. *Adv. Cancer Res.* **55,** 37–55.

DuBois, R. N., McLane, M. W., Ryder, K., Lau, L. F., and Nathans, D. (1990). A growth factor–inducible nuclear protein with a novel cysteine–histidine repetitive sequence. *J. Biol. Chem.* **265,** 19185–19191.

Dulbecco, R. (1970). Topoinhibition and serum requirement of transformed and untransformed cells. *Nature* **227,** 802–806.

Ebina, Y., Araki, E., Taira, M., Shimada, F., Mori, M., Craik, S. S., Siddle, K., Pierce, S. B., Roth, R. A., and Rutter, W. J. (1987). Replacement of lysine residue 1030 in the putative ATP-binding region of the insulin receptor abolishes insulin- and antibody-stimulated glucose uptake and receptor kinase activity. *Proc. Natl. Acad. Sci. U. S. A.* **84,** 704–708.

Edelman, A. M., Blumenthal, U. K., and Krebs, E. G. (1987). Protein serine/threonine kinases. *Annu. Rev. Biochem.* **56,** 567–613.

Edwards, D. R., Parfett, C. L., and Denhardt, D. T. (1985). Transcriptional regulation of two serum-induced RNAs in mouse fibroblasts: Equivalence of one species to B2 repetitive elements. *Mol. Cell. Biol.* **5,** 3280–3288.

Elder, P. K., French, C. L., Subramaniam, M., Schmidt, L. J., and Getz, M. J. (1988). Evidence that the functional beta-actin gene is single copy in most mice and is associated with 5′ sequences capable of conferring serum- and cycloheximide-dependent regulation. *Mol. Cell. Bio.* **8,** 480–485.

Ellis, C. E., Moran, M., McCormick, F., and Pawson, T. (1990). Phosphorylation of GAP and GAP-associated proteins by transforming and mitogenic tyrosine kinases. *Nature* **343,** 371–381.

Endemann, G., Yonezawa, K., and Roth, R. A. (1990). Phosphatidylinositol kinase or an associated protein is a substrate for the insulin receptor tyrosine kinase. *J. Biol. Chem.* **265,** 396–400.

Erisman, M. D., and Astrin, S. M. (1988). The *myc* oncogene. *In "The Oncogene Handbook"* (E. P. Reddy, A. M. Skalka and T. Curran, eds.), pp. 341–379. Elsevier, Amsterdam.

Escobeco, J., Barr, P. J., and Williams, L. T. (1988). Role of tyrosine kinase and membrane-spanning domains in signal transduction by the platelet-derived growth factor receptor. *Mol. Cell. Biol.* **8,** 5126–5131.

Evans, R. M. (1988). The steroid and thyroid hormone receptor superfamily. *Science* **240,** 889–895.

Fahrner, T. J., Carroll, S. L., and Milbrandt, J. (1990). The NGFI-B protein, an inducible member of the thyroid/steroid receptor family, is rapidly modified posttranslationlly. *Mol. Cell. Biol.* **10,** 6454–6459.

Fain, J., Wallace, M. A., and Wojcikiewicz, R. J. H. (1988). Evidence for involvement of guanine nucleotide-binding regulatory proteins in the activation of phospholipase by hormones. *FASEB J.* **20,** 2569–2574.

Ferrari, S., Battini, R., Kaczmarek, L., Rittling, S., Calabretta, B., De Riel, J. K., Philiponis, V., Wei,

J. F., and Baserga, R. (1986). Coding sequence and growth regulation of the human vimentin gene. *Mol. Cell. Biol.* **6,** 3614–3620.

Fields, S., and Jang, S. K. (1990). Presence of a potent transcription-activating sequence in the p53 protein. *Science* **249,** 1046–1049.

Fisch, T. M., Prywes, R., and Roeder, R. G. (1987). c-*fos* sequence necessary for basal expression and induction by epidermal growth factor, 12-*O*-tetradecanoyl phorbol-13-acetate and the calcium ionophore. *Mol. Cell. Biol.* **7,** 3490–3502.

Fisch, T. M., Prywes, R., Simon, M. C., and Roeder, R. G. (1989a). Multiple sequence elements in the c-*fos* promoter mediate induction by cAMP. *Genes Dev.* **3,** 198–211.

Fisch, T. M., Prywes, R., and Roeder, R. G. (1989b). An AP-1–binding site in the c-*fos* gene can mediate induction by epidermal growth factor and 12-*O*-tetradecanoyl phorbol-13-1acetate. *Mol. Cell. Biol.* **9,** 1327–1331.

Fornace, A. J., Jr., Nebert, D. W., Hollander, M. C., Luethy, J. D., Papthanasiou, M., Fargnoli, J., and Holbrook, N. J. (1989). Mammalian genes coordinately regulated by growth-arrest signal and DNA-damaging agents. *Mol. Cell. Biol.* **9,** 4196–4203.

Gauthire-Rouviere, C., Fernandez, A., and Lamb, N. J. C. (1990). *ras*-induced c-*fos* expression and proliferation in living rat fibroblasts involves C-kinase activation and the serum response element pathway. *EMBO J.* **9,** 171–180.

Gay, D. A., Sisodia, S. S., and Cleveland, D. W. (1989). Autoregulatory control of beta-tubulin mRNA stability is linked to translation elongation. *Proc. Natl. Acad. Sci. U.S.A.* **86,** 5763–5767.

Ghosh, S., and Baltimore, D. (1990). Activation *in vitro* of NF-kB by phosphorylation of its inhibitor IkB. *Nature* **344,** 678–682.

Ghosh, S., Gifford, A. M., Riviere, L. R., Tempst, P., Nolan, G. P., and Baltimore, D. (1990). Cloning of the p50 DNA-binding subunit of NF-kB: Homology to *rel* and *dorsal. Cell* **62,** 1019–1029.

Gilman, M. Z., Wilson, R. N., and Weinberg, R. A. (1986). Multiple protein-binding sites in the 5′-flanking region regulate c-*fos* expression. *Mol. Cell. Biol.* **6,** 4305–4316.

Gilman, M. Z. (1988). The c-*fos* serum-response element responds to protein kinase C-dependent and -independent signals but not to cyclic AMP. *Genes Dev.* **2,** 394–402.

Gius, D., Cao, X. M., Rauscher, F. J., III, Cohen, D. R., Curran, T., and Sukhatme, V. P. (1990). Transcriptional activation and repression by Fos are independent functions: The C terminus represses immediate-early gene expression via CArG elements. *Mol. Cell. Biol.* **10,** 4243–4255.

Gonzalez, G. A., and Montminy, M. R. (1989). Cyclic AMP stimulated somatostatin gene transcription by phosphorylation of CREB at serine 133. *Cell* **59,** 675–680.

Gonzalez, G. A., Yamamoto, K. K., Fischer, W. H., Karr, D., Menzel, P., Biggs, W., III, Vale, W. W., and Montminy, M. R. (1989). A cluster of phosphorylation sites on the cyclic AMP–regulated nuclear factor CREB predicted by its sequence. *Nature* **337,** 749–752.

Grasser, F. A., Scheidtman, K. H., Tuazon, P. T., Traugh, J. A., and Walter, G. (1988). *In vitro* phosphorylation of SV40 large-T antigen. *Virology* **165,** 13–22.

Greco, A., Gong, S. S., Ittmann, M., and Basilico, C. (1989). Organization and expression of the cell cycle gene, *ts11,* that encodes asparagine synthetase. *Mol. Cell. Biol.* **9,** 2350–2359.

Greenberg, M. E., Hermanowski, A. L., and Ziff, E. B. (1986). Effect of protein synthesis inhibitors on growth factor activation of c-*fos,* c-*myc,* and actin gene transcription. *Mol. Cell. Biol.* **6,** 1050–1057.

Greenberg, M. E., Siegfried, Z., and Ziff, E. B. (1987). Mutation of the c-*fos* gene dyad symmetry element inhibits serum inducibility of transcription *in vivo* and the nuclear regulatory factor binding *in vitro. Mol. Cell. Biol.* **7,** 1217–1225.

Greenberg, M. E., and Ziff, E. B. (1984). Stimulation of 3T3 cells induces transcription of the c-*fos* protooncogene. *Nature* **311,** 433–438.

Greene, L. A., and Tischler, A. S. (1982). Pheochromocytoma cultures in neurobiological research. *Adv. Cell. Neurobiol.* **3,** 373–414.

Grinstein, S., Smith, J. D., Onizuka, R., Cheung, R. K., Gelfand, E. W., and Benedict, S. (1988). Activation of Na^+/H^+ exchange and expression of cellular protooncogenes in mitogen- and phorbol ester-treated lymphocytes. *J. Biol. Chem.* **263,** 8658–8665.

Grinstein, S., Rotin, D., and Mason, M. J. (1989). Na^+/H^+ exchange and growth factor–induced cytosolic pH changes: Role in cellular proliferation. *Biochim. Biophys. Acta* **988,** 73–97.

Hall, A. (1990). ras and GAP—who's controlling whom? *Cell* **61,** 921–923.

Hartzell, S., Ryder, K., Lanahan, A., Lau, L. F., and Nathan, D. (1989). A growth factor–responsive gene of murine BALB/c 3T3 cells encodes a protein homologous to human tissue factor. *Mol. Cell. Biol.* **9,** 2567–2573.

Hattori, K., Angel, P., Le Beau, M. M., and Karin, M. (1988). Structure and chromosomal localization of the functional intronless human Jun protooncogene. *Proc. Natl. Acad. Sci. U.S.A.* **85,** 9148–9152.

Haycock, J. W. (1990). Phosphorylation of tyrosine hydrozylase *in situ* at serine 8, 19, 31, and 40. *J. Biol. Chem.* **265,** 11682–116791.

Hayes, T. E., Kitchen, A. M., and Cochran, B. H. (1987). Inducible binding of a factor to the c-*fos* regulatory region. *Proc. Natl. Acad. Sci. U.S.A.* **84,** 1272–1276.

Hayes, T. E., Sengupta, P., and Cochran, B. H. (1988). The human c-*fos* serum response factor and the yeast factors GRM/PRTF have related DNA-binding specificities. *Genes Dev.* **2,** 1713–1722.

Hazel, T. G., Nathans, D., and Lau, L. F. (1988). A gene inducible by serum growth factors encodes a member of the steroid and thyroid hormone receptor superfamily. *Proc. Natl. Acad. Sci. U.S.A.* **85,** 8444–8448.

Hazel, T. G., Misra, R., Davis, I. J., Greenberg, M. E., and Lau, L. F. (1991). Nur77 is differentially modified in PC12 cells upon growth-factor treatment and membrane depolarization. *Mol. Cell. Biol.* **11,** 3239–3246.

Heikkila, R., Schwab, G., Wickstrom, E., Loke, S. L., Pluznik, D. H., Watt, R., and Neckers, L. M. (1987). A c-*myc* antisense oligodeoxynucleotide inhibits entry into S phase but not progress from G_0 to G_1. *Nature* **328,** 445–449.

Heintz, N. (1989). Temporal regulation of gene expression during the mammalian cell cycle. *Curr. Opin. Cell Biol.* **1,** 275–278.

Heldin, C. H., Westermark, B., and Waterson, A. (1979). Platelet-derived growth factor: purification and partial characterization. *Proc. Natl. Acad. Sci U.S.A.* **76,** 3722–3726.

Herrera, R. E., Shaw, P. E., and Nordheim, A. (1989). Occupation of the c-*fos* serum response element *in vivo* by a multiprotein complex is unaltered by growth-factor induction. *Nature* **340,** 68–70.

Herschman, H. R. (1989). Extracellular signals, transcriptional responses, and cellular specificity. *Trends Biochem. Sci.* **14,** 455–458.

Hirai, S. I., Ryseck, R. P., Mechta, F., Bravo, R., and Yaniv, M. (1989). Characterization of *junD:* A new member of the *jun* protooncogene family. *EMBO J.* **8,** 1433–1439.

Hiraki, Y., Rosen, O. M., and Birnbaum, M. J. (1988). Growth factors rapidly induce expression of the glucose transporter gene. *J. Biol. Chem.* **263,** 13655–13662.

Hirschhorn, R. R., Aller, P., Yuan, Z. A., Gibson, C. W., and Baserga, R. (1984). Cell cycle–specific cDNAs from mammalian cells temperature-sensitive for growth. *Proc. Natl. Acad. Sci. U.S.A.* **81,** 6004–6008.

Holley, R. W., and Kiernan, J. A. (1968). Contact inhibition of cell division in 3T3 cells. *Proc. Natl. Acad. Sci. U.S.A.* **60,** 300–304.

Holt, J. T., Gopal, T. V., Moulton, A. D., and Nienhuis, A. W. (1986). Inducible production of c-*fos* antisense RNA inhibits 3T3 cell proliferation. *Proc. Natl. Acad. Sci. U.S.A.* **83,** 4794–4798.

Holt, J. T., Redner, R. L., and Nienhuis, A. W. (1988). An oligomer complementary to c-*myc* mRNA inhibits proliferation of HL-60 promyelocytic cells and induces differentiation. *Mol. Cell Biol.* **8,** 963–973.

Honegger, A. M., Szapary, D., Schmidt, A., Lyall, R., Van Obberghen, E., Dull, T. J., Ullrich, A., and Schlessinger, J. (1987). A mutant epidermal growth factor receptor with defective protein tyrosine kinase is unable to stimulate protooncogene expression and DNA synthesis. *Mol. Cell. Biol.* **7,** 4568–4571.

Jackson, S. P., MacDonald, J. J., Lees-Miller, S., and Tjian, R. (1990). GC box binding induces phosphorylation of Spl by a DNA-dependent protein kinase. *Cell* **63,** 155–165.

Jackson-Grusby, L. L., Swiergiel, J., and Linzer, D. I. H. (1987). A growth-related mRNA in cultured mouse cells encodes a placental calcium-binding protein. *Nucleic Acids Res.* **15,** 6677–6690.

Jamal, S., and Ziff, E. (1990). Transactivation of c-*fos* and beta-actin genes by *raf* as a step in early response to transmembrane signals. *Nature* **344,** 463–466.

Jamieson, G. A., Jr., Mayforth, R. D., Villereal, M. L., and Sukhatme, V. P. (1989). Multiple intracellular pathways induce expression of a zinc-finger-encoding gene (EGR1): Relationship to activation of the Na^+/H^+ exchanger *J. Cell. Physiol.* **139,** 262–268.

Janssen-Timmen, U., Lemaire, P., Mattéi, M. G., Revelant, O., and Charnay, P. (1989). Structure, chromosome mapping and regulation of the mouse zinc-finger gene *Krox-24;* evidence for a common regulatory pathway for immediate-early serum-response genes. *Gene* **80,** 325–336.

Jarvis, E. E., Clark, K. L., and Sprague, G. F. (1989). The yeast transcription activator PRTF, a homolog of the mammalian serum response factor, is encoded by the MCM1 gene. *Genes Dev.* **3,** 936–945.

Jonat, C., Rahmsdorf, H. J., Park, K. K., Cato, A. C., Gebel, S., Ponta, H., and Herrlich, P. (1990). Antitumor promotion and antiinflammation: Down-modulation of AP-1 (Fos/Jun) activity by glucocorticoid hormone. *Cell* **62,** 1189–1204.

Jones, R. H., and Jones, N. C. (1989). Mammalian cAMP-responsive element can activate transcription in yeast and binds a yeast factor(s) that resembles the mammalian transcription factor ATF. *Proc. Natl. Acad. Sci. U.S.A.* **86,** 2176–2180.

Jones, T. R., and Cole, M. D. (1987). Rapid cytoplasmic turnover of c-*myc* mRNA: Requirement of the 3′ untranslated sequences. *Mol. Cell. Biol.* **7,** 4513–4521.

Joseph, L. J., Le Beau, M. M., Jamieson, G. A., Jr., Acharya, S., Shows, T. B., Rowley, J. D., and Sukhatme, V. P. (1988). Molecular cloning, sequencing, and mapping of EGR2, a human early growth response gene encoding a protein with "zinc-binding finger" structure [published erratum appears in *Proc. Natl. Acad. Sci. U.S.A.* (1989). Jan **86**(2), 515]. *Proc. Natl. Acad. Sci. U.S.A.* **85,** 7164–7168.

Kabnick, K. S., and Housman, D. E. (1988). Determinants that contribute to cytoplasmic stability of human c-*fos* and beta-globin mRNAs are located at several sites in each mRNA. *Mol. Cell. Biol.* **8,** 3244–3250.

Kaczmarek, L., Hyland, J. K., Watt, R., Rosenberg, M., and Baserga, R. (1985). Microinjected c-Myc as a competence factor. *Science* **228,** 1313–1315.

Kaibuchi, K., Fukumoto, Y., Oku, N., Hori, Y., Yamamoto, T., Toyoshima, K., and Takai, Y. (1989). Activation of the serum response element and 12-*O*-tetradecanoyl phorbol-13-acetate response element by activated c-Raf-1 protein in a manner independent of protein kinase C. *J. Biol. Chem.* **264,** 20855–20858.

Kalauskas, A., Ellis, C., Pawson, T., and Cooper, J. A. (1990). Binding of GAP to activated PDGF receptor. *Science* **247,** 1578–1581.

Kato, G. J., Barrett, J., Villa-Garcia, M., and Dang, C. V. (1990). An amino-terminal c-Myc domain required for neoplastic transformation activates transcription. *Mol. Cell. Biol.* **10,** 5914–5920.

Katz, C., and Kahana, C. (1987). Transcriptional activation of mammalian ornithine decarboxylase during stimulated growth. *Mol. Cell. Biol.* **7,** 2641–2643.

Kelly, K., Cochran, B. H., Stiles, C. D., and Leder, P. (1983). Cell-specific regulation of the c-*myc* gene by lymphocyte mitogens and platelet-derived growth factor. *Cell* **35,** 603–610.

Kieran, M., Blank, V., Logeat, F., Vandekerckhove, J., Lottspeich, F., Le Ball, O., Urban, M. B., Kourilsky, P., Baeuerle, P. A., and Israel, A. (1990). The DNA-binding subunit of NF-kB is

identical to factor KBF1 and homologous to the *rel* oncogene product. *Cell* **62,** 1007–1018.
Kim, S. -J., Lafyatis, R., Kim, K. Y., Angel, P., Fukiki, H., Karin, M., Sporn, M. B., and Roberts, A. B. (1990). Regulation of collagenase gene expression by okadaic acid, an inhibitor of protein phosphatases. *Cell Reg.* **1,** 269–278.
König, H., Ponta, Rahmsdorf, U., Büscher, M., Schönthal, A., Rahmsdorf, H. J., and Herrlich, P. (1989). Autoregulation of *fos:* The dyad symmetry element as the major target of repression. *EMBO J.* **8,** 2559–2566.
Kornhauser, J. M., Nelson, D. E., Mayo, K. E., and Takahashi, J. S. (1990). Photic and circadian regulation of c-*fos* gene expression in the hamster suprachiasmatic nucleus. *Neuron* **5,** 127–134.
Kovacina, K. S., Yonezawa, K., Brautigan, D. L., Tonks, N. K., Rapp, U. R., and Roth, R. A. (1990). Insulin activates the kinase activity of the *raf*-1 protooncogene by increasing serine phosphorylation. *J. Biol. Chem.* **265,** 12115–12118.
Krebs, E. G., Eisenman, R. N., Kuenzel, E. A., Litchfield, D. W., Lozeman, F. J., Lüscher, B., and Sommercorn, J. (1988). Casein kinase II as a potentially important enzyme concerned with signal transduction. *Cold Spring Harb. Symp. Quant. Biol.* **53,** 77–84.
Kypta, R. M., Goldberg, Y., Ulug, E. T., and Coutneidge, S. A. (1990). Association between the PDGF receptor and members of the *src* family of tyrosine kinases. *Cell* **62,** 481–492.
L'Allemain, G., Franchi, A., Cragoe, E., Jr., and Pouysségur, J. (1984). Blockade of the Na^+/H^+ antiport abolishes growth factor–induced DNA synthesis in fibroblasts. *J. Biol. Chem.* **259,** 4313–4319.
Lamb, N. J. C., Fernandez, A., Tourkine, N., Jeanteur, P., and Blanchard, J. (1990). Demonstration in living cells of an intragenic negative regulatory element within the rodent c-*fos* gene. *Cell* **61,** 485–496.
Lamph, W. W., Wamsley, P., Sassone-Corsi, P., and Verma, I. M. (1988). Induction of protooncogene Jun/AP-1 by serum and TPA. *Nature* **334,** 629–631.
Landschulz, W. H., Johnson, P. F., and McKnight, S. L. (1989). The DNA-binding domain of the rat liver nuclear protein C/EBP is bipartite. *Science* **243,** 1681–1688.
Larrodera, P., Cornet, M. E., Diaz-Meco, M. T., Lopez-Barahonaz, M., Johansen, T., and Moscat, J. (1990). Phospholipase C–mediated hydrolysis of phosphatidylcholine is an important step in PDGF-stimulated DNA synthesis. *Cell* **61,** 1113–1120.
Latinkic, B. V., O'Brien, T. P., and Lau, L. F. (1991). Promoter function and structure of the growth factor–inducible immediate early gene *cyr61. Nucleic Acids Res.,* (in press).
Lau, L. F., and Nathans, D. (1985). Identification of a set of genes expressed during the G_0–G_1 transition of cultured mouse cells. *EMBO J.* **4,** 3145–3151.
Lau, L. F., and Nathans, D. (1987). Expression of a set of growth-related, immediate-early genes in BALB/c 3T3 cells: Coordinate regulation with c-*fos* or c-*myc. Proc. Natl. Acad. Sci. U.S.A.* **84,** 1182–1186.
Lau, L. F., and Nathans, D. (1991). Genes induced by serum growth factors. *In* "Molecular Aspects of Cellular Regulation, Vol. 6. Hormonal Regulation of Gene Transcription" (P. Cohen and J. G. Foulkes, eds.), pp. 165–202. Elsevier, Amsterdam.
Lemaire, P., Revelant, O., Bravo, R., and Charnay, P. (1988). Two mouse genes encoding potential transcription factors with identical DNA-binding domains are activated by growth factors in cultured cells. *Proc. Natl. Acad. Sci. U.S.A.* **85,** 4691–4695.
Lemaire, P., Vesque, C., Schmitt, J., Stunnenberg, H., Frank, R., and Charnay, P. (1990). The serum-inducible mouse gene *Krox-24* encodes a sequence-specific transcriptional activator. *Mol. Cell. Biol.* **10,** 3456–3467.
Lenardo, M. J., and Baltimore, D. (1989). NF-kB: A pleiotropic mediator of inducible and tissue-specific gene control. *Cell* **58,** 227–229.
Leung, S., and Miyamoto, N. G. (1989). Point mutational analysis of the human c-*fos* serum response–factor binding site. *Nucleic Acids Res.* **17,** 1177–1195.
Levine, A. J., and Momand, J. (1990). Tumor-suppressor genes; The p53 and retinoblastoma sensitivity

genes and gene products. *Biochim. Biophys. Acta* **1032,** 119–136.

Lim, R. W., Varnum, B. C., O'Brien, T. G., and Herschman, H. R. (1989). Induction of tumor promotor–inducible genes in murine 3T3 cell lines and tetradecanoyl phorbol acetate-nonproliferative 3T3 variants can occur through protein kinase C–dependent and –independent pathways. *Mol. Cell. Biol.* **9,** 1790–1793.

Linzer, D. I., and Nathans, D. (1983). Growth-related changes in specific mRNAs of cultured mouse cells. *Proc. Natl. Acad. Sci. U.S.A.* **80,** 4271–4275.

Linzer, D. I., Lee, S. J., Ogren, L., Talamantes, F., and Nathans, D. (1985). Identification of proliferin mRNA and protein in mouse placenta. *Proc. Natl. Acad. Sci. U.S.A.* **82,** 4356–4359.

Lopez-Rivas, A., Mendoza, S. A., Nanberg, E., Sinnett-Smith, J., and Rozengurt, E. (1987). Ca^{2+}-mobilizing actions of platelet-derived growth factor differ from those of bombesin and vasopressin in Swiss 3T3 mouse cells. *Proc. Natl. Acad. Sci. U. S. A.* **84,** 5768–5772.

Lu, K., Levine, R. A., and Campisi, J. (1989). c-*ras*-Ha gene expression is regulated by insulin or insulin-like growth factor and by epidermal growth factor in murine fibroblasts. *Mol. Cell. Biol.* **9,** 3411–3417.

Lucibello, F. C., Lowag, C., Neuberg, M., and Müller, R. (1989). Trans-repression of the mouse c-*fos* promoter: A novel mechanism of Fos-mediated transregulation. *Cell* **59,** 999–1007.

Lucibello, F. C., Slater, E. P., Jooss, K. U., Beato, M., and Müller, R. (1990). Mutual transrepression of Fos and the glucocorticoid receptor: Involvement of a functional domain in Fos which is absent in FosB. *EMBO J.* **9,** 2827–2834.

Lüscher, B., Kuenzel, E. A., Krebs, E. G., and Eisenman, R. N. (1989). Myc oncoproteins are phosphorylated by casein kinase II. *EMBO J.* **8,** 1111–1119.

Lüscher, B., Christenson, E., Litchfield, D. W., Krebs, E. G., and Eisenman, R. N. (1990). Myb DNA binding inhibited by phosphorylation at a site deleted during oncogenic activation. *Nature* **344,** 517–522.

Mahadevan, L. C., Targett, K., and Health, J. K. (1989). 2-Aminopurine abolishes epidermal growth factor–stimulated phosphorylation of complexed and chromatin-associated forms of a 33-kDa phosphoprotein. *Oncogene* **4,** 699–706.

Majerus, P. W., Ross, T. S., Cunningham, T. W., Caldwell, K. K., Jefferson, A. B., and Bansal, V. S. (1990). Recent insights into phosphatidylinositol signaling. *Cell* **63,** 459–465.

Malter, J. S. (1989). Identification of an AUUUA-specific messenger RNA-binding protein. *Science* **256,** 664–666.

Malviya, A. N., Rogue, P., Masmoudi, A., Labourdette, G., and Vincendon, G. (1990). Gene transcription: A role for nuclear protein kinase C? *Int. J. Cancer* **45,** 580–582.

Manak, J. R., de Bisschop, N., Kris, R. M., and Prywes, R. (1990). Casein kinase II enhances the DNA-binding activity of serum response factor. *Genes Dev.* **4,** 955–967.

Manfioletti, G., Ruaro, M. E., Sal, G. D., Philipson, L., and Schneider, C. (1990). A growth arrest–specific (*gas*) gene codes for a membrane protein. *Mol. Cell. Biol.* **10,** 2924–2930.

Matrisian, L. M. (1990). Metalloproteinases and their inhibitors in matrix remodeling. *Trends Gen.* **6,** 121–125.

Means, A. R., and Rasmussen, C. D. (1988). Calcium, calmodulin, and cell proliferation. *Cell Calcium* **9,** 313–319.

Meek, D. W., Simon, S., Kikkawa, U., and Eckhart, W. (1990). The p53 tumor-suppressor protein is phosphorylated at serine 389 by casein kinase II. *EMBO J.* **9,** 3253–3260.

Mehment, H., Morris, C., and Rozengurt, E. (1990). Multiple synergistic signal-transduction pathways regulate c-*fos* expression in Swiss 3T3 cells: The role of cyclic AMP. *Cell Growth & Differentiation* **1,** 293–298.

Meisenhelder, J., Suh, P.-G., Rhee, S. G., and Hunter, T. (1989). Phospholipase C-gamma is a substrate for the PDGF and EGF receptor protein tryosine kinase *in vivo* and *in vitro. Cell* **47,** 1109–1122.

Milbrandt, J. (1987). A nerve growth factor–induced gene encodes a possible transcriptional regulatory factor. *Science* **238,** 797–799.

Milbrandt, J. (1988). Nerve growth factor induces a gene homologous to the glucocorticoid receptor gene. *Neuron* **1,** 183–188.

Minty, A., and Kedes, L. (1986). Upstream regions of the human alpha cardiac actin gene that modulate its transcription in muscle cells: Presence of an evolutionarily conserved repeated motif. *Mol. Cell. Biol.* **6,** 2125–2136.

Mohn, K. L., Laz, T. M., Hsu, J.-C., Melby, A. E., Bravo, R. and Taub, R. (1991). The immediate-early growth response in regenerating liver and insulin-stimulated H-35 cells: Comparison with serum-stimulated 3T3 cells and identification of 41 novel immediate-early genes. *Mol. Cell. Biol.,* **11,** 381–390.

Mohun, T., Garrett, N., and Treisman, R. (1987). *Xenopus* cytoskeletal actin and human c-*fos* gene promoters share a conserved protein-binding site. *EMBO J.* **6,** 667–673.

Molloy, C. J., Bottaro, D. P., Flemming, T. P., Marshall, M. S., Gibbs, J. B., and Aaronson, S. A. (1989). PDGF induction of tyrosine phosphorylation of GTPase activating protein. *Nature* **342,** 711–714.

Moolenaar, W. H. (1986). Effects of growth factors on intracellular pH regulation. *Annu. Rev. Physiol.* **48,** 363–376.

Moore, J. P., Todd, J. A., Hesketh, T. R., and Metcalfe, J. C. (1986). c-*fos* and c-*myc* gene activation, ionic signals, and DNA synthesis in thymocytes. *J. Biol. Chem.* **261,** 8158–8162.

Morrison, D. K., Kaplan, D. R., Rapp, U., and Roberts, T. M. (1988). Signal transduction from membrane to cytoplasm: Growth factors and membrane-bound oncogene products increase Raf-1 phosphorylation and associated protein kinase activity. *Proc. Natl. Acad. Sci. U.S.A.* **85,** 8855–8859.

Morrison, D. K., Kaplan, D. R., Escobedo, J. A., Rapp, U. R., Roberts, T. M., and Williams, L. T. (1989). Direct activation of the serine/threnonine kinase activity of Raf-1 through tyrosine phosphorylation by the PDGF beta-receptor. *Cell* **58,** 649–657.

Mulcahy, L. S., Smith, M. R., and Stacey, D. W. (1985). Requirement for *ras* protooncogene function during serum-stimulated growth of NIH 3T3 cells. *Nature* **313,** 241–243.

Müller, R., Bravo, R., Burckhardt, J., and Curran, T. (1984). Induction of c-*fos* gene and protein by growth factors precedes activation of c-*myc. Nature* **312,** 716–720.

Müller, R., Bravo, R., Muller, D., Kurz, C., and Renz, M. (1987). Different types of modification in c-*fos* and its associated protein p39: Modulation of DNA binding by phosphorylation. *Oncogene Res.* **2,** 19–32.

Murre, C., McCaw, P. S., and Baltimore, D. (1989). A new DNA-binding and dimerization motif in immunoglobulin enhancer binding, *daughterless, MyoD,* and *myc* proteins. *Cell* **56,** 777–783.

Mylin, L. M., Bhat, J. P., and Hopper, J. E. (1989). Regulated phosphorylation and dephosphorylation of GAL4, a transcriptional activator. *Genes Dev.* **3,** 1157–1165.

Nakai, A., Kartha, S., Sakurai, A., Toback, F. G., and DeGroot, L. J. (1990). A human early-response gene homologous to murine *nur77* and rat *NGFI-B* and related to the nuclear receptor superfamily. *Mol. Endocrinol* **4,** 1438–1443.

Nathans, D., Lau, L. F., Christy, B., Hartzell, S., Nakabeppu, Y., and Ryder, K. (1988). Genomic response to growth factors. *Cold Spring Harb. Symp. Quant. Biol.* **53,** 893–900.

Nemerson, Y. (1988). Tissue factor and hemostasis. *Blood* **71,** 1–8.

Nishibe, S., Wahl, M. I., Wedegaertner, P. B., Kim, J. J., Rhee, S. G., and Carpenter, G. (1990a). Selectivity of phospholipase C phosphorylation by the epidermal growth factor receptor, the insulin receptor, and their cytoplasmic domains. *Proc. Natl. Acad. Sci. U.S.A.* **87,** 424–428.

Nishibe, S., Wahl, M. I., Hernandex-Sotomayor, S. M. T., Tonks, N. K., Rhee, S. G., Carpenter, G. (1990b). Increases of the catalytic activity of phospholipase c-γ tyrosine phosphorylation. *Science* **250,** 1253–1256.

Nishikura, K., and Murray, J. M. (1987). Antisense RNA of protonocogene c-*fos* blocks renewed growth of quiescent 3T3 cells. *Mol. Cell. Biol.* **7,** 639–649.

Nishina, H., Sato, H., Suzuki, T., Sato, M., and Iba, H. (1990). Isolation and characterization of *fra-2,* an additional member of the *fos* gene family. *Proc. Natl. Acad. Sci. U.S.A.* **87,** 3619–3623.
Nishizuka, Y. (1986). Studies and perspectives on protein kinase C. *Science* **233,** 305–312.
Norman, C., Runswick, M., Pollock, R., and Treisman, R. (1988). Isolation and properties of cDNA clones encoding SRF, a transcription factor that binds to the c-*fos* serum response element. *Cell* **55,** 989–1003.
Ny, T., Sawdey, M., Lawrence, D., Millan, J. L., and Loskutoff, D. J. (1986). Cloning and sequence of a cDNA coding for the human beta-migrating endothelial-cell-type plasminogen activator inhibitor. *Proc. Natl. Acad. Sci. U.S.A.* **83,** 6776–6780.
O'Brien, T. P., Yang, G. P., Sanders, L., and Lau, L. F. (1990). Expression of *cyr61,* a growth factor–inducible immediate-early gene. *Mol. Cell. Biol.* **10,** 3569–3577.
Ofir, R., Dwarki, V. J., Rashid, D., and Verma, I. M. (1990). Phosphorylation of the C-terminus of Fos protein is required for transcriptional transrepression of the c-*fos* promoter. *Nature* **348,** 80–82.
Oquendo, P., Alberta, J., Wen, D. Z., Graycar, J. L., Derynck, R., and Stiles, C. D. (1989). The platelet-derived growth factor-inducible *KC* gene encodes a secretory protein related to platelet alpha-granule proteins. *J. Biol. Chem.* **264,** 4133–4137.
Pardee, A. B., Dubrow, R., Hamlin, J. L., and Kletzien, R. F. (1978). Animal cell cycle. *Annu. Rev. Biochem.* **47,** 715–750.
Parsons, J. T. (1990). Closing the GAP in a signal transduction pathway. *Trends Gen.* **6,** 169–171.
Passmore, S., Elble, R., and Tye, B. -K. (1989). A protein involved in minichromosome maintenance in yeast binds a transcriptional enhancer conserved in eukaryotes. *Genes Dev.* **3,** 921–935.
Pessin, M. S., Baldassare, J. J., and Raben, D. M. (1990). Molecular species analysis of mitogen-stimulated 1,2-diglycerides in fibroblasts. *J. Biol. Chem.* **265,** 7959–7966.
Pouysségur, J., Sardet, C., Franchi, A., L'Allemain, G., and Paris, S. (1984). A specific mutation abolishing Na^+/H^+ antiport activity in hamster fibroblasts precludes growth at neutral and acidic pH. *Proc. Natl. Acad. Sci. U.S.A.* **81,** 4833–4837.
Prywes, R., and Roeder, R. G. (1986). Inducible binding of a factor to the c-*fos* enhancer. *Cell* **47,** 777–784.
Prywes, R., Dutta, A., Cromlish, J. A., and Roeder, R. G. (1988). Phosphorylation of serum-response factor, a factor that binds to the serum-response element of the c-FOS enhancer. *Proc. Natl. Acad. Sci. U.S.A.* **85,** 7206–7210.
Prywes, R., and Roeder, R. G. (1987). Purification of the c-*fos* enhancer-binding protein. *Mol. Cell Biol.* **7,** 3482–3489.
Quantin, B., and Breathnach, R. (1988). Epidermal growth factor stimulates transcription of the c-*jun* protooncogene in rat fibroblasts. *Nature* **334,** 538–539.
Ralph, R. K., Darkin-Rattray, S., and Schofield, P. (1990). Growth-related protein kinases. *Bioessays* **12,** 121–124.
Ran, W., Dean, M., Levine, R. A., Henkle, C., and Campisis, J. (1986). Induction of c-*fos* and c-*myc* mRNA by epidermal growth factor or calcium inophore is cAMP dependent. *Proc. Natl. Acad. Sci. U.S.A.* **83,** 8216–8220.
Rapp, U. R., Heidecker, G., Huleihel, M., Cleveland, J. L., Choi, W. C., Pawson, T., Ihle, J. N., and Anderson, W. B. (1988). *raf* family serine/threonine protein kinases in mitogen signal transduction. *Cold Spring Harb. Symp. Quant. Biol.* **53,** 173–184.
Rasmussen, C. D., and Means, A. R. (1989). Calmodulin, cell growth, and gene expression. *Trends Neurosci.* **12,** 433–438.
Rauscher, F. J., III, Morris, J. F., Tournay, O. E., Cook, D. M., and Curran, T. (1990). Binding of the Wilms' tumor locus zinc-finger protein to the EGR-1 consensus sequence. *Science* **250,** 1259–1262.
Raychaudhuri, P., Bagchi, S., and Nevins, J. R. (1989). DNA-binding activity of the adenovirus-induced E4F transcription factor is regulated by phosphorylation. *Genes Dev.* **3,** 620–627.

Raycroft, L., Wu, H. Y., and Lozano, G. (1990). Transcriptional activation by wild-type but not transforming mutants of the p53 antioncogene. *Science* **249,** 1049–1051.

Reddy, E. P., Skalka, A. M. and Curran, T. (1988). "*The Oncogen Handbook.*" Elsevier, Amsterdam.

Rice, N. P., and Gilden, R. V. (1988). The *rel* oncogene. *In* "*The Oncogene Handbook.*" (E. P. Reddy, A. M. Skalka, and T. Curran, eds.), pp. 495–512. Elsevier, Amsterdam.

Richmond, A., Balentien, E., Thomas, H. G., Flaggs, G., Barton, D. E., Spiess, J., Bordoni, R., Francke, U., and Derynck, R. (1988). Molecular characterization and chromosomal mapping of melanoma growth–stimulatory activity, a growth factor structurally related to beta-thromboglobulin. *EMBO J.* **7,** 2025–2033.

Rivera, V. M., and Greenberg, M. E. (1990). Growth factor–induced gene expression: The ups and downs of c-*fos* regulation. *New Biologist* **2,** 751–758.

Rivera, V. M., Sheng, M., and Greenberg, M. E. (1990). The inner core of the serum-response element mediates both the rapid induction and subsequent repression of c-*fos* transcription following serum stimulation. *Genes Dev.* **4,** 255–268.

Rodland, K. D., Muldoon, L. L., Lenormand, P., and Magun, B. E. (1990). Modulation of RNA expression by intracellular calcium. *J. Biol. Chem.* **265,** 11000–11007.

Rollins, B. J., Morrison, E. D., and Stiles, C. D. (1988). Cloning and expression of *JE,* a gene inducible by platelet-derived growth factor and whose product has cytokine-like properties. *Proc. Natl. Acad. Sci. U.S.A.* **85,** 3738–3742.

Rollins, B. J., Stier, P., Ernst, T., and Wong, G. G. (1989a). The human homolog of the *JE* gene encodes a monocyte secretory protein. *Mol. Cell. Biol.* **9,** 4687–4695.

Rollins, B. J., and Stiles, C. D. (1989b). Serum-inducible genes. *Adv. Cancer Res.* **53,** 1–32.

Rozengurt, E. (1986). Early signals in the mitogenic response. *Science* **234,** 161–166.

Ruderman, N. B., Kapeller, R., White, M. F., and Cantley, L. C. (1990). Activation of phosphatidylinositol 3-kinase by insulin. *Proc. Natl. Acad. Sci. U. S. A.* **87,** 1411–1415.

Rutherford, R. B., and Ross, R. (1976). Platelet factors stimulate fibroblasts and smooth muscle cells quiescent in plasma serum to proliferate. *J. Cell Biol.* **69,** 196–203.

Ryan, W. A., Jr., Franza, B. R., Jr., and Gilman, M. Z. (1989). Two distinct cellular phosphoproteins bind to the c-*fos* serum-response element. *EMBO J.* **8,** 1785–1792.

Ryder, K., and Nathans, D. (1988). Induction of protonocogene c-*jun* by serum growth factors. *Proc. Natl. Acad. Sci. U.S.A.* **85,** 8464–8467.

Ryder, K., Lau, L. F., and Nathans, D. (1988). A gene activated by growth factors is related to the oncogene v-*jun. Proc. Natl. Acad. Sci. U.S.A.* **85,** 1487–1491.

Ryder, K., Lanahan, A., Perez Albuerne, E., and Nathans, D. (1989). *jun-D:* A third member of the *jun* gene family. *Proc. Natl. Acad. Sci. U. S. A.* **86,** 1500–1503.

Ryseck, R. P., Hirai, S. I., Yaniv, M., and Bravo, R. (1988). Transcriptional activation of c-*jun* during the G_0–G_1 transition in mouse fibroblasts. *Nature* **334,** 535–537.

Ryseck, R. P., MacDonald Bravo, H., Zerial, M., and Bravo, R. (1989a). Coordinate induction of fibronectin, fibronectin receptor, tropomyosin, and actin genes in serum-stimulated fibroblasts. *Exp. Cell Res.* **180,** 537–545.

Ryseck, R. P., MacDonald Bravo, H., Mattéi, M. G., Ruppert, S., and Bravo, R. (1989b). Structure, mapping and expression of a growth factor–inducible gene encoding a putative nuclear hormonal-binding receptor. *EMBO J.* **8,** 3327–3335.

Sassone-Corsi, P., Sisson, J. C., and Verma, I. M. (1988). Transcriptional autoregulation of the protooncogene *fos. Nature* **334,** 314–319.

Sassone-Corsi, P., and Verma, I. M. (1987). Modulation of c-*fos* gene transcription by negative and positive cellular factors. *Nature* **326,** 507–510.

Schneider, C., King, R. M., and Phillipson, L. (1988). Genes specifically expressed at growth arrest of mammalian cells. *Cell* **54,** 787–793.

Schönthal, A., Herrlich, P., Rahmsdorf, H. J., and Ponta, H. (1988). Requirement for *fos* gene expression in the transcriptional activation of collagenase by other oncogenes and phorbol esters. *Cell* **54,** 325–334.

Schönthal, A., Buscher, M., Angel, P., Rahmsdorf, H. J., Ponta, H., Hattori, K., Chiu, R., Karin, M., and Herrlich, P. (1989). The Fos and Jun/AP-1 proteins are involved in the downregulation of *fos* transcription. *Oncogene* **4,** 629–636.

Schönthal, A. (1990). Nuclear protooncogene products: Fine-tuned components of signal-transduction pathways. *Cell. Signalling* **2,** 215–225.

Schröter, H., Shaw, P. E., and Nordheim, A. (1987). Purification of intercalator-released p67, a polypeptide that interacts specifically with the c-*fos* serum-response element. *Nucleic Acids Res.* **15,** 10145–10158.

Schröter, H., Mueller, C. G. F., Meese, K., and Nordheim, A. (1990). Synergism in ternary complex formation between the dimeric glycoprotein $p67^{SRF}$, polypeptide $p62^{TCF}$, and the c-*fos* serum-response element. *EMBO J.* **9,** 1123–1130.

Schüle, R., Rangarajan, P., Kliewer, S., Ransone, L. J., Bolado, J., Yang, N., Verma, I. M., and Evans, R. M. (1990). Functional antagonism between oncoprotein c-Jun and the glucocorticoid receptor. *Cell* **62,** 1217–1226.

Schutte, J., Viallet, J., Nau, M., Segal, S., Fedorko, J., and Minna, J. (1989). *junB* inhibits and c-*fos* stimulates the transforming and trans-activating activities of c-*jun. Cell* **59,** 987–997.

Severinsson, L., Ek, B., Mellstrom, K., Claesson-Welsh, L., and Heldin, C. -H. (1990). Deletion of the kinase insert sequence of the platelet-derived growth factor beta-receptor affects receptor kinase activity and signal transduction. *Mol. Cell. Biol.* **10,** 801–809.

Shaw, G., and Kamen, R. (1986). A conserved AU sequence from the 3′ untranslated region of GM-CSF mRNA mediates selective mRNA degradation. *Cell* **46,** 659–667.

Shaw, P. E., Schröter, H., and Nordheim, A. (1989a). The ability of a ternary complex to form over the serum-response element correlates with serum inducibility of the human c-*fos* promoter. *Cell* **56,** 563–572.

Shaw, P. E., Frasch, S., and Nordheim, A. (1989b). Repression of c-*fos* transcription is mediated through $p67^{SRF}$ bound to the SRE. *EMBO J.* **8,** 2567–2574.

Shen-Ong, G. L. (1990). The *myb* oncogene. *Biochim. Biophys. Acta* **1032,** 39–52.

Sheng, M., Dougan, S. T., McFadden, G., and Greenberg, M. E. (1988). Calcium and growth-factor pathways of c-*fos* transcriptional activation require distinct upstream regulatory sequences. *Mol. Cell. Biol.* **8,** 2787–2796.

Sheng, M., and Greenberg, M. E. (1990). The regulation and function of c-*fos* and other immediate-early genes in the nervous system. *Neuron* **4,** 477–485.

Sheng, M., McFadden, G., and Greenberg, M. E. (1990). Membrane depolarization and calcium induce c-*fos* transcription via phosphorylation of transcription factor CREB. *Neuron* **4,** 571–582.

Shurtleff, S. A., Downing, J. R., Rock, C. O., Hawkins, S. A., Roussel, M. F., and Sherr, C. J. (1990). Structural features of the colony-stimulating factor 1 receptor that affect its association with phosphatidylinositol 3-kinase. *EMBO J.* **9,** 2415–2421.

Shyu, A. B., Belasco, J. G., and Greenberg, M. E. (1991). Two distinct destablizing elements in the c-*fos* message trigger deadenylation as a first step in rapid mRNA decay. *Genes Dev.* **5,** 221–231.

Shyu, A. B., Greenberg, M. E., and Belasco, J. G. (1989). The c-*fos* transcript is targeted for rapid decay by two distinct mRNA degradation pathways. *Genes Dev.* **3,** 60–72.

Siegfried, Z., and Ziff, E. B. (1989). Transcription activation by serum, PDGF, and TPA through the c-*fos* DSE: Cell type–specific requirements for induction. *Oncogene* **4,** 3–11.

Simmons, D. L., Levy, D. B., Yannoni, Y., and Erikson, R. L. (1989). Identification of a phorbol ester–repressible v-*src*–inducible gene. *Proc. Natl. Acad. Sci. U.S.A.* **86,** 1178–1182.

Skup, D., Windass, J. D., Sor, F., George, H., Williams, B. R. G., Fukuhara, H., De Mayer-Guignard,

J., and De Mayer, E. (1982). Molecular Cloning of partial cDNA copies of two distinct mouse IFN-β mRNAs. *Nucleic Acids Res.* **10,** 3069–3084.

Sobczak, J., Mechti, N., Tournier, M. F., Blanchard, J. M., and Duguet, M. (1989). c-*myc* and c-*fos* gene regulation during mouse liver regeneration. *Oncogene* **4,** 1503–1508.

Spencer, C. A., and Groudine, M. (1990). Transcription elongation and eukaryotic gene regulation. *Oncogene* **5,** 777–785.

Sporn, M. B., and Roberts, A. B. (1990). "*Peptide Growth Factors and Their Receptors,*" Springer-Verlag, Berlin.

Stiles, C. D. (1983). The molecular biology of platelet-derived growth factor. *Cell* **33,** 653–655.

Stoeckle, M. Y., and Barker, K. A. (1990). Two burgeoning families of platelet factor 4–related proteins: Mediators of the inflammatory response. *New Biologist* **2,** 313–323.

Stumpo, D. J., Stewart, T. N., Gilman, M. Z., and Blackshear, P. J. (1988). Identification of c-*fos*sequences involved in induction by insulin and phorbol esters. *J. Biol. Chem.* **263,** 1611–1614.

Subramaniam, M., Schmidt, L. J., Crutchfield, C. E., and Getz, M. J. (1989). Negative regulation of serum-responsive enhancer elements. *Nature* **340,** 64–66.

Sukhatme, V. P., Kartha, S., Toback, F. G., Taub, R., Hoover, R. G., and Tsai-Morris, C. H. (1987). A novel early growth response gene rapidly induced by fibroblast, epithelial cell, and lymphocyte mitogens. *Oncogene Res.* **1,** 343–355.

Sukhatme, V. P., Cao, X. M., Chang, L. C., Tsai-Morris, C. H., Stamenkovich, D., Ferreira, P. C., Cohen, D. R., Edwards, S. A., Shows, T. B., Curran, T., Le Beau, M. M., and Adamson, E. D. (1988). A zinc finger–encoding gene coregulated with c-*fos* during growth and differentiation, and after cellular depolarization. *Cell* **53,** 37–43.

Tanaka, M., and Herr, W. (1990). Differential transcriptional activation by Oct-1 and Oct-2: Independent activation domains induce Oct-2 phosphorylation. *Cell* **60,** 375–386.

Temin, H. M. (1971). Stimulation by serum of multiplication of stationary chicken cells. *J. Cell Physiol.* **78,** 161–170.

Thomas, T. P., Talwar, H. S., and Anderson, W. B. (1988). Phorbol ester–mediated association of protein kinase C to the nuclear fraction in NIH 3T3 cells. *Cancer Res.* **48,** 1910–1919.

Thompson, C. B., Challoner, P. B., Neiman, P. E., and Groudine, M. (1985). Levels of c-*myc* oncogene mRNA are invariant throughout the cell cycle. *Nature* **314,** 363–366.

Thompson, C. B., Challoner, P. B., Neiman, P. E., and Groudine, M. (1986). Expression of the c-*myb* protooncogene during cellular proliferation. *Nature* **319,** 374–380.

Tirone, F., and Shooter, E. M. (1989). Early gene regulation by nerve growth factor in PC12 cells: Induction of an interferon-related gene. *Proc. Natl. Acad. Sci. U.S.A.* **86,** 2088–2092.

Todaro, G. J., and Green, H. (1963). Quantitative studies of the growth of mouse embryo cells in culture and their development into established lines. *J. Cell Biol.* **17,** 299–313.

Todaro, G. J., Lazar, G. K., and Green, H. (1965). The initiation of cell division in a contact-inhibited mammalian cell line. *J. Cell. Comp. Physiol.* **66,** 325–334.

Treisman, R. (1985). Transient accumulation of c-*fos* RNA following serum stimulation requires a conserved 5′ element and c-*fos* 3′ sequences. *Cell* **42,** 889–902.

Treisman, R. (1986). Identification of a protein-binding site that mediates transcriptional response of the c-*fos* gene to serum factors. *Cell* **46,** 567–574.

Treisman, R. (1987). Identification and purification of a polypeptide that binds to the c-*fos* serum-response element. *EMBO J.* **6,** 2711–2717.

Ullrich, A., and Schlessinger, J. (1990). Signal transduction by receptors with tryosine kinase activity. *Cell* **61,** 203–212.

Vairo, G., Argyriou, S., Bordun, A.-M., Gonda, T. J., Cragoe, E. J., and Hamilton, J. A. (1990). Na^+/H^+ exchange involvement in colony-stimulating-factor-1-stimulated macrophage proliferation. *J. Biol. Chem.* **265,** 16929–16939.

van Lancker, J. L. (1989). Molecular events in liver regeneration and repair. *Curr. Top. Pathol.* **79,** 205–254.

Varnum, B. C., Lim, R. W., Kujubu, D. A., Luner, S. J., Kaufman, S. E., Greenberger, J. S., Gasson, J. C., and Hershcman, H. R. (1989a). Granulocyte–macrophage colony-stimulating factor and tetradecanoyl phorbol acetate induce a distinct, restricted subset of primary-response TIS genes in both proliferating and terminally differentiated myeloid cells. *Mol. Cell. Biol.* **9,** 3580–3583.

Varnum, B. C., Lim, R. W., Sukhatme, V. P. and Herschman, H. R. (1989b). Nucleotide sequence of a cDNA encoding TIS11, a message induced in Swiss 3T3 cells by the tumor promoter tetradecanoyl phorbol acetate. *Oncogene* **4,** 119–120.

Varnum, B. C., Lim, R. W., and Herschman, H. R. (1989c). Characterization of TIS7, a gene induced in Swiss 3T3 cells by the tumor promoter tetradecanoyl phorbol acetate. *Oncogene* **4,** 1263–1265.

Vogt, P. K., and Bos, T. J. (1990). *jun:* Oncogene and transcription factor. *Adv. Cancer Res.* **55,** 1–35.

Walsh, K., and Schimmel, P. (1987). Two nuclear factors compete for the skeletal muscle actin promoter. *J. Biol. Chem.* **262,** 9492–9432.

Walsh, K. (1989). Cross-binding of factors to functionally different promoter elements in c-*fos* and skeletal actin genes. *Mol. Cell. Biol.* **9,** 2191–2201.

Watson, M. A., and Milbrandt, J. (1989). The NGFI-B gene, a transcriptionally inducible member of the steroid receptor gene superfamily: Genomic structure and expression in rat brain after seizure induction. *Mol. Cell. Biol.* **9,** 4213–4219.

Weinmaster, G., and Lemke, G. (1990). Cell-specific cyclic AMP–mediated induction of the PDGF receptor. *EMBO J.* **9,** 915–920.

Weiss, A., and Imboden, B. (1987). Cell-surface molecules and early events involved in human T-lymphocytes activation. *Adv. Immunol.* **41,** 1–38.

Whitman, M., Downes, C. P., Keeler, M., Keller, T., and Cantley, L. (1988). Type I phosphatidylinositol kinase makes a novel inositol phospholipid, phosphatidylinositol-3-phosphate. *Nature* **332,** 644–646.

Wilkinson, D. G., Bhatt, S., Chavrier, P., Bravo, R., and Charnay, P. (1989). Segment-specific expression of a zinc-finger gene in the developing nervous system of the mouse. *Nature* **337,** 461–464.

Williams, J. G., and Penman, S. (1975). The messenger RNA sequence in growing and resting mouse fibroblasts. *Cell* **6,** 197–206.

Williams, L. T. (1989). Signal transduction by the platelet-derived growth factor receptor. *Science* **243,** 1564–1570.

Wilson, T., and Treisman, R. (1988a). Removal of poly (A) and consequent degradation of c-*fos* mRNA facilitated by 3′ AU-rich sequences. *Nature* **336,** 396–399.

Wilson, T., and Treisman, R. (1988b). Fos C-terminal mutations block down-regulation of c-*fos* transcription following serum stimulation. *EMBO J.* **7,** 4193–4202.

Woodgett, J. R. (1990). Molecular cloning and expression of glycogen synthase kinase-3/Factor A. *EMBO J.* **9,** 2431–2438.

Wright, J. J., Gunter, K. C., Mitsuya, H., Irving, S. G., Kelly, K., and Siebenlist, U. (1990). Expression of a zinc-finger gene in HTLV-I and HTLV-II transformed cells. *Science* **248,** 588–591.

Yamamoto, K. K., Gonzalez, G. A., Biggs, W. H., III, and Montminy, M. R. (1988). Phosphorylation-induced binding and transcriptional efficacy of nuclear factor CREB. *Nature* **334,** 494–498.

Yang, G. P., and Lau, L. F. (1991). Cyr61, product of a growth factor–inducible immediate-early gene, is associated with the extracellular matrix and the cell surface. *Cell Growth & Differentiation* (in press).

Yang-Yen, H. F., Chambard, J. C., Sun, Y. L., Smeal. T., Schmidt, T. J., Drouin, J., and Karin, M. (1990). Transcriptional interference between c-Jun and the glucocorticoid receptor: Mutual inhibition of DNA binding due to direct protein–protein interaction. *Cell* **62,** 1205–1215.

Zerial, M., Toschi, L., Ryseck, R. P., Schuermann, M., Müller, R., and Bravo, R. (1989). The product of a novel growth factor–activated gene, Fos B, interacts with Jun proteins enhancing their DNA binding activity. *EMBO J.* **8,** 805–813.

Zetterberg, A., and Larsson, O. (1985). Kinetic analysis of regulatory events in G_1 leading to proliferation or quiescence of Swiss 3T3 cells. *Proc. Natl. Acad. Sci. U.S.A.* **82,** 5365–5369.

Ziff, E. B. (1990). Transcription factors: A new family gathers at the cAMP response site. *Trends Genet.* **6,** 69–72.

Zinn, K., Keller, A., Whittemore, L. A., and Maniatis, T. (1988). 2-Aminopurine selectively inhibits the induction of beta-interferon c-*fos,* and c-*myc* gene expression. *Science* **240,** 210–213.

Zipfel, P. F., Irving, S. G., Kelly, K., and Siebenlist, U. (1989). Complexity of the primary genetic response to mitogenic activation of human T cells. *Mol. Cell. Biol.* **9,** 1041–1048.

Section II

Cellular, Biochemical, and Molecular Parameters of *in vitro* Model Systems in Which Modifications in Cell-Growth Control Are Functionally Related to the Onset of Differentiation

6

Gene Expression during Development of the Osteoblast Phenotype: An Integrated Relationship of Cell Growth to Differentiation

JANE B. LIAN, GARY S. STEIN, THOMAS A. OWEN, STEVEN DWORETZKY, MELISSA S. TASSINARI, MICHAEL ARONOW, DAVID COLLART, VICTORIA SHALHOUB, SCOTT PEURA, LEESA BARONE, JOSEPH BIDWELL, AND SHIRWIN POCKWINSE

Department of Cell Biology
University of Massachusetts Medical Center
Worcester, Massachusetts 01655

I. INTRODUCTION

A functional relationship between cell growth and the initiation and progression of events associated with differentiation has been viewed as a fundamental question by developmental biologists for more than a century. In the case of bone, this relationship of growth and differentiation must be maintained and stringently regulated, both during development and throughout the life of the organism to support tissue remodeling.

Two distinct pathways are associated with bone formation, each necessitating a highly integrated series of developmental processes and signaling mechanisms, and both dependent on proliferation for increasing tissue mass as well as for regulation of key components of bone-cell phenotype expression. In one pathway, intramembranous bone development, bone tissue (e.g. the calvarium) forms directly from differentiation of mesenchymal progenitor cells to committed osteoblasts. These cells produce high levels of Type I collagen, alkaline phosphatase, and unique noncollagenous proteins that promote extracellular-matrix mineralization. The second is a more complex pathway, endochondral bone formation, which initially involves chondrogenesis from mesenchymal progenitors, resorption of calcified cartilage, vascular invasion, followed by osteogenesis (Marks and Popoff, 1988; Cowell *et al.,* 1987) and marrow formation (Fig. 1). Signals for osteoblast differentiation can therefore derive from multiple sources.

The complexity of bone-cell differentiation is further illustrated by the remodeling of all bone. Remodeling (replacing old bone with new) involves first the maturation and/or activation of osteoclasts (multinucleated resorbing cells of bone), which requires, in part, factors produced by osteoblasts (Rodan and Martin, 1981). Osteoblast proliferation and differentiation, in turn, is regulated by factors [growth factors, chemotactic proteins (see Table I)] released from the bone matrix undergoing resorption, thereby regulating tissue stability and turnover in response to changes in skeletal requirements or physiologic demands for calcium and phosphate (Rodan and Martin, 1981; Raisz and Kream, 1983a, b). Thus, another important component of osteoblast differentiation is acquisition of competency for responsiveness to hormones that regulate bone turnover.

Understanding bone-tissue organization is an essential element to defining mechanisms associated with the progressive differentiation of osteoblast as directly related to development of a mineralized extracellular matrix and to the capacity for influencing osteoclast differentiation (Fig. 1). Here it is necessary to consider location of cells within the bone-forming tissue, since this may contribute to concentration gradients of inductive factors and effector molecules. Within this context, the timing of skeletal tissue vascularization during osteogenesis is also an important consideration, since this provides for the availability and/or recruitment of bone marrow-derived hematopoietic stem cells, which are progenitors of osteoclasts. This concept is reflected by the initiation of cartilage resorption (when vascularization occurs), initiating the onset of osteogenesis during endochondral bone formation, and by the first evidence of calvarial bone resorption following vascularization. More recently, the potential of bone-marrow cells to differentiate to osteoblastic-like cells *in vitro* is another example of the *in vivo* mechanisms available to maintain coupling between bone resorption and bone formation (Owen and Friedenstein, 1988).

For many years, bone was defined anatomically and examined largely in a descriptive manner by ultrastructural analysis and by biochemical and histochemical methods. These studies provided the basis for our understanding of bone-

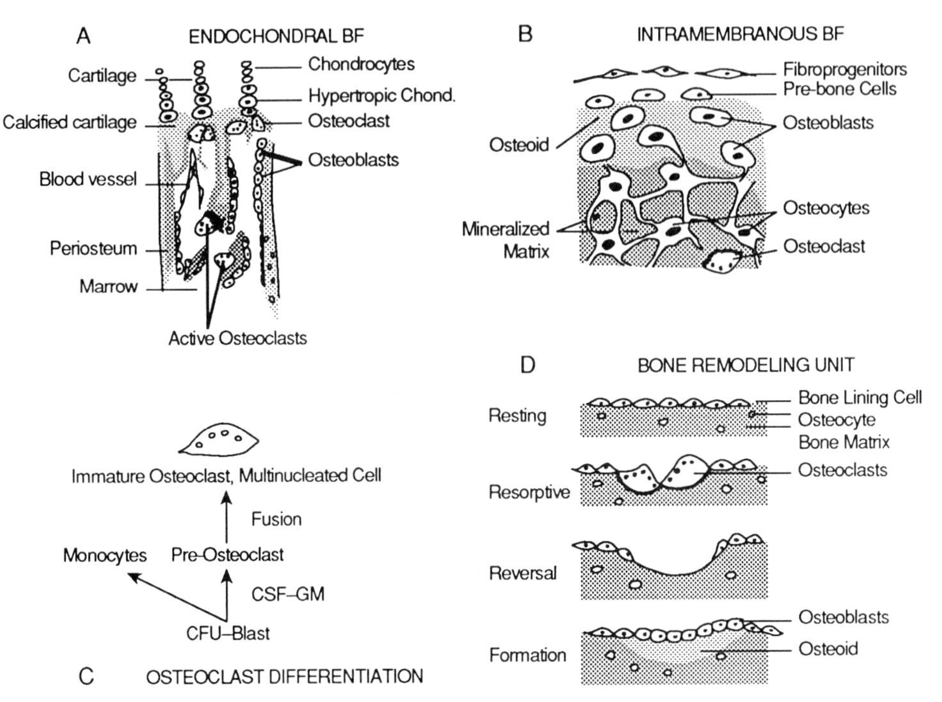

Fig. 1. Cellular aspects of bone formation and remodeling. (A) Events associated with endochondral bone formation, as occurs in growth of long bones are cartilage development, followed by hypertrophy of the chondrocyte and calcification of cartilage matrix, which is resorbed by osteoclasts after vascular invasion. Bone-marrow progenitors differentiate with bone formation from osteoblast deposition of bone matrix on cores of calcified cartilage. The calcified cartilage and initial, new-woven bone are resorbed and remodeled into a mature lamellar (layered) bone. (B) Intramembranous bone formation, as occurs in calvarium, or appositional growth of bone diaphysis, is direct. Osteoblasts differentiate from a periosteal progenitor cell (capable of proliferation) to cells producing osteoid (the nonmineralized matrix). The final stage of osteoblast maturation is the osteocyte surrounded by mineralized tissue. (C) The ontogeny of the osteoclast from its progenitor in the hemopoietic stem-cell system to the active resorbing cell is illustrated. (D) The remodeling sequence as it occurs in the removal of calcified cartilage, in trabecular bone, and formation of haversian systems is depicted. The osteoblast-lining cells on the bone surface (resting phase) must retract for osteoclast activation on mineralized surface (resorption phase). Mononuclear cells smooth out the resorbed surface (reversal phase) and are followed by osteoblasts, which form new bone (formation phase). Mineralization of the osteoid and formation of the bone-lining cell will complete the cycle.

tissue organization and orchestration of the progressive recruitment, proliferation, and differentiation of the various cellular components of bone tissue. Now, complemented by an increased knowledge of molecular mechanisms that are associated with and regulate expression of genes encoding phenotypic components of bone and those that may control the progressive development and maturation of the bone-cell phenotype, our understanding of bone cell and tissue differentiation is rapidly increasing.

TABLE I

Bone Cell Proteins

Protein	MW	Properties	Reference
Extracellular Matrix Proteins			
Collagen Type I	≈320,000	Principal structural component of the bone ECM	Glimcher, (1976)
Osteocalcin	5800	Binds Ca^{2+} to hydroxyapatite, contains γ-carboxyglutamic acid, vitamin D–responsive, mineralization related, involved in osteoclast recruitment–differentiation	Price (1988); Hauschka (1989)
Osteopontin (2ar)(SPPI)	32,600 (from sequence)	Phosphoprotein, cell attachment via RGD sequence, binds to hydroxylapatite	Franzen and Heingard (1985); Smith and Denhardt (1989); Smith and Denhardt (1987)
Matrix Gla protein	10,000	Contains γ-carboxyglutamic acid	Price (1988)
Bone sialoprotein	33,600 (from sequence)	Phosphoprotein binds to cells and hydroxyapatite through RGD sequence	Franzen and Heinegard (1985)
Bone acidic glycoprotein	75,000	Cell attachment	Gorski and Shimizu (1988)
Osteonectin (SPARC)	33,000 (from sequence)	Phosphorylated glycoprotein, binds calcium	Termine *et al.* (1981)
Thrombospondin	3 × 150,000	Contains RGD sequence, binds calcium and other proteins	Gehron *et al.* (1989)
Fibronectin	550,000	Cell attachment via RGD sequence, interacts with glycosaminoglycans	Ruoslahti (1988)
Bone proteoglycan I (biglycan)	75,000	ECM organization	Fisher *et al.* (1987)
Bone Proteoglycan II (Decorin)	120,000	ECM organization	Kinne and Fisher (1987)
BP-2	19,000	Mineralization	Mardon and Triffit (1987)
Important Cellular Proteins			
Alkaline phosphatase	140,000 (homodimer)	Mineralization	Rodan and Rodan (1984)
Plasminogen activator		Osteoblast catabolism	Hamilton *et al.* (1985)
Stromelysin		Osteoblast catabolism	Rodan and Noda (in press)
Collagenase	≈56,000 (non-glycosylated)	Osteoblast catabolism	Otsuka *et al.* (1984); Heath *et al.* (1984)

Protein	MW	Properties	Reference
Growth and Phenotype Modulators			
FGF	≈16,000	Stimulates proliferation	Hauschka (1986)
TGFβ	25,000 (Dimer)	Regulates growth (both stimulates and inhibits growth), regulates ECM biosynthesis	Antosz *et al.* (1989); Noda (1990)
β_2 Microglobulin		Mitogenic	Centrella *et al.* (1989)
PDGF	30,000 (Dimer)	Regulates growth	
IGFI	7650	Regulates growth	
LIF	26,000	Stimulates osteoblast proliferation	Abe *et al.* (1988); Noda *et al.* (1990)
Bone morphogenic proteins			
BMP1	730 amino acids	Bone–TGFβ family, induces cartilage formation *in vivo*	Wozney *et al.* (1988)
BMP2A	396 amino acids	Bone–TGFβ family, induces cartilage formation *in vivo*	Wozney *et al.* (1988); Wang *et al.* (1990)
BMP2B	408 amino acids	Bone–TGFβ family, induces cartilage formation *in vivo*	Wozney *et al.* (1988)
BMP3	472 amino acids	Bone–TGFβ family, induces cartilage formation *in vivo*	Wozney *et al.* (1988)
Osteogenin	28–43 kDa	Induces cartilage and bone formation *in vivo*	Sampath *et al.* (1987); Luyten *et al.* (1989)
Osteoinductive factor (OIF)	22–28 kDa (12 kDa deglycosylated)	Induces bone formation *in vivo* with TGFβ	Bentz *et al.* (1989); Madisen *et al.* (1990)
OP-1	15–18 kDa (reduced)	TGFβ family	Ozkaynak *et al.* (1990)

This chapter will review concepts and experimental approaches associated with the relationship of cellular proliferation to differentiation during development of the osteoblast phenotype, with the understanding that this relationship is of broad biological relevance. Consequently, it is reasonable to assume that analogous principles may apply to the regulation of phenotype expression and establishment of tissue organization in general. We will review several of the key elements of the sequence of events and regulatory mechanisms by which cell growth contributes to the onset and progression of osteoblast differentiation, whereby cells undergo a developmental maturation process, progressively acquiring phenotypic properties associated with the biosynthesis, organization, and mineralization of the extracellular matrix—the distinguishing characteristic and primary functional property of a specialized bone cell.

II. THE GROWTH–DIFFERENTIATION RELATIONSHIP

A. General Features

Although the concept of the relationship between proliferation and differentiation as a necessary component of the developmental process is long-standing, the experimental approaches have been largely descriptive, and the results, primarily correlative. However, the recent development of culture systems that support the differentiation of specialized cells has permitted the combined use of biochemical, molecular, and ultrastructural approaches to address the integrated relationships of cell growth to the expression of cell-and tissue-specific phenotypic properties. The application of *in situ* methodologies is particularly important in approaching questions related to differentiation, since these *in vitro* systems support the development of a tissue-like organization analogous to that occurring *in vivo,* allowing for an understanding of molecular signaling mechanisms at the single-cell level. Examples of such *in vitro* systems are schematically illustrated in Fig. 2 and include pluripotent promyelocytic leukemia cells, which develop the monocytic, macrophage, or granulocytic phenotype (Rovera *et al.,* 1979; Huberman and Callahan, 1979); adipocytes (Spiegelman and Green, 1980); myoblasts (Nadal-Ginard, 1978; Okazaki and Holtzer, 1966); keratinocytes (Rheinwald and Green, 1975; Rice and Green, 1979); and osteoblasts (Bellows *et al.,* 1986; Bhargava *et al.,* 1988; Ecarot-Charrier *et al.,* 1983; Gerstenfeld *et al.,* 1987; Aronow *et al.,* 1990; Owen *et al.,* 1990a). While this chapter will be restricted to a consideration of proliferation–differentiation relationships in osteoblasts, these relationships in

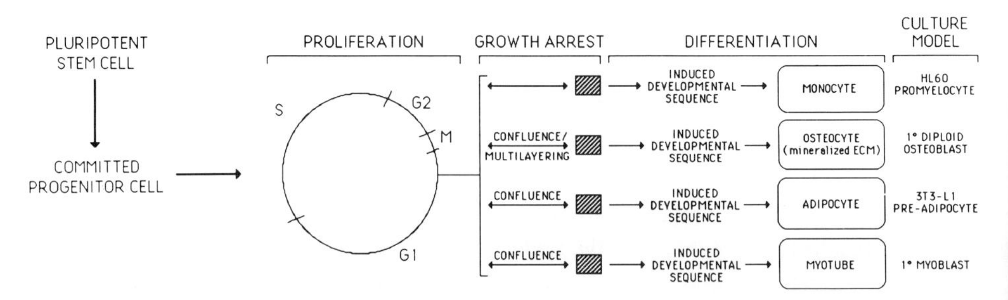

Fig. 2. Schematic illustration of growth and the progressive expression of differentiation of several phenotypes. Initially, the committed progenitor cell (for example, HL60 promyelocytic leukemia cells, primary cells of embryonic calvaria osteoblasts, 3T3-L1 preadipocytes or myoblasts) actively proliferate, expressing cell cycle and cell growth-regulated genes, as well as genes encoding proteins for extracellular-matrix formation. Following growth arrest, a developmental sequence involving the sequential and selective expression of genes that results in the differentiated cell and tissue phenotype occurs. Completion of the proliferation period marks an important transition point, where expression of tissue-specific genes, often functionally related to the down-regulation of proliferation, is initiated.

promyelocytic leukemia and hematopoietic stem cells, erythroleukemia cells, and melanocytes are discussed in other chapters of this volume.

There are several features of the proliferation–differentiation relationships exhibited by these *in vitro* systems that bear striking similarities to tissue development *in vivo,* validating the culture models. These relationships establish what appear to be important and possibly rate-limiting steps in the differentiation process. Initially, a pluripotent stem cell and/or committed progenitor cell undergoes active proliferation, increasing the pool of precursor cells and the tissue mass to accommodate the biosynthesis of the specialized cell products required for intercellular and extracellular structural and functional properties unique to a developing tissue. While actively proliferating, these cells express genes encoding cell-cycle and cell growth–related proteins that support the complex and interdependent events associated with the proliferative process. Additionally, gene expression in proliferating cells may suppress expression of genes for later events in the differentiation process.

Regulation of gene expression in proliferating cells that relate to cell-cycle control is largely mediated at several post-transcriptional levels (e.g., mRNA stability and phosphorylation), providing the basis for a rapid response to accommodate cellular events, which include DNA replication and mitotic division. In contrast, at the completion of proliferative activity, the down-regulation of cell cycle-related gene expression and the initiation of cell and tissue-specific gene expression is largely controlled transcriptionally. It should be emphasized that while the completion of proliferative activity in promyelocytes, osteoblasts, preadipocytes, and myoblasts is an important point in the development of the tissue-specific phenotype, a progressive series of events occurs postproliferatively that are necessary for the ordered development of the structural and functional characteristics of differentiated cell. Each system requires its own complex series of regulatory steps that can be mediated at multiple levels.

B. The Osteoblast Developmental Sequence

1. Composition and Development of Bone Tissue

For the past two decades, considerable effort has been directed toward identification and characterization of proteins synthesized by and associated with bone, providing the necessary foundation for addressing the regulation of bone-cell phenotype expression, both with respect to bone-cell differentiation and the influence of bone on other tissues and physiologic processes. Two examples that illustrate these relationships are the role of osteoblast-synthesized proteins in the biosynthesis, organization, and mineralization of the bone extracellular matrix and the potential utilization of some growth factors synthesized by bone in the modulation of osteoblast proliferation and growth control of nonskeletal tissues. Transforming growth factor β (TGFβ) is the most extensively studied example of an osteoblast-synthesized factor that influences growth and/or differentiation of both skeletal and nonskeletal cells (Massague, 1987; Centrella, *et al.,* 1988; Joyce, *et al.,* 1990; Sporn, *et al.,* 1983; Noda and Rodan, 1987; Rosen, *et al.,* 1988; Noda and Camilliere, 1989; Moses, *et al,* 1987).

The importance of having made significant progress toward establishing the protein composition of bone cells should not be underestimated. There is a tendency to label such endeavors as *descriptive.* However, without an in-depth and comprehensive understanding of cell and tissue protein composition and knowledge of when during the development of the cellular phenotype expression occurs, pursuit of molecular mechanisms that mediate control of gene expression cannot provide insight into biological relationships. Rather, such molecular studies pursued independently provide an inventory of gene-regulatory sequences and promoter-binding factors in the absence of a biological perspective, and thereby are also only *descriptive,* but at the gene level. Clearly, a biologically meaningful understanding necessitates the combined application of biochemical and molecular approaches within the context of cell and tissue structure.

a. Extracellular Matrix Proteins Associated with Bone. The proteins associated with bone can be assigned to several categories on the basis of structural and functional properties; those that constitute, organize, or modify the extracellular matrix: inductive factors; and growth modulators. It should be emphasized that an understanding of the functional properties of these proteins is incomplete and that a comprehensive characterization of a bone-cell protein requires a combined application of

1. biochemical and molecular approaches to determine when the proteins are synthesized and/or modified with respect to the progressive differentiation of the osteoblast phenotype, as well as to definitively establish the cells that synthesize the proteins;
2. *in situ* immunohistochemistry to localize the proteins within cells and tissue; and, where applicable
3. studies to assess potential functionality of the bone-cell proteins (for example, mineralization, or resorption, or morphogenic, mitogenic, or chemotactic activity).

Table I summarizes the properties of the principal bone-cell proteins that have been identified to date (reviewed by Heinegård and Oldberg, 1989; Boskey, 1989; Rodan and Noda, 1990).

Type I collagen is the principal component of the bone extracellular matrix serving as the highly organized three-dimensional scaffold for the ordered deposition of mineral. Several classes of proteins associated with the collagen extracellular matrix exhibit properties consistent with involvement in either the initiation and propagation of mineralization or regulation of mineral turnover (Heinegård and Oldberg, 1989; Boskey, 1989; Rodan and Noda, 1990) (Table I). These include alkaline phosphatase, osteocalcin, osteopontin, matrix Gla protein, osteonectin, thrombospondin, and several glycoproteins and proteoglycans. Interestingly, many of these proteins were initially reported to be bone specific and subsequently shown to be associated with and/or synthesized by other cell types and tissues except for osteocalcin. Common features shared by the tissues that many of these proteins are associated with is an extensive extracellular matrix, although only the bone-cell matrix is mineralized. Therefore the presence of alkaline phophatase, osteonectin, or osteopontin alone does not appear to constitute the rate-limiting steps that render the extracellular matrix competent for mineralization. This raises the possibility that either levels, local concentrations, and/or spatial organization of the proteins, together with other bone matrix-specific factors are functionally related to modifications in the extracellular matrix associated with mineralization.

Recently, analysis of the genes encoding two osteoblast-associated proteins, collagen Type I and alkaline phosphatase, has provided insight into potential molecular mechanisms that may account for the bone requirement for transcripts from these genes, which are expressed in several tissues. In the case of Type I collagen, the use of tissue-specific enhancer elements in the 5′ regulatory sequences of the gene as well as utilization of alternative transcription-intitiation sites have been reported to contribute to cell-and tissue-specific expression of Type I collagen (Kratochwil *et al.,* 1989). The normal expression (transcription, mRNA levels, and protein synthesis) of Type I collagen in odontoblasts, and the absence of Type I collagen transcription in fibroblasts of MOV13 transgenic mice with a retrovirus integrated in the first exon, is compelling evidence for alternative tissue-specific utilization of promoter regulatory elements. In the case of alkaline phosphatase, two mechanisms support tissue-specific expression and regulation in response to development and/or physiological requirements of cells and tissues. Two rodent genes have been identified: a bone, liver, kidney, and placental gene and an intestinal gene (Goldstein *et al.,* 1980; Zernik, 1990). In humans at least four genes exhibit tissue-specific expression (Henthorn *et al.,* 1988; Knoll *et al.,* 1988; Millan and Manes, 1988; Weiss *et al.,* 1988): a bone, liver, and kidney gene, a placental gene, an intestinal gene, and a germ-cell gene. Here additional tissue-specific regulation of expression is derived from the utilization of tissue-specific 5′ regulatory elements and by alternative splicing (Matsuura *et al.,* 1990).

b. Growth Modulators of the Bone-Cell Phenotype. The remodeling of bone throughout the life of an organism necessitates maintenance of a pool of osteoblasts to regenerate bone tissue, or osteoprogenitor cells that are competent to undergo proliferative activity and differentiate. A series of growth factors from bone cells and bone matrix have been isolated, and the growth-factor receptors, identified on osteoblasts (Canalis *et al.,* 1988). Furthermore, other evidence suggests that growth is modulated by both systemic (hormonal effects) and local factors (cytokines). Thus factors that influence osteoblast growth and differentiation are also synthesized by other skeletal and nonskeletal cells reflecting the involvement of endocrine, paracrine, and autocrine control Among the growth factors (Table I) that have been isolated to date from bone cells or osteosarcomas that effect osteoblast proliferation are TGFβ (Massague, 1987; Centralla *et al.*, 1988; Joyce *et al.*, 1990; Sporn *et al.*, 1983; Noda an Rodan, 1987; Rosen *et al.*, 1988; Noda and Camilliere, 1989; Moses *et al.*, 1987), insulin-like growth factor (IGF-I) (Ernst and Froesch, 1988; Zapf *et al.*, 1981); Gray *et al.*, 1989; Ernst *et al.*, 1989), β_2 microglobulin, and a bone-derived growth factor (BDGF) (Centrella *et al.*, 1989). TGFβ is not only mitogenic but has also been demonstrated to stimulate growth and differentiation of mesenchymal precursor cells in the periosteum to both cartilage and bone formation (Joyce *et al.*, 1990). Basic fibroblast growth factor (FGF) (Hauschka *et al.*, 1986), epidermal growth factor (EGF) (Ng*et al.*, 1983; Hiramatsu *et al.*, 1982), and platelet-derived growth factor (PDGF) (Graves *et al.*, 1989; Betsholtz *et al.*, 1986) are present in the bone extracellular matrix and also modulate the proliferative activity of osteoblasts. While unquestionably these factors are viable candidates for physiologic modulation of osteoblast growth and differentiation, definitive evidence for their biosysthesis by osteoblasts must be established by isolation of the mRNAs from the cells.

A source of ambiguity is reports of both stimulation and inhibition of osteoblast proliferation by the various growth factors. In part, this may reflect receptor levels. But perhaps to a greater extent, variations are associated with the extent to which cells are proliferating or expressing specific genes encoding osteoblast phenotype proteins. Additionally, cellular responses to growth-modulating factors differ significantly for normal diploid osteoblasts and transformed osteoblasts and osteosarcoma cells (Holthuis *et al.*, 1990).

c. Hormones That Modulate Osteoblast Growth and Differentiation. A series of polypeptide [parathyroid hormone (PTH), calcitonin, and insulin] and steroid hormones that include vitamin D (reviewed by Suda *et al.*, 1990; Minghetti and Norman, 1988; and DeLuca, 1988), estrogen (Komm *et al.*, 1988; Ernst *et al.*, 1988; Gray *et al.*, 1987; Ernst *et al.*, 1989), thyroid hormone, retinoic acid, and glucocorticoids (Chen *et al.*, 1983; Majeska *et al.*, 1985; Canalis, 1983) are physiologic mediators of bone-cell proliferation and differentiation that both up-regulate and down-regulate genes encoding cell growth and bone phenotypic proteins. Several of these hormones (e.g., PTH and vitamin D) exhibit pleiotropic effects on both bone formation (osteoblasts) and bone resorption (osteoclasts). Hormones, such as vitamin D and glucocorticoids, have been shown to have profound effects on formation of bone tissue, either by inhibiting or stimulating bone-cell proliferation and/or by enhancing or depressing expression of properties related to development of the osteoblast phenotype (Wong *et al.*, 1990, Owen *et* *al.*, 1991). The extent to which these hormones exhibit positive or negative effects on bone-cell gene expression is, in part, a reflection of the organization of the 5′ regulatory regions of genes expressed during osteoblast growth and differentiation, in which a series of steroid-responsive elements reside, and sequence-specific interactions of individual or multiple steroid-receptor complexes with their responsive elements may account for synergistic and antagonistic effects on transcription (Jantzen *et al.*, 1987; Grange *et al.*, 1989; Schüle *et al.*, 1988; Strähle *et al.*, 1988; Ankenbauer *et al.*, 1988). Recent results indicate the possibility of several steroid-receptor complexes recognizing and interacting with a single steroid-responsive element (e.g., recognition of the osteocalcin gene vitamin D–responsive element by the vitamin D receptor and the retinoic acid receptor) (Schüle *et al.*, 1990). These findings are intriguing, since earlier reports suggest that retinoic acid and 1,25-$(OH)_2D_3$ may impinge on a common pathway to modulate gene expression in osteoblastic cells (Nishimoto *et al.*, 1987b; Kim and Chen, 1989). However, the extent to which such dual recognitions of steroid-responsive elements by multiple steroid hormones occurs with physiological concentrations remains to be established.

A clinical situation directly resulting from an impairment in the transcription of vitamin D–responsive genes is vitamin D–resistant rickets (Type II) (Hughes *et al.*, 1988). This hereditary disorder is characterized by the classical features of vitamin D deficiency (rickets or osteomalacia), despite an adequate vitamin D intake and normal-to-high circulating levels of the active metabolite, 1,25-dihydroxyvitamin D. The molecular basis of this bone disorder is the alteration in sequences of the vitamin D–receptor gene encoding amino acids that bind to promoter elements of genes responsive to the hormone. This is an elegant example of a perturbation of the DNA-binding protein influencing vitamin D–mediated modifications in gene expression. The influence of steroid hormones on osteoblast growth and differentiation is also mediated at the level of mRNA stability, further complicating analysis of regulatory mechanisms, but providing the basis for a high resolution and rapid, responsive series of mechanisms to utilize steroid hormone-mediated signal transduction to rapidly modulate gene expression in osteoblasts in response to physiologic regulators.

2. Model Systems Supporting Development of the Bone-Cell Phenotype

a. In Vivo. Traditionally skeletal growth and differentiation has been addressed primarily by the utilization of histology and histochemistry to examine avian and rodent skeletal development with emphasis on limb and craniofacial morphogenesis (Pechak *et al.*, 1986a, b). These approaches have been complemented to some extent by bone-fracture studies in rat limbs and limb regeneration studies in amphibians. Such studies have provided an important foundation to our understanding of bone-cell growth and differentiation with respect to tissue organization, under conditions in which cell and tissue morphology and intercellular relationships are largely maintained. These results have been limited by the inability to address regulatory mechanisms other than at the protein and ultrastructural levels. However, several groups are generating a series of antibodies to osteoblasts at different stages of differentiation (Nijweide and Mulder, 1986; Bruder and Caplan, 1989a, b; Bruder and Caplan, 1990), which should be extremely instructive. With the development of high-resolution immunohistochemical (Heine *et al.,* 1987) and *in situ* hybridization methods, the localization of proteins and mRNAs in a semiquantitative manner can be detected at the single-cell level in tissue sections. These approaches afford the opportunity to definitively establish the expression of specific genes in bone cells at particular stages of osteogenesis as well as in relationship to the proliferative state of the osteoblast and expression of other osteoblast parameter simultaneously. *In situ* hybridization has been applied in studies of expression of osteopontin, osteonectin, and TGFβ-like genes during fetal development (Nomura *et al.,* 1988; Lyons *et al.,* 1989) and in investigations of gene expression in intramembranous bone and at the growth plate in endochondral bone (Weinreb *et al.,* 1990), with emphasis on the representation of mRNA transcripts from specific bone-related genes. Taken together, these studies indicate a pattern of gene expression that reflects anatomical location within the tissue and differentiated state of the cell. mRNA analysis from calvaria during fetal rat development (Yoon *et al.,* 1987) and during fracture healing (Jinguishi *et al.,* 1989; Bolander *et al.,* 1989) has also increased our understanding of gene expression during development of the osteoblast phenotype. However, the limitations of these approaches is that they offer minimal opportunity to modify growth and differentiation conditions and thereby experimentally address molecular mechanisms associated with the regulation of gene expression. A notable exception is the elegant studies of Eichler and colleagues (Thaller and Eichele, 1987; Wanger *et al.,* 1990; Thaller and Eichele, 1990), in which gradients of morphogenic factors associated with pattern formation in avian limbs have been examined.

An effective *in vivo* model to study cellular proliferation and differentiation related to endochondral bone formation is the subcutaneous implantation of demineralized bone particles into rats, which induces a series of events resulting in the production of an ectopic bone ossicle at the site of injection (Fig. 3). The sequence of chondro- and osteo-inductive events occurring in this bone-particle implant is directly comparable to that which takes place in the skeleton (Reddi, 1981). Morphological and biochemical events associated with endochondral bone formation in this model have been well characterized. Recently cellular levels of mRNA transcripts from cell growth-regulated and tissue-specific genes were examined and found to reflect the morphological and biochemical changes occurring during formation of the ossicle (Bortell *et al.,* 1990; Fig. 4). Several periods of proliferative activity occur corresponding to (1) the initial period of recruitment and production of fibroprogenitor cells (day 3); (2) onset of bone formation (day

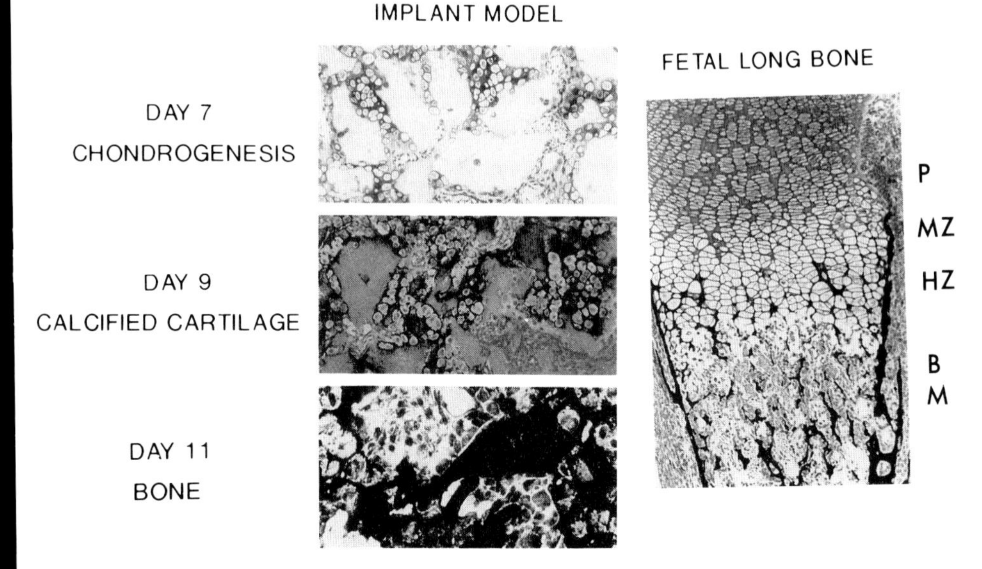

Fig. 3. The demineralized bone particle (DBP) implant model of endochondral bone formation. Histological sections of DBP implantation into subcutaneous pockets of young, growing rats, eliciting the recruitment and transformation of progenitor cells to chondrocytes (Days 0–7), which continue to undergo the normal sequelae of endochondral bone formation in a precisely timed program of events (left panel). Calcified cartilage follows (Days 7–9) and, concomitant with vascularization, becomes resorbed and replaced by bone tissue (Day 11), the sequence of events identical to endochondral bone formation in developing bone (right panel). The implant stages are compared to the cellular zones in development of bone. This longitudinal section of a rat 21-day fetal tibia shows proliferating chondrocytes (P), maturation zone (MZ) (column arrangement of chondrocytes), hypertrophic cells (HZ) undergoing calcification and the formation of bone spicules (B) and marrow (M).

9); and (3) formation of bone marrow (day 19). The mRNA levels of collagen Type II, a phenotypic marker of cartilage, peaks between days 7 and 9 post-implantation, corresponding to the appearance of chondrocytes in the implant, and rapidly declines on day 11 (to 5% of maximal value), when bone formation is observed. The peak mRNA levels of collagen Type I, found in fibroblasts and osteoblasts, occur first with the onset of bone formation (day 7–10) and again during formation of bone marrow (day 19).

Fig. 4. *(following page)* Expression of genes characterizing chondrogenesis and osteogenesis in the DBP bone implant model. Northern blot analyses, quantitated from densitometric measurements (percentage maximal expression), showing expression of Type II collagen unique to chondrocytes: histone H4, tightly coupled to DNA synthesis; collagen Type I, ubiquitous in connective tissue; and TGFβ throughout the experimental course. Type II is found only in the chondrogenic period. Type I is seen to increase at the onset of the calcified cartilage and the osteogenic period. The expression of TGFβI parallels the expression of Type I collagen.

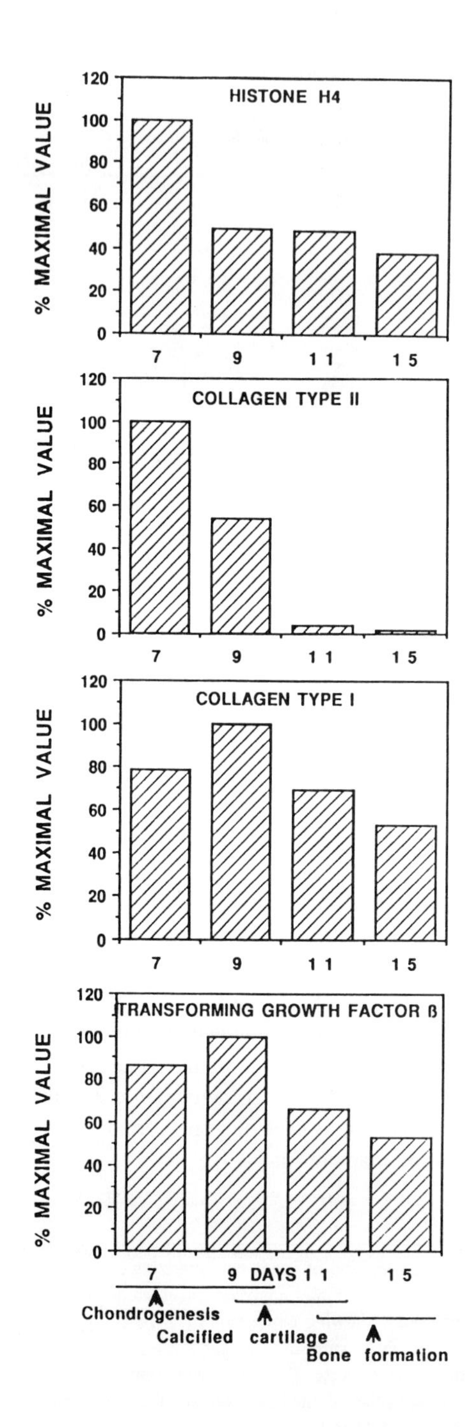
HISTONE H4
COLLAGEN TYPE II
COLLAGEN TYPE I
TRANSFORMING GROWTH FACTOR β
% MAXIMAL VALUE
120
100
80
60
40
20
0
7
9
1 1
1 5
DAYS
Chondrogenesis
Calcified cartilage
Bone formation

b. In Vitro. Normal diploid osteoblasts isolated from fetal rat calvaria (21-day rats) undergo an ordered developmental sequence during a 35-day culture period, resulting in the formation of multilayered nodules of cells with a mineralized Type I collagen extracellular matrix (Fig. 5) (Bellows *et al.*, 1986; Bhargava *et al.*, 1988; Ecarot-Charrier *et al.*, 1983; Gerstenfeld *et al.*, 1987; Aronow *et al.*, 1990; Ecarot-Charrier *et al.*, 1988; Stein *et al.*, 1989b; Owen *et al.*, 1990a). By the combined use of molecular, biochemical, and ultrastructural analysis, the expression of cell-growth and tissue-specific genes has been mapped during the progressive development of bone-cell phenotype within the context of the development of a bone tissue–like organization (Gerstenfeld *et al.*, 1987; Aronow *et al.*, 1990; Stein *et al.*, 1989b, Owen *et al.*, 1990a). The temporal sequence of expression of genes encoding osteoblast phenotype markers in culture follows a pattern of gene expression and tissue distribution determined by *in situ* hybridization observed in neonatal bones (Weinreb *et al.*, 1990; Sandberg, 1988; Lyons *et al.*, 1989) and during fetal calvarial development *in vivo* (Yoon *et al.*, 1987), supporting the biological relevance of the osteoblast culture system as a model for intermembranous bone differentiation. It is now generally accepted that the progenitor stem cell of the osteoblast is a pluripotent mesenchymal fibroblast that, before commitment, can develop as the osteoblast, chondrocyte, or other phenotypes (Friedenstein and Chailakhyan, 1987). However, the cells released from fetal calvaria by a sequential series of trypsin and collagenase digestions are cells that are committed to the osteoblast lineage, exhibiting such factors as PTH-hormone responsiveness and synthesis of bone-related proteins. While this precludes investigation of the process by which commitment to the osteoblast phenotype is regulated, the primary diploid osteoblast cultures offer a viable system in which to examine the relationship of cell growth to the initiation and progression of the complex of an interdependent series of events that result in development of the bone-cell phenotype—the osteoblast or osteocyte within a mineralized extracellular matrix.

The importance of experimentally addressing the regulatory events related to the recruitment and commitment of pluripotent mesenchymal fibroblasts into the developmental pathway (that initially restricts options for specialization to either the osteoblast and chondrocyte pathways and subsequently to solely the osteoblast phenotype) should not be underestimated. While to date our knowledge of this aspect of osteoblast differentiation is somewhat incomplete, this is largely owing to the limitations of model systems. However, recent studies utilizing bone-marrow cultures offer the possibility of examining events associated with stem-cell differentiation at the level of gene-regulatory mechanisms and relationships (Owen and Friedenstein, 1988; Leboy *et al.*, 1991; Long *et al.*, 1990; Benayahu *et al.*, 1989).

Osteosarcoma cells and transformed osteoblastic cell lines have been extensively utilized during the past several years to address questions related to factors

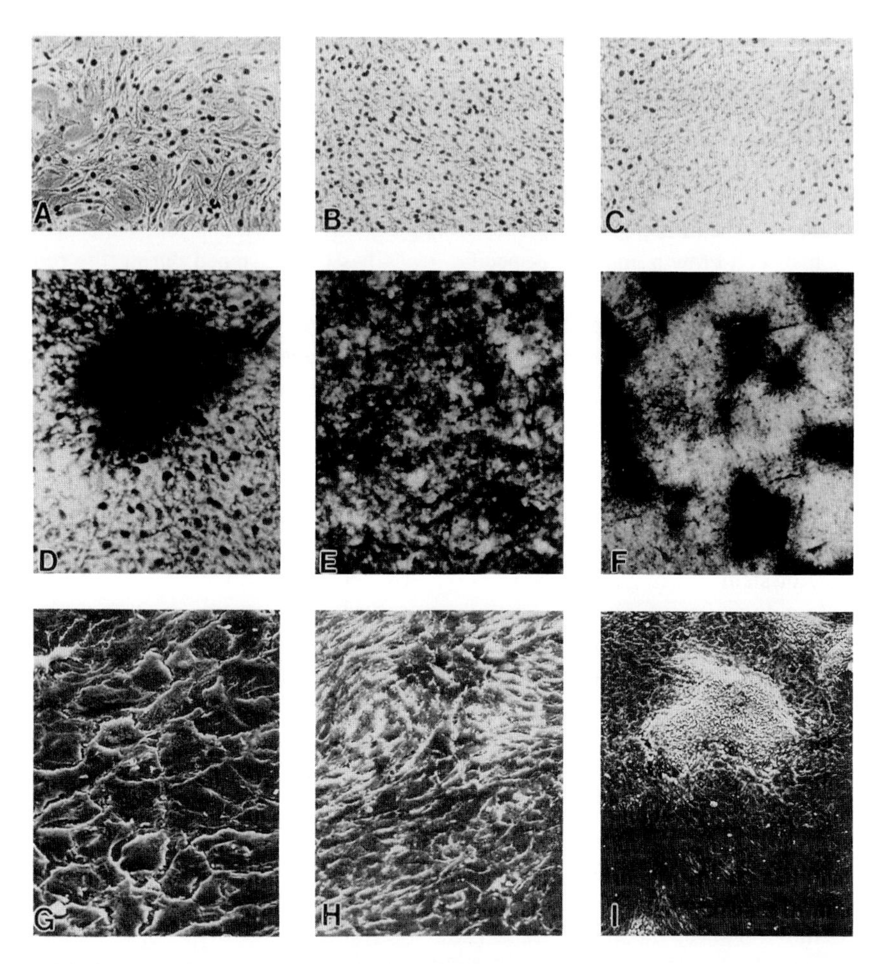

Fig. 5. Morphology of normal rat diploid osteoblast cultures. (A–F) Light microscopy of osteoblasts isolated from 21-day fetal calvarium bone by sequential collagenase digestion (Aronow *et al.*, 1990). (A–C) Cells labeled with tritiated thymidine show greater than 95% of the cells are proliferating after plating (A, Day 5; B, Day 9) at confluency. By Day 8, the cells begin to multilayer, forming nodules (C, Day 12) where cell proliferation ceases. The onset of expression of the osteoblast phenotype is indicated by positive staining of cells in the nodule for alkaline phosphatase (D, Day 12). By 2 weeks, every cell has ceased to proliferate, and greater than 90% of the cells are alkaline phosphatase-positive (E, Day 18). It is within this multilayered nodule that mineralization is initiated and, by Day 28 (F), von Kossa silver staining shows mineralized nodules throughout the dish. Scanning electron micrographs for periods representative of panels D, E, and F are shown in G, H, and I. A–F, photographed at 83×; G–I at 166×, 83×, 17×, respectively.

that influence expression of the osteoblast phenotype and to study the regulation of genes encoding osteoblast phenotype markers (Rodan and Noda, 1990). As has been demonstrated for the primary diploid osteoblasts, many of the tumor-derived osteoblastic cell lines have retained the principal properties exhibited by osteoblasts *in vivo,* including high alkaline phosphatase activity, growth-factor responsiveness, Type I collagen biosynthesis, responsiveness to hormones [1,25-$(OH)_2D_3$, parathyroid hormone, and glucocorticoids] as well as synthesis of bone-specific, noncollagenous proteins, such as osteocalcin. As such, both the diploid osteoblasts and osteosarcoma cell lines are valuable model systems; some, under the appropriate conditions, are competent to develop a mineralized extracellular matrix (Nishimoto *et al.,* 1987a). The ROS 17/2.8 cell line, which was isolated from a spontaneous rat tumor (Majeska *et al.,* 1978, 1980), expresses most genes that are expressed in mature osteoblasts, including osteocalcin. The UMR 106 line, which was isolated from a radiation-induced rat osteosarcoma, do not express osteocalcin, nor does an SV-40–transformed diploid osteoblast (Heath *et al.,* 1989), suggesting that they may represent less-differentiated osteoblasts. However, assignment of the developmental stage to a transformed osteoblast or osteosarcoma cell cannot be made definitively, since the deregulation of growth control uncouples the relationships between proliferation and differentiation that are operative in normal diploid cells. Furthermore, genes encoding osteoblast phenotype proteins that are expressed sequentially and independently during bone-cell differentiation (Owen *et al.,* 1990a) are expressed concomitantly following transformation (Majeska *et al.,* 1985; and Holthuis *et al.,* 1990).

Other *in vitro* models that have increased our understanding of the regulation of osteoblast growth and differentiation include cultured calvarial organ cultures (Canalis *et al.,* 1977) and outgrowth cultures from trabecular bone (Beresford *et al.,* 1984). These models support many of the *in vivo* phenotype properties and responses. The limitations are the quantity of cells available and, with the organ cultures, that the duration of cell viability is restricted.

3. A Temporal Pattern of Gene Expression during Development of the Osteoblast Phenotype

Expression of cell-growth genes and genes associated with the progressive development of the osteoblast phenotype is observed in cultured osteoblasts (Fig. 6). Initially, during the first 10–12 days following isolation of osteoblasts from calvaria, there is a period of active proliferation reflected by mitotic activity (Fig. 7) with expression of cell cycle (e.g., histone) and cell growth (e.g., c-*myc,* c-*fos* and AP-1 activity) regulated genes. These genes encode proteins that support proliferation by functioning as transactivation factors, in the case of c-*myc* and c-*fos,* and as proteins, which play the primary role in packaging newly replicated DNA into chromatin, in the case of histones. During this proliferation period, and

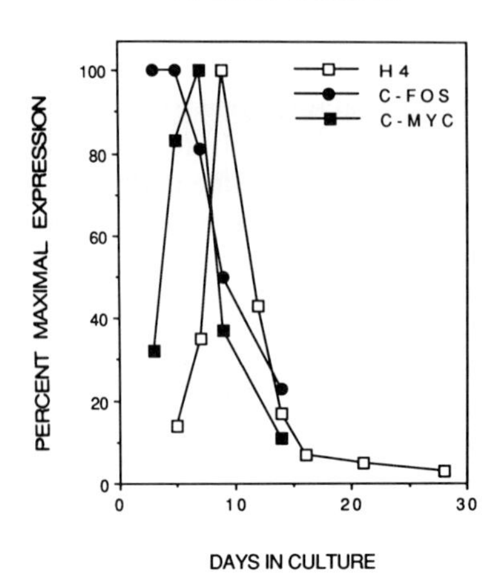

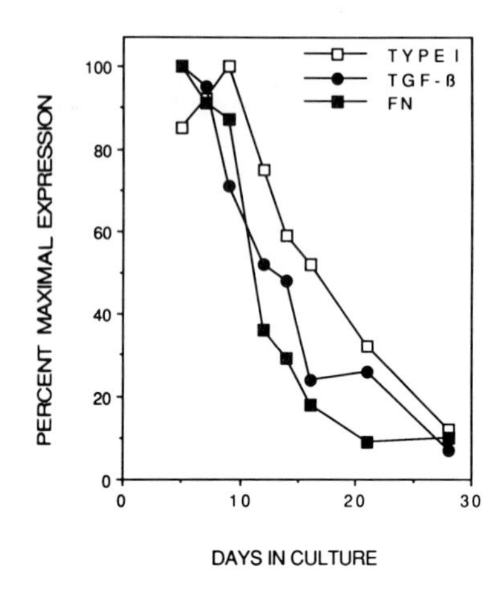

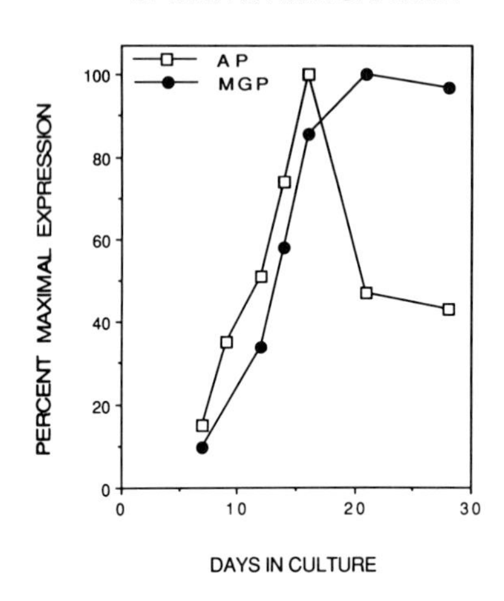

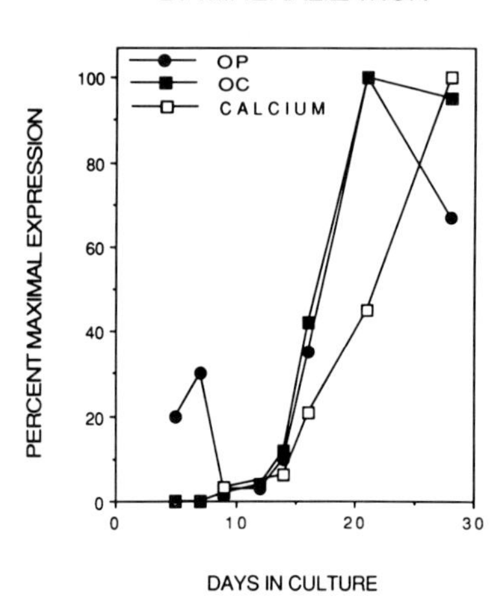

Fig. 6. Temporal expression of cell-growth and osteoblast-phenotype–related genes during the development of *in vitro* formed bone-like tissue by normal diploid rat osteoblasts. Temporal expression of cell growth (A), extracellular matrix (B), and osteoblast-phenotype–related genes (C, D) during the development of the osteoblast phenotype *in vitro.* Isolated primary cells were cultured after confluence in BGJb medium supplemented with 10% FCS, 50 μg/ml ascorbic acid and 10mM β-glycerol phosphate. Cellular RNA was isolated at the times indicated (3,5,7,10,12,14,16,20,28, and 35 days) during the differentiation time course and assayed for the steady-state levels of various transcripts by Northern blot analysis. The resulting blots were quantitated by scanning densitometry and the results, plotted relative to the maximal expression of each transcript. (A) Çell growth-related genes shown are H4 histone (reflects DNA synthesis), c-*myc* and c-*fos*. (B) Extracellular matrix-associated genes represented are Type I collagen, fibronectin (FN) and transforming growth factor β (TFG-β). (C) Genes associated with extracellular-matrix maturation shown are alkaline phosphatase (AP) and matrix Gla protein (MGP). (D) Genes induced with extracellular-matrix mineralization represented are osteopontin (OP), osteocalcin (OC), and calcium accumulation. Note the induction of alkaline phosphatase at the end of proliferation and the induction of osteocalcin and osteopontin with the onset of calcium deposition (Day 12).

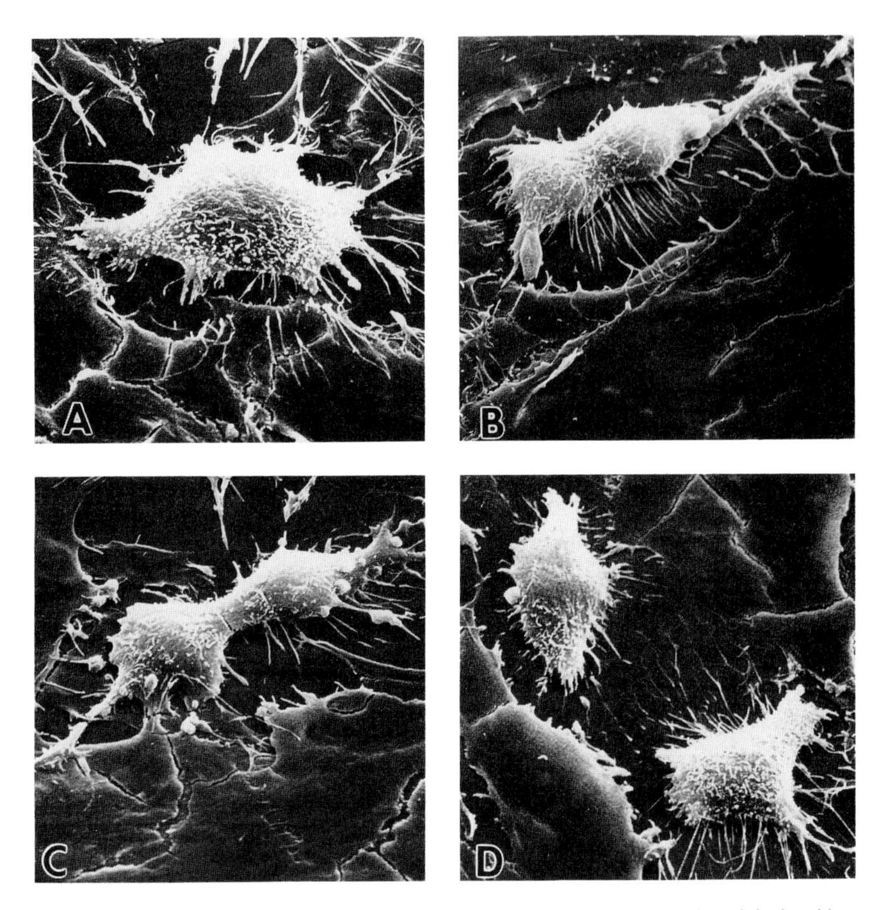

Fig. 7. Scanning electron microscopy of osteoblasts undergoing mitosis. Mitotic activity is evident in confluent cultures at Day 7, resulting in the formation of multilayered nodules. Panel A shows a rounded mitotic osteoblast over a monolayer of nondividing osteoblasts with a more flattened morphology. Panels B and C show osteoblasts completing a mitotic cleavage division, and Panel D shows two progeny cells immediately following mitosis. A–D magnification, approximately 665×.

fundamental to development of the bone-cell phenotype, several genes associated with formation of the extracellular matrix (Type I collagen, fibronectin, and TGFβ) are actively expressed. These genes are gradually down-regulated during subsequent stages of osteoblast differentiation. The parallel relationship of TGFβ with Type I collagen gene expression in cultured osteoblasts (Owen *et al.*, 1990a), as well as during endochondral bone formation *in vivo* (Carrington *et al.*, 1988; Bortell *et al.*, 1990; Nomura *et al.*, 1988; Lyons *et al.*, 1989), is consistent with a major role for TGFβ in regulating extracellular-matrix biosynthesis. It is primarily during the proliferation period that the activity of growth factors, their regulators, and the associated signal-transduction mechanisms influence the osteoblast parameters; this has recently been reviewed by Centrella *et al.*, (1988). Modifications of proliferation-related genes, e.g. c-*fos* expression *in vivo* (in transgenic animals), results in altered bone formation (Rüther *et al.*, 1987), reflecting the importance of expression during the proliferative period for control of cell growth, and for regulation of genes later during development of the osteoblast phenotype that directly involves bone-cell differentiation.

With the decline in DNA synthesis (^{3}H-thymidine incorporation and histone gene expression), the expression of alkaline phosphatase (enzyme activity and mRNA), a protein associated with the bone-cell phenotype, increases greater than 10-fold immediately following the down-regulation of proliferation (Fig. 5 and 6). During this period (from 12 to 18 days), the extracellular matrix undergoes a series of modifications in composition and organization that renders it competent for mineralization. Then, as the cultures progress into the mineralization stage, cellular levels of alkaline phosphatase mRNA decline. Two other bone-related genes, osteopontin (Nomura *et al.*, 1988; Oldberg *et al.*, 1986; Heinegård and Oldberg, 1989) and osteocalcin (Lian and Friedman, 1978; Lian *et al.*, 1985), exhibit a different pattern of expression. Osteopontin is also expressed during the period of active proliferation (at 25% of maximal levels), but reaches peak levels of expression during the mineralization period (days 16–20). This is not an unexpected result for two reasons. First, osteopontin expression during the proliferative period is consistent with the increased level of expression during the pre-replicative phase of the cell cycle following serum stimulation of quiescent fibroblasts and following oncogene transformation, or of phorbol ester treatment of fibroblasts (Craig *et al.*, 1989). Here one can speculate that the proliferation and tumorigenic-related function of osteopontin may be related to control of relationships between cells and extracellular matrices (Craig *et al.*, 1989). Indeed, the detection of osteopontin in the serum of both humans and rodents with metastatic tumors is consistent with a functional relationship of osteopontin to cell adhesion and/or to cell–cell interactions mediated by the extracellular matrix. Second, the induced expression of osteopontin coincident with mineralization may be related to physical properties of the protein. Osteopontin is a 60-Da acidic glycoprotein (Oldberg *et al.*, 1986) containing O-phosphoserine, thereby possessing several putative calcium-binding sites that are known to be important for cell proliferation and also for mineralization of the extracellular matrix in bone (Glimcher, 1989). It therefore appears that expression of the osteopontin gene early and late in the osteoblast developmental sequence may be mediated by alternative regulatory mechanisms. The lower levels of osteopontin mRNA observed during the proliferation period may in part reflect mRNA transcribed *in vivo* in osteoblasts undergoing matrix mineralization before the isolation from fetal calvaria. Additionally, recent results suggest that osteopontin may be preferentially localized on the surface of the mineralized bone extracellular matrix in the region of osteoclast contact, implicating this phosphoprotein in the resorption process (Reinholt *et al.,* 1990). The general structural and biological properties of noncollagenous proteins associated with the bone-cell extracellular matrix has recently been reviewed by Heinegård and Oldberg (1989) and Boskey (1989).

Yet another category of genes expressed during the osteoblast developmental sequence are represented by the vitamin K–dependent calcium-binding proteins, osteocalcin (Hauschka *et al.,* 1975) or bone Gla protein (Price *et al.,* 1976) a 5- to 7-kDa protein that binds tightly to hydroxyapatite, and the matrix Gla protein (Fraser and Price, 1988). Both proteins are characterized by γ-carboxyglutamic acid residues, and both are vitamin D–responsive proteins. While osteocalcin is bone specific, interestingly, matrix Gla protein is not found exclusively in osteoblasts, but is also abundant in chondrocytes and in several nonskeletal tissues with extensive extracellular matrices (e.g., lung and kidney) (Fraser and Price, 1988). Osteocalcin is not detectable during the proliferation period (prior to day 12) in culture and does not reach a significant level of expression until 16 to 20 days after isolation, when expression increases coordinately with total mineral accumulation. In contrast, the matrix Gla protein is initially expressed in proliferating osteoblasts and exhibits an increased expression immediately postproliferatively, which is sustained independent of mineralization (Barone *et al.,* 1991). Osteocalcin is positively regulated by vitamin D, but the addition of hormone to the cultures during proliferation will not induce osteocalcin gene expression (Owen *et al.,* 1991). Several manipulations of the culture system (Owen *et al.,* 1990a) demonstrate that like osteopontin, its induction of high levels of mRNA are dependent upon formation of a mineralized extracellular matrix. Osteocalcin has been shown to contribute to regulation of the mineral phase in bone both *in vitro,* as a potential inhibitor of mineral nucleation (Boskey, 1985; Romberg *et al.,* 1986), and *in vivo,* as a bone-matrix signal that promotes osteoclast differentiation and activation (Glowacki and Lian, 1987). Thus its expression late in the osteoblast developmental sequence indicates osteocalcin as a marker of the mature osteoblast and is consistent with a possible role for osteocalcin synthesized and bound to mineral and participating in the coupling of bone formation to bone resorption. Expression of these genes in a sequential fashion reflects development of the tissue organization, which can readily be seen by determining cellular levels of mRNA transcribed from genes for cell growth (e.g., H4) and a mature osteoblast phenotype marker (e.g., osteocalcin) at the single-cell level by *in situ* hybridization (Fig. 8).

The observed temporal expression of osteoblast phenotype properties during the developmental sequence is a reflection of functional activities necessary for the progressive formation of bone tissue. The expression of alkaline phosphatase mRNA and enzyme activity before the initiation of osteoblast mineralization suggests that alkaline phosphatase may be involved in preparation of the extracellular matrix for the ordered deposition of mineral, and that the coexpression of

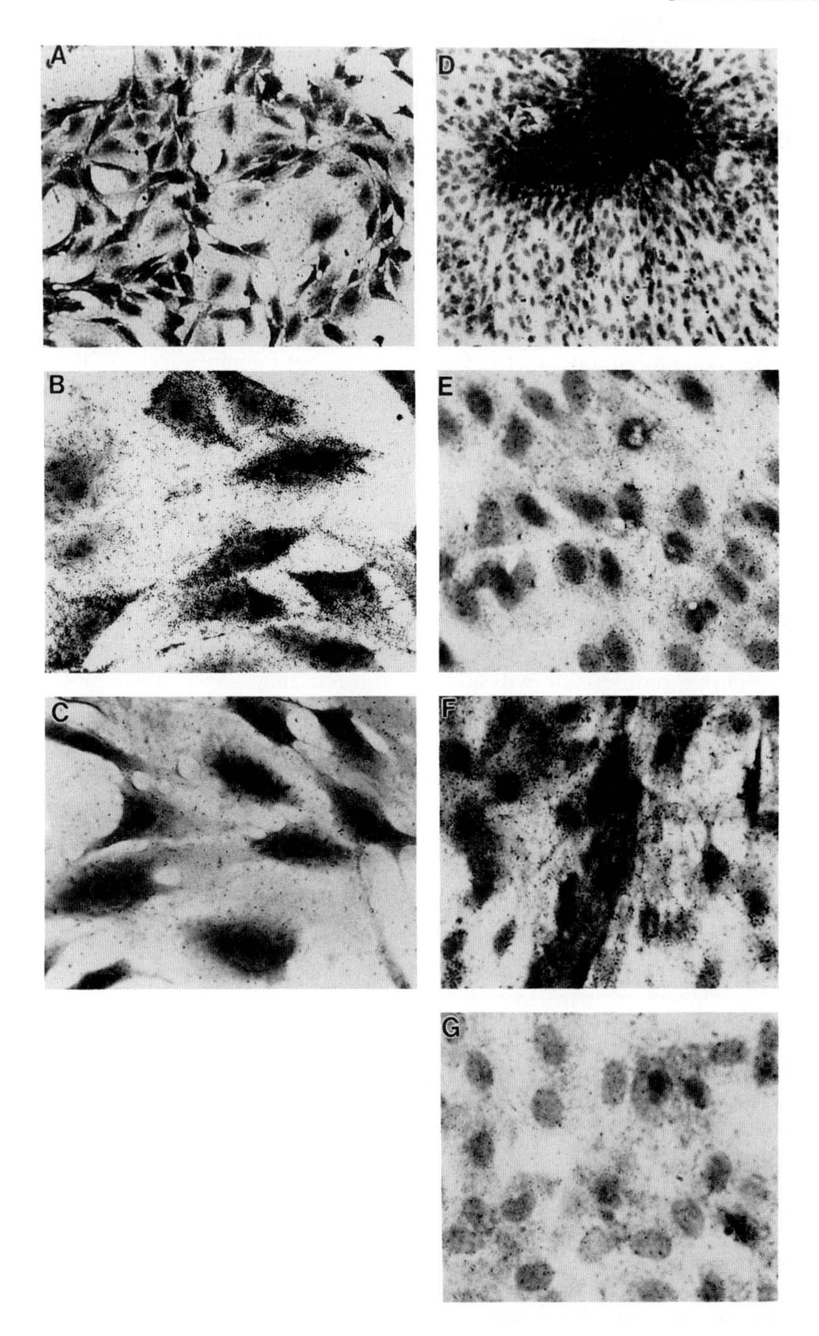
A
B
C
D
E
F
G

other genes such as osteocalcin and osteopontin may support the onset and progression of extracellular matrix mineralization. Alternatively, the induction of these mineralization-associated genes may reflect an acquisition of osteoblast properties necessary for signaling bone turnover *in vivo.* Taken together, the patterns of expression of these genes and the synthesis of the encoded proteins, determined biochemically and by histochemical staining, demonstrate that a temporal sequence of gene expression exists during the culture period, associated with development of the extracellular matrix, and reflects maturation of the osteoblast phenotype *in vitro.*

The biological relevance of the temporal expression of osteoblast parameters *in vitro* to bone formation *in vivo* is demonstrated by the fidelity of the tissue-like organization developed at the completion of the *in vitro* osteoblast developmental sequence. This is reflected by intense von Kossa silver staining of the mineralized nodules, indicating hydroxyapatite deposition (Fig. 5). The bone tissue-like organization in these cultures is further supported by comparison of the ultrastructure of the mineralized regions of the culture (Fig. 9). Sections through an intact 21-day fetal rat calvarium exhibit a similar ordered deposition of crystals within and between the orthogonally organized bundles of collagen fibrils. No evidence for cell necrosis or intracellular calcification is indicated in the cultures, particularly where mineralized matrix has enveloped the osteoblasts. The *in vivo* relevance of the temporal pattern of gene expression observed *in vitro* is further documented by a pattern of gene expression analogous to that of the mature osteocyte in osseous tissue. Alkaline phosphatase and collagen mRNA levels are almost nondetectable, and osteopontin and osteocalcin levels have declined; in heavily mineralized mature cultures (after day 35), *in vivo* alkaline phosphatase histochemistry and active collagen biosynthesis are not associated with mature osteocytes. *In situ* hybridization of osteoblast genes indicates much more reduced activity in mature bone than that found in osteoblasts at the mineralizing front.

Fig. 8. *In situ* hybridization of a proliferation marker gene (histone H4) and a differentiation-specific gene (osteocalcin) in rat osteoblast cultures. The distribution of cells selectively expressing histone and osteocalcin genes are shown in relation to the development and organization of the bone-like tissue organization. Panels A–C represent Day 7 (proliferation period) and Panels D–F represent Day 27 (mineralization period). Top row (A, D) shows overall organization of cells and tissue (82×). Panel B (322×) shows that on Day 7 50% of cells are labeled with H4 histone probe, consistent with all cells proliferating. An absence of detectable histone mRNA occurs at later times as shown in E (345×) demonstrating nonproliferating cells in the internodular region on Day 27. Panel C (322×) shows the absence of osteocalcin on Day 7, and the appearance of osteocalcin restricted to only the more mature osteoblast in an organized nodule (Panel F, 338×) while Panel G (338×) shows the absence of osteocalcin in the internodule areas on Day 27.

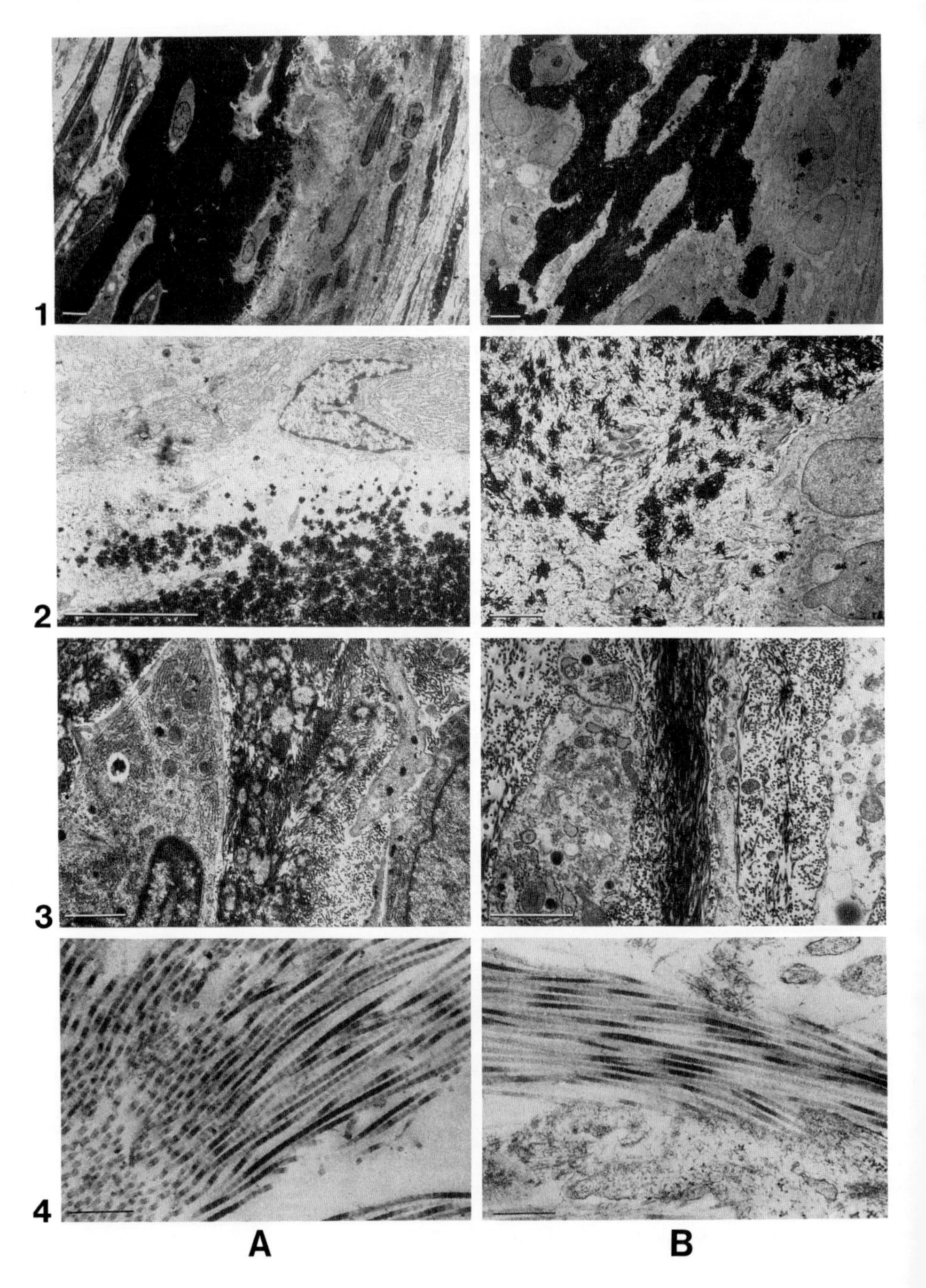
1
2
3
4
A
B

4. A Model and Supporting Evidence for a Functional Relationship of Cell Growth to Expression of the Osteoblast Phenotype

By combining ^{3}H-thymidine labeling and *in situ* autoradiography with alkaline phosphatase histochemistry, it has been possible to directly establish the relationship between proliferation and initiation of tissue-specific gene expression at the single-cell level during the osteoblast developmental sequence (Owen *et al.*, 1990a). As seen in Fig. 5, it is apparent that proliferation initially ceases in the discrete multilayered foci that form throughout the osteoblast cultures. It is these cells in the multilayered nodules that first express alkaline phosphatase. This is in contrast to proliferating cells in the internodular regions of the cultures, where alkaline phophatase activity is not observed until several days later when the entire cultures consist of multilayered non-proliferating cells.

These results confirm, on an individual-cell basis, that a temporal sequence of gene expression occurs and that at least some events (proliferation and alkaline phosphatase expression) appear to be sequential, mutually exclusive events in the same cell; i.e., proliferation must be down-regulated before the expression of alkaline phosphatase. This transition from a proliferating cell to one that can express an early marker of the osteoblast phenotype (alkaline phosphatase) represents the first restriction point at which cessation of proliferation appears to be required for initiation of tissue-specific gene expression associated with the distinctive characteristic features of bone—formation of the mineralized extracellular matrix.

Another molecular marker for the transition point between completion of proliferation and the onset of gene expression characteristic of the osteoblast phenotype (e.g., alkaline phosphatase) is expression of cell cycle-independent histone genes that encode high-molecular-weight poly (A)$^+$ mRNAs (Shalhoub *et al.*, 1989) (Fig. 10). Expression of poly (A)$^+$ histone mRNAs at the onset of tissue-specific gene expression has similarly been observed during monocytic differentiation in HL-60 promyelocytic leukemia cells (Stein *et al.*, 1989a) as well as during adipocyte differentiation (Fig. 10). The encoded histone proteins may be involved in the remodeling of chromatin architecture that occurs following the completion of proliferation, and may be necessary to support expression of genes

Fig. 9. Ultrastructural organization of fetal bone tissue (A) compared to mineralized nodule formed in a 35-day osteoblast culture (B). Row 1 shows general tissue organization of cells in the mineralized bone–extracellular matrix enveloping the osteoblast. Row 2 shows early stages of mineral deposition associated with collagen fibrils and the absence of intracellular calcification. Row 3 shows identical orthogonal organization of the collagen matrix in both the intact calvaria and in the mineralized nodule formed *in vitro.* For this micrograph in Column A, a demineralized calvarium was sectioned. Row 4 shows higher magnification of the collagen bundles with mineral deposition within the collagen fibrils. Bars in the lower micrograph represent 2.5 μm. Others represent 5 μm.

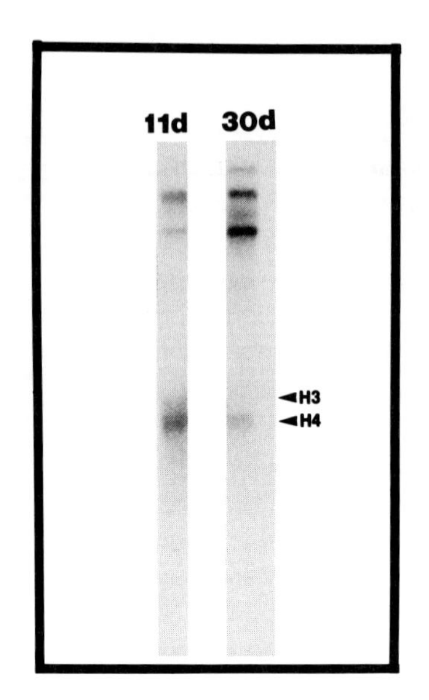
11d
30d
H3
H4

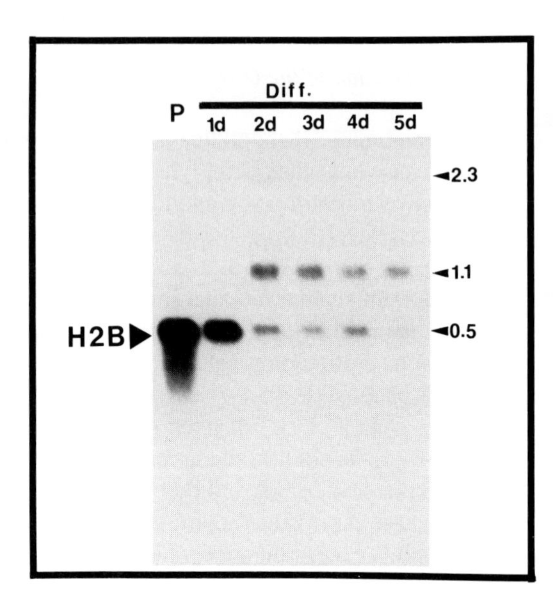
Diff.
P
1d
2d
3d
4d
5d
2.3
1.1
0.5
H2B

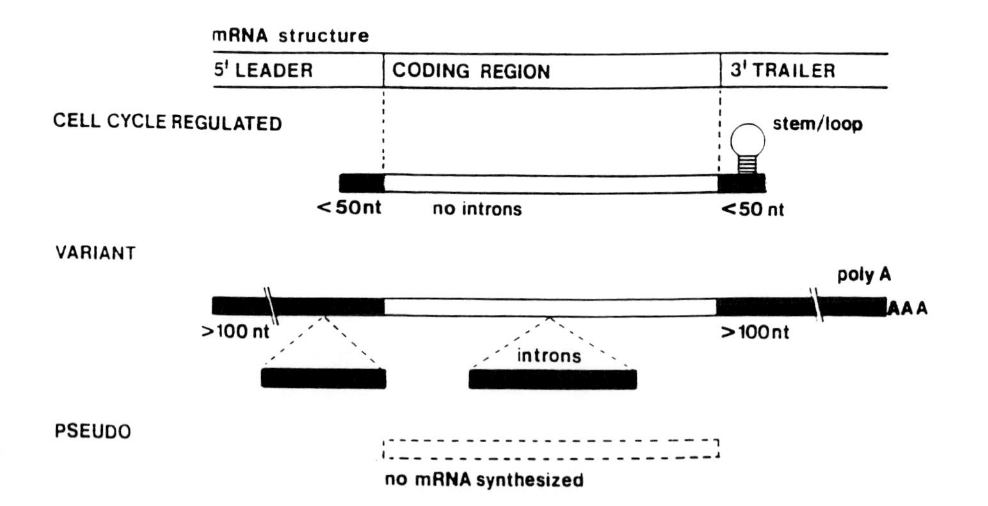
mRNA structure
5' LEADER
CODING REGION
3' TRAILER
CELL CYCLE REGULATED
stem/loop
<50nt
no introns
<50 nt
VARIANT
poly A
AAA
>100 nt
>100nt
introns
PSEUDO
no mRNA synthesized

associated with the development of tissue-specific phenotypic properties and the specialized functions of the differentiated cells.

Molecular, biochemical, and histochemical parameters are consistent with a reciprocal and functional relationship between proliferation and a sequential development of the osteoblast phenotype, which is schematically illustrated in Fig. 11. The progressive and interdependent series of biochemical events that characterizes the osteoblast developmental sequence reflects the selective expression, initially of genes encoding cell growth and extracellular matrix proteins, and subsequently, a series of tissue-specific genes (Owen *et al.,* 1990a, b). Such modifications in the temporal pattern of gene expression, reflected by both the activation and repression (e.g., suppression of phenotypic markers, alkaline phosphatase, and osteocalcin in proliferating osteoblasts), as well as by the extent to which specific genes are expressed, are the basis for proposing a developmental sequence with three distinct periods: proliferation, extracellular matrix maturation, and mineralization, and two principal transition points in the osteoblast developmental sequence where important regulatory signals may be required for the progressive expression of the bone-cell phenotype to proceed. The first of these points is when proliferation is down-regulated and gene expression associated with extracellular-matrix maturation is initiated, and the second is the transition at the onset of mineralization. Molecular mechanisms operative at the two principal transition points are described in Chapter 10.

Schematically illustrated in Fig. 11 are the interdependent relationships between proliferation, production of the extracellular matrix, and the progressive expression of genes reflecting osteoblast differentiation. A mechanism that may account for exclusion of expression of a subset of phenotypic genes (e.g., AP and OC)

Fig. 10. Appearance of the differentiation-specific histone gene in cultures of normal diploid rat osteoblasts and HL60 cells treated with TPA (12-*O*-tetradecanoylphorbol-13-acetate). Northern blot analysis of RNA from osteoblasts (left panel, top) or poly $(A)^+$ RNA from HL60 cells (right panel, top) were fractionated electrophoretically in 1.5% agarose–6% formaldehyde gels, transferred to nitrocellulose and hybridized to cloned genomic DNA containing an H3 and H4 histone mRNA coding sequence (osteoblast RNA) or hybridized to a 340 nucleotide ^{32}P-labeled probe from the cloned protein coding segment of a DNA replication-independent H2B histone gene (Collart *et al.,* 1991) (HL60 cells). The high-molecular-weight histone mRNAs represent transcripts from the differentiation-specific histone genes, which are expressed reciprocally with the low-molecular-weight mRNAs transcribed by the cell cycle-regulated histone genes. Lower panel illustrates schematically the three classes of histone genes: (1) cell cycle-regulated histone genes (expressed during proliferation), which encode mRNA that lacks introns and poly$(A)^+$ sites, containing nontranslated leader and trailer sequences of less than 50 nucleotides and terminated in a region of hyphenated dyad symmetry, which forms a stem-loop structure involved in processing; and (2) histone genes, which exhibit elevated levels of expression during the shut-down of proliferation and onset of differentiation. These genes encode mRNAs *containing introns,* polyadenylation sites, and extensive nontranslated leader and trailer sequences. A third class of histone sequences encode pseudogenes that are not transcribed.

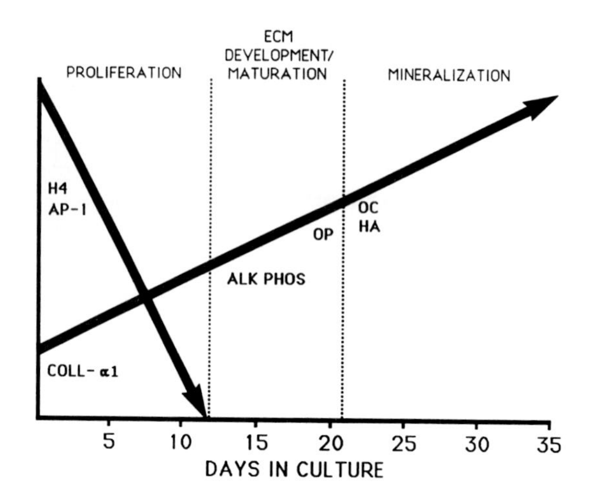

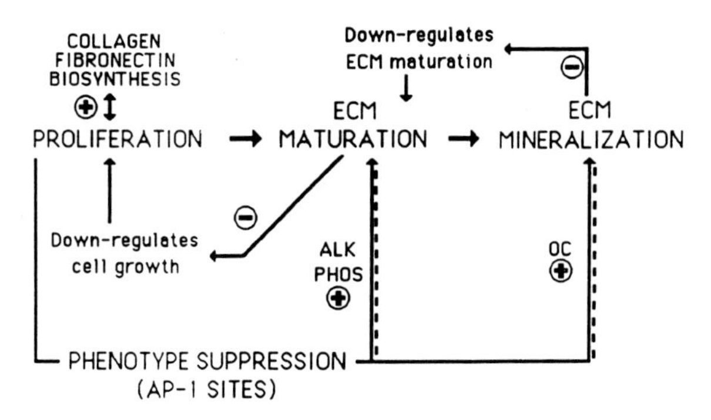

Fig. 11. Model of the relationship between proliferation and differentiation during the rat osteoblast developmental sequence. This relationship is schematically illustrated within the context of modifications in expression of cell cycle and cell growth-regulated genes, as well as genes associated with the maturation, development, and mineralization of the osteoblast extracellular matrix. The three principal periods of the osteoblast developmental sequence are designated within broken vertical lines (proliferation, matrix development and maturation, and mineralization). Commitment periods and restriction points indicated were experimentally established. A functional relationship between the down-regulation of proliferation and the initiation of extracellular-matrix development and maturation is based on stimulation of alkaline phosphatase and osteopontin gene expression when proliferation is inhibited, but the developmental sequence is induced only to the second transition point. Growth of the osteoblast under conditions that do not support mineralization confirms the Day 20 restriction point, since the developmental sequence proceeds through the proliferation and the extracellular-matrix development and maturation periods, but not further. The lower panel illustrates a series of signaling mechanisms whereby the proliferation period supports the synthesis of a type I collagen–fibronectin ECM, which continues to mature and mineralize. The formation of this matrix down-regulates proliferation, and matrix mineralization down-regulates the expression of genes associated with the formation–maturation period. The occupancy of the AP-1 sites in the OC box and VDRE of the osteocalcin gene and the alkaline phosphatase VDRE by Fos–Jun and/or related proteins suppresses the basal and vitamin D–induced expression of the alkaline phosphatase (Alk Phos) and osteocalcin (OC) genes before the initiation of basal expression.

during the proliferation period, but not collagen, is the binding of the Fos–Jun protein complex to AP-1 sites that reside in the 5′ regulatory elements in these genes (Owen *et al.,* 1990c; Lian *et al.,* 1991). Occupancy of AP-1 sites by the Fos–Jun complex can provide a general mechanism for *phenotype suppression* of genes expressed postproliferatively, which then become selectively activated by a gene-specific mechanism at appropriate times during the osteoblast developmental sequence. See Chapter 10 for details of this postulated mechanism.

Central to understanding the molecular and cellular mechanisms that mediate relationships between events occurring during the progressive development of the osteoblast phenotype is to distinguish the extent to which specific gene expression is the component of a temporal sequence and the extent to which specific genes are functionally coupled, i.e., causally related. In particular, the manner in which proliferation supports expression of certain osteoblast-related genes, while the completion of proliferative activity is required for expression of other osteoblast phenotype markers, is necessary for functionally defining growth–differentiation interactions within the context of bone-cell differentiation. This relationship between cell growth and expression of osteoblast-related genes can be addressed experimentally (schematically illustrated in Fig. 12) by abbreviating or extending the proliferation period of the osteoblast developmental sequence and determining which genes associated with extracellular-matrix biosynthesis, maturation, or mineralization are prematurely up-regulated or delayed in initiation of expression. The growth–differentiation relationship can be further approached in studies in which the onset and progression of differentiation are systematically modified, and observing consequent changes in both expression of genes during proliferation and the influence on proliferative activity. Taken together, these approaches will lead to a more comprehensive appreciation of the distinction between functionally coupled events and temporal events associated with the growth–differentiation relationship that are operative during development of the osteoblast phenotype. Several examples of this experimental approach follow.

One direct demonstration that the down-regulation of proliferation induces the expression of some genes (normally expressed later in the osteoblast developmental sequence) is derived from experiments that establish that inhibition of DNA synthesis in actively proliferating osteoblasts results in a rapid and selective down-regulation of cell-growth genes (Owen *et al.,* 1990a) (Fig. 13). This is paralleled by a fourfold increase in alkaline phosphatase mRNA levels, indicating that the premature down-regulation of proliferation induces the expression of an early marker for the extracellular-matrix maturation period of the osteoblast developmental sequence. Increased expression of alkaline phosphatase with decreased proliferative activity has similarly been observed in ROS 17/2.8 cells (Majeska *et al.,* 1985). With inhibition of DNA synthesis, levels of osteopontin mRNA also increase to levels that approximate those present during the extracellular-matrix mineralization period of cultured osteoblasts. These data suggest

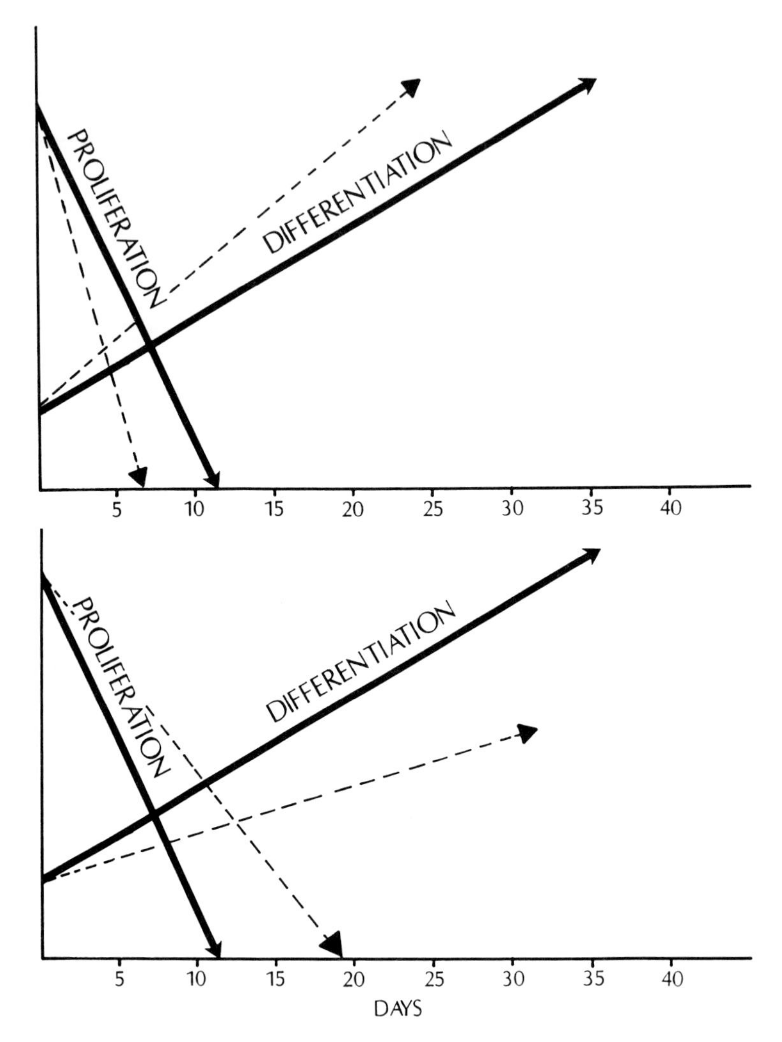

Fig. 12. Illustration of experimental approaches to address the growth and differentiation relationships. The top profile illustrates an experimental approach to assess the influence of proliferation on expression of genes associated with osteoblast differentiation by determining consequences of growth inhibition on differentiation parameters, i.e., establishing which are accelerated. This will facilitate distinguishing changes in the levels of expression of osteoblast (differentiation) genes that are coupled to the proliferative state of the cell versus those genes that become temporally expressed, dependent on the formation and organization of the extracellular matrix, following the completion of proliferative activity. The lower panel illustrates the potential consequences of extending the proliferation period or delaying the onset of the differentiation. In this way, one can further establish which differentiation-related genes are expressed, dependent upon the shut-down of proliferation, i.e., which proliferation-related events are dependent on formation of the bone extracellular matrix.

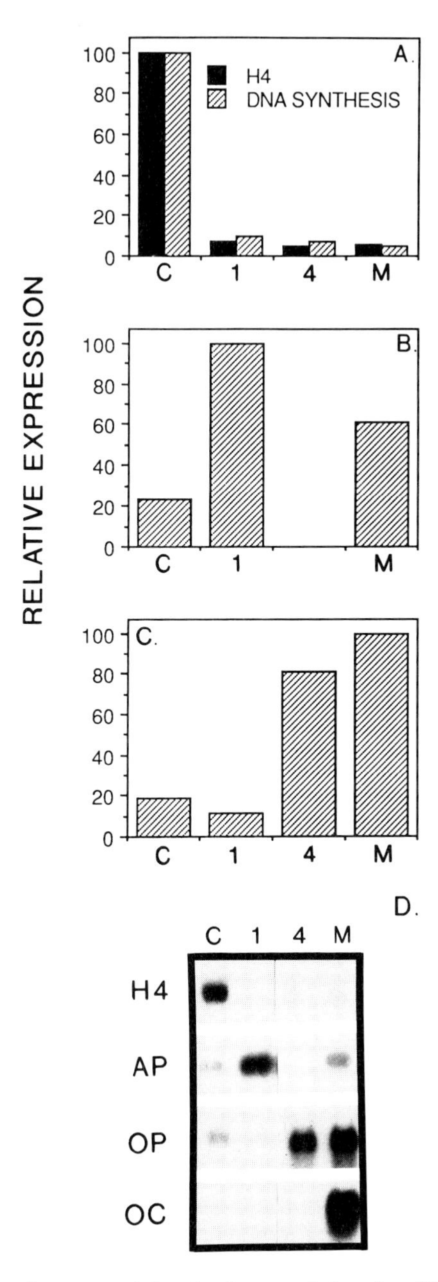

Fig. 13. Expression of genes coupled to the down-regulation of proliferation. Proliferation was inhibited in actively growing osteoblasts (Day 5) by addition of 5 m*M* hydroxyurea. Cells were harvested at 1 and 4 hr., assayed for DNA synthesis and cellular levels of H4 histone mRNA (A), alkaline phosphatase mRNA (AP) (B) and osteopontin mRNA (OP) (C). Hydroxyurea-treated cells were compared to two controls, the nonhydroxyurea-treated cells (C, Day 5), and cells from mineralized cultures (M, 30 days after plating). Both alkaline phosphatase and osteopontin, but not osteocalcin (OC), are induced following inhibition of proliferation by hydroxyurea; the induction is to a level similar to that found in mature cells expressing the phenotypic markers at their maximal levels (D).

a direct functional coupling of the down-regulation of proliferation at the first transition point early during the osteoblast developmental sequence, with the preferential expression of genes normally induced in cells with a mature extracellular matrix. However, osteocalcin, which is expressed in osteoblasts late during he period of extracellular-matrix mineralization, is not induced by simply inhibiting proliferative activity (Owen *et al.,* 1990a). The absence of osteocalcin induction is consistent with the concept that there is at least a second set of genes, whose expression is not directly coupled to the down-regulation of proliferation, but rather to development of the more-differentiated osteoblasts in a mineralized matrix. Thus, inhibition of proliferation supports expression of genes that are expressed only during progression of the developmental sequence, up to the stage where mineralization is initiated. These experiments in which premature differentiation has been promoted by inhibition of proliferation reveal a second transition point, the onset of mineralization. Mineral deposition may be reqauired to signal expression of a subset of osteoblast-phenotype genes, such as osteocalcin. Additionally, other genes expressed earlier than the onset of mineralization, which may be required to render the matrix competent for mineralization, may also not be induced by inhibition of proliferation.

The biological significance of the second transition point in the osteoblast developmental sequence at the onset of mineralization is further supported by studies that directly demonstrate a relationship between mineralization and the sequential expression of genes during the progressive development of the osteoblast phenotype. When cells are maintained under conditions that support extracellular-matrix mineralization, osteocalcin and osteopontin mRNA and biosynthesis increase steadily beginning at day 15, in parallel with calcium accumulation in the cell layer (Fig. 14). In contrast, when cultures are maintained under nonmineralizing conditions, calcium does not begin to accumulate in the cell layer until approximately day 25. At this time, osteocalcin gene expression occurs, and higher levels of osteopontin mRNA are induced. Notably, the normal down-regulation of alkaline phosphatase observed in mineralizing cultures is not as marked. The presence or absence of mineralization has no effect on the expression of genes occurring during the proliferative period, on passage through the first transition point, or the onset of alkaline phosphatase expression. These experiments provide additional evidence to support the existence of the second transition point, since the cells can progress through the proliferation and extracellular-matrix maturation sequence to the onset of mineralization, but cannot initiate expression of genes related to the mineralization stage unless mineral accumulation occurs. Genes such as osteocalcin are not only temporally expressed late in the osteoblast developmental sequence, but are also *coupled* to deposition of hydroxyapatite.

An example of an experimental approach to assess the consequences of accelerated differentiation on expression of cell growth and tissue-specific genes related

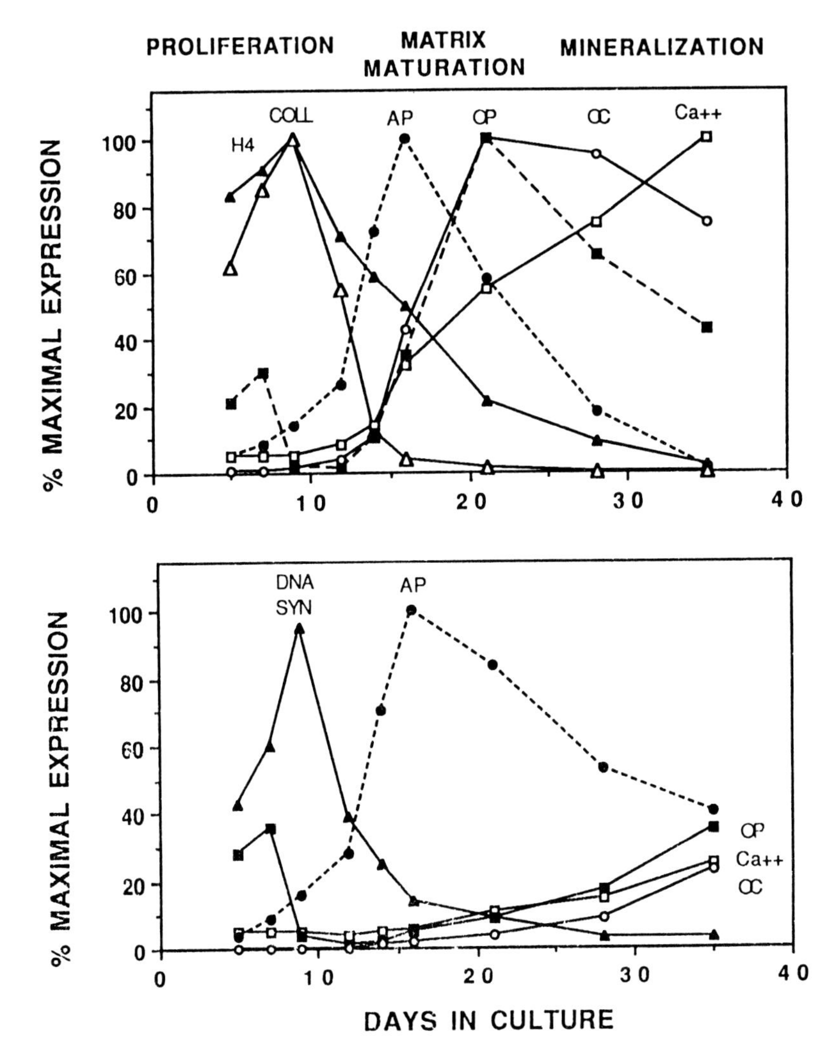

Fig. 14. Effect of delayed differentiation (mineralization) on expression of osteoblast phenotype markers. Top panel shows, in primary diploid rat osteoblast cultures, the expression of cell growth histone (H4) and phenotype marker genes [collagen (COLL), alkaline phosphatase (AP), osteopontin (OP), osteocalcin (OC), and calcium (Ca^{2+})] under normal differentiation conditions and, in the lower panel, expression of genes under conditions where the onset of mineralization is delayed by deleting 10 m*M* β-glycerol phosphate from the culture medium (BGJb supplemented with 10% FCS serum and 50 mg/ml ascorbic acid). This delay in the onset of mineralization does not affect proliferation period (DNA synthesis) or the onset of matrix maturation, as indicated by AP expression. However, with the delay of the onset of mineralization, AP expression is not as rapidly down-regulated, and OP and OC are not induced until calcium deposition occurs after Day 20.

to development of the osteoblast phenotype is chronic treatment of osteoblast cultures with dexamethasone, a synthetic potent glucocorticoid. In subcultivated cells, the effect is even more pronounced, in that the initiation of the osteoblast developmental sequence in the absence of 10^{-7} *M* dexamethasone is significantly delayed compared to progression of the developmental sequence in primary osteoblast-cultured cells (Aronow *et al.*, 1990; Wong *et al.*, 1990). Glucocorticoids maintain the extent of expression of differentiation parameters to that found in primary cell cultures. Here one observes an increase in the number of bone nodules (Fig. 15), paralleled by an increase in alkaline phosphatase activity and earlier induction of osteocalcin, a marker of the mature osteoblast phenotype. Thus expression of these phenotypic markers reflects the normal temporal sequence, but over a shorter period. Whether some of the relationships between genes that are coupled or functionally related have been altered owing to direct modulation by glococorticoids, or whether the accelerated differentiation is secondary to other modifications in osteoblast developmental parameters, remains to be established. However, the presence of glucocorticoid-responsive elements in several genes expressed during osteoblast growth and differentiation and the effects of glococorticoids on messenger RNA stability are consistent with the observed modifications in physiological control of the osteoblast phenotype.

Although unquestionably a simplification of an extremely complex series of biological interactions, the temporal pattern of expression suggests a working model for the relationship between growth and differentiation, whereby genes involved in the production and deposition of the extracellular matrix must be expressed during the proliferative period for differentiation to occur. One can postulate that proliferation is functionally related to the synthesis of a bone-specific extracellular matrix and that the maturation and organization of the extracellular matrix contributes to the shutdown of proliferation, which then promotes expression of genes that render the matrix competent for mineralization, a final process essential for complete expression of the mature osteoblast phenotype (Fig. 11). The onset of extracellular-matrix mineralization and/or events early during the mineralization period may be responsible for the down-regulation of genes expressed during extracellular-matrix maturation and organization. Clearly, in this model the development of an extracellular matrix is integrally related to the differentiation stages. Numerous studies have shown modifications of osteoblast phenotype properties (summarized in Aronow *et al.*, 1990 and Owen *et al.*, 1990a) under a variety of conditions that promote or alter extracellular-matrix biosynthesis and organization. This working model provides a basis for addressing whether particular stages of osteoblast differentiation exhibit selective responsiveness to actions of hormones and other physiologic factors that influence osteoblast activity, as well as other questions related to the molecular mechanisms associated with bone formation.

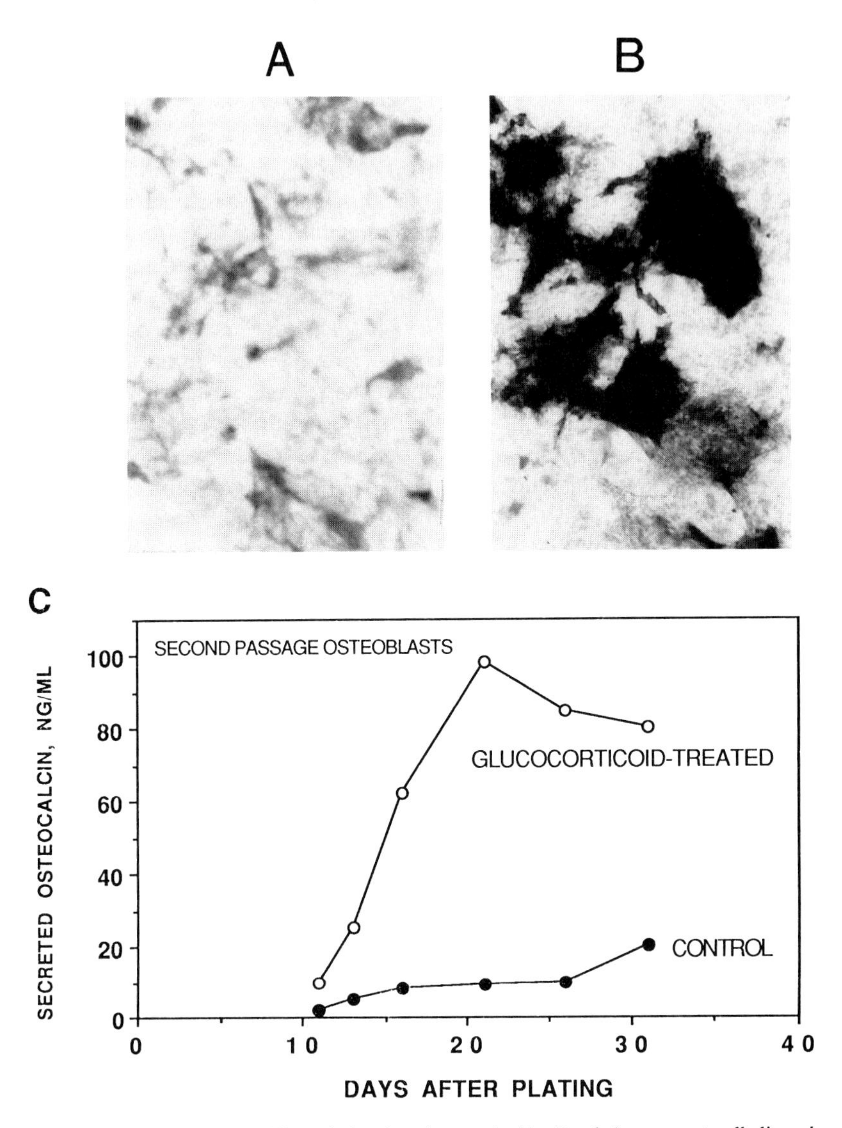

Fig. 15. Acceleration of differentiation by glucocorticoids. Panel A represents alkaline phosphatase histochemical staining of rat osteoblast cultures, second-passage cells, Day 10 compared to 10^{-7} *M* dexamethasone–treated cultures initiated from Day 3 after plating (Panel B). Note the formation of nodules in B as compared to the diffuse and weak alkaline phosphatase staining of the nontreated cultures. Panel C shows the increased synthesis of osteocalcin, a marker of a mature osteoblast phenotype [measured by radioimmunoassay (RIA) of secreted protein in media], in the dexamethasone-treated cultures compared to the untreated, subcultivated cells that did not develop nodules.

C. Cell Structure and Regulation of Osteoblast Gene Expression

It is becoming increasingly apparent that cell structure is intimately involved in the development of the osteoblast phenotype from several standpoints. First, the organization of the extracellular matrix is necessary to support the ordered deposition of hydroxyapatite and possibly the binding and/or localization of a series of proteins associated with extracellular-matrix mineralization (e.g., osteocalcin, osteopontin). Additionally, the changes in cell shape that are observed during the progressive differentiation of the osteoblast are striking. In cultured osteoblasts, a transition in the cellular morphology coincides with, and appears to be related to, the multilayering of cells during the initial stages of nodule formation. A series of analogous modifications in overall cell morphology accompanies osteoblast differentiation in skeletal tissue. Then, with the more advanced stages of extracellular-matrix mineralization, both in cultured osteoblasts and *in vivo,* extended cellular processes throughout the extracellular matrix support the nutriation of the bone mineral and may additionally facilitate cell–cell communication via the exchange of informational macromolecules. The intracellular architecture also contributes significantly to the structural and functional properties of the osteoblast, both the modifications that accompany developmental changes, and those required by the mature osteoblast for maintenance of cellular function in a mineralized extracellular matrix.

Unquestionably, the cytoarchitecture (cytoskeleton and associated macromolecules) and the nuclear architecture (nuclear matrix) are elements that modulate overall cellular morphology. Particularly exciting are several lines of evidence that implicate the cytoarchitecture and the nuclear architecture of the osteoblast in the regulation of gene expression that supports both phenotype development and maintenance of bone-cell function. Recent results are consistent with involvement of the nuclear matrix in transcriptional-level control through mechanisms related to gene localization and concentration of transcription factors within the nucleus. Modifications in the cytoskeleton during osteoblast differentiation have been observed (Egan *et al.,* 1991a, b), suggesting a potentially important role of the cytoskeleton in posttranscriptional regulation by mediating mRNA stability and/or translation. Support for such a cytoskeleton-mediated regulatory mechanism is provided by evidence for the cytoskeleton association of mRNA on both free and membrane-bound polyribosomes (Zambetti *et al.,* 1987), the asymmetric distribution of mRNAs within the cytoplasm (Lawrence *et al.,* 1988), and results that directly demonstrate that mRNA stability is dramatically modified when the subcellular localization within the cytoplasm of mRNAs is modified (Zambetti *et al.,* 1987).

1. The Extracellular Matrix and Osteoblast Gene Expression

The contribution of collagenous extracellular matrix formation to changes in cell structure, osteoblast differentiation, and gene expression has been effectively addressed in cultured osteoblasts. First, with formation of the collagen matrix, changes in cell shape are observed, with the cuboidal osteoblasts becoming markedly smaller as the collagen extracellular matrix accumulates and organizes (Bellows *et al.,* 1986; Aronow *et al.,* 1990). Osteoblasts *in vitro* do not form a contact-inhibited monolayer; rather, they form multilayered nodules, with each cell surrounded by a highly ordered collagen extracellular matrix (Bellows *et al.,* 1986; Aronow *et al.,* 1990; Bhargava, 1988; Owen *et al.,* 1990a) (Fig. 9). The differentiation pathway of the cuboidal osteoblast synthesizing matrix to the mature osteocyte, with long processes for communication and nutriation as the matrix mineralized, is highly dependent on collagen. Shown in Fig. 16 are the effects of ascorbic acid, required for collagen synthesis, on the osteoblast developmental sequence (Owen, *et al.,* 1990a). Formation of the collagenous extracellular matrix contributes to the shut-down of proliferation requisite for the induction of genes characteristic of the mature osteoblast phenotype. This relationship is clearly evident from results of these studies in which cells were cultured at various concentrations of ascorbic acid. With higher levels of collagen synthesis and accumulation in the extracellular matrix, proliferation ceases at a lower cell density; and, coordinately, alkaline phosphatase mRNA levels and enzyme activity per cell are greater. Thus there is a contribution of signals from the extracellular matrix that promotes the progressive differentiation of the osteoblast. This expression of genes for bone cell-phenotype proteins is significantly dampened with decreased synthesis and organization of the extracellular matrix.

Consistent with a coupled relationship between cell growth, extracellular-matrix biosynthesis and development of the osteoblast phenotype is provided by results from studies in which growth arrest is initiated early during the proliferation period and maintained subsequently by chronic treatment with DNA-synthesis inhibitor, hydroxyurea. As indicated in Fig. 16, the cell growth–extracellular matrix–biosynthesis relationship is reflected by inihibition of proliferation, resulting in cessation of collagen biosynthesis and accumulation. The involvement of the extracellular matrix in promotion of gene expression, associated with the subsequent development of the osteoblast phenotype, is reflected by the low level of alkaline phosphatase activity and the absence of osteocalcin gene expression, coupled with no evidence for extracellular-matrix mineralization. Further experimental results that support the active involvement of the collagen extracellular matrix in the progressive expression of the osteoblast phenotype are shown in Fig. 17, where chronic vitamin D treatment is initiated early during the osteoblast developmental sequence. The resulting inhibition of both collagen biosynthesis and proliferation reflects the cell growth–extracellular matrix–biosynthesis relationship with the absence of subsequent expression of genes related to extracellular-matrix maturation, organization, and mineralization. Fig. 18 clearly demonstrates the inhibition of organization and formation of the nodules in calvarial-derived osteoblast cul-

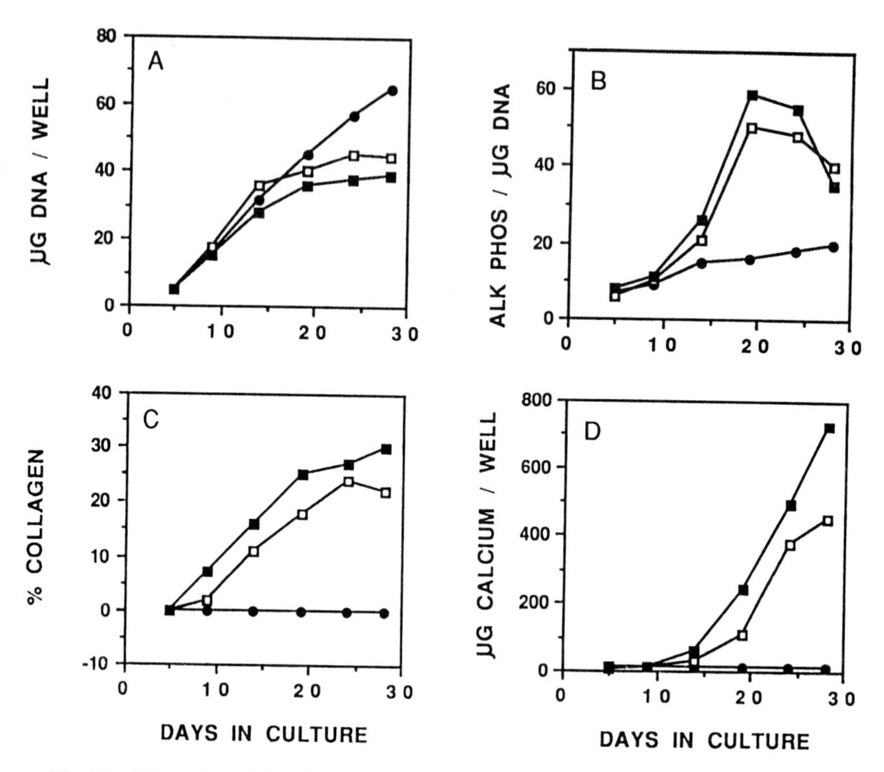

Fig. 16. Effect of ascorbic acid and collagen accumulation on osteoblast proliferation (cell number) and differentiation parameters (collagen content, alkaline phosphatase activity, and calcium deposition). Cells were maintained during a 30-day culture period, first in MEM medium (until Day 7) and then in BGJb medium supplemented with 10% FCS, 10m*M* β-glycerol phosphate and containing 0 (•), 25 (□), or 50 (■) µg/ml ascorbic acid. Values plotted represent the mean of three sample wells per time point. (A) DNA was determined by a fluorometric assay. (B) Alkaline phosphatase is expressed as n*M* *p*-nitrophenol (released from the substrate *p*-nitrophenol phosphate) per microgram DNA. (C) Percentage collagen accumulated in the extracellular matrix was determined from the hydroxyproline concentration in total amino acid analysis of the cell layers. (D) Calcium was determined by atomic absorption spectroscopy in the same samples hydrolyzed for the amino acid analysis. With increasing ascorbic acid concentrations, the cells reached confluency at a lower density, and there was an increase in collagen accumulation associated with a parallel increase in alkaline phosphatase activity. In the absence of a collagen matrix, no accumulation of calcium occurred, reflecting the absence of mineralization.

tures when vitamin D treatment is initiated late during the proliferative period. When vitamin D treatment is initiated during the osteoblast developmental sequence, following the onset of mineralization, nodule growth is arrested. This further confirms the involvement of the osteoblast Type I collagen extracellular matrix not only in initiation of extracellular matrix maturation and mineralization, but equally important, in supporting the progression and completion of mine-

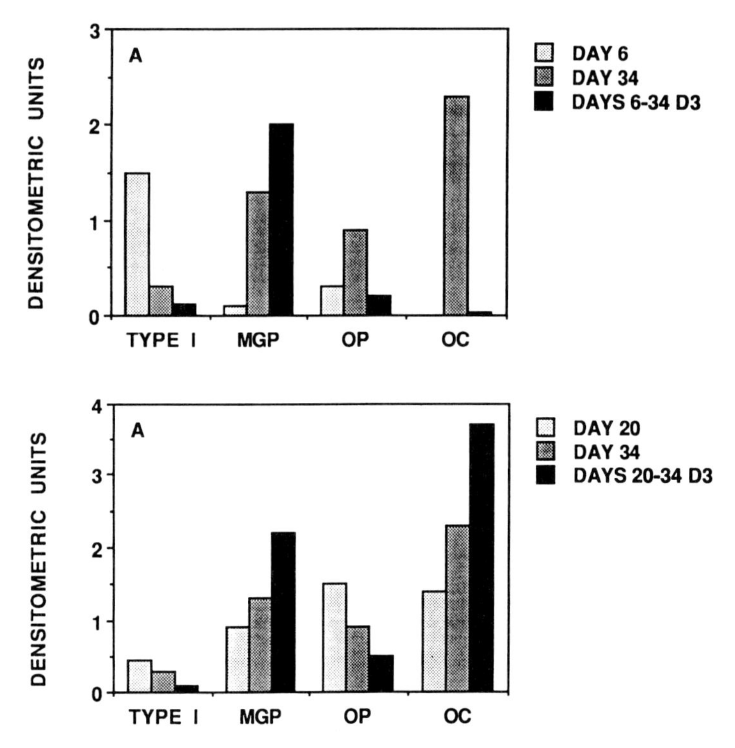

Fig. 17. Inhibition of Type I collagen gene expression by vitamin D and subsequent arrest of differentiation in rate osteoblast cultures. Cellular levels of mRNA transcribed from vitamin D–regulated genes after chronic treatment of the rat osteoblast cultures with $10^{-8}M$ 1,25-$(OH)_2D_3$ was initiated from Day 6 (Panel A) or from Day 20 (Panel B) of the developmental sequence. Two control values, the day of initiation of hormone at either Day 6 or Day 20 (gray bars) and the last day (Day 34, hatch bar) of the experiment, are compared to the hormone-treated cultures to Day 34 (solid bar). Note partial inhibition of osteopontin (OP) by chronic vitamin D initiated at either time (Panel A and B) and, for osteocalcin (OC), the absence of its expression on Day 6 in control, and the inability of vitamin D treatment to induce its expression from Day 6 to 34 vitamin D–treated cultures in panel A, compared to OC peak expression in Day 34 control. In contrast, OC is stimulated by vitamin D under chronic conditions from Days 20 to 34 (Panel B) after the formation of a mineralized matrix. MGP is stimulated by hormone during both periods.

ralization and other related events that accompany expression of the mature osteoblast phenotype.

A complementary experimental approach substantiating the supportive and/or inductive effect of the collagen extracellular matrix on osteoblast differentiation is culturing primary diploid osteoblasts on Type I collagen film or in a collagen gel (Fig. 19). Here we observe an accelerated progression of the development of the osteoblast phenotype by early and enhanced expression of alkaline phosphatase activity.

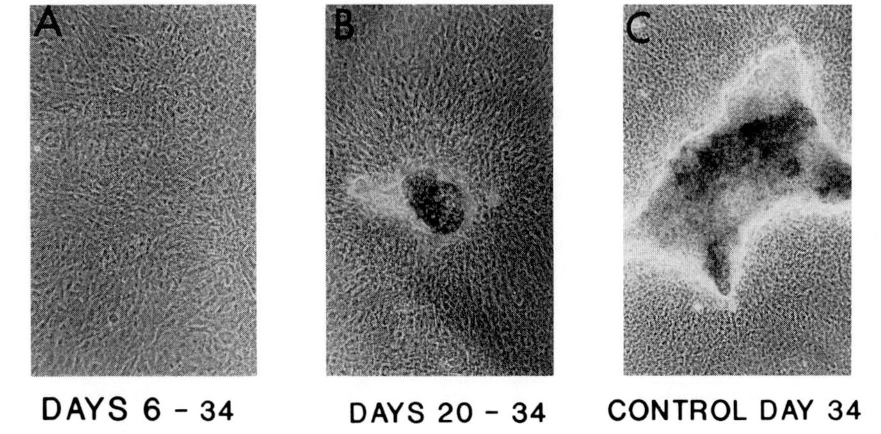

Fig. 18. Microscopy of vitamin D–treated osteoblast cultures. Phase-contrast micrographs of rat osteoblast cultures show the inhibition of nodule formation (Panel A) and nodule growth (Panel B) by intitiation of chronic hormone treatment from day 6 (A) or day 20–34 (B) compared to control cultues shown in Panel C.

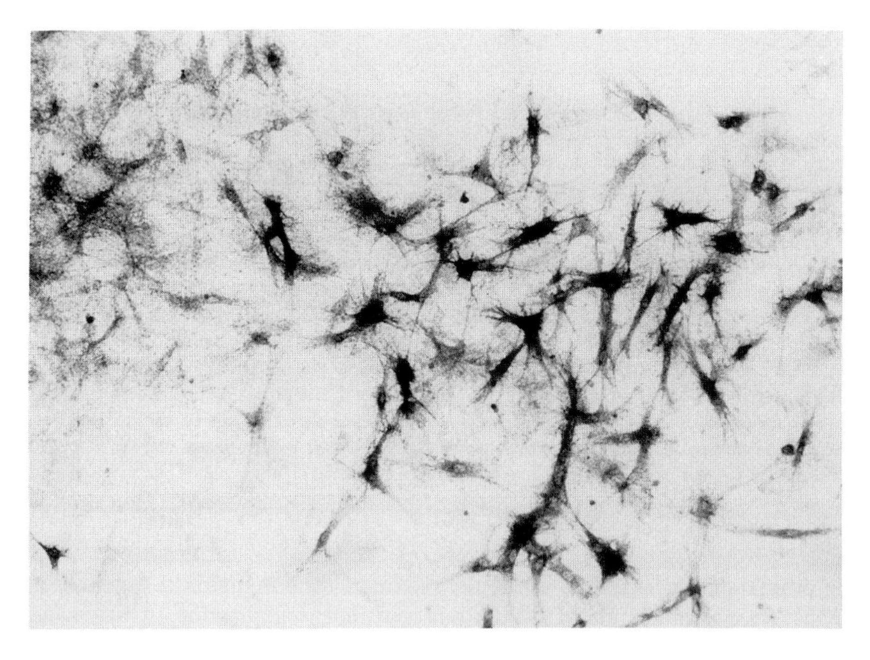

Fig. 19. Promotion of osteoblast differentiation by collagen matrix. Fetal calvaria-derived rat osteoblasts cultured on a Type I collagen gel compared to control culture plate. Histochemical staining for alkaline phosphatase demonstrates the ability of the collagen ECM to promote expression of the differentiated phenotype marker. Activity of alkaline phophatase–DNA determined histochemically confirms collagen-induced differentiation. Note the more osteocytic-like appearance of the cells with extensive cellular processes.

2. The Nuclear Matrix and Transcriptional Control

A key question to be addressed in order to understand regulation of gene expression at the trasnscriptional level is how, with a low representation of promoter-binding factors and sequence-specific regulatory elements in the nucleus, a threshold concentration can be achieved to initiate RNA synthesis. Here the answers may, at least in part, reside in the three-dimensional organization of the cell nucleus, where the nuclear matrix exhibits properties consistent with a functional involvement in the control of transcription.

The nuclear matrix is operationally defined as the proteinaceous nuclear substructure that resists both nuclease digestion and salt extractions. While considerable controversy exists regarding the extent to which various nuclear-matrix preparations retain *in vivo* fidelity of composition and organization (Fey and Penman, 1988), the existence of the nuclear matrix as a network of polymorphic anastomosing filaments within the nucleus is undeniable. Indications of a role for the nuclear matrix in the regulation of gene expression include sites for DNA replication (Pardoll *et al.,* 1980); preferential association with actively transcribed genes (Nelkin *et al.,* 1980; Robinson *et al.,* 1982; Schaack *et al.,* 1990; Stief *et al.,* 1989; Zenk *et al.,* 1990); association with HN-RNA (van Eeklen and van Venrooij, 1981); RNA synthesis at fixed transcriptional complexes; pre-mRNA splicing (Zeitlin *et al.,* 1987); and specific association of steroid receptors with the nuclear matrix (Barrack and Coffey, 1983; Kumara-Siri *et al.,* 1986). Involvement of the nuclear matrix in the regulation of cell- and tissue-specific gene expression is further suggested by recent demonstrations of variations in the nuclear-matrix protein composition of different cells and tissues (Capco *et al.,* 1982; Fey *et al.,* 1984; He *et al.,* 1990; Fey *et al.,* 1986; Marashi *et al.,* 1982).

Additional support for participation of the nuclear matrix in transcriptional control is provided by the results presented in Fig. 20, which displays the two-dimensional electrophoretic profiles of nuclear-matrix proteins analyzed throughout the osteoblast developmental sequence. It is evident that changes in the protein composition of the nuclear matrix parallel sequential expression of genes during the progressive of the osteoblast phenotype. The composition of the nuclear matrix is constant within each of the three principal periods of osteoblast differentiation, but is modified dramatically at the two key transition points: at the completion of the proliferation period, and at the onset of extracellular-matrix mineralization. This relationship between nuclear-matrix protein composition and expression of specific genes is further supported by retention of the characteristic stage-specific representation of nuclear-matrix proteins when the osteoblast developmental sequence is delayed (Dworetzky *et al.,* 1990).

More direct evidence linking the nuclear matrix with the regulation of gene expression is provided by the recent demonstration that the histone gene is associated with the nuclear matrix only during the proliferative period of the osteoblast

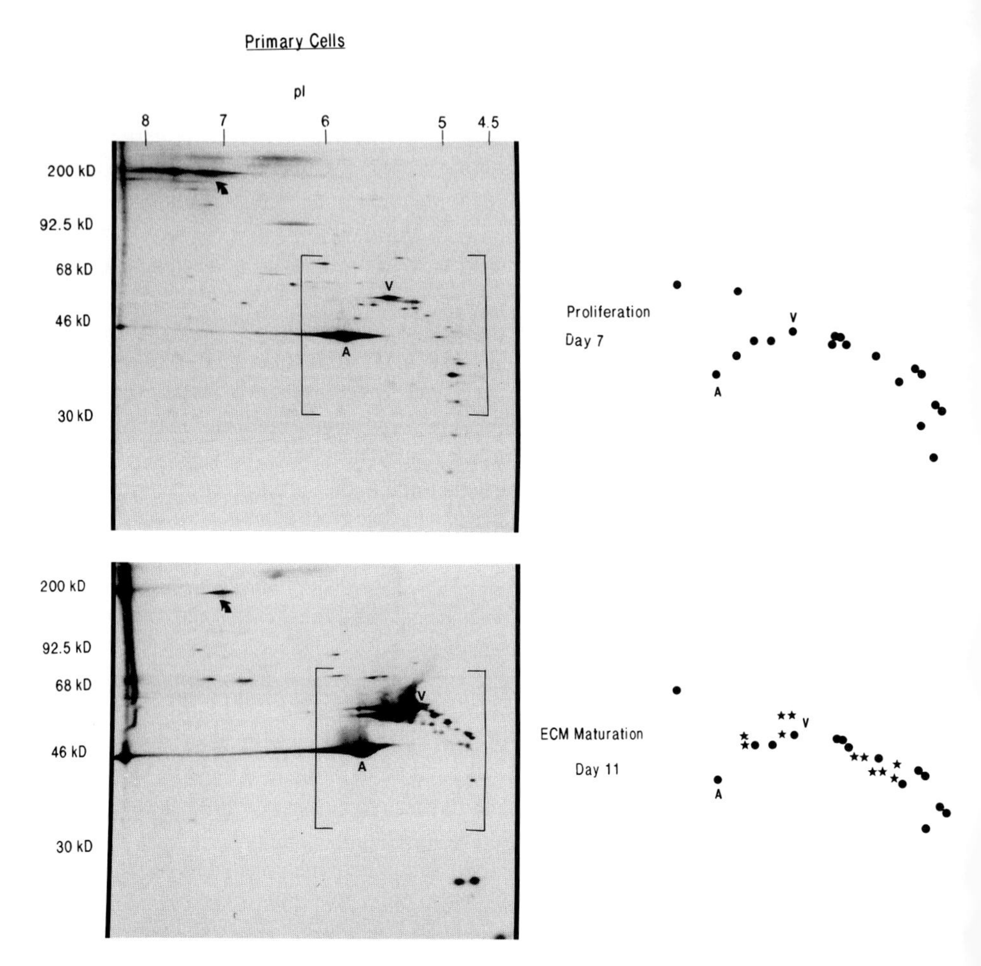

Fig. 20. Progressive changes in protein composition of the nuclear matrix during osteoblast differentiation. The fluorographs represent two-dimensional gel electrophoresis of ^{35}S-methionine–labeled, nuclear-matrix proteins isolated from Days 7, 11, 15, and 23 primary rat osteoblast cultures. Schematically illustrated in the bracketed region are stage-specific changes that occur in the nuclear-matrix composition. The arrow points to a large 190 kD protein that completely disappears as the cells differentiate. Symbols represent proteins that appear on that particular day. •, Day 7; ★, Day 11,⊗, Day 15, ❍, Day 23. The symbols are removed with the given protein is no longer present.

developmental sequence, when actively transcribed, and that the H4 histone gene distal promoter–binding factor NMP-1 (Figs. 21–26) is a component of the nuclear matrix (Dworetzky, *et al.*, in press). Results presented in Fig. 21, where a sequential series of radiolabeled segments of an H4 histone gene promoter were assayed for sequence-specific binding to nuclear-matrix proteins, provide an initial indication that a 141-bp element resides between −589 and −730 upstream from

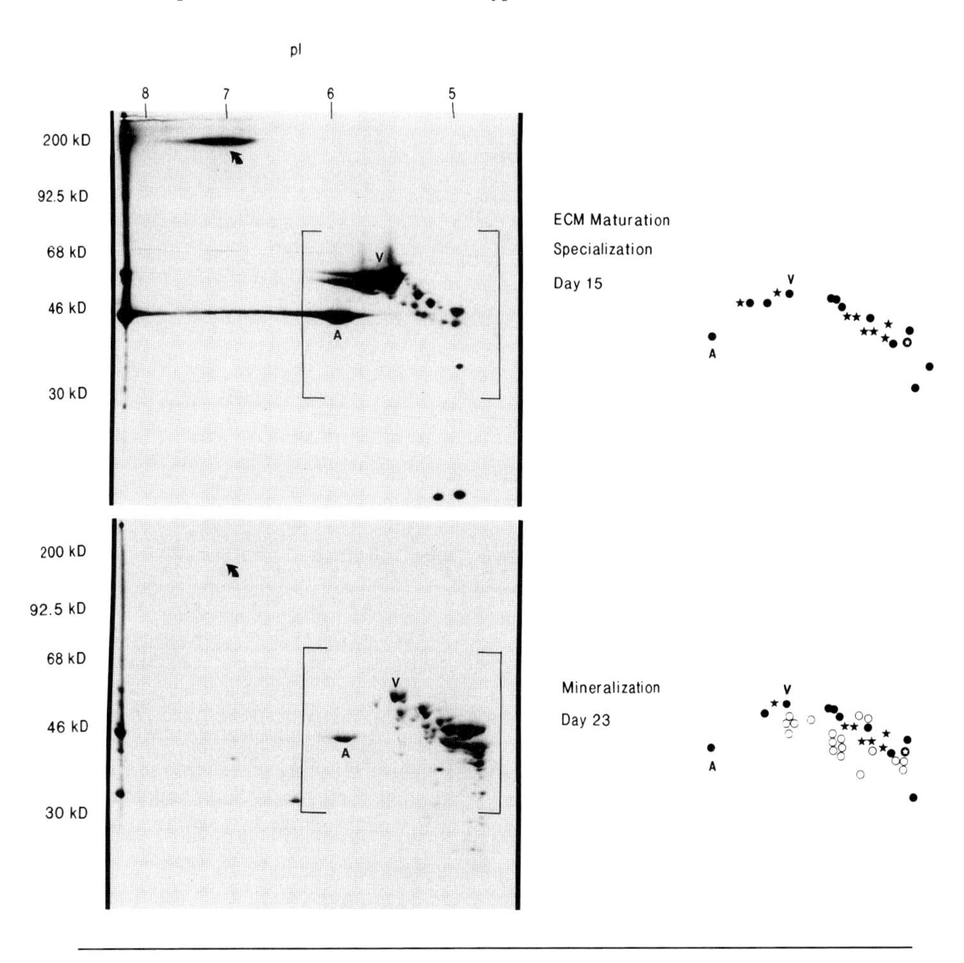

the cap site, which supports sequence-specific interactions of nuclear-matrix proteins with a histone gene-regulatory sequence. These sequence-specific protein–DNA interactions were confirmed and further localized by gel mobility shift analysis (Fig. 22) and established at single nucleotide resolution by copper-phenanthroline (OP-CU) footprint analysis (Fig. 23A), dimethyl sulfate DMS protection (Fig. 23B), and competition analysis (Fig. 23C). The representation of the NMP-1 nuclear-matrix protein in transcription factor extracts by gel mobility shift analysis is shown in Fig. 24 and characterized with respect to molecular weight by ultraviolet crosslinking studies (Fig. 25) and DNA-binding–site recognition by NMP-1 oligonucleotide affinity chromatography. Further support for involvement of the nuclear matrix in transcriptional control during development of the bone-cell phenotype is provided by our recent observation that a homologous nuclear-matrix attachment site resides in the osteocalcin gene promoter, and

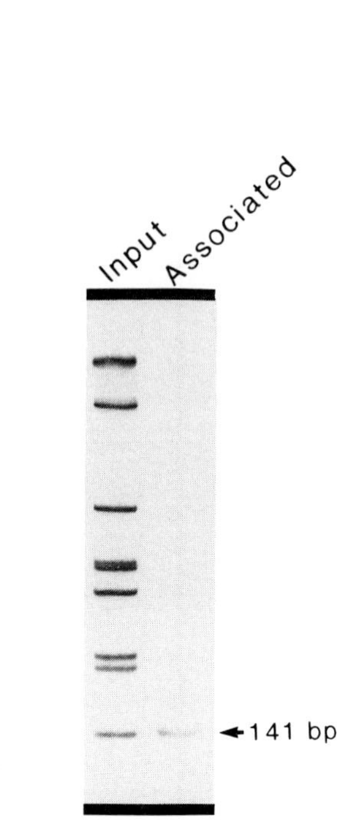

Fig. 21. Sequence-specific interactions of the nuclear-matrix protein NMP-1 with a regulatory element in the H4 histone gene promoter. An *in vitro* matrix-associated region was identified within the H4 histone gene promoter. The *input* lane represents electrophoretically fractionated, ^{32}P-labeled segments of an H4 histone gene promoter. A mixture of radiolabeled cloned histone gene promoter segments and pUC19 fragments were incubated in the presence of unlabeled E. coli competitor DNA and a nuclear-matrix preparation. After isolation of the protein–DNA complex followed by deproteinization, a specific 141-bp DNA fragment is seen to be preferentially associated with the nuclear matrix (associated late). This fragment is located between −589 and −730 upstream from the cap site.

that the sequence-specific DNA-binding proteins are associated with the nuclear matrix only when the osteocalcin gene is expressed (J. Bidwell, J. Stein, J. Lian and G. Stein, unpublished results). These results are consistent with a role for the nuclear matrix in the concentration and localization of both actively transcribed genes and transcription factors. This may explain how, with low representation of factors and sequence-specific regulatory elements in the nucleus, a threshold to initiate transcription can be obtained. In a broader biological context, these results provide a basis for understanding transcriptional control mediated by the nuclear architecture.

3. *Transduction and Integration of Cellular Signals for Regulating Osteoblast Proliferation and Differentiation*

Understanding control of the osteoblast developmental sequence necessitates defining a complex series of regulatory mechanisms mediated by the integration of both extracellular and intracellular signaling pathways. These modulate expression

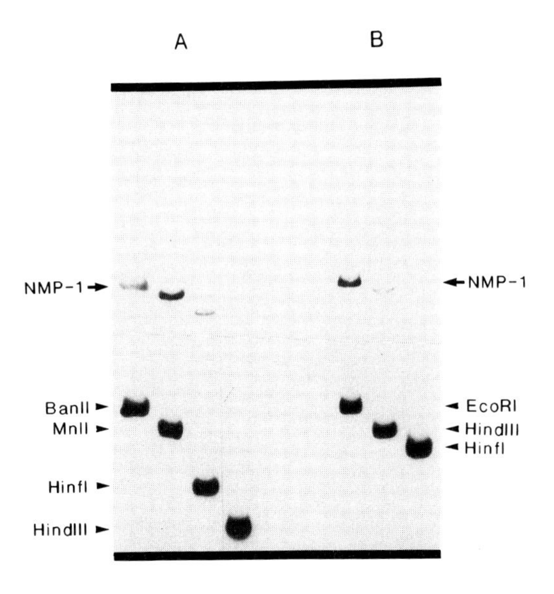

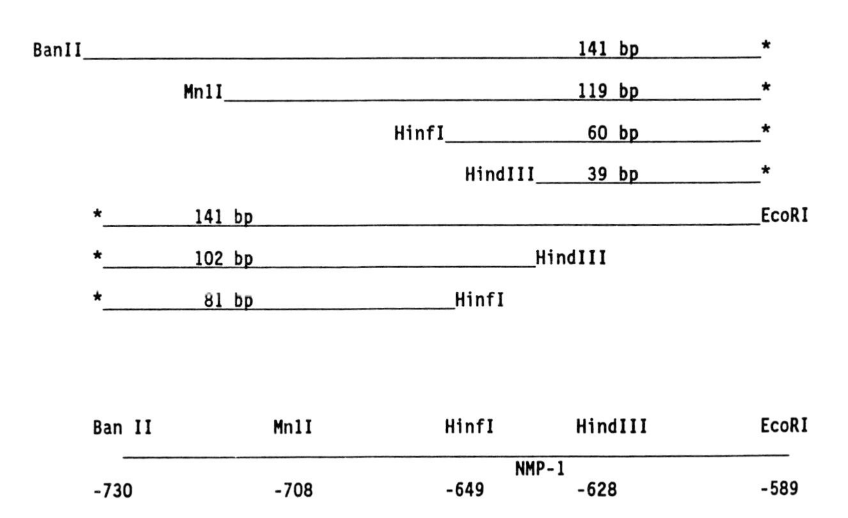

Fig. 22. Nuclear-matrix protein–DNA interaction within the 141-bp distall promoter element of an H4 histone gene. A gel mobility shift assay in conjunction with a bidirectional deletion analysis was carried out to determine the region within the 141-bp fragment of the H4 histone gene promoter where the site of protein–DNA interaction occurs. It is demonstrated that nuclear matrix proteins(s), NMP-1, bind to this distal element between the *Hin*f1 and *Hin*dIII restriction sites (−649 to −628). The lower panel shows schematically the deletion fragments used in the gel shift assays.

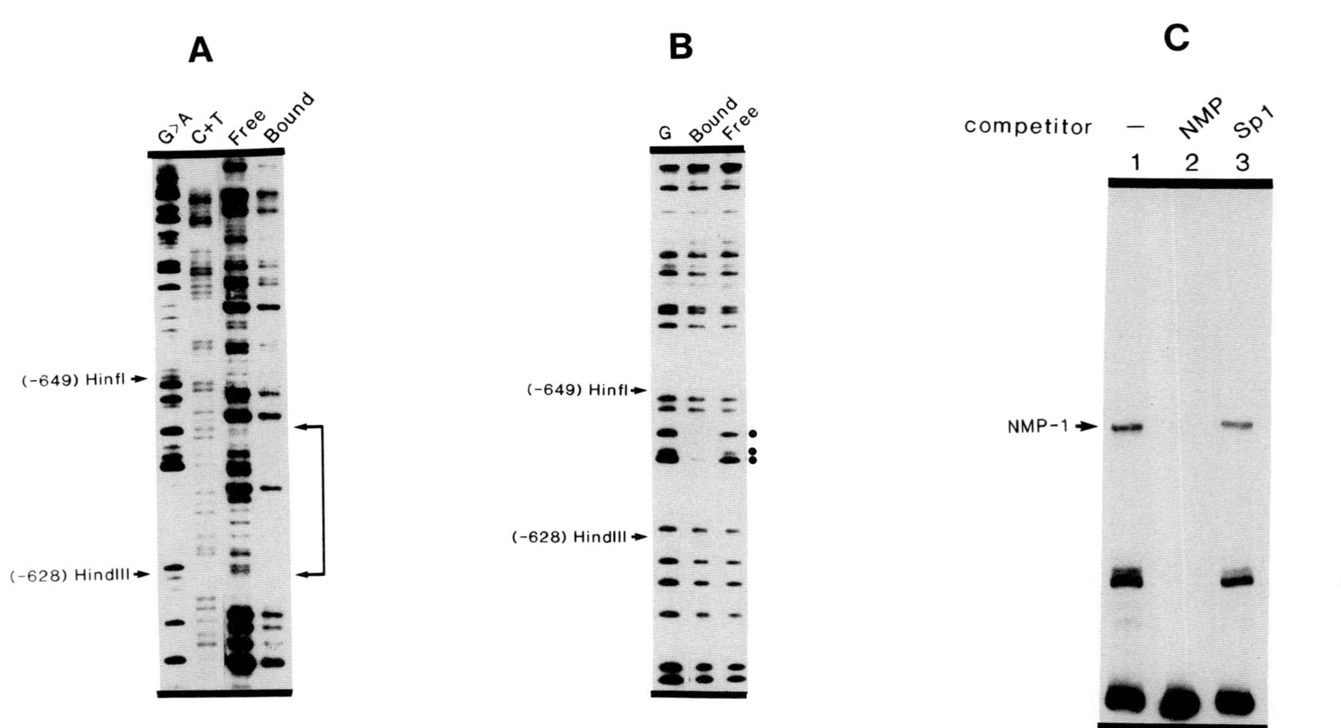

Fig. 23. Mapping of a nuclear-matrix attachment site to H4 histone gene promoter. A. Op-Cu footprint analysis of NMP-1 binding in the H4 histone gene promoter. The sequence-specific interactions of NMP-1 were determined by footprint analysis. Comparison of the free and bound lanes shows that NMP-1 protects the DNA between the *Hin*f1 and *Hin*dIII restriction sites (indicated by arrows). B. Identification of specific guanine residue contacts in the H4 histone gene promoter with NMP-1. Specific nucleotide contacts of NMP-1 were established by DMS protection of the gel shifted protein–DNA complex. The recognition-binding sequence is GGA*CGTCA*. The CGTCA motif contains the core recognition binding sequence for the ATF family of transcription factors. Free, free probe; Bound, bound probe; G, G ladder; •, protected G residues. C. Competition analysis of the NMP-1–DNA complex. Oligonucleotides synthesized for the NMP-1 and Sp1 binding sequences were used as competitors in the gel mobility shift assay. The oligonucleotides were added in 250 molar excess. The NMP oligonucleotide specifically competes out the NMP-1 interaction, whereas the Sp1 oligonucleotide does not have any effect.

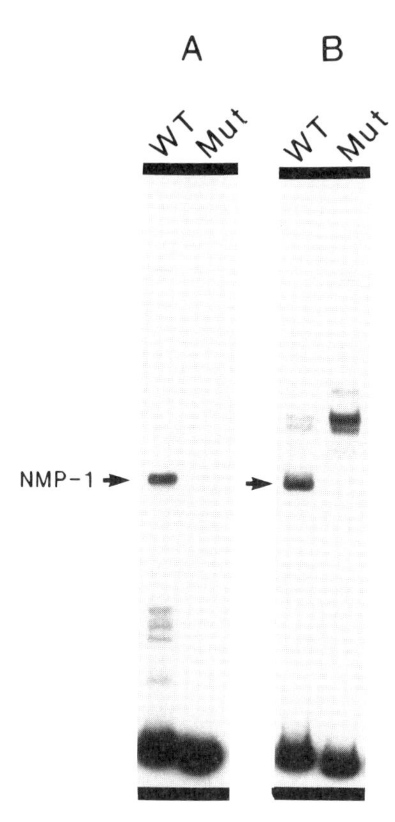

Fig. 24. Representation of NMP-1 in nuclear-matrix versus transcription-factor extracts. The NMP-1 interaction can be found in transcription-factor extracts. The mobility of the shift DNA appears to be the same; sequence-specific binding is further established by site-directed mutagenesis. The mut lanes have the CG changed to GA within the CGTCA binding sequence. It is clearly shown that the NMP-1 interaction is completely abolished when using these probes with both extracts.

of cell growth and tissue-specific genes that support progressive development of the osteoblast phenotype. Historically signal transduction was defined as the amplification of extracellular regulatory information at the level of the cell membrane, primarily mediated by cyclic nucleotides. However, it is becoming increasingly apparent that regulation of gene expression associated with control of osteoblast growth and differentiation involves the processing of regulatory signals at multiple levels. Unquestionably, the well-established cyclic AMP–adenylate cyclase and inositol phosphate–cytosolic calcium signal-transduction pathways (Donahue *et al.,* 1988; Reid *et al.,* 1987; Yamaguchi *et al.,* 1987) are operative. The rapid and pronounced parathyroid hormone effects on intracellular calcium levels in osteoblasts is a striking example of the combined use of both signal-transduction mechanisms (Bidwell *et al.,* 1991). But the mechanisms by which physiological mediators of osteoblast growth and differentiation, such as steroid hormones (e.g., vitamin D, glucocorticoids, estrogen) convey regulatory signals to influence the composition and organization of promoter-binding factors at gene regulatory elements (that determine the extent to which specific genes are transcribed and the extent to which mRNA transcripts serve as templates for synthesis of cellular proteins) are challenging questions that must be addressed. Here an exchange of information via regulatory macromolecules between the extracellular environment and the cytoplasm, as well as between the nucleus and the cytoplasm, must be considered. Involvement of the extracellular matrix, the cytoskeleton, and the nuclear matrix in the concentration of regulatory macromolecules, as well as in the localization of elements of the transcription and translation machinery, are timely and biologically important questions. The potential for the extracellular and intracellular matrices to provide solid-state, signal-transduction pathways engaged in the bidirectional exchange of regulatory information is a viable concept that will significantly enhance our understanding of the control of osteoblast-phenotype development.

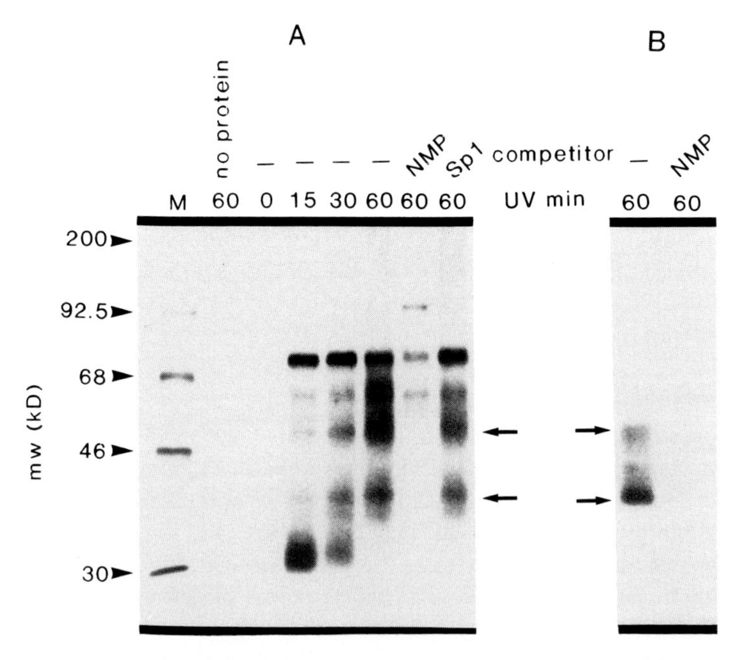

Fig. 25 UV cross-linking experiments to determine the molecular weight of NMP-1. Panel A represents the UV cross-linking of total nuclear matrix extracts with the distal element, and Panel B shows UV cross-linking of the protein fraction obtained by NMP oligonucleotide affinity chromatography. The lower two bands in Panel A (arrows) are the only bands competed out by the NMP oligonucleotide. Interestingly, only these bands appear when purified fractions are used (Panel B, arrows). These proteins represent proteins of 43,000 and 54,000 molecular weight.

III. CONCLUSION

By the combined application of molecular, biochemical, histochemical, and ultrastructural approaches, a temporal sequence of gene expression associated with development of bone-cell phenotype has been defined *in vivo* and in primary cultures of diploid osteoblasts. Multiple lines of evidence indicate that the observed modifications in gene expression reflect a developmental sequence that has (1) three principal periods: proliferation, extracellular matrix maturation, and mineralization; and (2) two restriction points to which the cells can progress, but cannot pass without further signals: the first when proliferation is down-regulated, and gene expression associated with extracellular-matrix maturation is induced; and the second, when mineralization occurs. Initially, actively proliferating cells expressing cell cycle and cell growth-regulated genes produce a fibronectin–Type I collagen extracellular matrix. Several lines of evidence support a reciprocal and functionally coupled relationship between the decline in proliferative activity and the subsequent induction of genes associated with extracellular-matrix maturation and mineralization. The extracellular matrix contributes to both the shutdown of proliferation and the development of the mature osteoblast phenotype. Recent results are consistent with involvement of the nuclear matrix in regulating the expression of genes that mediate osteoblast growth and differentiation within the three-dimensional context of cellular architecture.

ACKNOWLEDGMENTS

Studies from the authors' laboratories described in this review were supported by grants from the National Institutes of Health (GM32010, GM32381, AR33920, AR35166, AR39122, AR39588, HD22400), the National Science Foundation (BMS 88-19989, DCB88-96116), the March of Dimes Birth Defects Foundation (1-813), the International Life Sciences Institute (Washington, D.C.), and the Northeast Osteogenesis Imperfecta Society. We thank Christine Dunshee for photographic assistance, Donna Conlan and Mary Beth Kennedy for assistance with cell cultures. The editorial assistance of Patricia Jamieson in the preparation of the manuscript is most appreciated.

REFERENCES

Abe, E., Ishimi, Y, Takahashi, N., Akatsu, T., Ozawa, H., Yamana, H., Yoshiki, S., and Suda, T. (1988). A differentiation-inducing factor produced by the osteoblastic cell line MC3T3E1 stimulates bone resorption by promoting osteoclast formation. *J. Bone Miner. Res.* **3,** 635–645.

Ankenbauer, W., Stråhle, U., and Schütz, G. (1988). Synergistic action of glucocorticoid- and estradiol-responsive elements. *Proc. Natl. Acad. Sci. U.S.A.* **85,** 7526–7530.

Antosz, M. E., Bellows, C. G., and Aubin, J. E. (1989). Effects of transforming growth factor β and epidermal growth factor on cell proliferation and the formation of bone nodules in isolated rat calvaria cells. *J. Cell. Physiol.* **140,** 386–395.

Aronow, M. A., Gerstenfeld, L. C., Owen, T. A., Tassinari, M. S., Stein, G. S., and Lian, J. B. (1990). Factors that promote progressive development of the osteoblast phenotype in cultured fetal rat calvaria cells. *J. Cell. Physiol.* **143,** 213–221.

Barone, L. M., Owen, T. A., Tassinari, M. S., Bortell, R., Stein, G. S. and, Lian, J. B. (1991) Developmental expression and hormonal regulation of the rat matrix Gla protein (MGP) gene in chondrogenesis and osteogenesis. *J. Cell. Biochem.* (in press).

Barrack, E. R., and Coffey, D. S. (1983). Hormone receptors and the nuclear matrix *In* Gene Regulation by Steroid Hormones II (A. K. Roy and J. H. Clark, eds.), pp. 239–266, Springer-Verlag, N.Y., N.Y.

Bellows, C. G., Aubin, J. E., Heersche, H. N. M., and Antosz, M. E. (1986). Mineralized bone nodules formed *in vitro* from enzymatically released rat calvaria cell populations. *Calcif. Tissue Int.* **38,** 143–154.

Benayahu, D., Kletter, Y., Zipori, D., and Weintraub, S. (1989). Bone marrow–derived stromal cell line expressing osteoblastic phenotype *in vitro* and osteogenic capacity *in vivo*. *J. Cell. Physiol.* **140,** 1–7.

Bentz, H., Nathan, R. M., Rosen, D. M., Armstrong, R. M., Thompson, A. Y., Segarini, P. R., Mathews, M. C., Dasch, J. R., Piez, K. A., and Seyedin, S. M. (1989). Purification and characterization of a unique osteoinductive factor from bovine bone. *J. Biol. Chem.* **264,** 20805–20810.

Beresford, J. N., Gallagher, J. A., Poser, J. W., and Russell, R. G. G. (1984). Production of osteocalcin by human bone cells *in vitro*. Effects of 1,25$(OH)_2D_3$, 24, 25$(OH)_2D_3$, parathyroid hormone, and glucocorticoids. *Metab. Bone Dis. and Rel. Res.* **5,** 229–234.

Betsholtz, C., Johnsson, A., Heldin, C.-H., Westermark, B., Lind, P., Urdea, M. S., Eddy, R., Shows, T. B., Philpott, K., Mellor, A. L., Knott, T. J., and Scott, J. (1986). cDNA sequence and chromosomal localization of human platelet-derived growth factor A-chain and its expression in tumor cell lines. *Nature* **320,** 695–699.

Bhargava, U., Bar-Lev, M., Bellows, C. G., and Aubin, J. E. (1988). Ultrastructural analysis of bone nodules formed *in vitro* by isolated fetal rat calvaria cells. *Bone* **9,** 155–163.

Bidwell, J. P., Fryer, M. J., Firek, A. F., Donahue, H. J., Heath, H., III. (1991). Desensitization of rat osteoblast-like cells (ROS 17/2.8) to parathyroid hormone uncouples the adenosine 3′, 5′-mononophosphate and cytosolic ionized calcium response limbs. *Endocrinology* **128:** 1021–1028.

Bolander, M. E., Joyce, J. E., Boden, S. D., Oliver, B., and Heydemann, A. (1989). Estrogen receptor mRNA expression during fracture healing in the rat detected by polymerase chain-reaction amplification. *J. Bone Miner. Res.* **4,** S259.

Bortell, R., Barone, L. M., Tassinari, M. S., Lian, J. B., and Stein, G. S. (1990). Gene expression during endochondral bone development: Evidence for coordinate expression of transforming growth factor β and collagen type I. *J. Cell. Biochem.* **44,** 81–91.

Boskey, A. L. (1989). Noncollagenous matrix proteins and their role in mineralization. *Bone Miner.* **6,** 111–123.

Boskey, A. L., Wians, F. H., Jr., and Hauschka, P. V. (1985). The effect of osteocalcin on *in vitro* lipid-induced hydroxyapatite formation and seeded hydroxyapatite growth. *Calcif. Tissue. Int.* **37,** 57–62.

Bruder, S. P., and Caplan, A. (1989a). First bone formation and the dissection of an osteogenic lineage in the embryonic chick tibia is revealed by monoclonal antibodies against osteoblasts. *Bone* **11,** 359–375.

Bruder, S. P., and Caplan, A. I. (1989b). Discrete stages within the osteogenic lineage are revealed by alterations in the cell-surface architecture of embryonic bone cells. *Connect. Tissue Res.* **20,** 73–79.

Bruder, S. P., and Caplan, A. I. (1990). Terminal differentiation of osteogenic cells in the embryonic chick tibia is revealed by a monoclonal antibody against osteocytes. *Bone* **11,** 189–198.

Canalis, E. (1983). Effect of glucocorticoids on type I collagen synthesis, alkaline phosphatase activity, and deoxyribonucleic acid content in cultured rat calvariae. *Endocrinology* **112,** 931–939.

Canalis, E. M., Dietrich, J. W., Maina, D. M., and Raisz, L. G. (1977). Hormonal control of bone collagen synthesis *in vitro,* effects of insulin and glucagon. *Endocrinology* **100,** 668–674.

Canalis, E., Centrella, M., and McCarthy, T. (1988). Effects of basic fibroblast growth factor on bone formation *in vitro*. *J. Clin. Invest.* **81,** 1572–1577.

Capco, D. G., Wan, K. M., and Penman, S. (1982). The nuclear matrix: Three-dimensional architecture and protein composition. *Cell* **29,** 847–858.

Carrington, J. L., Roberts, A. B., Flanders, K. C., Roche, N. S., and Reddi, A. H. (1988). Accumulation, localization, and compartmentation of transforming growth factor β during endochondral bone development. *J. Cell Biol.* **107,** 1969–1975.

Centrella, M., McCarthy, T. L., and Canalis, E. (1988). Skeletal tissue and transforming growth factor β. *FASEB J.* **2,** 3066-3073.

Centrella, M., McCarthy, T. L., and Canalis, E. (1989). β_2 microglobulin enhances insulin-like growth factor I receptor levels and synthesis in bone cell cultures. *J. Biol. Chem.* **264,** 18268–18271.

Chen, T.L., Cone, C. M., and Feldman, D. (1983). Glucocorticoid modulation of cell proliferation in cultured osteoblast-like bone cells: Differences between rat and mouse. *Endocrinology* **112,** 1739–1746.

Collart, D., Ramsey-Ewing, A., Bortell, R., Lian, J., Stein, J., and Stein, G. (1991). Isolation and characterization of a cDNA from a human histone H2B gene, which is reciprocally expressed in relation to replication-dependent H2B histone genes during HL60 cell differentiation. *Biochemistry* **30:** 1610–1617.

Cowell, H. R., Hunziker, E. B., and Rosenberg, L. (1987). The role of hypertrophic chondrocytes in endochondral ossification and in the development of secondary centers of ossification. *J. Bone Joint Surg.* **69-A,** 159–162.

Craig, A. M., Smith, J. H., and Denhardt, D. T. (1989). Osteopontin, a transformation-associated, cell-adhesion phosphoprotein, is induced by 12-O-tetradecanoylphorbol 13-acetate in mouse epidermis. *J. Biol. Chem.* **264,** 9682–9689.

DeLuca, H. F. (1988). The vitamin D story: A collaborative effort of basic science and clinical medicine. *FASEB J.* **2,** 224–236.

Donahue, H. J., Fryer, M. J., Eriksen, E. F., Heath, H., III. (1988). Differential effects of parathyroid hormone and its analogies on cytosolic calcium ion and cAMP levels in cultured rat osteoblast-like cells. *J. Biol. Chem.* **263,** 13522–13527.

Dworetzky, S. I., Fey, E. G., Penman, S., Lian, J. B., Stein, J. L., and Stein, G. S. (1990). Progressive changes in the protein composition of the nuclear matrix during rat osteoblast differentiation. *Proc. Natl. Acad. Sci. U.S.A.* **87,** 4605–4609.

Dworetzky, S. I., Wright, K. L., Fey, E. G., Penman, S., Lian, J. B., Stein, J. L., and Stein, G. S. (1991). Sequence-specific DNA binding proteins are components of a nuclear matrix attachment site. (ms submitted).

Egan, J. J., Gronowicz, G., and Rodan, G. A. (1991a). Cell density–dependent decrease in cytoskeletal actin and myosin in cultured osteoblastic cells: Correlation with cyclic AMP changes. *J. Cell. Biochem.* **45,** 93–100.

Egan, J. J., Gronowicz, G., and Rodan, G. A. (1991b). Parathyroid hormone promotes the disassembly of cytoskeletal actin and myosin in cultured osteoblastic cells: Mediation by cyclic AMP. *J. Cell. Biochem.* **45,** 101–111.

Ernst, M., and Froesch, E. R. (1988). Growth hormone–dependent stimulation of osteoblast-like cells in serum-free cultures via local synthesis of insulin-like growth factor I. *Biochem. Biophys. Res. Commun.* **151,** 142–148.

Ernst, M., Schmid., C. H., and Froesch, E. R. (1988). Enhanced osteoblast proliferation and collagen gene expression by estradiol. *Proc. Natl. Acad. Sci. U.S.A.* **85,** 2307–2310.

Ernst, M., Heath, J. K., and Rodan, G. A. (1989). Estradiol effects on proliferation, messenger ribonucleic acid for collagen and insulin-like growth factor-I, and parathyroid hormone–stimulated adenylate cyclase activity in osteoblastic cells from calvariae and long bones. *Endocrinology* **125,** 825–833.

Ecarot-Charrier, B., Glorieux, F. H., van der Rest, M., and Pereira, G. (1983). Osteoblasts isolated from mouse calvaria initiate matrix mineralization in culture. *J. Cell Biol.* **96,** 639–643.

Ecarot-Charrier, B., Shepard, N., Charette, G., Grynpas, M., and Glorieux, F. H. (1988). Mineralization in osteoblast cultures: A light and electron microscopic study. *Bone* **9,** 147–154.

Fey, E. G., Wan, K. M., and Penman, S. (1984). Epithelial cytoskeletal framework and nuclear matrix–intermediate filament scaffold: Three-dimensional organization and protein composition. *J. Cell Biol.* **98,** 1973–1984.

Fey, E. G., Krochmalnic, G., and Penman, S. (1986). The nonchromatin substructures of the nucleus: The ribonucleoprotein (RNP)-containing and RNP-depleted matrices analyzed by sequential fractionation and resinless section electron microscopy. *J. Cell Biol.* **102,** 1654–1665.

Fey, E. G., and Penman, S. (1988). Nuclear-matrix proteins reflect cell type of origin in cultured human cells. *Proc. Natl. Acad. Sci. U.S.A.* **85,** 121–125.

Fisher, L. W., Hawins, G. R., Turcoss, N., and Termine, J. D. (1987). Purification and partial characterization of small proteoglycans I and II, bone sialoproteins I and II, and osteonectin from the mineral compartment of developing human bone. *J. Biol. Chem.* **262,** 9702–9708.

Franzen, A., and Heinegard, D. (1985). Isolation and characterization of two sialoproteins present only in bone calcified matrix. *Biochem. J.* **232,** 715–724.

Fraser, J. D., and Price, P. A. (1988). Lung, heart, and kidney express high levels of mRNA for the vitamin K–dependent matrix Gla protein: Implications for the possible functions of matrix Gla protein and for the tissue distribution of the γ-carboxylase. *J. Biol. Chem.* **263,** 11033–11036.

Friedenstein, A. J., and Chailakhyan, R. K. (1987). Bone marrow osteogenic stem cells: *In vitro* cultivation and transplantation in diffusion chambers. *Cell Tissue Kinet.* **20**, 263–272.

Gehron-Robey, P., Young, M., Fischer, L., and McClain, T. (1989). Thrombospondin is an osteoblast-derived component of mineralized extracellular matrix. *J. Cell. Biol.* **108,** 719–727.

Gerstenfeld, L. C., Chipman, S. D., Glowacki, J., and Lian, J. B. (1987). Expression of differentiated function by mineralizing cultures of chicken osteoblasts. *Dev. Biol.* **122,** 49–60.

Glimcher, M. J. (1976). Composition, structure and organization of bone and other mineralized tissues and the mechanism of calcification. *In* "Handbook of Physiology," Vol. 7 (R. O. Greep and E. B. Astwood, eds.), pp. 25–116. Williams & Wilkins, Baltimore, Maryland.

Glimcher, M. J. (1989). Mechanism of calcification: Role of collagen fibrils and collagen–phosphoprotein complexes *in vitro* and *in vivo. Anat. Rec.* **224,** 139–153.

Glowacki, J., and Lian, J. (1987). Impaired recruitment of osteoclast progenitors in deficient osteocalcin-depleted bone implants. *Cell Differ. Dev.* **21,** 247–254.

Goldstein, D. J., Rogers, C. E., and Harris, H. (1980). Expression of alkaline phosphatase loci in mammalian tissues. *Proc. Natl. Acad. Sci. U.S.A.* **77,** 2857–2860.

Gorski, J. P., and Shimizu, K. (1988). Isolation of new phosphorylated glycoprotein from mineralized phase of bone that exhibits limited homology to adhesive protein osteopontin. *J. Biol. Chem.* **263,** 15938–15945.

Grange, T., Roux, J., Rigeaud, G., and Pictet, R. (1989). Two remote glucocorticoid-responsive units interact cooperatively to promote glucocorticoid induction of rat tyrosine aminotransferase gene expression. *Nucleic Acids Res.* **17,** 8695–8709.

Graves, D. T., Valentin-Opran, A., Delgado, R., Valente, A. J., Mundy, G., and Piche, J. (1989). The potential role of platelet-derived growth factor as an autocrine or paracrine factor for human bone cells. *Connect. Tissue Res.* **23,** 209–218.

Gray, T. K., Flynn, T. C., Gray, K. M., and Nabell, L. M. (1987). 17β-Estradiol acts directly on the clonal osteoblast cell line UMR 106. *Proc. Natl. Acad. Sci. U.S.A.* **84,** 6267–6271.

Gray, T. K., Mohan, S., Linkhard, T. A., and Baylink, D. (1989). Estradiol stimulates *in vitro* the secretion of insulin-like growth factors by the clonal osteoblastic cell line UMR 106. *Biochem. Biophys. Res. Commun.* **158,** 407–411.

Hamilton, J. A., Lingelbach, S., Partridge, N. C., and Martin, T. J. (1985). Regulation of plasminogen activator production by bone-resorbing hormones in normal and malignant osteoblasts. *Endocrinology* **116,** 2186–2191.

Hauschka, P. V., Lian, J. B., and Gallop, P. M. (1975). Direct identification of the calcium-binding amino acid, γ-carboxyglutamate, in mineralized tissue. *Proc. Natl. Acad. Sci. U.S.A.* **72,** 3925–3929.

Hauschka, P. V., Mavrakos, A. E., Iafrati, M. D., Doleman, S. E., and Klagsbrun, M. (1986). Growth factors in bone matrix. Isolation of multiple types by affinity chromatography on heparin-sepharose. *J. Biol. Chem.* **261,** 12665–12675.

Hauschka, P. V., Lian, J. B., Cole, D. E. C., and Gundberg, C. M. (1989). Osteocalcin and matrix GLA protein—vitamin K-dependent proteins in bone. *Physiol. Rev.* **69,** 990–1947.

He, D., Nickerson, J. A., and Penman, S. (1990). Core filaments of the nuclear matrix. *J. Cell Biol.* **110,** 569.

Heath, J. K., Atkinson, S. J., Meikle, M. C., and Reynolds, J. J. (1984). Mouse osteoblasts synthesize collagenase in response to bone-resorbing agents. *Biochim. Biophys. Acta* **802,** 151–154.

Heath, J. K., Rodan, S. B., Yoon, K., Rodan, G. A. (1989). Rat calvarial cell lines immortalized with SV-40 large-T antigen: Constitutive and retinoic acid–inducible expression of osteoblastic features. *Endocrinology* **124,** 3060–3068.

Heine, U. I., Munoz, E. F., Flanders, K. C., Ellingsworth, L. R., Lam, H.-Y. P., Thompson, N. L., Roberts, A. B., and Sporn, M. B. (1987). Role of transforming growth factor-β in the development of the mouse embryo. *J. Cell. Biol.* **105,** 2861–2876.

Heinegård, D., and Oldberg, Å. (1989). Structure and biology of cartilage and bone matrix noncollagenous macromolecules. *FASEB J.* **3,** 2042–2051.

Henthorn, P. S., Raducha, M., Kadesch, T., Weiss, M. J., and Harris, H. (1988). Sequence and characterization of the human intestinal alkaline phosphatase gene. *J. Biol. Chem.* **263,** 12011–12019.

Hiramatsu, M., Kumegawa, M., Hatakeyama, K., Yajima, T., Minami, N., and Kodama, H. (1982). Effect of epidermal growth factor on collagen synthesis in osteoblastic cells derived from newborn mouse calvaria. *Endocrinology* **111,** 1810–1817.

Holthuis, J., Owen, T. A., van Wijnen, A. J., Wright, K. L., Ramsey-Ewing, A., Kennedy, M. B., Carter, R., Cosenza, S. C., Soprano, K. J., Lian, J. B., Stein, J. L., and Stein, G. S. (1990). Tumor cells exhibit deregulation of the cell-cycle histone gene-promoter factor HiNF-D. *Science* **247:** 1454–1457.

Huberman, E., and Callahan, M. F. (1979). Induction of terminal differentiation in human promyelocytic leukemia cells by tumor-promoting agents. *Proc. Natl. Acad. Sci. U.S.A.* **76,** 1293–1297.

Hughes, M. R., Malloy, P. J., Kiebach, D. G., Kesterson, R. A., Pike, J. W., Feldman, D., and O'Malley, B. W. (1988). Point mutations in the human vitamin D–receptor gene associated with hypocalcemic rickets. *Science* **242,** 1702–1705.

Jantzen, M. -M., Strähle, U., Gloss, B., Stewart, F., Schmid, W., Boshart, M., Miksicek, R., and Schütz, G. (1987). Cooperativity of glucocorticoid response elements located far upstream of the tyrosine aminotransferase gene. *Cell* **49,** 29–38.

Jinguishi, S., Heydemann, A., Joyce, M., Soltero, R., and Bolander, M. E. (1989). Acidic fibroblast growth factor is synthesized in fracture callus by inflammatory cells around the blood clot and regulated cartilage formation during fracture healing. *J. Bone Miner. Res.* **4,** S285.

Joyce, M. E., Roberts, A. B., Sporn, M. B., and Bolander, M. E. (1990). Transforming growth factor-βand the initiation of chondrogenesis and osteogenesis in the rat femur. *J. Cell Biol.* **110,** 2195–2207.

Kim, H. T., and Chen, T. L. (1989). 1,25-Dihydroxyvitamin D_3 interaction with dexamethasone and retinoic acid: Effects of procollagen messenger ribonucleic acid levels in rat osteoblast-like cells. *Mol. Encocrinol.* **3,** 97–104.

Kinne, R. W., and Fisher, L. W. (1987). Keratin Sulfate proteoglycan in rabbit compact bone is sialoprotein II. *J. Biol. Chem.* **262,** 10206–10211.

Knoll, B. J., Rothblum, K. N., and Longley, M. (1988). Nucleotide sequence of the human placental alkaline phosphatase gene: Evolution of the 5′ flanking region by deletion–substitution. *J. Biol. Chem.* **263,** 12020–12027.

Komm, B. S., Terpening, C. M., Benz, D. J., Graeme, K. A., Gallegos, A., Korc, M., Greene, G. L., O'Malley, B. W., and Haussler, M. R. (1988). Estrogen binding, receptor mRNA, and biological response in osteoblast-like osteosarcoma cells. *Science* **241,** 81–83.

Kratochwil, K., von der Mark, K., Kollar, E. J., Jaenisch, R., Mooslehner, K., Schwarz, M., Haase, K., Gmachi, I., and Harbers, K. (1989). Retrovirus-induced insertional mutation in *Mov13* mice affects collagen I expression in a tissue-specific manner. *Cell* **57**:807–816.

Kumara-Siri, M. H., Shapiro, L. E., and Surks, M. I. (1986). Association of the 3,5,3′-triiodo-L-thyronine nuclear receptor with the nuclear matrix of cultured growth hormone–producing rate pituitary tumor cells (GC cells). *J. Biol. Chem.* **261,** 2844–2852.

Lawrence, J. B., Singer, R. H., Villnave, C. A., Stein, J. L., and Stein, G. S. (1988). Intracellular distribution of histone mRNAs in human fibroblasts studied by *in situ* hybridization. *Proc. Natl. Acad. Sci. U.S.A.* **85,** 463–467.

Leboy, P. S., Beresford, J. N., Devlin, C., and Owen, M. E. (1991). Dexamethasone induction of osteoblast mRNAs in rat marrow stromal cell cultures. *J. Cell. Physiol.* **146:** 370–378.

Lian, J. B., and Friedman, P. A. (1978). The vitamin K–dependent synthesis of gamma-carboxy-glutamic acid by bone microsomes. *J. Biol. Chem.* **253,** 6623–6626.

Lian, J. B., Couttes, M. C., and Canalis, E. (1985). Studies of hormonal regulation of osteocalcin synthesis in cultured fetal rat calvariae. *J. Biol. Chem.* **60**, 8706–8710.

Lian, J. B., Stein, G. S., Bortell, R., and Owen, T. A. (1991). Phenotype suppression: A postulated molecular mechanism for mediating the relationship of proliferation and differentiation by *fos–jun* interactions at AP-1 sites in steroid-responsive promoter elements of tissue-specific genes. *J. Cell. Biochem.* **45,** 9–14.

Long, M. W., Williams, J. L., and Mann, K. G. (1990). Expression of human bone–related proteins in the hematopoietic microenvironment. *J. Clin. Invest.* **86,** 1387–1395.

Luyten, F. P., Cunningham, N. S., Ma, S., Muthukumaran, N., Hammonds, R. G., Nevins, W. B., Wood, W. I., and Reddi, A. H. (1989). Purification and partial amino acid sequence of osteogenin, a protein initiating bone differentiation. *J. Biol. Chem.* **264,** 13377–13380.

Lyons, K. M., Pelton, R. W., and Hogan, B. L. M. (1989). Patterns of expression of murine Vgr-1 and BMP-2a RNA suggest that transforming growth factor-β–like genes coordinately regulate aspects of embryonic development. *Genes Dev.,* **1:**1657–1668.

Madisen, L., Neubauer, M., Plowman, G., Rosen, D., Segarini, P., Dasch, J., Thompson, A., Ziman, J., Betz, H., and Purchio, A. F. (1990). Molecular cloning of a novel bone-forming compound: Osteoinductive factor *DNA Cell Biol.* **9,** 303–309.

Majeska, R. J., Rodan, S. B., and Rodan, G. A. (1978). Maintenance of parathyroid hormone response in clonal rat osteosarcoma lines. *Exp. Cell Res.* **111,** 465–472.

Majeska, R. J., Rodan, S. B., and Rodan, G. A. (1980). Parathyroid hormone–responsive clonal cell lines from rat osteosarcoma. *Endocrinology* **107,** 1494.

Majeska, R. J., Nair, B. C., and Rodan, G. A. (1985). Glucocorticoid regulation of alkaline phosphatase in the osteoblastic osteosarcoma cell line ROS 17/2.8. *Endocrinology* **116,** 170–179.

Marashi, F., Baumbach, L., Rickles, R., Sierra, F., Stein, J. L., and Stein, G. S. (1982). Histone proteins in HeLa S_3 cells are synthesized in a cell cycle stage–specific manner. *Science* **215,** 683-685.

Mardon, H., and Triffit, J. T. (1987). A tissue-specific protein in rat osteogenic tissues. *J. Bone Miner. Res.* **3,** 191–199.

Marks, S. C., Jr., and Popoff, S. N. (1988). Bone cell biology: The regulation of development, structure, and function in the skeleton. *Am. J. Anat.* **183,** 1-144.

Massagué, J. (1987). The TGFβ family of growth and differentiation factors. *Cell* **49,** 437–438.

Matsuura, S., Kishi, F., and Kajii, T. (1990). Characterization of a 5′-flanking region of the human liver/bone/kidney alkaline phosphatase gene: Two kinds of mRNA from a single gene. *Biochem. Biophys. Res. Commun.* **168,** 993–1000.

Millan, J. L., and Manes, T. (1988). Seminoma-derived Nagao isozyme is encoded by a germ-cell alkaline phosphatase gene. *Proc. Natl. Acad. Sci. U.S.A.* **85,** 3024–3028.

Minghetti, P. P., and Norman, A. W. (1988). $1,25(OH)_2$-vitamin D_3 receptors: Gene regulation and genetic circuitry. *FASEB J.* **2,** 3043–3053.

Moses, H. L., Shipley, G. D., Leof, E. B., Halper, J., Coffey, R. J., Jr., and Tucker, R. F. (1987). Transforming growth factors. *In* "Control of Animal Cell Proliferation" (A. L. Boynton and H. L. Leffert eds.), pp. 75–92. Academic Press, New York.

Nadal-Ginard, B. (1978). Commitment, fusion, and biochemical differentiation of a myogenic cell line in the absence of DNA synthesis. *Cell* **15,** 855–864.

Nelkin, B. D., Pardoll, D. M., and Vogelstein, B. (1980). Localization of SV40 genes with supercoiled loop domains. *Nucleic Acids Res.* **8,** 5623–5633.

Ng, K. W., Partridge, N. C., Niall, M., and Martin, T. J. (1983). Epidermal growth receptors of clonal lines of a rat osteogenic sarcoma and osteoblast-rich rat bone cells. *Calcif. Tissue Int.* **35,** 298–309.

Nijweide, P. J., and Mulder, R. J. P. (1986). Identification of osteocytes in osteoblast-like cell cultures using a monoclonal antibody specifically directed against osteocytes. *Histochemistry* **84,** 342–347.

Nishimoto, S. K., Stryker, W. F., and Nimni, M. E. (1987a). Calcification of osteoblast-like rat osteosarcoma cells in agarose suspension cultures. *Calcif. Tissue Int.* **41,** 274–280.

Nishimoto, S. K., Salka, C., and Nimni, M. E. (1987b). Retinoic acid and glucocorticoids enhance the effect of 1,25-dihydroxyvitamin D_3 on bone γ-carboxyglutamic acid protein synthesis by rat osteosarcoma cells. *J. Bone Miner. Res.* **2,** 571–577.

Noda, M. (1989). Transcriptional regulation of osteocalcin production by transforming growth factor β in rat osteoblast-like cells. *Endocrinology* **124,** 612–617.

Noda, M., and Camilliere, J. J. (1989). *In vivo* stimulation of bone formation by transforming growth factor β. *Endocrinology* **124,** 2991–2994.

Noda, M., and Rodan, G. A. (1987). Type β transforming growth factor (TGFβ) regulation of alkaline phosphatase expression and other phenotype-related mRNAs in osteoblastic rat osteosarcoma cells. *J. Cell. Physiol.* **133,** 426–437.

Noda, M., Vogel, R. L., Hasson, D. M., and Rodan, G. A. (1990). Leukemia inhibitory factor suppresses proliferation, alkaline phosphatase activity, and type I collagen messenger ribonucleic acid level and enhances osteopontin mRNA level in murine osteoblast-like (MC3T3E1) cells. *Endocrinology* **127,** 185–190.

Nomura, S., Wills, A. J., Edwards, D. R., Heath, J. K., and Hogan, B. L. M. (1988). Developmental expression of 2ar (osteopontin) and SPARC (osteonectin) RNA as revealed by *in situ* hybridization. *J. Cell Biol.* **106,** 441–450.

Okazaki, K., and Holtzer, H. (1966). Myogenesis: Fusion, myosin synthesis, and the mitotic cycle. *Proc. Natl. Acad. Sci. U.S.A.* **56,** 1484–1488.

Oldberg, Å., Franzén, A., and Heinegård, D. (1986). Cloning and sequence analysis of rat bone sialoprotein (osteopontin) cDNA reveals an Arg-Gly-Asp cell-binding sequence. *Proc. Natl. Acad. Sci. U.S.A.* **83,** 8819–8823.

Otsuka, K., Sodek, J., and Limeback, H. (1984). Synthesis of collagenase and collagenase inhibitors by osteoblast-like cells in culture. *Eur. J. Biochem.* **145,** 123–129.

Owen, M., and Friedenstein, A. J. (1988). Stromal stem cells: Marrow-derived osteogenic precursors, *In* "Cell and Molecular Biology and Vertebrate Hard Tissues" (D. Evered and S. Harnett, eds.), pp. 42–60. John Wiley, Chichester, England.

Owen T. A., Aronow, M., Shalhoub, V., Barone, L. M., Wilming, L., Tassinari, M. S., Kennedy, M. B., Pockwinse, S., Lian, J. B., and Stein, G. S. (1990a). Progressive development of the rat osteoblast phenotype *in vitro:* Reciprocal relationships in expression of genes associated with osteoblast proliferation and differentiation during formation of the bone extracellular matrix. *J. Cell. Physiol.* **143,** 420–430.

Owen, T. A., Holthuis, J., Markose, E., van Wijnen, A. K., Wolfe, S. A., Grimes, S. E., Lian, J. B., and Stein, G. S. (1990b). Modifications of protein–DNA interactions in the proximal promoter of a cell-growth–regulated histone gene during onset and progression of osteoblast differentiation. *Proc. Natl. Acad. Sci. U.S.A.* **87,** 5129–5133.

Owen, T. A., Bortell, R., Yocum, S. A., Smock, S. L., Zhang, M., Abate, C., Shalhoub, V., Aronin, N., Wright, K. L., van Wijnen, A. J., Stein, J. L., Curran, T., Lian, J. B., and Stein, G. S. (1990c).

Coordinate occupancy of AP-1 sites in the vitamin D–responsive and CCAAT box elements by Fos–Jun in the osteocalcin gene: A model for phenotype suppression of transcription. *Proc. Natl. Acad. Sci. U.S.A.,* **87,** 9990–9994.

Owen, T. A., Aronow, M. A., Barone, L. M., Bettencourt, B., Stein, G., and Lian, J. B. (1991). Pleiotropic effects of vitamin D on osteoblast gene expression are related to the proliferative and differentiated state of the bone-cell phenotype: Dependency upon basal levels of gene expression, duration of exposure, and bone–matrix competency in normal rat osteoblast cultures. *Endocrinology,* **128:** 1496–1504.

Ozkaynak, E., Rueger, D. C., Drier, E. A., Corbett, C., Ridge, R. J., Sampath, T. K., and Oppermann, H. (1990). OP-1 cDNA encodes an osteogenic protein in the TGFβ family. *The EMBO J.* **9,** 2085–2093.

Pardoll, D. M., Vogelstein, B., and Coffey, D. S. (1980). A fixed site of DNA replication in eukaryotic cells. *Cell* **19,** 527–536.

Pechak, D. G., Kujawa, J. J., and Caplan, A. I. (1986a). Morphological and histochemical events during first bone formation in embryonic chick limbs. *Bone* **7,** 441–458.

Pechak, D. G., Kujawa, J. J., and Caplan, A. I. (1986b). Morphology of bone development and bone remodeling in embryonic chick limbs. *Bone* **7,** 459–472.

Price, P. A. (1988). Role of vitamin K–dependent proteins in bone metabolism. *Ann. Rev. Nutr.* **8,** 865–883.

Price, P. A., Otsuka, A. S., Poser, J. W., Kristaponis, J., and Raman, N. (1976). Characterization of a γ-carboxyglutamic acid–containing protein from bone. *Proc. Natl. Acad. Sci. U.S.A.* **73,** 1447–1451.

Raisz, L. G., and Kream, B. E. (1983a). Regulation of bone formation (Part 1). *N. Engl. J. Med.* **309,** 29–35.

Raisz, L. G., and Kream, B. E. (1983b). Regulation of bone formation (Part 2). *N. Engl. J. Med.* **309,** 83–89.

Reddi, A. H. (1981). Cell biology and biochemistry of endochondral bone development. *Coll. Relat. Res. Vol.* **1,** 209–226.

Reid, I. R., Civitelli, R., Halstead, L. R., Avioli, L. V., and Hruska, K. A. (1987). Parathyroid hormone acutely elevates intracellular calcium in osteoblast-like cells, *Am J. Physiol.* **253,** E45–E51.

Reinholt, F. P., Hultenby, K., Oldberg, Å., and Heinegård, D. (1990). Osteopontin—-a possible anchor of osteoclasts to bone. *Proc. Natl. Acad. Sci. U.S.A.* **87,** 4473–4475.

Rheinwald, J. G., and Green, H. (1975). Serial cultivation of strains of human epidermal keratinocytes: The formation of keratinizing colonies from single cells. *Cell* **6,** 331–344.

Rice, R. H., and Green, H. (1979). Presence in human epidermal cells of soluble protein precursor of the cross-linked envelope: Activation of the cross-linking by calcium ions. *Cell* **18,** 681–694.

Robinson, S. I., Nelkin, B. D., and Vogelstein, B. (1982). The ovalbumin gene is associated with the nuclear matrix of chicken oviduct cells. *Cell* **28,** 99–106.

Rodan, G. A., and Martin, T. J. (1981). Role of osteoblasts in hormonal control of bone resorption—a hypothesis. *Calcif. Tissue Int.* **33,** 349–351.

Rodan, G. A., and Rodan, S. P. (1984). Expression of the osteoblastic phenotype. *In* "Bone and Mineral Research," Vol 2. (W. A. Peck, ed.), pp. 244–285. Elsevier Science Publishers, Amsterdam.

Rodan, G. A., and Noda, M. (1991). Gene expression in osteoblastic cells. *CRC Crit. Rev. Eukaryotic Gene Expression* **1:** 85–98.

Romberg, R. W., Werness, P. G., Riggs, B. L., and Mann, K. G. (1986). Inhibition of hydroxyapatite crystal growth by bone-specific and other calcium-binding proteins. *Biochemistry* **25,** 1176–1180.

Rosen, D. M, Stempien, S. A., Thompson, A. Y., and Seyedin, S. M. (1988). Transforming growth factor β modulates the expression of osteoblast and chondrocyte phenotypes *in vivo. J. Cell. Physiol.* **134,** 337–346.

Rovera, G., Santoli, D., and Damsky, C. (1979). Human promyelocytic leukemia cells in culture differentiate into macrophage-like cells when treated with a phorbol diester. *Proc. Natl. Acad. Sci. U.S.A.* **76,** 2779–2783.

Ruoslahti, E. (1988). Structure and biology of proteoglycans. *Annu. Rev. Cell. Biol.* **4,** 229–255.

Rüther, U., Garber, C., Komitowski, D., Müller, R., and Wagner, E. F. (1987). Deregulated c-*fos* expression interferes with normal bone development in transgenic mice. *Nature* **325,** 412–416.

Sampath, T. K., Muthukumaran, N., and Reddi, A. H. (1987). Isolation of osteogenin, an extracellular matrix–associated, bone-inductive protein, by heparin affinity chromatrography. *Proc. Natl. Acad. Sci. U.S.A.* **84,** 7109–7113.

Sandberg, M., Autio-Harmainen, H., and Vuorio, E. (1988). Localization of the expression of types I, III, and IV collagen, TGFβ1 and c-*fos* genes in developing human calvarial bones. *Dev. Biol.* **130,** 324–334.

Schaack, J., Ho, W. Y-W., Freimuth, P., and Shenk, T. (1990). Adenovirus terminal protein mediates both nuclear-matrix association and efficient transcription of adenovirus DNA. *Genes Dev.* **4,** 1197–1208.

Schüle, R., Muller, M., Kaltschmidt, C., and Renkawitz, R. (1988). Many transcription factors interact synergistically with steriod receptors. *Science* **241,** 1418–1420.

Schüle, R., Kazuhiko, U., Mangelsdorf, D. J., Bolado, J., Pike, J. W., and Evans, R. M. (1990). Jun–Fos and receptors for vitamins A and D recognize a common response element in the human osteocalcin gene. *Cell* **61,** 497–504.

Shalhoub, V., Gerstenfeld, L. C., Lian, J. B., Stein, J. L., and Stein, G. S. (1989). Down-regulation of cell growth and cell cycle–regulated genes during chick osteoblast differentiation with the reciprocal expression of histone gene variants. *Biochemistry* **28,** 5318–5322.

Smith, J. H., and Denhardt, D. T. (1987). Molecular cloning of a tumor-promoter–inducible mRNA found in JB6 mouse epidermal cells: Induction is stable at high, but not at low, cell densities. *J. Cell. Biochem* **34,** 13–22.

Smith, J. H., and Denhardt, D. T. (1989). Evidence for two pathways of protein kinase C induction of 2AR expression: Correlation with mitogenesis. *J. Cell. Physiol.* **139,** 189–195.

Spiegelman, B. M., and Green, H. (1980). Control of specific protein biosynthesis during the adipose conversion of 3T3 cells. *J. Biol. Chem.* **255,** 8811–8818.

Sporn, M. B., Roberts, A. B., Shull, J. H., Smith, J. M., Ward, J. M., and Sodek, J. (1983). Polypeptide transforming growth factors isolated from bovine sources and used for would healing *in vivo. Science* **219,** 1329–1331.

Stein, G., Lian, J., Stein, J., Briggs, R., Shalhoub, V., Wright, K., Pauli, U., and van Wijnen, A. J. (1989a). Altered binding of human histone gene transcription factors during the shutdown of proliferation and onset of differentiation in HL-60 cells. *Proc. Natl. Acad. Sci. U.S.A.* **86,** 1865–1869.

Stein, G. S., Lian, J. B., Gerstenfeld, L. C., Shalhoub, V., Aranow, M., Owen, T., and Markose, E. (1989b). The onset and progression of osteoblast differentiation is functionally related to cellular proliferation. *Connect. Tissue Res.* **20,** 3–13.

Stief, A., Winter, D. M., Strätling, W. H., and Sippel, A. E. (1989). A nuclear DNA attachment element mediates elevated and position-independent gene activity. *Nature* **341,** 343–345.

Strähle, U., Schmid, W., and Schütz, G. (1988). Synergistic action of the glucocorticoid receptor with transcription factors. *EMBO J.* **7,** 3389–3395.

Suda, T., Shinki, T., and Takahashi, N. (1990). The role of vitamin D in bone and intestinal cell differentiation. *Annu. Rev. Nutr.* **10,** 195–211.

Termine, J. D., Kleinman, H. K., Whitson, S. W., Conn, K. M., McGarvey, M. L., and Martin, G. R. (1981). Osteonectin, a bone-specific protein linking mineral to collagen. *Cell* **26,** 99–105.

Thaller, C., and Eichele, G. (1987). Identification and spatial distribution of retinoids in the developing chick limb bud. *Nature* **327,** 625–628.

Thaller, C., and Eichele, G. (1990). Isolation of 3,4-didehydroretinoic acid, a novel morphogenetic signal in the chick wing bud. *Nature* **345,** 815–819.

van Eeklen, C. A. G., and van Venrooij, W. J. (1981). hnRNA and its attachment to a nuclear-protein matrix. *J. Cell Biol.* **88,** 554–563.

Wagner, M., Thaller, C., Jessell, T., and Eichele, G. (1990). Polarizing activity and retinoid synthesis in the floor plate of the neural tube. *Nature* **345,** 819–822.

Wang, E. A., Rosen, V., D'Alessandro, J. S., Bauduy, M., Cordes, P., Harada, T., Israel, D. I., Hewick, R. M., Kerns, K. M., LaPan, P., Luxenberg, D. P., McQuaid, D., Moutsatsos, I. K., Nove, J., and Wozney, J. M. (1990). Recombinant human bone morphogenetic protein induces bone formation. *Proc. Natl. Acad. Sci. U.S.A.* **87,** 2220–2224.

Weinreb, M., Shinar, D., and Rodan, G. A. (1990). Different pattern of alkaline phosphatase, osteopontin, and osteocalcin expression in developing rat bone visualized by *in situ* hybridization. *J. Bone Miner. Res.* **5,** 831–842.

Weiss, M. J., Ray, K., Henthorn, P. S., Lamb, B., Kadesch, T., and Harris, H. (1988). Structure of the human liver/bone/kidney alkaline phosphatase gene. *J. Biol. Chem.* **263,** 12002–12010.

Wong, M.-M., Rao, L. G., Ly, H., Hamilton, L., Tong, J., Sturtridge, W., McBroom, R., Aubin, J. E., and Murray, T. M. (1990). Long-term effects of physiologic concentrations of dexamethasone on human bone–derived cells. *J. Bone Miner. Res.* **5,** 803–813.

Wozney, J. M., Rosen, V., Celeste, A. J., Mitsock, L. M., Whitters, M. J., Kriz, R. W., Hewick, R. M., and Wang, E. A. (1988). Novel regulators of bone formation: Molecular clones and activities. *Science* **242,** 1528–1533.

Yamaguchi, D. T., Hahn, T. J., Iida-Klein, A., Kleeman, C. R., and Muallem, S. (1987). Parathyroid hormone–activated calcium channels in an osteoblast-like clonal rat osteosarcoma cell line. *J. Biol. Chem.* **262,** 7711–7718.

Yoon, K., Buenaga, R., and Rodan, G. A. (1987). Tissue specificity and developmental expression of osteopontin. *Biochem. Biophys. Res. Commun.* **148,** 1129–1136.

Zambetti, G., Stein, J., and Stein, G. (1987). Targeting of a chimeric human histone fusion mRNA to membrane-bound polysomes in HeLa cells. *Proc. Natl. Acad. Sci. U.S.A.* **84,** 2683–2687.

Zapf, J., Walter, H., and Froesch, E. R. (1981). Radioimmunological determination of insulin-like growth factor I and II in normal subjects and in patients with growth disorders and extrapancreatic tumor hypoglycemia. *J. Clin. Invest.* **68,** 1321–1327.

Zeitlin, S., Parent, A., Silverstein, S., and Efstratiadis, A. (1987). Pre-mRNA splicing and the nuclear matrix. *Mol. Cell. Biol.* **7,** 111–120.

Zenk, D. W., Ginder, G. D., and Brotherton, T. W. (1990). A nuclear-matrix protein binds very tightly to DNA in the avian β-globin gene enhancer. *Biochemistry* **29,** 5221–5226.

Zernik, J., Thiede, M. A., Twarog, K., Stover, M. L., Rodan, G. A., Upholt, W. B., and Rowe, D. W. (1990). Cloning and analysis of the 5′ region of the rat bone/liver/kidney/placenta alkaline phosphatase gene, a dual-function promoter. *Matrix,* **10:** 38–47.

7

Growth and Differentiation of Myelomonocytic Cells

BRENT L. KREIDER AND GIOVANNI ROVERA

The Wistar Institute of Anatomy and Biology
Philadelphia, Pennsylvania 19104

I. INTRODUCTION

Both proliferation and differentiation are controlled mainly by (1) cytokines and their receptors, (2) intracellular signal transduction pathways, and (3) the eventual gene activation that occurs within the nucleus of the responding cell. In the first level of control, cytokines interact with their corresponding receptors on the cell. The most simplistic scenario is that in which a single cytokine binds to its corresponding receptor and initiates a cascade of events that eventually direct the cell to proliferate or differentiate. This initial step can be easily regulated by mechanisms that control cytokine production or receptor expression. However, the successful differentiation of a given cell often involves many factors working together in a strict sequential order, priming, committing, and then completing the differentiation program. Thus, the levels of control can become progressively more complex. In addition, certain cytokines are capable of controlling the surface

MOLECULAR AND CELLULAR APPROACHES TO THE CONTROL
OF PROLIFERATION AND DIFFERENTIATION

expression of receptors for other factors and can further complicate receptor regulation in the presence of multiple factors, which is usually the case *in vivo.* A study by Walker *et al.* (1985) showed that the cytokines interleukin 3 (IL-3), granulocyte–macrophage colony-stimulating factor (GM-CSF), granulocyte (G)-CSF, and macrophage (M)-CSF were all capable of down-regulating the receptors for one another on purified bone-marrow cells. The relative degree of down-modulation varied for these cytokines, with the most striking effects seen using IL-3. Finally, soluble forms of some of these surface receptors might exist as well as cytokine analogs that compete with the native ligand for receptor binding but do not transduce a signal.

Once the ligand–receptor interaction has successfully occurred, the signal must be transduced from the cell membrane to the nucleus. For this, the cell utilizes one or more of the second-messenger pathways designed to both transduce and amplify the extracellular signal. These signal-transduction pathways include the phosphoinositide-related calcium messenger system, the various phospholipases (A_2, C, and D), the large family of GTP-binding proteins, protein kinase C (PKC), and other protein kinases. There are numerous studies on the biochemistry of phosphoinositide metabolism and the manner in which this pathway transduces and amplifies extracellular signals (for review see Berridge and Irvine, 1984, 1989; Altmann, 1988). In addition, the guanine nucleotide-binding regulatory proteins (G proteins) and their specific capabilities as mediators between the cell membrane and nucleus have been well studied (Gilman, 1987; Neer and Clapham, 1988; Freissmuth *et al.,* 1989). PKC and the cyclic AMP-dependent protein kinase (PKA) have also been investigated extensively for their role in second-messenger pathways (Kishimoto *et al.,* 1980; Kikkawa *et al.,* 1982, 1983, 1987; Nishizuka, 1986; Mellon *et al.,* 1989).

Intracellular signal transduction entails a complex series of conversions and modifications involving not only different compounds but also different subunits of very similar components. Common to most signaling pathways are the G proteins, whose activation can transduce the extracellular signal in two general ways. By coupling with adenylate cyclase (AC), PKA is activated on conversion of ATP into cyclic AMP. PKA then directs the phosphorylation of various protein targets within the cell. Alternatively, coupling with phospholipase C (PLC) leads to the formation of inositol triphosphate (IP_3) and diacylglycerol (DAG). IP_3 mediates the release of intracellular Ca^{2+} stores, which in turn can activate Ca^{2+}-dependent protein kinase, and DAG activates PKC directly. These kinases then phosphorylate proteins within the cell, either activating or inactivating their targets. A given receptor might also have intrinsic kinase activity (see Ullrich and Schlessinger, 1990) and not only activate the pathways upon ligand binding, but also phosphorylate other substrates directly. Obviously, specific phosphatases are needed to dephosphorylate the protein targets acted on by the many kinases. Thus, there are clearly many points within this intracellular signaling mechanism at which both proliferation and differentiation may be controlled.

Once the signal initiated by the growth factor–receptor interaction is received by the nucleus, through a series of relatively ill-defined interactions, a specific set or sets of genes become activated. This gene activation can be divided into two categories, an immediate-early (IE) response and a late response. The IE gene response refers to the induction of a certain set of genes, in the absence of protein synthesis, immediately following growth-factor exposure, usually within the first 2 hr. Late-response elements include the genes associated with the differentiative phenotype of the specific cell. These genes may help direct differentiation or may be activated merely as a consequence of the differentiative program of the cell, without any direct involvement in commitment or completion of the differentiative process.

By far the most work has focused on IE gene activation, which was first characterized by identifying genes induced by serum in quiescent cells (Cochran *et al.,* 1983; Lau and Nathans, 1985, 1987; Almendral *et al.,* 1988). This IE response varies from cell type to cell type and includes *fos* and the *fos*-related antigens (fra) (Greenberg and Ziff, 1984; Muller *et al.,* 1984; Cohen and Curran, 1988; Curran, 1988; Distel *et al.,* 1987), the Jun family (c-*jun, jun-B* and *jun-D*) (Rauscher *et al.,* 1988a; Ryder and Nathans, 1988; Ryder *et al.,* 1988, 1989), and the TPA-induced sequences (TIS) genes (TIS-1, -7, -8, -10, -11) (Lim *et al.,* 1987), which also include *zif/268* (Christy *et al.,* 1988), *nur77* (Hazel *et al.,* 1988), PC-4 (Tirone and Shooter, 1989), KROX-24 (LeMaire *et al.,* 1988), *egr-1* (Sukhatme *et al.,* 1988), NGF1A (Milbrandt, 1987), and NGF1B (Watson and Milbrandt, 1989). The protein products encoded by this large family of IE genes represent DNA-binding proteins (Bohman *et al.,* 1987; Franza *et al.,* 1988; Rauscher *et al.,* 1988b, c), zinc fingers (Milbrandt, 1987; Sukhatme *et al.,* 1988; Chavrier *et al.,* 1989; Cao *et al.,* 1990), members of the steroid and thyroid hormone–receptor superfamily (Hazel *et al.,* 1988; Ryseck *et al.,* 1989; Watson and Milbrandt, 1989), and cytokine-like proteins (Rollins *et al.,* 1988; Tirone and Shooter, 1989). With such a diverse repertoire of gene products, the IE response induced by a given growth factor could easily initiate a complex differentiation program or direct the proliferation of the responding cell. The controlled induction of these genes is therefore probably critical for the proper cellular response to a differentiative or proliferative signal.

A number of cellular genes play such crucial roles in growth or differentiation that malignant transformation can occur upon their deregulation. These genes are the various protooncogenes; it was the initial studies on their mechanism of action that helped to dissect the general levels of proliferative and differentiative control described above. Protooncogenes encode growth factors, growth-factor receptors, second-messenger components, and putative transcriptional modulators, all of which have been shown to alter proper growth and/or differentiation of various cell types. Many of these protooncogenes fall into the category of IE genes.

In this chapter, we shall address the control of myelomonocytic growth and differentiation with regard to cytokine–receptor interactions, intracellular signal transduction, and gene activation. We shall examine cell lines only as a model to dissect these steps. Even though the use of cell lines has a drawback in that some of the mechanisms identified may not necessarily occur *in vivo,* they do have the distinct advantage of providing a homogeneous cell population and an unlimited number of cells. We attempt to provide a general overview of the possible mechanisms involved in the control of growth and differentiation, and refer the reader to the most recent reviews or reports concerning specific points for detailed explanation.

II. CELL LINES AS MODELS OF MYELOMONOCYTIC DIFFERENTIATION

Two of the first-established cell lines shown to be capable of differentiation *in vitro* are the M1 and murine erythroleukemia (MEL) or Friend cell lines. MEL cells can be induced to differentiate into orthochromatic normoblasts, and therefore represent a unique model system for studying erythroid differentiation and globin gene regulation (reviewed in Tsiftsoglou and Wong, 1985). M1 cells were first isolated as a myeloid leukemia cell line (Fibach and Sachs, 1976). A variant of the parent cell line, termed M1D+, is capable of differentiation in response to IL-6 (Liebermann *et al.,* 1982; Lord *et al.,* 1990) and has subsequently been used as a model of *in vitro* myelomonocytic differentiation. In M1D+ cells, proliferation is uncoupled from differentiation such that growth is immediately arrested upon stimulation with IL-6. The establishment of these two cell lines generated great interest in isolating additional lines capable of *in vitro* differentiation representing as many lineages as possible. Thus a number of human leukemic cell lines were established, including K562, established from a patient with chronic myeloid leukemia in blast crisis (Lozzio and Lozzio, 1975); KG1, derived from a patient with erythroleukemia (Koeffler and Golde, 1978); U937, a histiocytic lymphoma-derived line (Sundstrom and Nilsson, 1976); and HL-60, a myelogenous cell line isolated from a promyelocytic leukemia patient (Collins *et al.,* 1977). The properties of HL-60, KG1, and K562 cells have been reviewed (Koeffler and Golde, 1980), and a more recent article focuses on HL-60 and U937 as models of myelomonocytic development (Harris and Ralph, 1985). Finally, Collins (1987) has recently reviewed the proliferation, differentiation, and cellular oncogene expression of the HL-60 cell line.

Although all of these lines have been useful in investigations of growth and differentiation control, the HL-60 cell line has undergone the most extensive experimental scrutiny and generated a great abundance of information. As noted by Collins, the reason for the great research interest in HL-60 cells stems from their ability to differentiate *in vitro* into granulocytes or monocytes, depending on the stimulus used. These cells normally proliferate continuously in culture and are blocked at the promyelocytic stage of differentiation. A wide variety of agents remove this differentiation block and result in growth arrest and maturation of these cells to either granulocytes (dimethyl sulfoxide and retinoic acid) or monocytes (tetradecanoyl phorbol acetate, interferon-γ, vitamin D_3, tumor necrosis factor-α, and sodium butyrate). In addition, these cells have a number of oncogenic alterations. HL-60 cells contain a 15- to 30-fold genomic amplification of the c-*myc* protooncogene when compared to normal cells (Collins and Groudine, 1982), and this amplification results in high basal levels of c-*myc* RNA in HL-60 cells. Two other genomic mutations, alterations in the N-*ras* and *p53* genes, are also found in these cells; the N-*ras* gene can be *activated* by specific point mutations. One such mutation, that of the second nucleotide of codon 61, has been documented in the HL-60 cells (Bos *et al.*, 1985); the *p53* gene has been deleted in HL-60 cells, resulting in no expression of this mRNA or protein (Wolf and Rotter, 1985).

These alterations have implications for the leukemic phenotype of the HL-60 cells. First, it is known that *myc* and *ras* can act cooperatively to malignantly transform primary rodent fibroblasts (Land *et al.*, 1983). Second, *p53* has been shown to act as a growth-suppressor gene, and its absence in HL-60 cells could contribute significantly to their phenotype. In light of the above characteristics, it is obvious that the HL-60 model system is very valuable for the investigation of the mechanism responsible for the block in myelomonocytic differentiation seen in these cells and exactly what triggers the release of this block upon chemical induction.

The $M1D^+$, HL-60, KG1, K562, and U937 cell lines all share two features: (1) they are leukemic, and (2) when it can be induced, immediate growth arrest accompanies differentiation. With the exception of TPA and retinoic acid, the various chemical agents cause differentiation of these lines by either unknown or poorly defined mechanisms. Thus, it becomes important to have available a system to analyze "normal" cell growth and differentiation, that is, a growth factor-dependent cell line with a nonleukemic phenotype.

Two nonleukemic, growth factor-dependent murine cell lines have been established: FDC-P1 (Dexter *et al.*, 1980) and 32DC1-23 (Greenberger *et al.*, 1983), both bone marrow-derived, IL-3-dependent myeloid cells that have been used extensively for the investigation of growth control mediated by IL-3. The FDC-P1 cells also grow in the presence of GM-CSF. Although both of these lines are nonleukemic and growth-factor dependent, neither is capable of differentiating past the blast-like phenotype. As for growth factor-dependent hematopoietic cell lines that are nonleukemic and capable of differentiation *in vitro*, only the 32DC13(G) cell line is presently available as a model system. The 32D cell line was originally established from C3H/HeJ long-term, bone-marrow cultures as a nonadherent, IL-3–dependent basophil–mast cell line. It was later shown that upon removal from IL-3 and culture in the presence of G-CSF, a subclone of the parental cell line, termed 32DC13(G), could terminally differentiate into neutrophilic granulocytes (Valtieri *et al.*, 1987; Ohta *et al.*, 1989). This cell line represents the only nonleukemic *in vitro* model of myeloid differentiation presently available and has been used extensively to study the regulation of myelomonocytic stem-cell differentiation. Very recently, a human hematopoietic cell line, KMT-2, was established from umbilical cord blood (Tamura *et al.*, 1990). These cells are IL-3–dependent and express a large number of IL-3 receptors; thus KMT-2 promises to be a useful tool in analyses of signal transduction by human IL-3 and for the biochemistry of the IL-3 receptor.

In the following sections, we present some of the information obtained through studies with cell lines and what these findings imply concerning mechanisms of growth and differentiation control. We focus on the HL-60 (growth factor-independent; differentiation-competent; leukemic), 32DC13(G) (growth factor-dependent; differentiation-competent; nonleukemic), and FDC-P1 (growth factor-dependent; differentiation-incompetent; nonleukemic) cell systems since they represent an excellent mix of cell types with specific proliferative and differentiative potentials.

III. CYTOKINES AND THEIR RECEPTORS

Very early studies by Metcalf (reviewed in Metcalf, 1984) used bone-marrow cells in colony-forming assays and laid the groundwork for the involvement of the various cytokines in myelomonocytic differentiation. Those studies showed that the effects of IL-3 and GM-CSF are not lineage restricted and are focused on more immature cells, whereas G-CSF and M-CSF appear to act in a lineage-restricted manner on more mature, committed cells. A major area of investigation turned toward the mechanisms by which an alteration or complete abrogation of a cell's growth-factor requirements might lead to a leukemic phenotype. As expected, cells able to produce the ligand required for their growth often grew autonomously. One of the early studies reporting this phenomenon involved the introduction of a retroviral vector expressing GM-CSF into FDC-P1 cells (Lang *et al.*, 1985). The infected cells secreted GM-CSF, grew independent of exogenous GM-CSF, and produced tumors in syngeneic mice. Similarly, FDC-P1 or 32DC123 cells transfected with a retroviral vector carrying the IL-3 gene were able to proliferate in the absence of IL-3 and became leukemogenic (Browder *et al.*, 1989; Hapel *et al.*, 1986).

However, the use of antibodies against IL-3 led to conflicting conclusions about the mechanisms by which this growth-factor autonomy was achieved. In the study by Hapel *et al.* (1986) addition of a neutralizing antibody against IL-3 blocked the autocrine stimulation of their IL-3–transfected FDC-P1 cells, suggesting that IL-3 was secreted from the cell and then interacted with surface receptors to stimulate the growth of the secreting cell. In contrast, Browder *et al.* (1989) found that exogenously added antibodies had no effect on the growth-factor independence of their transfected 32DC123 cells, supporting a much more powerful mechanism of autocrine stimulation, wherein IL-3 can interact with its receptor intracellularly, before it is expressed on the cell surface. Either of these autocrine loops can render a cell growth-factor independent and may be an important mechanism by which growth control is altered. In fact, a study by Ostertag's group (Laker *et al.,* 1987) suggests that these two mechanisms may be sequential steps toward tumorigenicity. They found that transfected FDC-P1 cells first secrete the ligand, in this case GM-CSF, and respond through an interaction at the outer membrane. They then presumably acquire a second mutation that abrogates the need for secretion of GM-CSF for autonomous growth. This second step is believed to irreversibly transform the cell into a tumorigenic clone.

How do these experimentally induced autocrine growth models reflect possible *in vivo* transforming events? This question remains unanswered, but there is some evidence supporting the possible role of autocrine stimulation in leukemogenesis. In one study, two of three cases of acute myeloblastic leukemia (AML), which provided clones capable of autonomous proliferation *in vitro,* were found to secrete GM-CSF (Young and Griffin, 1986). Although the constitutive expression of GM-CSF was not shown to be responsible for the disease state in these two AML cases, it does suggest the involvement of such mechanisms *in vivo.* More recently it has been shown that one third of *in vivo*-derived, growth factor-independent leukemic FDC-P1 variants express either GM-CSF or IL-3 (Duhrsen *et al.,* 1990). In all cases, this was due to the insertion of intracisternal A-particle DNA in the 5′-flanking regions of the CSF genes. If further studies can provide additional evidence for autocrine growth of cancer cells *in vivo,* the information from studies using the *in vitro* model systems mentioned may well lead to better therapeutic approaches (reviewed in Sporn and Roberts, 1985).

An additional strategy for the investigation of specific growth factor–receptor interactions involves the introduction of a foreign receptor into a naive cell in order to monitor the ability of the transfected cell to respond to the corresponding ligand. Introduction of the murinc c-*fms* gene, which encodes the M-CSF or CSF-1 receptor (Sherr *et al.,* 1985), into FDC-P1 cells, or transfection of the human c-*fms* gene into 32DC13 cells, results in M-CSF–dependent growth of both of these cell lines (Rohrschneider and Metcalf, 1989; Kato and Sherr, 1990). In addition, transfection of 32DC13 cells with the human epidermal growth factor (EGF) receptor resulted in the ability of EGF to deliver a proliferative signal to these cells (Pierce *et al.,* 1988). Similar results were obtained with the IL-3–dependent pre-mast cell line IC2 (Wang *et al.,* 1989).

These types of studies are important for two reasons. First, such artificial systems allow analyses of specific growth factor-receptor signaling mechanisms and the importance of the integrity of such pathways. Second, they indicate that in many cases, naive cells possess all of the components necessary to transduce the signal mediated through a receptor they normally do not express. That is, cellular nonresponsiveness to a specific factor rests solely in the lack of appropriate receptor expression. This last point supports the concept of growth or differentiation control at the level of genetic programming, which determines the type of receptor to be expressed on the cell surface. Such restricted receptor expression might well represent a more efficient control mechanism than restricted ligand production in determining differentiation of progenitor cells, considering the wide variety of growth factors usually present in the microenvironment of the bone marrow. Receptor control could be at two levels: (1) differentiation versus growth, or (2) differentiation along one lineage versus differentiation along a different lineage.

Several studies suggest that this type of control is actually used by hematopoietic cells. It has long been known that interaction of a ligand with its receptor on the cell surface is followed by the internalization of the ligand–receptor complex and eventual recycling of the same or a newly synthesized receptor to the cell surface. This finding led to the concept of down-regulation of a receptor by its own ligand. Moreover, some growth factors are able to down-regulate the receptors for other growth factors as well. This complex pattern of receptor down-modulation (Walker *et al.,* 1985) suggested an intricate system of cellular-response control. It seems that IL-3 not only directs the proliferation of an immature progenitor cell through interactions with its own receptor, but may also prevent terminal differentiation of the cell by neutralizing the functional receptors for another factor that would induce differentiation. Indeed, studies using the 32DC13(G) system showed that even very small concentrations of IL-3 effectively block the G-CSF–induced differentiation of this cell line into neutrophilic granulocytes (Valtieri *et al.,* 1987). Although no receptor studies have been published concerning the down-modulation of the G-CSF receptor by IL-3 in this cell line, the biological responses seen, i. e., IL-3 antagonism of G-CSF–induced differentiation, fully support the findings of Walker *et al.* (1985). This same cell system also supports the notion of receptor modulation as a regulatory mechanism of the differentiation of a multipotential progenitor cell along a specific lineage.

32DC13(G) cells have no detectable GM-CSF receptors on their surface and do not normally respond to GM-CSF. However, upon exposure to G-CSF, these cells become responsive to GM-CSF (Kreider *et al.,* 1990). Growth in GM-CSF results in the generation of granulocytes and monocyte–macrophages, with readily detectable GM-CSF receptors on the cell surface, while maintaining a self-renewing progenitor pool. The ability of G-CSF to induce GM-CSF responsiveness in 32DC13(G) cells through induction of the GM-CSF receptor is an example of lineage determination based on receptor control. The 32DC13(G) cells cannot form monocyte–macrophages unless the GM-CSF receptor is present on the cell surface. A second growth factor, G-CSF, induces expression of these receptors, resulting in maturation along an additional lineage. Thus, the potential of one growth factor to down-regulate a specific receptor, while another growth factor can up-regulate the same or a different receptor, represents a delicate but precise mechanism by which these factors control a cellular response.

A final point concerns the strong oncogenic potential of a number of growth-factor receptors. In many cases, the alteration of a specific receptor by truncation, point mutation, or even overexpression of the intact protein results in transformation. The v-*erbB* oncogene represents a truncated epidermal growth-factor receptor (EGF-R), which lacks the extracellular-binding domain, is constitutively activated, and transforms cells *in vitro* (for review, see Vennstrom and Damm, 1988). The v-fms protein, unlike its nononcogenic counterpart c-fms, is constitutively phosphorylated on tyrosine and mediates the phosphorylation of a number of protein targets in the absence of M-CSF (Woolford *et al.*, 1985; Tamura *et al.*, 1986; Morrison *et al.*, 1988). The two proteins differ due to a number of point mutations, but more importantly, the carboxy-terminal 40 amino acids of c-fms are replaced by 11 unrelated amino acids in v-fms, such that a tyrosine residue near the carboxy-terminus is deleted in the receptor. A number of studies have indicated that this tyrosine residue negatively regulates signal transduction through the CSF-1 receptor. However, substitution of this residue alone is not sufficient to confer transforming ability to the c-fms protein, indicating that additional alterations may be required for full transforming capabilities (see Browning *et al.*, 1986 and Roussel *et al.*, 1987, 1988a). In contrast, a single residue substitution in the extracellular domain of the CSF-1 receptor results in transformation of NIH 3T3 cells and, in combination with the carboxy-terminus tyrosine substitution, transforming efficiency is increased (Roussel *et al.*, 1988b). It has been proposed that the extracellular mutation causes a conformational change in the receptor that mimics ligand binding, allowing signaling in the absence of exogenous M-CSF.

Such alterations, either by mimicking ligand binding or removing negative regulatory regions, result in the deregulation of a component of normal cell growth and differentiation and result in malignancy. Thus, while growth factors are critical for proper control of cell maturation and proliferation, the receptors for these factors may play an even more crucial role, since very minor alterations of the receptor protein can induce the aberrant signaling that releases the transformed cell from its dependence on the microenvironment for growth-stimulatory factors.

IV. SIGNAL TRANSDUCTION

Immediately following ligand–receptor interaction, a variety of second-messenger pathways may be activated, directly or indirectly, by the receptor–ligand complex. Despite the detailed biochemical characterization of the various G proteins and of the pathways by which phosphatidylinositol forms inositol triphosphate and diacylglycerol, the importance of these processes in cellular proliferation or differentiation remains unclear, since such analyses do not necessarily provide clues as to the sites that determine deregulation of the responding cell. An alternative to classic biochemistry and a more direct way to study the roles of these signaling mechanisms involves the alteration of one or more of the components of these pathways and an evaluation of effects. Such an approach has been used to investigate the mode of G protein action and the functional role of these proteins in the cell response to external stimuli.

Bacterial toxins from *Vibrio cholera* and *Bordetella pertussis* can covalently modify and irreversibly alter the activity of the G proteins owing to ADP-ribosylation of their α subunits (Gilman, 1987). Thus, an alteration of a response normally elicited by a specific stimulus in cells pretreated with these agents, suggests the involvement of G proteins in such a signaling pathway. In this way, a role for G-protein signaling has been established for neutrophils, monocytes, macrophages, and mast cells (Harnett and Klaus, 1988). In addition, nonhydrolyzable analogs of GTP such as Gpp(NH)p or GTP-γS result in constant activation of the α subunit of G proteins (Gilman, 1987), and have been used in conjunction with the bacterial toxins to further support the involvement of these transducers in the myeloid cell response.

These studies shed light on the involvement of the various G proteins in specific signal-transduction networks or eventual gene activation in myeloid cells, but did not address the possible correlation between deregulation of these components and malignancy. To investigate this concept, many studies focused on the *ras* oncogene, whose product has GTP-binding properties indicative of a G protein and localizes to the inner surface of the cytoplasmic membrane (see Barbacid, 1987; Sigal, 1988). Unlike the G proteins, the *ras* gene product does not rely on a receptor–ligand interaction for activation, and is capable of transforming a number of cell types, including those of the myeloid lineage. For this reason, the *ras* oncogene can be considered a G protein *mutant.* 32DC13(G) cells transformed by *ras* (32D-ras cells) do not achieve growth-factor independence, but are blocked in their ability to differentiate properly (Mavilio *et al.,* 1989), suggesting that the mechanism by which G-CSF stimulates 32DC13(G) cells to undergo terminal differentiation is altered by the presence of the *ras* oncogene. In fact, the G-CSF signal has been perturbed in such a way that the 32D-ras cells respond to G-CSF by proliferating instead of terminally differentiating. Interestingly, HL-60 cells carry a N-*ras* point mutation. Since both the 32D-ras and HL-60 cells are blocked at the promyelocytic stage of differentiation, the activated ras protein must be capable of initiating and/or maintaining this block either alone or in concert with other components. Given the highly abnormal genetic background in HL-60 cells, it seems likely that a number of alterations contribute to the leukemic phenotype, of which ras is one. In the case of 32DC13(G) cells, the 32D-ras phenotype suggests that G-CSF–mediated differentiation normally utilizes a pathway with which *ras* interacts and can interfere, most likely that of G protein-mediated transduction.

Although the *ras* oncogene has been shown to alter normal myeloid cell responses, the mechanisms underlying these alterations remain unclear. However, a recent report has shed light on the mechanism responsible for the functional uncoupling between intracellular PLC and external signals observed in *ras*-transformed cells (Alonso *et al.,* 1990). The normal coupling of G proteins to PLC is significantly impaired in cells transformed by *ras* and other membrane-associated oncogenes, suggesting that G proteins in the transformed cells are modified such that they are unable to interact with PLC. In fact, changes in isoelectric focusing patterns have suggested covalent modifications of the G proteins in *ras*-transformed cells (Gilman, 1987). These data support the potential for cell transformation when one or more of these components is altered or constitutively activated.

As with G proteins, there are also agents capable of altering the normal constraints of PKC. For example, 12-*O*-tetradecanoylphorbol-13-acetate (TPA) directly stimulates PKC, bypassing the receptor–ligand interaction step that normally precedes its activation. Use of TPA has allowed the biochemical dissection of the PKC signaling mechanism and has provided a tool for studying aberrant signaling through this protein kinase. One interesting phenomenon is the macrophage-like differentiation of HL-60 cells treated with TPA, a response also seen in most acute myeloid leukemia and chronic myelogenous leukemia blast-crisis cells when exposed to TPA (see Collins, 1987 and references therein). The finding that the differentiation block in the HL-60 cells, and perhaps in leukemic cells *in vivo,* can be bypassed by direct overexpression of PKC suggests that this pathway somehow overrides the leukemic phenotype by inducing differentiation. Since the downstream effect of kinase activation is protein phosphorylation, the pattern of phosphorylation seen upon TPA induction of HL-60 cells might provide clues as to the identity of the protein products involved in the differentiation process. However, the TPA-induced differentiation of HL-60 cells is associated with the increased phosphorylation of a number of different proteins (Feurstein and Cooper, 1983) and, since phosphorylation can either activate or inactivate the modified protein, sorting out the role of these various proteins represents a difficult task.

Protein phosphorylation has also been investigated under more physiological conditions, that is, in response to growth factors associated with the myeloid lineage. A number of studies have demonstrated the phosphorylation of specific protein products in response to IL-2, IL-3, G-CSF, GM-CSF, TNF-α and γ-IFN in established cell lines (Evans *et al.,* 1987; Morla *et al.,* 1988; Evans *et al.,* 1990). Interestingly, a few proteins have been shown to be phosphorylated by more than one of these factors, indicating the possible existence of a common intermediate in these signaling pathways. Information about the role of these proteins in the cellular response to specific ligands, as well as the mechanism leading to their phosphorylation, is still limited, but will provide invaluable insight into the signal transduction in normal or altered hematopoietic cell responses.

A final note involving protein phosphorylation is the effect of the kinase oncogenes on myeloid cell lines. Two powerful kinases, the v-*src* and v-*abl* oncogenes, are capable of abrogating the growth-factor requirements of a number of myeloid cell lines (Cook *et al.,* 1985; Pierce *et al.,* 1985; Rovera *et al.,* 1987; Cleveland *et al.,* 1989; Anderson *et al.,* 1990), and in the cases when it could be tested, blocked the differentiative potential of these cells (Rovera *et al.,* 1987; Kruger and Anderson, 1990). These proteins are associated with the plasma membrane, and constitutively phosphorylate targets within the cell, resulting in strong proliferative signals without the need for exogenous ligand. Again, the deregulation and overexpression of components similar to those involved in intracellular signal transduction results in a drastically altered phenotype.

V. GENE ACTIVATION

As mentioned previously, the gene activation that occurs in response to a growth factor can be categorized as an immediate-early (IE) or a late response. Unfortunately, analyses of the transcriptional activation of specific IE genes in response to differentiative, proliferative, or generalized inducers have not yet served to identify genes consistently associated with differentiation or proliferation. The PC-12 pheochromocytoma-derived cell line differentiates along the neuronal lineage in response to nerve growth factor (NGF) but proliferates only when exposed to EGF or TPA. However, a number of IE genes are induced by all three of these agents, including c-*fos,* TIS-1, -7, -8, and -11 (Kujubu *et al.,* 1987; Tirone and Shooter, 1989) and *egr-1* (Cao *et al.,* 1990). Similarly, TPA, EGF, or fibroblast growth factor (FGF) treatment of NIH 3T3 cells induces expression of c-*fos* and all five of the TIS genes (Lim *et al.,* 1989). *egr-1* activation has also been seen during neuronal, glial, and cardiac muscle cell differentiation using an embryonal carcinoma cell line (Sukhatme *et al.,* 1988).

The HL-60, M1D+ and 32D cell lines have provided valuable information about gene activation in the myeloid lineage. A rapid, transient induction of c-fos mRNA occurs upon TPA induction of HL-60 cells (Mitchell *et al.,* 1985; Muller *et al.,* 1985), which ultimately results in macrophage-like differentiation of these cells. IL-6 treatment of M1D+ cells results in the induction of c-myc, c-myb, c-fos, c-fes and c-fms mRNA (Liebermann and Hoffman-Liebermann, 1989) as well as a number of unique mRNA species, termed MyD, which have not yet been fully characterized (Lord *et al.,* 1990). A subclone of 32DC13 cells that proliferate in response to GM-CSF show a rapid, transient induction of c-fos, TIS-7, -8, -10, and -11 mRNAs when treated with either GM-CSF or TPA (Varnum *et al.,* 1989).

32DC13(G) cells that differentiate in response to G-CSF induce c-fos, c-jun, jun-B, TIS-7, -8, and -11 mRNAs in response to this factor (Kreider and Rovera, unpublished results). Clearly, the available induction data obtained using both mitogenic and differentiative agents as well as generalized inducers such as TPA, do not allow assignment of any of the IE genes as specific triggers for the differentiative or the proliferative response. In fact, a role for these genes in both processes seems likely.

What then is the exact role of the IE response in myeloid cell maintenance? The IE response is probably better viewed as the induction of a specific set of genes that work in concert to initiate the proper cellular response instead of the induction of individual genes. Survey of the Fos and Jun gene families illustrates the potential complexity of this response. The Fos protein can form heterodimers with any of the Jun proteins, and the Jun proteins themselves are capable of forming Jun–Jun homodimers or Jun–Jun heterodimers. All of these dimers may have different gene targets, based on subtle differences in sequence recognition, binding affinities, or interaction with proteins other than Jun or Fos. In addition, the ratio of the various dimer species is most likely controlled by the absolute concentrations of the different Fos and Jun proteins as well as their affinities for one another for dimer formation. Last, some of these dimers may actually serve as repressors rather than activators. Thus, induction of the mRNA for one of these genes in response to a specific factor may be misleading if taken alone, and when more than one of these genes are induced, the great number of possible interactions occurring at the protein level suggests a complicated mechanism of subsequent gene activation. The possible regulatory effects of these dimers on the transcription of the specific IE genes, which encode the proteins that form these dimers, adds to this complexity. Obviously, much work remains to be done in evaluating the importance of IE gene induction in response to both proliferative and differentiative signals.

Late gene activation in response to various growth factors has not been so intensely studied as the IE response, mainly because these genes are quite specific to the differentiation lineage of the responding cell. That is, no common set of genes is induced as there is for the IE response. The late genes associated with the myeloid cell response include myeloperoxidase, lactoferrin, lysozyme, and a number of surface antigens, including specific growth-factor receptors. The induction of these mRNAs, or the presence of the proteins encoded by them, has long been used as a molecular marker of differentiation along a specific lineage, but their role in the differentiation process remains uncharacterized. To date, it is not entirely clear whether the induction of these genes occurs solely as a result of differentiation commitment and progression or whether they actually help effect the differentiation program in some way. Since the proteins encoded by these late genes most often serve a functional role in the mature cell, it is likely that induction of these genes occurs after the differentiation program has already been initiated, and that they represent activated genes with no real regulatory role in this process. Instead, their induction provides the terminally differentiated cell with the necessary components required for its eventual function.

VI. CONCLUDING REMARKS

In this chapter we have outlined the progressive steps involved in the control of myeloid cell growth and differentiation and, where available, the data to support such controlling processes. Clearly, the cell response to differentiative or proliferative signals emerges as a complex picture, involving myriad components.

Understanding the mechanisms controlling myeloid cell growth and differentiation has two major goals. First is an understanding of the events that progressively signal a *normal* cell to either enter and complete the cell cycle or initiate a specific differentiation program. These two fundamental processes control myeloid cell development, as well as hematopoietic cell maintenance in general, and are critical for the overall balance of immature and mature blood cells within the body. The second goal is dependent on understanding the first. When a cell escapes the normal constraints of growth control or is in some manner blocked in its ability to differentiate, malignant transformation may result. Understanding how a cell is normally controlled at these two levels may lead to identification of the controlling step that has been bypassed in the malignantly transformed cell, and to clues about strategies to restore this controlling element. The controlling elements discussed in this chapter must be further documented and described in detail for such goals to be reached.

REFERENCES

Almendral, J. M., Sommer, D., MacDonald-Bravo, H., Burckhardt, J., Perera, J., and Bravo, R. (1988). Complexity of the early genetic response to growth factors in mouse fibroblasts. *Mol. Cell. Biol.* **8,** 2140–2148.

Alonso, T., Srivastava, S., and Santos, E. (1990). Alterations of G-protein coupling function in phosphoinositide signaling pathways of cells transformed by *ras* and other membrane-associated and cytoplasmic oncogenes. *Mol. Cell. Biol.* **10,** 3117–3124.

Altman, J. (1988). Ins and outs of cell signalling. *Nature (London)* **331,** 119–120.

Anderson, S. M., Carroll, P. M., and Lee, F. D. (1990). Abrogation of IL-3 dependent growth requires a functional v-*src* gene product: Evidence for an autocrine growth cycle. *Oncogene* **5,** 317–325.

Barbacid, M. (1987). ras genes. *Annu. Rev. Biochem.* **56,** 779–827.

Berridge, M. J., and Irvine, R. F. (1984). Inositol triphosphate, a novel second messenger in cellular signal transduction. *Nature (London)* **312,** 315–321.

Berridge, M. J., and Irvine, R. F. (1989). Inositol phosphates and cell signalling. *Nature (London)* **341,** 197–205.

Bohmann, D., Bos, T. J., Admon, A., Nishimura, T., Vogt, P.K., and Tjian, R. (1987). Human proto-oncogene c-*jun* encodes a DNA-binding protein with structural and functional properties of transcription factor AP-1. *Science* **238,** 1386–1392.

Bories, D., Raynal, M-C., Solomon, D. H., Darzynkiewics, Z., and Cayre, Y. E. (1989). Down-regulation of a serine protease, myeloblastin, causes growth arrest and differentiation of pro-myelocytic leukemia cells. *Cell* **59,** 959–968.

Bos, J., Verlaan-de Vries, M., Jansen, A., Veeneman, G., van-Boom, J., and vanderEb, A. (1985). Three different mutations in codon 61 of the human N-ras gene detected by synthetic oligoucleotide hybridization. *Nucleic Acids Res.* **12,** 9155–9163.

Browder, T. M., Abrams, J. S., Wong, P. M., and Nienhuis, A. W. (1989). Mechanism of autocrine stimulation in hematopoietic cells producing interleukin-3 after retrovirus-mediated gene transfer. *Mol. Cell. Biol.* **9,** 204–213.

Browning, P. J., Bunn, H. F., Cline, A., Shuman, M., and Nienhuis, A. W. (1986). "Replacement" of COOH-terminal truncation of v-*fms* with c-*fms* sequences markedly reduces transformation potential. *Proc. Natl. Acad. Sci. U.S.A.* **83,** 7800–7804.

Cao, X., Koski, R. A., Gashler, A., McKiernan, M., Morris, C. F., Gaffney, R., Hay, R. V., and Sukhatme, V. P. (1990). Identification and characterization of the Egr-1 gene product, a DNA-binding zinc-finger protein induced by differentiation and growth signals. *Mol. Cell. Biol.* **10,** 1931–1939.

Chavrier, P., Janssen-Timmen, U., Mattei, M. G., Zerial, M., Bravo, R., and Charnay, P. (1989). Structure, chromosome location, and expression of the mouse zinc-finger gene *Krox-20:* Multiple gene products and coregulation with the protooncogene c-*fos. Mol. Cell. Biol.* **9,** 787–797.

Christy, B. A., Lau, L. F., and Nathans, D. (1988). A gene activated in mouse 3T3 cells by serum growth factors encodes a protein with "zinc-finger" sequences. *Proc. Natl. Acad. Sci. U.S.A.* **85,** 7857–7861.

Cleveland, J. L., Dean, M., Rosenberg, N., Wang, J. Y., and Rapp, U. R. (1989). Tyrosine kinase oncogenes abrogate interleukin-3 dependence of murine myeloid cells through signaling pathways invovling c-*myc:* Conditional regulation of c-*myc* transcription by temperature-sensitive v-*abl. Mol. Cell. Biol.* **9,** 5685–5695.

Cochran, B. H., Reffel, A. C., and Stiles, C. D. (1983). Molecular cloning of gene sequences regulated by platelet-derived growth factor. *Cell* **33,** 939–947.

Cohen, D. R., and Curran, T. (1988). *fra-1:* A serum-inducible, cellular immediate-early gene that encodes a *fos*-related antigen. *Mol. Cell. Biol.* **8,** 2063–2069.

Collins, S. J., Gallo, R. C., and Gallagher, R. E. (1977). Continuous growth and differentiation of human myeloid leukaemic cells in suspension culture. *Nature (London)* **270,** 347–349.

Collins, S., and Groudine, M. (1982). Amplification of endogenous *myc*-related DNA sequences in a human myeloid leukemia cell line. *Nature (London)* **298,** 679–681.

Collins, S. J. (1987). The HL-60 promyelocytic leukemia cell line: Proliferation, differentiation, and cellular oncogene expression. *Blood* **70,** 1233–1244.

Cook, W. D., Metcalf, D., Nicola, N. A., Burgess, A. W., and Walker, F. (1985). Malignant transformation of a growth factor–dependent myeloid cell line by Abelson virus without evidence of an autocrine mechanism. *Cell* **41,** 677–683.

Curran, T. (1988). The *fos* oncogene. *In* "The Oncogene Handbook" (E. P. Reddy, A. M. Skalka, and T. Curran, eds.), Elsevier Biomedical Press, Amsterdam.

Dexter, T. M., Garland, J., Scott, D., Scolnick, E., and Metcalf, D. (1980). Growth of factor-dependent hematopoietic precursor cell lines. *J. Exp. Med.* **152,** 1036–1047.

Distel, R. J., Ro, H. S., Rosen, B. S., Groves, D. L., and Spiegelman, B. M. (1987). Nucleoprotein complexes that regulate gene expression in adipocyte differentiation: Direct participation of c-fos. *Cell* **49,** 835–844.

Duhrsen, U., Stahl, J., and Gough, N. M. (1990). *In vivo* transformation of factor-dependent hemopoietic cells: Role of intracisternal A-particle transposition for growth-factor gene activation. *EMBO J.* **9,** 1087–1096.

Evans, S. W., Rennick, D., and Farrar, W. L. (1987). Identification of a signal-transduction pathway shared by hematopoietic growth factors with diverse biological specificity. *Biochem. J.* **244,** 683–691.

Evans, J. P. M., Mire-Sluis, A. R., Hoffbrand, V., and Wickremasinghe, R. G. (1990). Binding of G-CSF, GM-CSF, tumor necrosis factor α and γ-interferon to cell-surface receptors on human myeloid leukemia cells triggers rapid tyrosine and serine phosphorylation of 75-kDa protein. *Blood* **75,** 88–95.

Feurstein, N., and Cooper, H. L. (1983). Rapid protein phosphorylation induced by phorbol ester in HL60 cells. *J. Biol. Chem.* **258,** 10786–10793.

Fibach, E., and Sachs, L. (1976). Control of normal differentiation of myeloid leukemic cells. *J. Cell. Physiol.* **89,** 259–266.

Franza, B. R., Jr., Rauscher, F. J., III, Josephs, S. F., and Curran, T. (1988). The fos complex and fos-related antigens recognize sequence elements that contain AP-1 binding sites. *Science* **239,** 1150–1153.

Freissmuth, M., Casey, P. J., and Gilman, A. G. (1989). G proteins control diverse pathways of transmembrane signaling. *FASEB J.* **3,** 2125–2131.

Gilman, A. G. (1987). G proteins: Transducers of receptor-generated signals. *Annu. Rev. Biochem.* **56,** 615–649.

Greenberg, M. E., and Ziff, E. M. (1984). Stimulation of 3T3 cells induces transcription of the c-*fos* protooncogene. *Nature (London)* **311,** 433–438.

Greenberger, J. S., Sakakeeny, M. A., Humphries, R. K., Eaves, C. J., and Eckner, R. J. (1983). Demonstration of permanent factor-dependent multipotential (erythroid/neutrophil/basophil) hematopoietic progenitor cell lines. *Proc. Natl. Acad. Sci. U.S.A.* **80,** 2931–2935.

Hapel, A. J., Vande-Woude, G., Campbell, H. D., Young, I. G., and Robins, T. (1986). Generation of an autocrine leukaemia using a retroviral expression vector carrying the interleukin-3 gene. *Lymphokine Res.* **5,** 249–254.

Harnett, M. M., and Klaus, G. G. B. (1988). G protein regulation of receptor signalling. *Immunol. Today* **9,** 315–320.

Harris, P., and Ralph, P. (1985). Human leukemic models of myelomonocytic development: A review of the HL-60 and U937 cell lines. *J. Leukoc. Biol.* **37,** 407–422.

Hazel, T. G., Nathans, D., and Lau, L. F. (1988). A gene inducible by serum growth factors encodes a member of the steroid and thyroid hormone receptor superfamily. *Proc. Natl. Acad. Sci. U.S.A.* **85,** 8444–8448.

Kato, J., and Sherr, C. J. (1990). Human colony-stimulating factor 1 (CSF-1) receptor confers CSF-1 responsiveness to interleukin-3–dependent 32DCL3 mouse myeloid cells and abrogates differentiation in response to granulocyte CSF. *Blood* **75,** 1780–1787.

Kikkawa, U., Takai, Y., Minakuchi, R., Inohara, S., and Nishizuka, Y. (1982). Calcium-activated, phospholipid-dependent protein kinase from rat brain. *J. Biol. Chem.* **257,** 13341–13348.

Kikkawa, U., Takai, Y., Tanaka, Y., Miyake, R., and Nishizuka, Y. (1983). Protein kinase C as a possible receptor protein of phorbol esters. *J. Biol. Chem.* **258,** 11442–11445.

Kikkawa, U., Ogita, K., Ono, Y., Asano, Y., Shearman, M. S., Fujii, T., Aso, K., Sekiguchi, K., Igarashi, K., and Nishizuka, Y. (1987). The common structure and activities of four subspecies of rat brain protein kinase C family. *FEBS Lett.* **223,** 212–216.

Kishimoto, A., Takai, Y., Mori, T., Kikkawa, U., Nishizuka, Y. (1980). Activation of calcium and phosholipid-dependent protein kinase by diacylglycerol: Its possible relation to phosphatidyl inositol turnover. *J. Biol. Chem.* **255,** 2273–2276.

Koeffler, H. P., and Golde, D. W. (1978). Acute myelogenous leukemia: A human cell line responsive to colony-stimulating activity. *Science* **200,** 1153–1154.

Koeffler, H. P., and Golde, D. W. (1980). Human myeloid leukemia cell lines: A review. *Blood* **56,** 344–350.

Kreider, B. L., Phillips, P. D., Prystowsky, M. B., Shirsat, N., Pierce, J. H., Tushinski, R., and Rovera, G. (1990). Induction of the granuloctye-macrophage colony-stimulating factor (CSF) receptor by granulocyte CSF increases the differentiative options of a murine hematopoietic progenitor cell. *Mol. Cell. Biol.* **10,** 4846–4853.

Kruger, A., and Anderson, S. M. (1990). The v-*src* oncogene blocks the differentiation of a murine myeloid progenitor cell line and induces a tumorigenic phenotype. *Oncogene* (in press).

Kujubu, D. A., Lim, R. W., Varnum, B. C., and Herschman, H. R. (1987). Induction of transiently expressed genes in PC-12 pheochromocytoma cells. *Oncogene* **1,** 257–262.

Laker, C., Stocking, C., Bergholz, U., Hess, N., DeLamarter, J. F., and Ostertag, W. (1987). Autocrine stimulation after transfer of the granulocyte/macrophage colony-stimulating factor gene and autonomous growth are distinct but interdependent steps in the oncogenic pathway. *Proc. Natl. Acad. Sci. U.S.A.* **84,** 8458–8462.

Land, H., Parada, L., Weinberg, R. (1983). Tumorigenic conversion of primary embryo fibroblasts requires at least two cooperating oncogenes. *Nature (London)* **304,** 596–602.

Lang, R. A., Metcalf, D., Gough, N. M., Dunn, A. R., and Gonda, T. J. (1985). Expression of a hemopoietic growth factor cDNA in a factor-dependent cell line results in autonomous growth and tumorigenicity. *Cell* **43,** 531–542.

Lau, L. F., and Nathans, D. (1985). Identification of a set of genes expressed during the G0–G1 transition of cultured mouse cells. *EMBO J.* **4,** 3145–3151.

Lau, L. F., and Nathans, D. (1987). Expression of a set of growth-related immediate-early genes in BALB/c 3T3 cells: Coordinate regulation with c-*fos* or c-*myc*. *Proc. Natl. Acad. Sci. U.S.A.* **84,** 1182–1186.

LeMaire, P., Revelant, O., Bravo, R., and Charnay, P. (1988). Two mouse genes encoding potential transcription factors with identical DNA-binding domains are activated by growth factors in cultured cells. *Proc. Natl. Acad. Sci. U.S.A.* **85,** 4691–4695.

Liebermann, D., Hoffman-Liebermann, B., and Sachs, L. (1982). Regulation and role of different macrophage and granulocyte-inducing proteins in normal and leukemic myeloid cells. *Int. J. Cancer* **29,** 159–161.

Liebermann, D. A., and Hoffman-Liebermann, B. (1989). Protooncogene expression and dissection of the myeloid growth to differentiation developmental cascade. *Oncogene* **4,** 583–592.

Lim, R. W., Varnum, B. C., and Herschman, H. R. (1987). Cloning of tetradecanoyl phorbol ester–induced primary response sequences and their expression in density-arrested Swiss 3T3 cells and a TPA nonproliferative variant. *Oncogene* **1,** 263–270.

Lim, R. W., Varnum, B. C., O'Brien, T. G., and Herschman, H. R. (1989). Induction of tumor promotor–inducible genes in murine 3T3 cell lines and tetradecanoyl phorbol acetate-nonproliferative 3T3 variants can occur through protein kinase C–dependent and –independent pathways. *Mol. Cell. Biol.* **9,** 1790–1793.

Lord, K. A., Hoffman-Liebermann, B., and Liebermann, D. A. (1990). Complexity of the immediate-early response of myeloid cells to terminal differentiation and growth arrest includes ICAM-1, jun-B and histone variants. *Oncogene* **5,** 387–396.

Lozzio, C. B., and Lozzio, B. B. (1975). Human chronic myelogenous leukemia cell-line with positive Philadelphia chromosome. *Blood* **45,** 321–334.

Mavilio, F., Kreider, B. L., Valtieri, M., Naso, G., Shirsat, N., Venturelli, D., Reddy, E. P., and Rovera, G. (1989). Alteration of growth and differentiation factors response by Kirsten and Harvey sarcoma viruses in IL-3–dependent murine hematopoietic cell line 32DC13 (G). *Oncogene* **4,** 301–308.

Mellon, P. L., Clegg, C. H., Correll, L. A., and McKnight, S. G. (1989). Regulation of transcription by cyclic AMP–dependent protein kinase. *Proc. Natl. Acad. Sci. U.S.A.* **86,** 4887–4891.

Metcalf, D. (1984). *In* "The Hemopoietic Colony-Stimulating Factors." Elsevier Biomedical Press, Amsterdam.

Milbrandt, J. (1987). A nerve growth factor–induced gene encodes a possible transcriptional regulatory factor. *Science* **238,** 797–799.

Mitchell, R., Zokas, L., Schreiber, R., and Verma, I. (1985). Rapid induction of the expression of protooncogene *fos* during human monocytic differentiation. *Cell* **40,** 209–217.

Morla, A. O., Schreurs, J., Miyajima, A., and Wang, J. Y. J. (1988). Hematopoietic growth factors activate the tyrosine phosphorylation of distinct sets of proteins in interleukin-3–dependent murine cell lines. *Mol. Cell. Biol.* **8,** 2214–2218.

Morrison, D. K., Browning, P. J., White, M. F., and Roberts, T. M. (1988). Tyrosine phosphorylations *in vivo* associated with v-*fms* transformation. *Mol. Cell. Biol.* **8,** 176–185.

Muller, R., Bravo, R., Burckhardt, J., and Curran, T. (1984). Induction of c-*fos* gene and protein by growth factors precedes activation of c-*myc. Nature (London)* **312,** 716–720.

Muller, R., Curran, T., Muller, D., and Guilbert, L. (1985). Induction of c-*fos* during myelomonocytic differentiation and macrophage proliferation. *Nature (London)* **314,** 546–548.

Murray, M., Cunningham, J., Parada, L., Dautry, F., Lebowitz, P., and Weinberg, R. (1983). The HL-60 transforming sequence: A *ras* oncogene coexisting with altered *myc* genes in hematopoietic tumors. *Cell* **33,** 749–757.

Neer, E. J., and Clapham, D. E. (1988). Roles of G protein subunits in transmembrane signalling. *Nature (London)* **333,** 129–134.

Nishizuka, Y. (1986). Studies and perspectives of protein kinase C. *Science* **233,** 305–312.

Ohta, M., Anklesaria, P., Fitzgerald, T. J., Kase, K., Leif, J., Delamarter, J., Farber, N., Wright, E., and Greenberger, J. S. (1989). Long-term bone marrow cultures: Recent studies with clonal hematopoietic and stromal cell lines. *Pathol. Immunopathol. Res.* **8,** 1–20.

Pierce, J. H., DiFiore, P. P., Aaronson, S. A., Potter, M., Pumphrey, J., Scott, A., and Ihle, J. N. (1985). Neoplastic transformation of mast cells by Abelson-MuLV: Abrogation of IL-3 dependence by a nonautocrine mechanism. *Cell* **41,** 685–693.

Pierce, J. H., Ruggiero, M., Fleming, T. P., DiFiore, P. P., Greenberger, J. S., Varticovski, L., Schlessinger, J., Rovera, G., and Aaronson, S. A. (1988). Signal transduction through the EGF receptor transfected in IL-3–dependent hematopoietic cells. *Science* **239,** 628–631.

Rauscher, F., III, Cohen, D., Curran, T., Bos, T., Vogt, P., Bohmann, D., Tjian, R., and Franza, B., Jr. (1988a). Fos-associated protein p39 is the product of the *jun* protooncogene. *Science* **240,** 1010–1016.

Rauscher, F. J., III, Sambucetti, L. C., Curran, T., Distel, R. J., and Spiegelman, B. M. (1988b). A common DNA-binding site for fos protein complexes and transcription factor AP-1. *Cell* **52,** 471–480.

Rauscher, F. J., III, Voulalas, P. J., Franza, B. R., Jr., and Curran, T. (1988c). *Fos* and *jun* bind cooperatively to the AP-1 site: Reconstitution *in vitro. Genes Dev.* **2,** 1687–1699.

Rohrschneider, L. R., and Metcalf, D. (1989). Induction of macrophage colony-stimulating factor–dependent growth and differentiation after introduction of the murine c-*fms* gene into FDC-P1 cells. *Mol. Cell. Biol.* **9,** 5081–5092.

Rollins, B. J., Morrison, E. D., and Stiles, C. D. (1988). Cloning and expression of JE, a gene inducible by platelet-derived growth factor and whose product has cytokine-like properties. *Proc. Natl. Acad. Sci. U.S.A.* **85,** 3738–3742.

Roussel, M. F., Dull, T. J., Rettenmier, C. W., Ralph, P., Ulrich, A., and Sherr, C. J. (1987). Transforming potential of the c-*fms* protooncogene (CSF-1 receptor). *Nature (London)* **325,** 549–552.

Roussel, M. F., Downing, J. R., Ashmun, R. A., Rettenmier, C. W., and Sherr, C. J. (1988a). Colony-stimulating factor-1–mediated regulation of a chimeric c-*fms*/v-*fms* receptor containing the v-*fms*–encoded tyrosine kinase domain. *Proc. Natl. Acad. Sci. U.S.A.* **85,** 5903–5907.

Roussel, M. F., Downing, J. R., Rettenmier, C. W., and Sherr, C. J. (1988b). A point mutation in the extracellular domain of the human CSF-1 receptor (c-*fms* protooncogene product) activates its transforming potential. *Cell* **55,** 979–988.

Rovera, G., Valtieri, M., Mavilio, F., and Reddy, E. P. (1987). Effect of Abelson murine leukemia virus on granulocytic differentiation and interleukin-3 dependence of a murine progenitor cell line. *Oncogene* **1,** 29–35.

Ryder, K., and Nathans, D. (1988). Induction of protooncogene c-*jun* by serum growth factors. *Proc. Natl. Acad. Sci. U.S.A.* **85,** 8464–8467.

Ryder, K., Lau, L., and Nathans, D. (1988). A gene activated by growth factors is related to the oncogene v-*jun*. *Proc. Natl. Acad. Sci. U.S.A.* **85,** 1487–1491.

Ryder, K., Lanahan, A., Perez-Albuerne, E., and Nathans, D. (1989). Jun-D: A third member of the *jun* gene family. *Proc. Natl. Acad. Sci. U.S.A.* **86,** 1500–1503.

Ryseck, R-P., Macdonald-Bravo, H., Mattei, M-G., Siegfried, R., and Bravo, R. (1989). Structure, mapping, and expression of a growth factor–inducible gene encoding a putative nuclear hormonal binding receptor. *EMBO J.* **8,** 3327–3335.

Sachs, L. (1982). Control of growth and normal differentiation in leukemic cells: Regulation of the developmental program and restoration of the normal phenotype in myeloid leukemia. *J. Cell. Physiol. (Suppl.)* **1,** 151–164.

Sherr, C. J., Rettenmier, C. W., Sacca, R., Roussel, M. F., Look, A. T., and Stanley, E. R. (1985). The c-*fms* protooncogene product is related to the receptor for the mononuclear phagocyte growth factor, CSF-1. *Cell* **41,** 665–676.

Sigal, I. S. (1988). The *ras* oncogene: A structure and some function. *Nature (London)* **332,** 485–486.

Sporn, M. B., and Roberts, A. B. (1985). Autocrine growth factors and cancer. *Nature (London)* **313,** 745–747.

Sukhatme, V. P., Cao, X., Chang, L. C., Tsai-Morris, C. H., Stamenkovich, D., Ferreira, P. C. P., Cohen, D. R., Edwards, S. A., Shows, T. B., Curran, T., LeBeau, M. M., and Adamson, E. D. (1988). A zinc finger–encoding gene coregulated with c-*fos* during growth and differentiation, and after cellular depolarization. *Cell* **53,** 37–43.

Sundstrom, C., and Nilsson, K. (1976). Establishment and characterization of a human histiocytic lymphoma cell line (U937). *Int. J. Cancer* **17,** 565–577.

Tamura, T., Simon, E., Niemann, H., Snoek, G. T., and Bauer, H. (1986). Gp140$^{v\text{-}fms}$ molecules expressed at the surface of cells transformed by the McDonough strain of feline sarcoma virus are phosphorylated in tyrosine and serine. *Mol. Cell. Biol.* **6,** 4745–4748.

Tamura, S., Sugawara, M., Tanaka, H., Tezuka, E., Nihira, S., Miyamoto, C., Suda, T., and Ohta, Y. (1990). A new hematopoietic cell line, KMT-2, having human interleukin-3 receptors. *Blood* **76,** 501–507.

Tirone, F., and Shooter, E. M. (1989). Early gene regulation by nerve growth factor in PC12 cells: Induction of an interferon-related gene. *Proc. Natl. Acad. Sci. U.S.A.* **86,** 2088–2092.

Tsiftsoglou, A. S., and Wong, W. (1985). Molecular and cellular mechanisms of leukemic hemopoietic cell differentiation: An analysis of the Friend system. *Anticancer Res.* **5,** 81–99.

Ullrich, A., and Schlessinger, J. (1990). Signal transduction by receptors with tyrosine kinase activity. *Cell* **61,** 203–212.

Valtieri, M., Tweardy, D. J., Caracciolo, D., Johnson, K., Mavilio, F., Altmann, S., Santoli, D., and Rovera, G. (1987). Cytokine-dependent granulocytic differentiation: Regulation of proliferative and differentiative responses in a murine progenitor cell line. *J. Immunol.* **138,** 3829–3835.

Varnum, B. C., Lim, R. W., Kujubu, D. A., Luner, S. J., Kaufman, S. E., Greenberger, J. S., Gasson, J. C., and Herschman, H. R. (1989). Granulocyte–macrophage colony-stimulating factor and tetradecanoyl phorbol acetate induce a distinct, restricted subset of primary-response TIS genes in both proliferating and terminally differentiated myeloid cells. *Mol. Cell. Biol.* **9,** 3580–3583.

Vennstrom, B., and Damm, K. (1988). The *erbA* and *erbB* oncogenes. *In* "The Oncogene Handbook" (E. P. Reddy, A. M. Skalka, and T. Curran, eds.), Elsevier Biomedical Press, Amsterdam.

Walker, F., Nicola, N. A., Metcalf, D., and Burgess, A. W. (1985). Hierarchical down-modulation of hemopoietic growth factor receptors. *Cell* **43,** 269–276.

Wang, H.-M., Collins, M., Arai, K.-I., and Miyajima, A. (1989). EGF induces differentiation of an IL-3–dependent cell line expressing the EGF receptor. *EMBO J.* **8,** 3677–3684.

Watson, M. A., and Milbrandt, J. (1989). The NGFI-B gene, a transcriptionally inducible member of the steroid receptor gene superfamily: Genomic structure and expression in rat brain after seizure induction. *Mol. Cell. Biol.* **9,** 4213–4219.

Wolf, D., and Rotter, V. (1985). Major deletions in the gene encoding the p53 tumor antigen cause lack of p53 expression in HL-60 cells. *Proc. Natl. Acad. Sci. U.S.A.* **82,** 790–794.

Woolford, J., Rothwell, V., and Rohrschneider, L. R. (1985). Characterization of the human c-fms gene product and its expression in cells of the monocyte–macrophage lineage. *Mol. Cell. Biol.* **5,** 3458–3466.

Young, D. C., and Griffin, J. D. (1986). Autocrine secretion of GM-CSF in acute myeloblastic leukemia. *Blood* **68,** 1178–1181.

8

Erythroleukemia Cells

VICTORIA M. RICHON, JOSEPH MICHAELI,
RICHARD A. RIFKIND, AND PAUL A. MARKS

DeWitt Wallace Research Laboratories, Memorial Sloan-Kettering Cancer Center and the
Sloan-Kettering Division of the Graduate School of Medical Sciences,
Cornell University,
New York, New York, 10021

I. OVERVIEW OF MURINE ERYTHROLEUKEMIA DIFFERENTIATION

Murine erythroleukemia cells (MELC) provide a useful model system for investigation of the mechanism by which proliferating precursor cells in a differ-

MOLECULAR AND CELLULAR APPROACHES TO THE CONTROL
OF PROLIFERATION AND DIFFERENTIATION

entiation lineage withdraw from the cell-division cycle and express the genes characteristic of the differentiated phenotype. MELC are virus-transformed erythroid precursor cells, which can be induced to differentiate by a variety of chemical inducers (Friend *et al.,* 1971; Marks and Rifkind, 1978). The induction of differentiation is a multistep process and involves the regulation of expression of a number of genes. It is characterized by an initial latent period (10 to 12 hr), during which there are a number of changes in gene expression, but commitment to terminal cell division and expression of differentiation-related genes, such as globin genes, has not yet occurred. Commitment to terminal differentiation is defined as the irreversible capacity to express characteristics of the differentiated erythroid phenotype, including globin gene expression and loss of proliferative capacity, despite removal of the inducer (Gusella *et al.,* 1976; Fibach *et al.,* 1977). Upon continued exposure to inducer, there is a period of progressive (stochastic) recruitment of an increasing proportion of MELC until approximately 48 hr, when more than 90% of the cells are committed to terminal erythroid differentiation.

MELC lines were originally derived from spleens of susceptible mice infected with the Friend virus complex. The cells were subsequently established in continuous culture (Friend *et al.*, 1966). The transforming Friend virus complex includes at least two viruses; a defective spleen focus-forming virus (SFFV), and a murine leukemia helper virus (MuLV). One effect of the Friend virus transformation of this erythroid precursor cell, probably a cell at the CFU-e stage of erythropoiesis (Marks and Rifkind, 1978; Tsiftsoglou and Robinson, 1985), is to make cell proliferation independent of erythropoietin, the physiological regulator of erythroid precursor replication. Li *et al.,* (1990) have demonstrated that the SFFV *env* gene encodes a membrane glycoprotein, gp55, which binds to the erythropoietin receptor and stimulates or activates the receptor, bypassing the normal requirement for erythropoietin. Friend erythroleukemia cell lines continue to express gp55 in culture even when expression of other virus-encoded proteins has ceased (Anand *et al.,* 1981). Binding of gp55 to the erythropoietin receptor may explain the erythropoietin independence of MELC. The virus-transformed cells continuously proliferate and display a low rate of commitment to terminal erythroid-cell differentiation. The rate of MELC differentiation can be enhanced by a variety of inducing agents. Thus, the virus-induced block does not irreversibly destroy the potential of these cells to undergo terminal erythroid differentiation.

This chapter reviews recent work on erythroleukemia cell differentiation, including (1) studies on hexamethylene bisacetamide (HMBA)-induced MELC differentiation to the terminal erythroid phenotype, including recent findings characterizing variant cell lines with altered kinetics of HMBA-induced differentiation; (2) the development of new polar–apolar inducers of differentiation; and (3) early and late events involved in the control of cell division and exit from the cell cycle in induced MELC.

II. EFFECT OF HMBA ON CELL CYCLE PROGRESSION AND GLOBIN TRANSCRIPTION

A. Characteristics of HMBA-Induced Differentiation

Terminal differentiation of MELC (745A-DS19) is characterized by accumulation of high levels of hemoglobin and cessation of cell division. Induced differentiation is a complex, multistep process involving an early latent period (10 to 12 hr) before any cells are committed, followed by progressive recruitment of committed cells and expression of the differentiated phenotype (Fig. 1). During the latent period, a number of changes occur. These include alterations in membrane ion permeability (Bernstein *et al.*, 1979; Chapman, 1980), alterations in membrane fluidity (Lyman *et al.*, 1976; Muller *et al.*, 1980), changes in cell volume (Gazitt *et al.*, 1978), a prompt (< 1 hr) translocation of protein kinase C (PKC) activity to the membrane (Melloni *et al.*, 1987; L. Leng, unpublished observation), and a transient increase in cyclic adenosine monophosphate (cAMP) concentration (Gazitt *et al.*, 1978a), as well as changes in the expression of certain protooncogenes: c-*myb*, c-*myc*, and the p53 gene (Shen *et al.*, 1983; Lachman and Skoultchi, 1984; Ramsay *et al.*, 1986; Watson, 1988; Richon *et al.*, 1989). The changes in

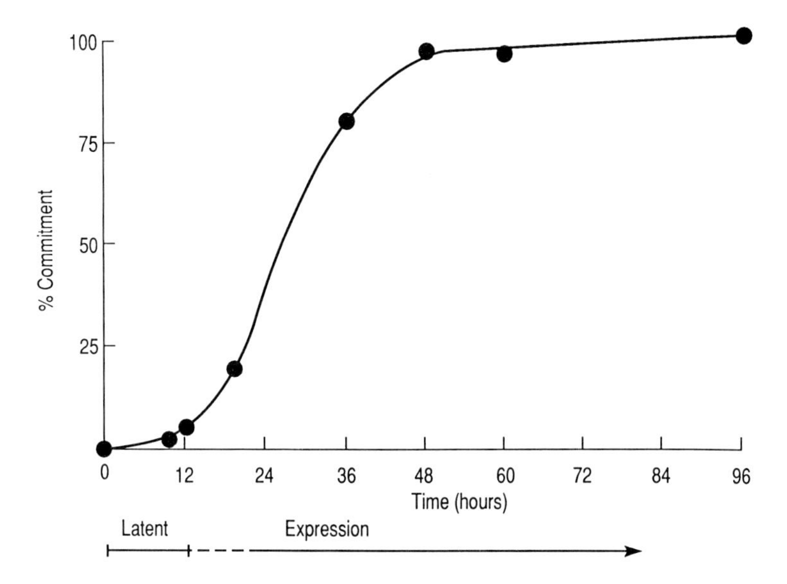

Fig. 1. MELC (DS19) in culture with 5 m*M* HMBA are induced to commit to terminal differentiation. An initial latent period of about 12 hr is followed by progressive recruitment (expression) of cells to differentiate and cease proliferation. Commitment was determined as previously described (Fibach *et al.*, 1977). MELC were cultured with HMBA as described (Reuben *et al.*, 1976).

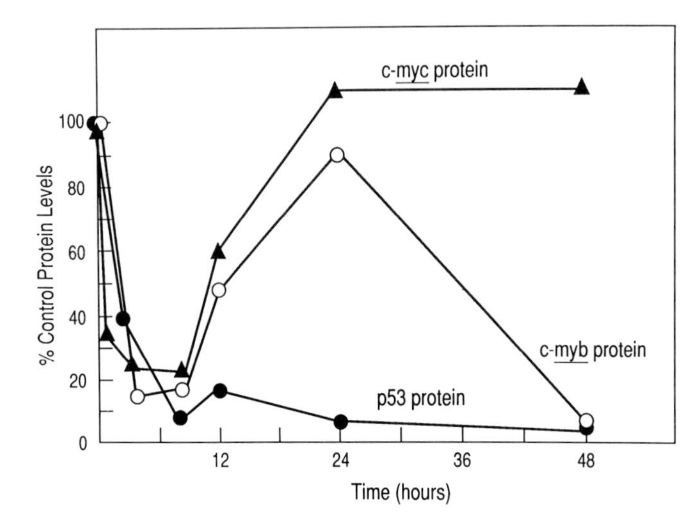

Fig. 2. Expression of c-*myc,* c-*myb* and p53 proteins during HMBA-induced differentiation in MELC. Protein levels were determined by laser densitometric quantification of the immunoprecipitation–Western blot analysis as described (Richon *et al.,* 1989). The levels were normalized to the level of protein in untreated MELC.

expression of protooncogenes include decreases in c-*myb* mRNA and protein (within 4 hr), in c-*myc* mRNA and protein (within 1 hr) and in p53 mRNA and protein (within 8 hr) (Fig. 2). Although these early changes are not themselves sufficient for irreversible induction of differentiation, they do appear to be necessary steps (see Section IV).

During the period of progressive recruitment of MELC to differentiate, coordinated modulation of expression of a number of genes can be detected, including those for α and β globins, the heme synthetic enyzmes, erythroid-specific membrane proteins and rRNA synthesis (Marks and Rifkind, 1978). By 27 hr, structural changes in DNA can be demonstrated by a decrease in sedimentation rate on an alkaline sucrose gradient, which may reflect an accumulation of single-strand breaks in the DNA of cells induced to erythroid differentiation (Terada *et al.,* 1978; Scher and Friend, 1978). The c-*myb* and c-*myc* mRNA and protein levels return toward control values by 12 to 24 hr; then the c-*myb* mRNA and protein levels decrease again, and remain thereafter at almost undetectable levels.

B. Characteristics of Cell Cycle Progression during Differentiation

A number of studies have suggested that the action of inducers of MELC differentiation may be cell cycle related. McClintock and Papaconstantinou (1974) and Levy *et al.* (1975) demonstrated a requirement for exposure to inducer during at least one S phase, in order to achieve induction of differentiation. Geller *et al.* (1978) have shown that commitment to differentiate is more rapid in cultures of MELC initiated with cells synchronized in the G_2 or G_1 phase of the cell cycle than in mid or late S, suggesting that passage through G_1 and or early S in the presence of inducer may be important for the induction process.

Beckman *et al.* (1987) have shown that blocking MELC in late G_1 to S phase with the DNA synthesis inhibitor, aphidicolin, enhances HMBA-induced differentiation. They suggest that DNA synthesis inhibitors may enhance differentiation by lengthening a particular phase of the cell cycle (G_1 or early S), in which the cell is receptive to differentiation stimuli.

In terms of cell proliferation, unsynchronized MELC, exposed to HMBA, lag behind an uninduced culture for approximately 12 hr, which is about the time required for one transit through the cell cycle. Terada *et al.* (1977) identified an inducer-mediated transient prolongation of G_1 in MELC undergoing differentiation induced by HMBA or dimethyl sulfoxide (DMSO), which appears to account for the cell growth lag. Gambari *et al.* (1979) demonstrated that an effect of HMBA during late G_1 or early S phase of the cell cycle is required for the change in cell-cycle control that results in the subsequent prolonged G_1 phase. Cells synchronized in late S–G_2 by centrifugal elutriation, then exposed to HMBA, do not display G_1 prolongation until after they have traversed late G_1–early S phase and reach the subsequent G_1. Following this prolonged G_1, HMBA-treated cultures progress through the cell cycle for about two to five generations (cycle time of 10 to 12 hr) until they arrest permanently in G_1–G_0 and express the characteristics of terminal erythroid differentiation.

The accumulated evidence suggests that there is a critical time in late G_1 or early S phase when inducer must be present in order to induce differentiation. This is followed by prolongation of the subsequent G_1 phase, during which there is the first detectable increase in the rate of globin gene transcription (see below). It is not known whether G_1 prolongation itself is critical to the induction of differentiation, or whether factors controlling the transient G_1 prolongation are also involved in the permanent G_1 arrest characteristic of terminal stages of erythroid differentiation.

C. Regulation of Globin Gene Expression

During HMBA-induced MELC differentiation there is a 10- to 30-fold increase in the level of globin mRNA owing, primarily, to an increase in globin gene transcription (Profous-Juchelka *et al.,* 1983; Sheffery *et al.,* 1984; Ganguly and Skoultchi, 1985). In MELC, two β-like globin genes are expressed, βmaj and βmin, and two α globin genes (Popp *et al.,* 1981; Smith *et al.,* 1984). In uninduced

MELC, there is a low level of constitutive expression of both the α and βmin globin genes (Nudel *et al.,* 1977; Curtis *et al.,* 1980). HMBA initiates marked activation of transcription of the α and βmaj genes. Increased transcription of the α globin gene is detected within 12 to 24 hr of exposure to HMBA and achieves a 20- to 30-fold increase in rate of transcription by approximately 36 hr (Sheffery *et al.,* 1984). Activation of βmaj globin gene transcription is detected only by 24 to 36 hr and requires about 48 hr to reach a 30-fold increase (Salditt-Georgieff *et al.,* 1984). An increment in induced globin mRNA synthesis is initially detected in the first transiently prolonged G_1 phase after a complete S phase traversed in the presence of HMBA. Once accelerated globin mRNA synthesis is initiated, transcription continues through both S and G_1 phases of the subsequent cell cycles (Gambari *et al.,* 1979).

Several features of DNA and chromatin structure in the globin gene domains have been examined in relation to inducer-mediated modulation of expression of the α and βmaj globin genes. HMBA causes a number of specific changes in chromatin configuration about the globin gene domains, including disappearance of the second intron DNase I hypersensitive site in the βmaj globin gene and appearance of novel DNase I hypersensitive sites within 100 bp 5′ of both the α and βmaj genes (Marks *et al.,* 1987).

In addition to characterizing the changes in chromatin structure around the globin genes during HMBA induction, the transcription factors required to form active α and β globin gene transcription complexes have been intensively investigated. Tsai *et al.* (1989) and Evans and Felsenfeld (1989) have cloned and characterized an erythroid-specific factor that recognizes and binds to a core DNA sequence [(A/T)GATA(A/G)] that has been observed in both the mouse α and βmaj genes, as well as globin genes from other species including human and chicken. This factor has been termed GF-1 (Tsai *et al.,* 1989), EF-1 (Plumb *et al.,* 1989), Eryf 1 (Evans *et al.,* 1988), NF-E1 (Wall *et al.,* 1988), or EFγa (Gumucio *et al.,* 1988). This factor may also play a role in regulating the transcription of other erythroid-specific genes, since the core DNA-binding sequence is found near the promoters of other genes whose expression is induced in erythroid cells, such as glutathione peroxidase (Chambers *et al.,* 1986) and porphobilinogen deaminase (Chretien *et al.,* 1988). This factor is produced at high levels in MELC and does not increase during induction of these genes during differentiation (Tsai *et al.,* 1989). This suggests that additional factors may be required for the activation of these genes during differentiation.

Although not erythroid specific, several additional transcription factors that interact with the α and/or βmaj promoter regions and are involved in their transcription have been characterized (Barnhart *et al.,* 1989; Kim and Sheffery, 1990; Galson and Housman, 1988). Kim *et al.* (1990) have characterized four factors that regulate α globin transcription. These factors include the erythroid-specific NF-E1 described above, α-IRP, α-CP1, and α-CP2. None of the factors, except for NF-

E1, is erythroid specific. The α-CP2 binding activity increases during MELC differentiation and α-CP2 may therefore play a role in stimulating the α-promoter during inducer differentiation.

Regulation of the α and β globin genes during HMBA-induced differentiation is complex, requiring a number of specific changes in chromatin configuration about the globin gene domains and the coordinated regulation of several transcription factors.

D. Variant Cell Lines

Several variant MELC lines that display a range of inducer-mediated differentiation properties have been developed. These cell lines include inducer non-responsive lines (Marks *et al.*, 1983), cell lines that display accelerated kinetics of induced differentiation (Melloni *et al.*, 1987), and an erythropoietin-responsive cell line (Todokoro *et al.*, 1987).

Variant MELC lines with accelerated commitment to terminal differentiation, attributable largely to elimination of the latent period, and increased sensitivity to induction by lower concentrations of HMBA, have been used to gain insight into early inducer-mediated events. One cell line, V3.17, was selected from parental DS19 cells for resistance to low levels of vincristine (twofold to fivefold) (Melloni *et al.*, 1988). Several observations suggest that the development of vincristine resistance is related to alterations in the commitment phenotype. Four independently derived cell lines resistant to low levels of vincristine display the accelerated-commitment phenotype. The association of low-level vincristine resistance and accelerated inducer-mediated differentiation is stably inherited, even when cultures are maintained for prolonged periods in the absence of vincristine. MELC lines selected for higher levels of resistance to vincristine show a somewhat more accelerated rate of HMBA-mediated commitment but no further reduction in the optimal concentration of inducer (Richon *et al.*, 1991).

Cells resistant to low levels of vincristine display several characteristics of the multidrug resistance (MDR) phenotype, including cross-resistance to colchicine and vinblastine (but not adriamycin), reversal of vincristine resistance by verapamil, and decreased accumulation of [^{3}H]vincristine (Richon *et al.*, 1991). However, they lack other features of MDR; most notably they do not overexpress P-glycoprotein or *mdr1, mdr2,* or *mdr3* mRNA. Although overexpression of P-glycoprotein can be detected in MELC lines made resistant to high levels of vincristine, there is no clear relationship between the level of P-glycoprotein expression and the accelerated response to HMBA. Furthermore, DS19 cells induced to differentiate do not modulate their levels of P-glycoprotein. Therefore, although there appears to be a relationship between the development of vincristine resistance and accelerated differentiation, increased expression of P-glycoprotein is not a component of the mechanism altering the differentiation phenotype. There does appear to be a relationship between an increased level of protein kinase Cβ (PKCβ) activity found in V3.17 cells and their accelerated rate of inducer-mediated differentiation (See Section IV,G; Melloni *et al.,* 1989).

Vincristine-resistant MELC with accelerated inducer-mediated differentiation display cell cycle kinetics upon induction that differs from that observed with induced parental vincristine-sensitive cells. The G_1 prolongation following passage through S in the presence of HMBA is more pronounced in the vincristine-resistant cells, and a substantially higher proportion of cells become committed during the first G_1 than do the parental vincristine-sensitive cells (V. Richon, unpublished observations). The induction of differentiation in vincristine-resistant cells, as in vincristine-sensitive variants, is cell cycle–related.

Development of an inducer-nonresponsive cell line, R1, has also contributed to our understanding of inducer-mediated terminal differentiation (Marks *et al.,* 1983). Although R1 cells are resistant to induced differentiation by HMBA or other polar–apolar inducers of differentiation, HMBA does trigger a decrease in the levels of c-*myc* and p53 proteins. The level of c-*myb* protein, which undergoes a biphasic decrease in induced parental DS19 cells, is not suppressed either early (1 to 4 hr) or late (48 hr) during exposure to HMBA (unpublished observations). R1 cells made resistant to vincristine [cell line R1(VCR)] become sensitive to HMBA and can then be induced to differentiate (Michaeli *et al.,* 1990). Induced R1(VCR) cells display the characteristic decrease in c-*myb,* suggesting the developmental block in R1 cells involves an inability to suppress c-*myb* expression.

Although characteristically MELC grow independent of and are insensitive to the erythropoietic growth factor, erythropoietin (Epo), an Epo-responsive MELC was isolated by Todokoro *et al.* (1987) by dilutional cloning of an Epo nonresponsive line. This cell line, SKT6, also retains its sensitivity to induction by DMSO. Although both Epo and DMSO induce terminal erythroid differentiation, their effects upon the regulation of protooncogene expression are different (Todokoro *et al.,* 1988). Epo-mediated induction of SKT6 results in decreased c-*myb* expression, and increased c-*myc* expression, whereas DMSO decreases the expression of both protooncogenes. Todokoro *et al.* (1988) conclude that early down-regulation of c-*myb,* but not of c-*myc,* is essential for induced erythroid differentiation.

III. AGENTS ACTIVE AS INDUCERS OF DIFFERENTIATION AND INHIBITORS OF DIFFERENTIATION

A. Inducers of Differentiation

Since the discovery almost two decades ago (Friend *et al.,* 1971) that DMSO induces differentiation in MELC, a wide variety of differentiation-inducing agents have been identified or synthesized (Marks and Rifkind, 1978; Tsiftsoglou and Robinson, 1985). The observation that a wide variety of naturally occurring and synthetic compounds will induce differentiation in MELC lends weight to the hypothesis that induction may involve multiple, coordinated regulatory processes leading to a final common pathway of differentiation. Different inducing agents may interact with different components of the regulatory process.

Polar–apolar chemical inducers are among the most potent inducing agents (Marks *et al.,* 1989). Their mechanism of action is unknown. Discovery of this family of inducers resulted from attempts to improve the effectiveness of the polar compound, DMSO. (Tanaka *et al.,* 1975). The hypothesis was developed that a high concentration (280 m*M*) of DMSO was required because more than one solvent molecule must bind or interact with the cellular target site, and that the target might interact more efficiently with a single molecule carrying two or more of the active chemical groups. This stragegy has generated a series of active polar–apolar inducers, one of the most effective being HMBA (Reuben *et al.,* 1976). Structure–activity studies have demonstrated that the apolar–polar compounds must be relatively flexible, and that a 5- to 8-methylene bridge between polar groups is optimal for inducing activity (Marks and Rifkind, 1978). Although far more active than DMSO, the optimal effective concentration of HMBA required to induce MELC differentiation is still quite high (5 m*M*). The search for polar–apolar inducers with lower effective molar concentrations continues, and agents effective at much lower concentrations have been identified (Marks *et al.,* 1989). The compounds include flexible analogs of suberic acid bisdimethylamide, which are approximately 100 times more active than HMBA on a molar basis.

Certain purines and purine analogs form another class of inducing agents (Gusella and Housman, 1976; Kerr, 1990). Hypoxanthine, 6-thioguanine, and monomethylated derivatives of guanine, hypoxanthine, and xanthine are active inducers, whereas guanine and xanthine are inactive. Kerr (1990) has suggested that some of these compounds act as inducers, at least in part, by inhibiting adenosine diphosphate (ADP)-ribosylation. Other compounds that inhibit ADP-ribosylation, such as nicotinamide and benzamide, also induce MELC differentiation (Terada *et al.,* 1979); HMBA-induced differentiation is accompanied by a decrease in endogenous ADP-ribosyltransferase activity (Morioka *et al.,* 1979).

The purine analog 6-thioguanine is believed to act as an inducer by interfering with the structure and synthesis of DNA rather than by inhibiting ADP-ribosylation. Other differentiation inducers known to cause structural changes in DNA include actinomycin D, X-rays and UV irradiation. Still other agents, such as ouabain, the polar–apolar compounds, and certain fatty acids have effects on membrane function, but the relationship of these effects to induction remains unclear (Marks and Rifkind, 1978).

Activin A, a peptide homologous to transforming growth factor β (TGFβ), induces MELC differentiation (Eto *et al.,* 1987; Yamashita *et al.,* 1990). Eto *et al.* (1987) purified a homodimeric protein from a human leukemia cell culture supernatant that induced erythroid differentiation in MELC. This protein appears to be identical to activin A (Murata *et al.*, 1988). The mechanism of induction by activin A appears to differ from that of HMBA (Yamashita *et al.*, 1990). Increasing cell density suppresses the response to activin A in a range that only minimally affects HMBA-mediated differentiation.

B. Inhibitors of Differentiation

Studies with inhibitors of HMBA-induced terminal differentiation have provided additional evidence with respect to the multistep process involved in inducer-mediated differentiation. The glucocorticoid, dexamethasone, suppresses the expression of HMBA-mediated MELC terminal cell division and the accumulation of globin mRNAs (Chen *et al.*, 1982; Kaneda *et al.*, 1985). Suppression of differentiation is mediated by binding to the glucocorticoid receptor (Mayeux *et al.*, 1985), which blocks a late step in the induction process. Dexamethasone does not inhibit inducer-mediated early changes during the precommitment latent period, such as the fall in c-*myc* and p53 protein, and it can be added as late as 12 hr after inducer and still suppress induction. Dexamethasone does inhibit the late (48 hr) HMBA-induced suppression of c-*myb* mRNA and protein, but not the continued suppression of p53 protein (Richon *et al.*, 1989).

The activators of PKC, diacylglycerols and the tumor promoter 12-*O*-tetradecanoylphorbol-13-acetate (TPA), are also inhibitors of HMBA-induced differentiation (Fibach *et al.*, 1979; Pincus *et al.*, 1984). HMBA itself causes a decrease in the level of phosphatidylinositol metabolites (Faleto *et al.*, 1985). Inositol triphosphate and diacylglycerol decrease within 2 hr of exposure to HMBA. Diacylglycerol activates the calcium and phospholipid-dependent PKC activity, which may play a role in HMBA-induced differentiation (see Section IV,G).

IV. ROLE OF PROTEINS AFFECTING CELL CYCLE PROGRESSION IN THE INDUCTION OF MELC DIFFERENTIATION

A. c-*myc*

The protooncogene c-*myc* has been implicated in controlling cell proliferation, and a reduction in c-*myc* expression may be necessary for terminal differentiation in MELC (Cole, 1986; Dmitrovsky *et al.*, 1986; Coppola and Cole, 1986; Prochownik and Kukowska, 1986). The murine c-*myc* gene encodes two major proteins that are phosphorylated and localized in the nucleus (Abrams *et al.*, 1982; Donner *et al.*, 1982). The c-*myc* protein is believed to be involved in the passage from G_1 into S, and the amount of protein increases upon mitogen stimulation of quiescent cells (Kelly *et al.*, 1983; Kaczmarek *et al.*, 1985). c-*myc* has also been implicated in the control of DNA replication (Studzinski *et al.*, 1986), but its role remains controversial (Gutierrez *et al.*, 1987). In uninduced MELC, c-*myc* mRNA and protein levels are unchanged throughout the cell cycle (Hann *et al.*, 1985; Thompson *et al.*, 1986). However, after 12 to 24 hr of exposure to inducer, expression of c-*myc* mRNA becomes restricted to late G_1 (Lachman *et al.*, 1985).

Both c-*myc* mRNA and protein undergo a biphasic decline during inducer-mediated induction of MELC (Lachman and Skoultchi, 1984; Ramsay *et al.*, 1986; Richon *et al.*, 1989; Spotts and Hann, 1990). These changes in c-*myc* expression appear to be due to a complex pattern of regulation involving both transcriptional and posttranscriptional mechanisms. Immediately after addition of inducer, c-*myc* mRNA and protein levels fall (within 1 to 4 hr), owing to an enhanced block of elongation of transcription within the first exon, rather than to inhibition of transcriptional initiation (Nepveu *et al.*, 1987). Despite return of the rate of transcription to control levels, the levels of accumulated mRNA and protein remain low until approximately 8 to 12 hr, when the mRNA and protein levels approach those in untreated cultures. As c-*myc* mRNA and protein levels recover, the rate of c-*myc* protein synthesis dissociates from the rates of c-*myc* mRNA and protein accumulation; relatively high levels of c-*myc* protein are synthesized from moderate levels of c-*myc* mRNA (Spotts and Hann, 1990). After 60 to 72 hr of induction, c-*myc* protein levels again decrease, despite high levels of c-*myc* protein synthesis. This decrease corresponds with an increase in c-*myc* protein turnover (Spotts and Hann, 1990).

Several reports demonstrate that the regulation of c-*myc* is necessary for the induction of MELC differentiation. Deregulated expression of c-*myc,* by constitutive expression of an exogenous c-*myc* cDNA, inhibits MELC differentiation (Coppola and Cole, 1986; Dmitrovsky *et al.*, 1986; Prochownik and Kukowska, 1986), while constitutive expression of c-*myc* antisense RNA accelerates MELC differentiation (Prochownik *et al.*, 1988). The precise role of c-*myc* is unknown, but recent studies have examined the role of c-*myc* in the control of cell cycle progression. Prochownik *et al.* (1988) have studied the relationships between traversal of the cell cycle, differentiation, and c-*myc* levels in MELC expressing high levels of c-*myc* antisense transcripts. In response to DMSO, MELC that express the antisense c-*myc* transcripts differentiate faster and to a greater degree than do control MELC. The authors postulate that minimal *myc* levels are necessary for cells to traverse G_1. Cells with insufficient levels of c-*myc* due to expression of c-*myc* antisense RNA are unable to traverse G_1, and arrest at an earlier point in G_1 when serum deprived, compared to serum-deprived control MELC. This early point of arrest in G_1 may reflect a decision point between proliferation and differentiation. Thus, MELC expressing antisense c-*myc* transcripts and low levels of c-*myc* may have an increased proportion of cells in the early G_1 state and an increased proportion of cells entering the differentiation pathway.

Coppola *et al.* (1989) have investigated the relationships between constitutive expression of c-*myc*, inhibition of induced differentiation of MELC, and cell cycle events. In MELC that do not overexpress c-*myc* and that differentiate in response to inducer, a temporary decrease in c-*myc* is followed by a transient prolongation of G_1 (see Section II,B). MELC that constitutively overexpress c-*myc* and do not differentiate, also undergo a transient G_1 prolongation. Present evidence suggests that decreased levels of c-*myc* make MELC more likely to enter the differentiation pathway, while overexpression of c-*myc* appears to decrease the probability of entering the differentiation-responsive state.

B. c-*myb*

The c-*myb* protooncogene encodes a short-lived nuclear protein (Klempnauer *et al.*, 1984) apparently involved in regulation of proliferation and differentiation. Unlike c-*myc*, c-*myb* is not involved in the immediate response to serum growth factors, but c-*myb* does undergo a transient increase in expression, which correlates with cell cycle progression in chicken embryo fibroblasts. In mammals, c-*myb* is expressed in immature cells of the lymphoid, erythroid, and myeloid lineages (Duprey and Boetigger, 1985; Gonda and Metcalf 1984; Sheiness and Gardinier, 1984) and in a number of other tissues (Thompson *et al.*, 1986; Thiele *et al.*, 1988). Expression of c-*myb* decreases dramatically as cells differentiate, and the decrease precedes suppression of cell proliferation. Recently, Venturelli *et al.* (1990) have found that inhibition of c-*myb* protein synthesis in human peripheral blood mononuclear cells, by antisense oligodeoxynucleotides, prevents entry into S phase and cell proliferation. They further demonstrate that inhibition of c-*myb* protein synthesis leads to a decrease in DNA polymerase α mRNA, and suggest a functional link between c-*myb* and DNA polymerase α mRNA expression. The c-*myb* protein has DNA-binding activity (Moelling *et al.*, 1985) and binds specifically to the nucleotide sequence pyAAC$^{G}/_{T}$G (Biedenkapp *et al.*, 1988). c-*myb* can serve as a transcriptional activator (Weston and Bishop, 1989), but the cellular genes whose transcription is directly stimulated by c-*myb* have not been identified.

Both c-*myb* mRNA and protein undergo biphasic modulation of expression during induced differentiation in MELC (Ramsay *et al.*, 1986; Watson, 1988; Richon *et al.*, 1989). Initially (within 4 hr), the levels of c-*myb* mRNA and protein fall sharply. mRNA and protein levels then rise transiently between 12 and 24 hr, and by 48 hr, are once again markedly suppressed. Both the early and late decline in c-*myb* mRNA and protein appear to be due to a combination of an enhanced block in transcriptional elongation and a decrease in mRNA half-life (Ramsay *et al.*, 1986; Watson, 1988). The levels of mRNA and protein parallel one another throughout induction, and, unlike those of c-*myc*, c-*myb* protein levels do not seem to be controlled at a translational level (Richon *et al.,* 1989; Spotts and Hann, 1990).

Suppression of c-*myb* protein expression may be a critical regulatory step during induced differentiation. Dexamethasone inhibits HMBA-induced differentiation and blocks the late suppression of c-*myb* mRNA and protein (Ramsay *et al.,* 1986; Richon *et al.,* 1989). The inducer-resistant cell line, R1, does not differentiate in response to HMBA, and there is no HMBA-mediated suppression of c-*myb,* although c-*myc* expression is suppressed (V. Richon *et al.,* unpublished observations).

Weber *et al.* (1990) found that constitutive expression of an altenatively spliced c-*myb* mRNA species (*mbm2*) results in enhanced differentiation in MELC. *mbm2* contains the DNA-binding region and nuclear-localization signal present in the c-*myb* protein, but does not contain the transcriptional regulatory regions. The authors suggest that the *mbm2* protein may be functioning as a competitive inhibitor of c-*myb* protein by occupying the same DNA-binding sites without regulatory activity.

Several investigators have demonstrated that constitutive expression of exogenous c-*myb* cDNA inhibits MELC differentiation (Clarke *et al.,* 1988; McMahon *et al.,* 1988; McClinton *et al.,* 1990). Clarke *et al.* (1988) show that MELC clones expressing high levels of human c-*myb* mRNA differentiate poorly in response to DMSO. Clones that initially express low levels of human c-*myb* transcripts and differentiate normally were inhibited in their ability to differentiate after enhanced expression of the expression vector sequences by amplification by selection in methotrexate. The inhibitory effect of c-*myb* on MELC differentiation does not prevent the early decline in c-*myc* expression. McClinton *et al.* (1990) show that the inhibition of HMBA-induced differentiation by c-*myb* transfection is proportional to the level of exogenous c-*myb* mRNA expression. The c-*myb* transfectants display a number of properties upon HMBA induction. First, the initial suppression and subsequent recovery of endogenous c-*myb* and c-*myc* mRNA levels are similar for control cells and *myb* transfectants. Second, in contrast to control cells, terminal down-regulation of endogenous c-*myb* and c-*myc* mRNA levels does not occur for the *myb* transfectants. Third, activation of α-globin transcription is substantially blocked in *myb* transfectants as is hemoglobin expression. Constitutive expression of c-*myb* also inhibits differentiation in the Epo-responsive MELC line, SKT6 (Todokoro *et al.,* 1988).

C. p53

The p53 protein appears to play a role in the control of normal cellular proliferation. Wild-type p53 may function as a tumor-suppressor gene, since murine wild-type p53 inhibits transformation of rat cells in culture (Finlay *et al.,* 1989).

Both elevation of p53 level and inactivation of p53 protein by mutation have been associated with cellular transformation (Crawford *et al.*, 1981; Rotter, 1983; Munroe *et al.*, 1990). p53 mRNA expression is up-regulated by serum growth factors that induce the G_0 to S transition of fibroblasts (Reich and Levine, 1984). Cells microinjected with anti-p53 monoclonal antibodies (Mercer *et al.*, 1982) or transfected with a p53 antisense mRNA-expressing vector (Shohat *et al.*, 1987) fail to undergo cell proliferation and appear to be blocked in the G_1–S transition. p53 protein displaces DNA polymerase α bound to large T antigen, suggesting that p53 might be directly involved in the initiation or maintenance of replicative DNA synthesis (Gannon and Lane, 1987).

Rearrangement of the p53 gene occurs in a high proportion of MELC lines (Mowat *et al.*, 1985; Munroe *et al.*, 1990). Abnormalities in p53 include complete extinction of p53 expression, synthesis of truncated p53 polypeptides, or synthesis of altered forms of p53 that differ by amino acid substitutions. Loss of expression, truncation, or point mutation may all be functionally equivalent, suggesting a tumor-suppressor role for p53 in the development of MELC. Alternatively, the mutations and deletions in p53 found in MELC may not destroy the function of p53 but, rather, may alter its activity or activate novel properties of the protein.

In MELC, HMBA induction causes an initial decrease in p53 mRNA and protein levels (Shen *et al.*, 1983; Ben-Dori *et al.*, 1983). By 10 hr of induction, p53 mRNA has reaccumulated to about 50% of the level in untreated MELC (Khochbin *et al.*, 1988a; Richon *et al.*, 1989). The protein level, however, remains low during the remainder of the induction period. A change in transcription rate does not account for the down-regulation of p53 mRNA, and the half-lives of both the protein and mRNA do not vary during induction (Shen *et al.*, 1983; Khochbin *et al.*, 1988a). Therefore, it appears that there is posttranscriptional control of the mRNA and protein levels. Khochbin *et al.* (1988b) have examined the regulation of p53 protein accumulation throughout the cell cycle. The amount of p53 is constant during G_1, increases at the G_1–S interface, and then progressively rises through S. At G_2–M it is at approximately twice the level as in G_1. Induction of differentiation does not generate a change in this cell cycle-related pattern.

The role that p53 plays in the induction of MELC differentiation is not known, and defining its role is complicated by the high frequency of mutations of p53 that arise during the development of Friend virus-induced MELC lines. It is possible that the function of p53 during differentiation will depend upon the form of p53 expressed in a particular erythroleukemia cell line.

D. Retinoblastoma Protein

Retinoblastoma protein (pRB) has been implicated in the regulation of differentiation and the control of cell cycle progression; pRB is an approximately 105,000 dalton nuclear phosphoprotein (Lee *et al.*, 1987; Bernards *et al.*, 1989).

Robbins *et al.* (1990) show that pRB can repress c-*fos* expression, and propose that pRB regulates c-*fos* expression indirectly by regulating the activity of other transcription factors, rather than through direct binding to the c-*fos* promoter region. While the amount of pRB does not change during progression of the cell cycle, the phosphorylation state of pRB is regulated (DeCaprio *et al.*, 1989; Buchkovich *et al.*, 1989; Chen *et al.*, 1989). The least-phosphorylated form is present in G_1, and as cells progress through S and G_2, pRB becomes more highly phosphorylated. Pietenpol *et al.* (1990) implicate pRB as part of a pathway leading to growth suppression. TGFβ1 inhibits keratinocyte growth through down-regulation of c-*myc*, and the down-regulation of c-*myc* may be mediated by the underphosphorylated form of pRB. Induction of differentiation in several human leukemic cell lines leads to dephosphorylation of pRB (Chen *et al.*, 1989; Mihara *et al.*, 1989; Akiyama and Toyoshima, 1990). In MELC, the phosphorylation of pRB, as in other cells, is cell cycle regulated. During HMBA-induced differentiation, an increase in underphosphorylated pRB accompanies the transient prolongation of G_1. As MELC resume progression through the cell cycle, pRB is again phosphorylated, and the total pRB level increases by approximately threefold (unpublished observations). Although still preliminary, these data suggest that pRB plays a role in the control of cell cycle progression during HMBA-induced MELC differentiation.

E. Histones

The histones play a role in organizing eukaryotic chromatin and may also contribute to the reorganization of chromatin during differentiation. In the mouse there are 10 to 20 copies of the genes for each histone subtype (Jacob, 1976). The functional significance of histone protein heterogeneity and the multiplicity of genes encoding each subtype are not established. Nucleosomes, composed of different combinations of histone subtypes, display different physical properties (Simpson, 1981), and alterations in histone composition may contribute to the changes in chromatin conformation associated with differentiation (von Holt *et al.*, 1984).

The changes in mouse core histones that occur during MELC differentiation have been identified (Grove and Zweidler, 1984; Brown *et al.*, 1985; Brown *et al.*, 1988). The fully replication-dependent (FRD) histones predominate in rapidly dividing tissues; their synthesis is linked to DNA synthesis. The synthesis of partially replication-dependent (PRD) histones is associated with DNA synthesis in continuously dividing cells. The PRD genes are the principal types expressed in MELC. The replication-independent (RID) histones are constitutively expressed at a low level, independent of the cell cycle (Brown *et al.*, 1988). Between 24 and 60 hr of HMBA-induced differentiation, there is a switch in histone gene expression. The ratio of PRD to FRD transcripts increases severalfold over that found in uninduced MELC. This histone switch is tightly coupled to the differentiation process. MELC arrested at a precommitment stage of differentiation (by induction with HMBA in the presence of dexamethasone) do not exhibit histone switching. When dexamethasone alone is removed and the cells resume commitment, the expected change in histone gene switching occurs. If both dexamethasone and HMBA are removed, neither commitment nor the histone gene switch is observed (Brown *et al.,* 1988).

The histone $H1^0$ accumulates in the nuclei of a number of cell types when they are induced to differentiate, and the accumulation of $H1^0$ level accompanies loss of proliferative capacity (Pehrson and Cole, 1980; Osborne and Chabanas, 1984). $H1^0$ protein accumulation is characteristic of HMBA-induced differentiation in MELC (Keppel *et al.,* 1977; Chen *et al.,* 1982; Osborne and Chabanas, 1984), and precedes the onset of commitment. The accumulation of $H1^0$ is initiated in the G_2 phase of the cell cycle and is subsequently found in cells in all phases of the cell cycle (Osborne and Chabanas, 1984).

F. Topoisomerases

Two major types of DNA topoisomerase have been characterized based on function. DNA topoisomerase I makes a reversible single-strand break in the DNA double helix, whereas DNA topoisomerase II makes a transient double-strand break before resealing the DNA double helix. DNA topoisomerase II activity is affected by the growth state of the cell, whereas DNA topoisomerase I activity is unaffected (Tricoli *et al.,* 1985). DNA topoisomerase II protein is abundant in proliferating cells and decreases during density arrest. DNA topoisomerase II protein levels decrease gradually during MELC differentiation. By Day 5 of HMBA-mediated differentiation, only 5 to 10% of the topoisomerase protein remains (Bodley *et al.,* 1987). DNA topoisomerase I levels remain relatively constant throughout growth and differentiation in MELC. The decrease in DNA topoisomerase II may be a requirement for MELC differentiation or, alternatively, may be a characteristic of the differentiated state.

G. Protein Kinase C

Protein kinase C (PKC) is an important regulatory enzyme in signal-transduction pathways controlling cell proliferation and differentiation (Nishizuka, 1988). The enzyme is activated by calcium- and diacylglycerol-dependent binding to membrane phospholipids (Nishizuka, 1988). Of the several isozymes of PKC, MELC contain at least two, PKCα, which is the major isozyme form and, a minor component, PKC epsilon (Powell *et al.*, 1991) The minor isozyme form had previously been reported to be PKCβ (Melloni *et al.*, 1989). A role for PKC in

MELC differentiation is supported by several observations. The phorbol ester, TPA, depletes MELC PKC activity and suppresses HMBA-induced differentiation in MELC (Yamasaki *et al.,* 1977; Fibach *et al.,* 1979). Removal of TPA permits progressive reaccumulation of PKC and restoration of sensitivity to HMBA induction (Melloni *et al.,*1987). During induction, PKC is translocated, activated at the plasma membrane, then down-regulated (Melloni *et al.,* 1987; L. Leng, unpublished observation). There appears to be a relationship between the level of PKCε activity and responsiveness to HMBA (Melloni *et al.,* 1989). MELC variants that are relatively resistant to induction have relatively low levels of PKCε, while those with an accelerated response display higher PKCε activity. During induction with HMBA there is a fall in PKC activity, largely owing to a decline in PKCε activity. Cells responding more rapidly to HMBA more rapidly down-regulate their PKCε activity. The introduction of exogenous PKCε protein purified from MELC, but not PKCα, into permeabilized MELC results in an increased rate of HMBA-induced differentiation (Melloni *et al.,* 1990). Taken together, the data suggest that PKCε plays an important role in the signaling mechanisms by which HMBA induces cell differentiation.

V. SUMMARY: RELATIONSHIP OF THE CELL CYCLE TO COMMITMENT TO TERMINAL ERYTHROID DIFFERENTIATION

The mechanism by which agents induce MELC differentiation appears to be cell cycle related and to involve the modulation in expression of several proteins, which are themselves either cell cycle regulated or control cell cycle progression. There appears to be a critical time in late G_1 or early S phase when inducer must be present in order to cause differentiation (McClintock and Papaconstantinou, 1974; Levy *et al.,* 1975; Geller *et al.,* 1978). This is followed by transient prolongation of the next G_1, then a return to apparently normal cell cycling, before cell division is finally arrested in G_1 (Terada *et al.,* 1977). Inducers of differentiation cause a decrease in the expression of several proteins that are necessary for progression from G_1 into S. Decreased expression of p53, c-*myc,* and c-*myb* precedes the induced cessation of cell division. (Shen *et al.,* 1983; Lachman and Skoultchi, 1984; Ramsay *et al.,* 1986; Watson, 1988; Richon *et al.,* 1989). Proteins involved in the conformation and replication of DNA are also regulated during induced differentiation, including the core histones (Grove and Zweidler, 1984; Brown *et al.,* 1985), histone $H1^0$ (Keppel *et al.,* 1977; Chen *et al.,* 1982; Osborne and Chabanas, 1984), and DNA topoisomerase II (Bodley *et al.,* 1987). PKC (Melloni *et al.,* 1987) and pRB (unpublished observations) are protein components of signal-transduction pathways that mediate growth and differentiation, and are regulated during MELC differentiation. The complex regulation of all of these proteins during terminal erythroid differentiation may reflect the induction of a single major developmental pathway or several concurrent pathways. Understanding the roles of these proteins that are regulated in the differentiation process is the next challenge. It has been demonstrated that pRB may mediate TGFβ-induced growth suppression in keratinocytes, by suppressing expression of c-*myc* (Pietenpol *et al.,* 1990). It is not yet known whether pRB is involved in suppression of c-*myc* protein or other proteins critical to MELC differentiation.

Understanding the mechanism of action of differentiation inducers and their role in the control of the cell cycle may aid in the development of clinically more effective cytodifferentiation agents for the treatment of cancer. Phase I and phase II studies have demonstrated clinical activity of HMBA against a variety of malignancies, including metastatic lesions of breast, colon, and lung (Young *et al.,* 1988), and myelodysplastic syndrome (Andreeff *et al.,* 1988). The effectiveness of HMBA is limited by toxic side-effects, in particular thrombocytopenia. A better understanding of the biological basis for inducer activity may assist in the design of new differentiation inducers with improved therapeutic index.

ACKNOWLEDGMENTS

Studies performed in our laboratories were supported, in part, by grants from the National Cancer Institute (CA–31768–07), the Japanese Foundation for the Promotion of Cancer Research, and the Roberta Rudin Leukemia Research Fund.

REFERENCES

Abrams, H. D., Rohrschneider, L. R., and Eisenman, R. N. (1982). Nuclear location of the putative transforming protein of avian myelocytomatosis virus. *Cell* **29,** 427–439.

Akiyama, T., and Toyoshima, K. (1990). Marked alteration in phosphorylation of the RB protein during differentiation of human promyelocytic HL-60 cells. *Oncogene* **5,** 179–183.

Anand, R., Lilly, F., and Ruscetti, S. (1981). Viral protein expression in producer and nonproducer clones of Friend erythroleukemia cell lines. *J. Virol.* **37,** 654–660.

Andreeff, M., Young, C., Clarkson, B., Fetten, J., Rifkind, R. A., and Marks, P. A. (1988). Treatment of myelodysplastic syndromes (MDS) with hexamethylene bisacetamide. *Blood* **72,** 186a.

Barnhart, K. M., Kim, C. G., and Sheffery, M. (1989). Purification and characterization of an erythroid cell-specific factor that binds the murine α- and β-globin genes. *Mol. Cell. Biol.* **9,** 2606–2614.

Beckman, B. S., Kopfler, W., Koury, P., and Jeter, J. R., Jr. (1987). Effect of aphidicolin on Friend erythroleukemia cell maturation. *Exp. Cell Res.* **169,** 223–232.

Ben-Dori, R., Resnitzki, D., and Kimchi, A. (1983). Reduction in p53 synthesis during differentiation of Friend erythroleukemia cells. *FEBS Lett.* **162,** 384–389.

Bernards, R., Schackleford, G. M., Berber, M. R., Horowitz, J. M., Friend, S. H., Schartl, M., Bogenmann, E., Rapaport, J. M., McGee, T., Dryja, T. P., and Weinberg, R. A. (1989). Structure and expression of the murine retinoblastoma gene and characterization of its encoded protein. *Proc. Natl. Acad. Sci. U.S.A.* **86,** 6474–6478.

Bernstein, A., Hunt, H. D., Crichley, V., and Mak, T. W. (1979). Induction by ouabain of hemoglobin synthesis in cultured Friend erythroleukemia cells. *Cell* **9,** 375–381.

Biedenkapp, H., Borgmeyer, U., Sippel, A. E., and Klempnauer, K.-H. (1988). Viral *myb* oncogene encodes a sequence-specific DNA-binding activity. *Nature* **335,** 835–837.

Bodley, A. L., Wu, H.-Y., and Liu, L. F. (1987). Regulation of DNA topoisomerases during cellular differentiation. *NCI Monogr.* **4,** 31–35.

Brown, D. T., Wellman, S. E., and Sittman, D. B. (1985). Changes in the level of three different classes of histone mRNA during murine erythroleukemia cell differentiation. *Mol. Cell. Biol.* **5,** 2879–2886.

Brown, D. T., Yang, Y.-S., and Sittman, D. B. (1988). Histone gene switching in murine erythroleukemia cells is differentiation specific and occurs without loss of cell-cycle regulation. *Mol. Cell. Biol.* **8,** 4406–4415.

Buchkovich, K., Duffy, L. A., and Harlow, E. (1989). The retinoblastoma protein is phosphorylated during specific phases of the cell cycle. *Cell* **58,** 1097–1105.

Chambers, I., Frampton, J., Goldfarb, P., Affara, N., McBain, W., and Harrison, P. R. (1986). The structure of the mouse glutathione peroxidase gene: The selenocysteine in the active site is encoded by the termination codon TGA. *EMBO J.* **5,** 1221–1227.

Chapman, L. F. (1980). Effect of calcium on differentiation of Friend leukemia cells. *Dev. Biol.* **79,** 243–246.

Chen, Z., Banks, J., Rifkind, R. A., and Marks, P. A. (1982). Inducer-mediated commitment of murine erythroleukemia cells to differentiation: A multistep process. *Proc. Natl. Acad. Sci. U.S.A.* **79,** 471–475.

Chen, P.-L., Scully, P., Shew, J.-Y., Wang, J. Y. J., and Lee, W.-H. (1989). Phosphorylation of the retinoblastoma gene product is modulated during the cell cycle and cellular differentiation. *Cell* **58,** 1193–1198.

Chretien, S., Dubart, A., Beaupain, D., Raich, N., Grandchanp, B., Rosa, J., Goossens, M., and Romeo, P. H. (1988). Alternative transcription and splicing of the human porphobilinogen deaminase gene result either in tissue-specific or in housekeeping expression. *Proc. Natl. Acad. Sci. U.S.A.* **85,** 6–10.

Clarke, M. F., Kukowska-Latallo, J. L., Westin, E., Smith, M., and Prochownik, E. V. (1988). Constitutive expression of a c-*myb* cDNA blocks Friend murine erythroleukemia cell differentiation. *Mol. Cell. Biol.* **8,** 884–892.

Cole, M. D. (1986). The *myc* oncogene: Its role in transformation and differentiation. *Annu. Rev. Genet.* **20,** 361–384.

Coppola, J. A., Parker, J. M., Schuler, G. D., and Cole, M. D. (1989). Continued withdrawal from the cell cycle and regulation of cellular genes in mouse erythroleukemia cells blocked in differentiation by the c-*myc* oncogene. *Mol. Cell. Biol.* **9,** 1714–1720.

Coppola, J. A., and Cole, M. D. (1986). Constitutive c-*myc* oncogene expression blocks mouse erythroleukaemia cell differentiation but not commitment. *Nature (London)* **320,** 760–763.

Crawford, L. V., Pim, D. C., Gurney, E. G., Goodfellow, P., and Taylor-Papadimitriou, J. (1981). Detection of a common feature in several human tumor cell lines—a 53,000-dalton protein. *Proc. Natl. Acad. Sci. U.S.A.* **78,** 41–45.

Curtis, P., Finnigan, A. C., and Rovera, G. (1980). The β major and β minor globin nuclear transcripts of Friend erythroleukemia cells induced to differentiate in culture. *J. Biol. Chem.* **255,** 8971.

DeCaprio, J. A., Ludlow, J. W., Lynch, D., Furukawa, Y., Griffin, J., Piwnica-Worms, H., Huang, C.-M., and Livingston, D. M. (1989). The product of the retinoblastoma susceptibility gene has properties of a cell cycle–regulatory element. *Cell* **58,** 1085–1095.

Dmitrovsky, E., Kuehl, W. M., Hollis, G. F., Kirsch, I. R., Bender, T. P., and Segal, S. (1986). Expression of a transfected human c-*myc* oncogene inhibits differentiation of a mouse erythroleukaemia cell line. *Natue (London)* **322,** 748–750.

Donner, P., Greiser-Wilke, I., and Moeling, K. (1982). Nuclear localization and DNA binding of the transforming gene product of avian myelocytomatosis virus. *Nature (London)* **296,** 262–266.

Duprey, S. P., and Boetigger, D. (1985). Developmental regulation of c-*myb* in normal myeloid progenitor cells. *Proc. Natl. Acad. Sci. U.S.A.* **78,** 3600–3604.

Eto, Y., Tuji, T., Takezawa, M., Takano, S., Yokogawa, Y., and Shibai, H. (1987). Purification and characterization of erythroid differentiation factor (EDF) isolated from human leukemia cell line THP-1. *Biochem. Biophys. Res. Commun.* **142,** 1095–1103.

Evans, T., Reitman, M., and Felsenfeld, G., (1988). An erythrocyte-specific DNA-binding factor recognizes a regulatory sequence common to all chicken globin genes. *Proc. Natl. Acad. Sci. U.S.A.* **85,** 5976–5980.

Evans, T., and Felsenfeld, G. (1989). The erythroid-specific transcription factor Eryf1: A new finger protein. *Cell* **58,** 877–885.

Faleto, D. L., Arrow, A. S., and Macara, I. G. (1985). An early decrease in phosphatidylinositol turnover occurs on induction of Friend-cell differentiation and precedes the decrease in c-*myc* expression. *Cell* **43,** 315–325.

Fibach, E., Reuben, R. C., Rifkind, R. A., and Marks, P. A. (1977). Effect of hexamethylene bis-acetamide on the commitment to differentiation of murine erythroleukemia cells. *Cancer Res.* **37,** 440–444.

Fibach, E., Gambari, R., Shaw, P. A., Maniatis, G., Reuben, R. C., Sassa, S., Rifkind, R. A., and Marks, P. A. (1979). Tumor promoter–mediated inhibition of cell differentiation: Suppression of the expression of erythroid functions in murine erythroleukemia cells. *Proc. Natl. Acad. Sci. U.S.A.* **76,** 1906–1910.

Finley, C. A., Hinds, P. W., and Levine, A. J. (1989). The p53 protooncogene can act as a suppressor of transformation. *Cell* **57,** 1083–1093.

Friend, C., Patuleia, M. C., and DeHarvan, E. (1966). Erythrocyte maturation *in vitro* of murine (Friend) virus–induced leukemia cells. *Natl. Cancer Inst. Monogr.* **22,** 505–520.

Friend, C., Scher, W., Holland, J., and Sato, T. (1971). Hemoglobin synthesis in murine erythroleukemia cells *in vitro*: Stimulation of erythroid differentiation by dimethylsulfoxide. *Proc. Natl. Acad. Sci. U.S.A.* **68,** 378–382.

Galson, D. L., and Housman, D. E. (1988). Detection of two tissue-specific DNA-binding proteins with affinity for sites in the mouse β-globin intervening sequence 2. *Mol. Cell. Biol.* **8,** 381–392.

Gambari, R., Marks, P. A., and Rifkind, R. A. (1979). Murine erythroleukemia cell differentiation: Relationship of globin gene expression and of prolongation of G_1 to inducer effects during G_1–early S. *Proc. Natl. Acad. Sci. U.S.A.* **76,** 4511–4515.

Ganguly, S., and Skoultchi, A. I. (1985). Absolute rates of globin gene transcription and mRNA formation during differentiation of cultured mouse erythroleukemia cells. *J. Biol. Chem.* **260,** 12167–12173.

Gannon, J. V., and Lane, D. P. (1987). p53 and DNA polymerase α compete for binding to SV40 T antigen. *Nature (London)* **329,** 456–460.

Gazitt, Y., Deitch, A. D., Marks, P. A., and Rifkind, R. A. (1978a). Cell volume changes in relation to the cell cycle of differentiating erythroleukemia cells. *Exp. Cell Res.* **117,** 413–420.

Gazitt, Y., Reuben, R. C., Deitch, A. D., Marks, P. A., and Rifkind, R. A. (1978b). Changes in cyclic adenosine 3′:5′-monophosphate levels during induction of differentiation in murine erythroleukemic cells. *Cancer Res.* **38,** 3779–3783.

Geller, R., Levenson, R., and Housman, D. (1978). Significance of the cell cycle in commitment of murine erythroleukemia cells to erythroid differentiation. *J. Cell. Physiol.* **95,** 213–222.

Grove, G. W., and Zweidler, A. (1984). Regulation of nucleosomal core histone variants in differentiating murine erythroleukemia cells. *Biochemistry* **23,** 4436–4443.

Gonda, T. J., and Metcalf, D. (1984). Expression of *myb, myc,* and *fos* protooncogenes during the differentiation of a murine myeloid leukaemia. *Nature (London)* **310,** 249–251.

Gumucio, D. L., Rood, K. L., Gray, T. A., Blanchard, K. L., and Collins, F. S. (1988). Nuclear proteins that bind the human gamma-globin gene promoter: Alterations in binding produced by point mutations associated with hereditary persistence of fetal hemoglobin. *Mol. Cell. Biol.* **8,** 5310–5322.

Gusella, J. F., Geller, R., Clarke, B., Weeks, V., and Housman, D. (1976). Commitment to erythroid differentiation by Friend erythroleukemia cells: A stochastic analysis. *Cell* **9,** 221–229.

Gusella, J. F., and Housman, D. (1976). Induction of erythroid differentiation *in vitro* by purines and purine analogs. *Cell* **8,** 263–269.

Gutierrez, C., Guo, Z.-S., Farrell-Towt, J., Ju, G., and DePamphilis, M. L. (1987). c-*myc* Protein and DNA replication: Separation of c-*myc* antibodies from an inhibitor of DNA synthesis. *Mol. Cell. Biol.* **7,** 4594–4598.

Hann, S. R., Thompson, C. B., and Eisenman, R. N. (1985). c-*myc* oncogene protein synthesis is independent of the cell cycle in human and avian cells. *Nature (London)* **314,** 366–369.

Jacob, E. (1976). Histone-gene reiteration in the genome of mouse. *Eur. J. Biochem.* **65,** 275–284.

Kaczmarek, L., Hyland, J., Watt, R., Rosenberg, M., and Baserga, R. (1985). Microinjected c-*myc* as a competence factor. *Science* **228,** 1313–1315.

Kaneda, T., Murate, T., Sheffery, M., Brown, K., Rifkind, R. A., and Marks, P. A. (1985). Gene expression during terminal differentiation: Dexamethasone suppression of inducer-mediated α1 and βmaj-globin gene expression. *Proc. Natl. Acad. Sci. U.S.A.* **82,** 5020–5024.

Kelly, K., Cochran, B. H., Stiles, C. D., and Leder, P. (1983). Cell-specific regulation of the c-*myc* gene by lymphocyte mitogens and platelet-derived growth factor. *Cell* **35,** 603–610.

Keppel, F., Albert, B., and Eisen, H. (1977). Appearance of a chromatin protein during the erythroid differentiation of Friend virus–transformed cells. *Proc. Natl. Acad. Sci. U.S.A.* **74,** 653–659.

Kerr, S. J. (1990). Methylated oxypurines and induction of differentiation of murine erythroleukemia cells. *Mol. Cell. Biochem.* **92,** 37–44.

Khochbin, S., Principaud, E., Chabanas, A., and Lawrence, J.-J. (1988a). Early events in murine erythroleukemia cells induced to differentiate: Accumulation and gene expression of the transformation-associated cellular protein p53. *J. Mol. Biol.* **200,** 55–64.

Khochbin, S., Chabanas, A., and Lawrence, J.-J. (1988b). Early events in murine erythroleukemia cells induced to differentiate: Variation of the cell cycle parameters in relation to p53 accumulation. *Exp. Cell Res.* **179,** 565–574.

Kim, C. G., Swendeman, S. L., Barnhart, K. M., and Sheffery, M. (1990). Promoter elements and erythroid cell nuclear factors that regulate α-globin gene transcription *in vitro*. *Mol. Cell. Biol.* **10** 5958–5966.

Kim, C. G., and Sheffery, M. (1990). Physical characterization of the purified CCAAT transcription factor, α-CP1. *J. Biol. Chem.* **265,** 13362–13369.

Klempnauer, K. H., Symonds, G., Evan, G. I., and Bishop, J. M. (1984). Subcellular localization of proteins encoded by oncogenes of avian myeloblastosis virus and avian leukemia virus E26 and by the chicken c-*myb* gene. *Cell* **37,** 537–547.

Lachman, H. M., Hatton, K. S., Skoultchi, A. I., and Schildkrat, C. H. (1985). c-myc mRNA levels in the cell cycle change in mouse erythroleukemia cells following inducer treatment. *Proc. Natl. Acad. Sci. U.S.A.* **82,** 5323–5327.

Lachman, H. M., and Skoultchi, A. I. (1984). Expression of c-*myc* changes during differentiation of mouse erythroleukaemia cells. *Nature (London)* **310,** 592–594.

Lee, W.-H., Shew, J.-Y., Hong, F., Sery, T. W., Donoso, L. A., Young, L.-J., Bookstein, R., and Lee, E. Y.-H. P. (1987). The retinoblastoma-susceptibility gene encodes a nuclear phosphoprotein associated with DNA-binding activity. *Nature (London)* **329,** 642–645.

Levy, J., Terada, M., Rifkind, R. A., and Marks, P. A. (1975). Induction of erythroid differentiation by dimethylsulfoxide in cells infected with Friend virus: Relationship to the cell cycle. *Proc. Natl. Acad. Sci. U.S.A.* **72,** 28–32.

Li, J.-P., D'Andrea, A., Lodish, H. F., and Baltimore, D. (1990). Activation of cell growth by binding of Friend spleen focus-forming virus gp55 glycoprotein to the erythropoietin receptor. *Nature (London)* **343,** 762–764.

Lyman, G., Papahajopoulos, D., and Preisler, H. (1976). Phospholipid membrane stabilization by dimethylsulfoxide and other inducers of Friend leukemic cell differentiation. *Biochim. Biophys. Acta* **448,** 460–473.

Marks, P. A., Chen, Z. X., Banks, J., and Rifkind, R. A. (1983). Erythroleukemia cells: Variants inducible for hemoglobin synthesis without commitment to terminal cell differentiation. *Proc. Natl. Acad. Sci. U.S.A.* **80,** 2281–2284.

Marks, P. A., Sheffery, M., and Rifkind, R. A. (1987). Induction of transformed cells to terminal differentiation and the modulation of gene expression. *Cancer Res.* **47,** 659–666.

Marks, P. A., Breslow, R., Rifkind, R. A., Ngo, L., and Singh, R. (1989). Polar/apolar chemical inducers of differentiation of transformed cells: Strategies to improve therapeutic potential. *Proc. Natl. Acad. Sci. U.S.A.* **86,** 6358–6362.

Marks, P. A., and Rifkind, R. A. (1978). Erythroleukemic differentiation. *Annu. Rev. Biochem.* **47,** 419–448.

Mayeux, P., Felix, J.-M., Billat, C., and Jacquot, R. (1985). Effect of the antiglucocorticoid agent RU 38486 on the dexamethasone inhibition of Friend-cell differentiation. *Biochim. Biophys. Acta* **846,** 413–417.

McClintock, P. R., and Papaconstantinou, J. (1974). Regulation of hemoglobin synthesis in a murine erythroblastic leukemic cell: The requirement for replication to induce hemoglobin synthesis. *Proc. Natl. Acad. Sci. U.S.A.* **71,** 4551–4555.

McClinton, D., Stafford, J., Brents, L., Bender, T. P., and Kuehl, M. (1990). Differentiation of mouse erythroleukemia cells is blocked by late up-regulation of a c-*myb* transgene. *Mol. Cell. Biol.* **10,** 705–710.

McMahon, J., Howe, K. M., and Watson, R. J. (1988). The induction of Friend erythroleukaemia differentiation is markedly affected by expression of a transfected c-*myb* cDNA. *Oncogene* **3,** 717–720.

Melloni, E., Pontremoli, S., Michetti, M., Sacco, O., Cakiroglu, A. G., Jackson, J. F., Rifkind, R. A., and Marks, P. A. (1987). Protein kinase C activity and hexamethylene bisacetamide–induced erythroleukemia cell differentiation. *Proc. Natl. Acad. Sci. U.S.A.* **84,** 5282–5286.

Melloni, E., Pontremoli, S., Damiani, G., Viotti, P., Weich, N., Rifkind, R. A., and Marks, P. A. (1988). Vincristine-resistant erythroleukemia cells have marked increased sensitivity to hexamethylene bisacetamide–induced differentiation. *Proc. Natl. Acad. Sci. U.S.A.* **85,** 3835–3839.

Melloni, E., Pontremoli, S., Viotti, P. L., Patrone, M., Marks, P. A., and Rifkind, R. A. (1989). Differential expression of protein kinase C isozymes and erythroleukemia differentiation. *J. Biol. Chem.* **264,** 18414–18418.

Melloni, E., Pontremoli, S., Sparatore, B., Patrone, M., Grossi, F., Marks, P. A., and Rifkind, R. A. (1990). Introduction of the β isozyme of protein kinase c accelerates induced differentiation of murine erythroleukemia cells. *Proc. Natl. Acad. Sci. U.S.A.* **87,** 4417–4420.

Mercer, W. E., Nelson, D., DeLeo, A., Old, L. J., and Baserga, R. (1982). Microinjection of monoclonal antibody to protein p53 inhibits serum-induced DNA synthesis in 3T3 cells. *Proc. Natl. Acad. Sci. U.S.A.* **79,** 6309–6312.

Michaeli, J., Lebedev, Y. B., Richon, V. M., Chen, Z. X., Marks, P. A., and Rifkind, R. A. (1990). Conversion of differentiation-inducer resistance to differentiation-inducer sensitivity in erythroleukemia cells. *Mol. Cell. Biol.* **10,** 3535–3540.

Mihara, K., Cao, X.-R., Yen, A., Chandler, S., Driscoll, B., Murphree, A. L., T'Ang, A., and Fung, Y.-K. T. (1989). Cell cycle dependent regulation of phosphorylation of the human retinoblastoma gene product. *Science* **246,** 1300–1303.

Moelling, K., Pfaff, E., Beug, H., Beiming, P., Bunte, T., Schaller, H. E., and Graf, T. (1985). DNA-binding activity is associated with purified Myb proteins from AMV and E26 viruses and is temperature-sensitive for E26 to mutants. *Cell* **40,** 983–990.

Morioka, K., Tanaka, K., Nokuo, T., Ishizawa, M., and Ono, T. (1979). Erythroid differentiation and poly(ADP-ribose) synthesis in Friend leukemia cells. *Gann* **70,** 37–46.

Mowat, M., Cheng, A., Kimura, N., Bernstein, A., and Benchimol, S. (1985). Rearrangements of the cellular p53 gene in erythroleukaemic cells transformed by Friend virus. *Nature (London)* **314,** 633–636.

Muller, C. P., Volloch, Z., and Shinitzky, M. (1980). Correlation between cell density, membrane fluidity, and the availability of transferrin receptors in Friend erythroleukemia cells. *Cell Biophys.* **2,** 233–240.

Munroe, D. G., Peacock, J. W., and Benchimol, S. (1990). Inactivation of the cellular p53 gene is a common feature of Friend virus–induced erythroleukemia: Relationship of inactivation to dominant transforming alleles. *Mol. Cell. Biol.* **10,** 3307–3313.

Murata, M., Eto, Y., Shibai, H., Sakai, M., and Muramatsu, M. (1988). Erythroid differentiation factor is encoded by the same mRNA as that of the inhibin βa chain. *Proc. Natl. Acad. Sci. U.S.A.* **85,** 2434–2438.

Nepveu, A., Marcu, K. B., Skoultchi, A. I., and Lachman, H. M. (1987). Contributions of transcriptional and posttranscriptional mechanisms to the regulation of c-*myc* expression in mouse erythroleukemia cells. *Genes Dev.* **1,** 938–945.

Nishizuka, Y. (1988). The molecular heterogeneity of protein kinase C and its implications for cellular regulation. *Nature (London)* **334,** 661–665.

Nudel, U., Salmon, J., Terada, M., Bank, A., Rifkind, R. A., and Marks, P. A. (1977). Differential effects of chemical inducers on the expression of β globin genes in murine erythroleukemia cells. *Proc. Natl. Acad. Sci. U.S.A.* **74,** 1100–1104.

Osborne, H. B., and Chabanas, A. (1984). Kinetics of histone $H1^0$ accumulation and commitment to differentiation in murine erythroleukemia cells. *Exp. Cell Res.* **152,** 449–458.

Pehrson, J., and Cole, R. D. (1980). Histone $H1^0$ accumulates in growth-inhibited cultured cells. *Nature (London)* **285,** 43–44.

Pietenpol, J. A., Stein, R. W., Moran, E., Yaciuk, P., Schlegel, R., Lyons, R. M., Pittelkow, M. R., Munger, K., Howley, P. M., and Moses, H. L. (1990). TGF-β1 inhibition of c-*myc* transcription and growth in keratinocytes is abrogated by viral transforming proteins with pRB binding domains. *Cell* **61,** 777–785.

Pincus, S. M., Beckman, B. S., and George, W. J. (1984). Inhibition of dimethylsulfoxide-induced differentiation in Friend erythroleukemic cells by diacylglycerols and phospholipase C. *Biochem. Biophys. Res. Commun.* **125,** 491–499.

Plumb, M., Frampton, J., Wainwright, H., Walker, M., Macleod, K., Goodwin, G., and Harrison, P. (1989). GATAAG; a cis-acting control region binding an erythroid-specific nuclear factor with a role in globin and nonglobin gene expression. *Nucleic Acids Res.* **17,** 73–92.

Popp, R. A., Lalley, P. A., Whitney, J. B., III, and Anderson, W. F. (1981). Mouse α-globin genes and α-globin–like pseudogenes are not synthenic. *Proc. Natl. Acad. Sci. U.S.A.* **78,** 6362–6366.

Powell, C. T., Leng, L., Rifkind, R. A., and Marks, P. A. (1991). Identification of protein kinase C isozymes in murine erythroleukemia cells: Protein kinase alpha and epsilon. *Proc. Natl. Acad. Sci.* (USA) (In press).

Prochownik, E. V., Kukowska, J., and Rodgers, C. (1988). c-*myc* antisense transcripts accelerate differentiation and inhibit G_1 progression in murine erythroleukemia cells. *Mol. Cell. Biol.* **8,** 3683–3695.

Prochownik, E. V., and Kukowska, J. (1986). Deregulated expression of c-*myc* by murine erythroleukaemia cells prevents differentiation. *Nature (London)* **322,** 848–851.

Profous-Juchelka, H. R., Rueben, R. C., Marks, P. A., and Rifkind, R. A. (1983). Transcriptional and posttranscriptional regulation of globin gene accumulation in induced murine erythroleukemia cells. *Mol. Cell. Biol.* **3,** 229–232.

Ramsay, R. G., Ikeda, K., Rifkind, R. A., Marks, P. A. (1986). Changes in gene expression associated with induced differentiation of erythroleukemia: Protooncogenes, globin genes, and cell division. *Proc. Natl. Acad. Sci. U.S.A.* **83,** 6849–6853.

Reich, N. C., and Levine, A. J. (1984). Growth regulation of a cellular tumour antigen, p53, in nontransformed cells. *Nature (London)* **308,** 199–201.

Reuben, R. C., Wife, R. L., Breslow, R., Rifkind, R. A., and Marks, P. A. (1976). A new group of potent inducers of differentiation in murine erythroleukemia cells. *Proc. Natl. Acad. Sci. U.S.A.* **73,** 862–866.

Richon, V. M., Ramsay, R. G., Rifkind, R. A., and Marks, P. A. (1989). Modulation of the c-*myb,* c-*myc,* and p53 mRNA and protein levels during induced murine erythroleukemia cell differentiation. *Oncogene* **4,** 165–173.

Robbins, P. D., Horowitz, J. M., and Mulligan, R. C. (1990). Negative regulation of human c-fos expression by the retinoblastoma gene product. *Nature (London)* **346,** 668–671.

Rotter, V. (1983). p53, a transformation-related cellular-encoded protein, can be used as a biochemical marker for detection of primary mouse tumor cells. *Proc. Natl. Acad. Sci. U.S.A.* **80,** 2613–2617.

Salditt-Georgieff, M., Sheffery, M., Krauter, K., Darnell, J. E., Rifkind, R. A., and Marks, P. A. (1984). Induced transcription of the mouse β-globin transcription unit in erythroleukemia cells: Time course of induction and changes in chromatin structure. *J. Mol. Biol.* **172,** 437–450.

Scher, W., and Friend, C. (1978). Breakage of DNA and alterations in folded genomes by inducers of differentiation in Friend erythroleukemic cells. *Cancer Res.* **38,** 841–849.

Sheffery, M., Marks, P. A., and Rifkind, R. A. (1984). Gene expression in murine erythroleukemia cells: Transcriptional control and chromatin structure of the α_1-globin gene. *J. Mol. Biol.* **172,** 417–436.

Sheiness, D., and Gardinier, M. (1984). Expression of a protooncogene (proto-*myb*) in hemopoietic tissues of mice. *Mol. Cell. Biol.* **4,** 1206–1212.

Shen, D. W., Real, F. X., Deleo, A. B., Old, L. J., Marks, P. A., and Rifkind, R. A. (1983). Protein p53 and inducer-mediated erythroleukemia cell commitment to terminal cell division. *Proc. Natl. Acad. Sci. U.S.A.* **80,** 5919–5922.

Shohat, O., Greenberg, M., Resiman, D., Oren, M., and Rotter, V. (1987). Inhibition of cell growth mediated by plasmids encoding p53 antisense. *Oncogene* **1,** 277–283.

Simpson, R. T. (1981). Modulation of nucleosome structure by histone subtypes in sea urchin embryos. *Proc. Natl. Acad. Sci. U.S.A.* **78,** 6803–6807.

Smith, R. D., Yu, J., and Seale, R. L. (1984). Chromatin structure of the β-globin gene family in murine erythroleukemia cells. *Biochemistry* **23,** 785–790.

Spotts, G. D., and Hann, S. R. (1990). Enhanced translation and increased turnover of c-*myc* proteins occur during differentiation of murine erythroleukemia cells. *Mol. Cell. Biol.* **10,** 3952–3964.

Studzinski, G. P., Brelvi, Z. S., Feldman, S. C., and Watt, R. A. (1986). Participation of c-*myc* protein in DNA synthesis of human cells. *Science* **214,** 467–470.

Tanaka, M., Levy, J., Terada, M., Breslow, R., Rifkind, R. A., and Marks, P. A. (1975). Induction of erythroid differentiation in murine virus–infected erythroleukemia cells by highly polar compounds. *Proc. Natl. Acad. Sci. U.S.A.* **72,** 1003–1006.

Terada, M., Fried, J., Nudel, U., Rifkind, R. A., and Marks, P. A. (1977). Transient inhibition of initiation of S phase associated with dimethylsulfoxide induction of murine erythroleukemia cells to erythroid differentiation. *Proc. Natl. Acad. Sci. U.S.A.* **74,** 248–252.

Terada, M., Nudel, U., Fibach, E., Rifkind, R. A., and Marks, P. A. (1978). Changes in DNA associated with induction of erythroid differentiation by dimethylsulfoxide in murine erythroleukemia cells. *Cancer Res.* **38,** 835–840.

Terada, M., Fujiki, H., Marks, P. A., and Sugimura, T. (1979). Induction of erythroid differentiation of murine erythroleukemia cells by nicotinamide and related compounds. *Proc. Natl. Acad. Sci. U.S.A.* **76,** 6411–6414.

Thiele, C. J., Cohen, P. S., and Israel, M. A. (1988). Regulation of c-*myb* expression in human neuroblastoma cell during retinoic acid–induced differentiation. *Mol. Cell. Biol.* **8,** 1677–1683.

Thompson, C. B., Challoner, P. B., Nieman, P. E., and Groudine, M. (1986). Expression of the c-*myb* protooncogene during cellular proliferation. *Nature (London)* **319,** 374–380.

Todokoro, K., Kanazawa, S., Amanuma, H., and Ikawa, Y. (1987). Specific binding of erythropoietin to its receptor on responsive mouse erythroleukemia cells. *Proc. Natl. Acad. Sci. U.S.A.* **84,** 4126–4130.

Todokoro, K., Watson, R. J., Higo, H., Amanuma, H., Kuramochi, S., Yanagisawa, H., and Ikawa, Y. (1988). Down-regulation of c-*myb* gene expression is a prerequisite for erythropoietin-induced erythroid differentiation. *Proc. Natl. Acad. Sci. U.S.A.* **85,** 8900–8904.

Tricoli, J. V., Sahai, B. M., McCormick, P. J., Jarlinski, S. J., Bertram, J. S., and Kowalski, D. (1985). DNA topoisomerase I and II activities during cell proliferation and the cell cycle in cultured mouse embryo fibroblast (C3H 10T1/2) cells. *Exp. Cell Res.* **158,** 1–14.

Tsai, S.-F., Martin, D. I., Zon, L. I., D'Andrea, D., Wong, G. G., and Orkin, S. H. (1989). Cloning of cDNA for the major DNA-binding protein of the erythroid lineage through expression in mammalian cells. *Nature (London)* **339,** 446–451.

Tsiftsoglou, A. S., and Robinson, S. H. (1985). Differentiation of leukemic cell lines: Review focusing on murine erythroleukemia and human HL-60 cells. *Int. J. Cell Cloning* **3,** 349–366.

Venturelli, D., Travali, S., and Calabretta, B. (1990). Inhibition of T-cell proliferation by a *myb* antisense oligomer is accompanied by selective down-regulation of DNA polymerase α expression. *Proc. Natl. Acad. Sci. U.S.A.* **87,** 5963–5967.

von Holt, C., deGroot, P., Schwager, S., and Brandt, W. F. (1984). The structure of sea urchin histones and considerations on their function. *In* "Histone genes: Structure, Organization, and Regulation." (G. S. Stein, J. L. Stein, and W. F. Marzluff (eds.), pp. 65–105. John Wiley, New York.

Wall, L., deBoer, E., and Grosveld, F. (1988). The human β-globin gene 3′ enhancer contains multiple binding sites for an erythroid-specific protein. *Genes Dev.* **2,** 1089–1100.

Watson, R. J. (1988). Expression of the c-*myb* and the c-*myc* genes is regulated independently in differentiating mouse erythroleukemia cells by common processes of premature transcription arrest and increased mRNA turnover. *Mol. Cell. Biol.* **8,** 3938–3942.

Weber, B. L., Westin, E. H., and Clarke, M. F. (1990). Differentiation of mouse erythroleukemia cells enhanced by alternatively spliced c-*myb* mRNA. *Science* **249,** 1291–1293.

Weston, K., and Bishop, J. M. (1989). Transcriptional activation by the v-*myb* oncogene and its cellular progenitor, c-*myb*. *Cell* **58,** 85–93.

Yamasaki, H., Fibach, E., Nudel, U., Weinstein, I. B., Rifkind, R. A., and Marks, P. A. (1977). Tumor promoters inhibit spontaneous and induced differentiation of murine erythroleukemia cells in culture. *Proc. Natl. Acad. Sci. U.S.A.* **74,** 3451–3455.

Yamashita, T., Eto, Y., Shibai, H., and Ogata, E. (1990). Synergistic action of Activin A and hexamethylene bisacetamide in differentiation of murine erythroleukemia cells. *Cancer Res.* **50,** 3182–3185.

Young, C. W., Fanucchi, M. P., Walsh, T. B., Blatzer, L., Yaldaie, S., Stevens, Y. W., Gordon, C., Tong, W., Rifkind, R. A., and Marks, P. A. (1988). Phase I trial and clinical pharmacologic evaluation of hexamethylene bisacetamide by 10-day continuous intravenous infusion at 28-day intervals. *Cancer Res.* **48,** 7304–7309.

9

Growth and Differentiation in Melanocytes

ANN RICHMOND

Department of Veterans Affairs and Departments of Cell Biology and Medicine
Vanderbilt University School of Medicine
Nashville, Tennessee 37232

I. INTRODUCTION

Melanocytes of the skin are derived from the embryonic neural crest (see Fig. 1). It is thought that pigment cells do not differentiate until after the migration of the neural crest from the neural tube has been completed (Rawles, 1947, 1948). While fish and amphibians localize their pigment in connective tissue (dermis and epidermis of the skin, peritoneum, walls of blood vessels, etc.), birds and mammals concentrate pigment in ectodermal derivatives (hair and feathers). Pigmented

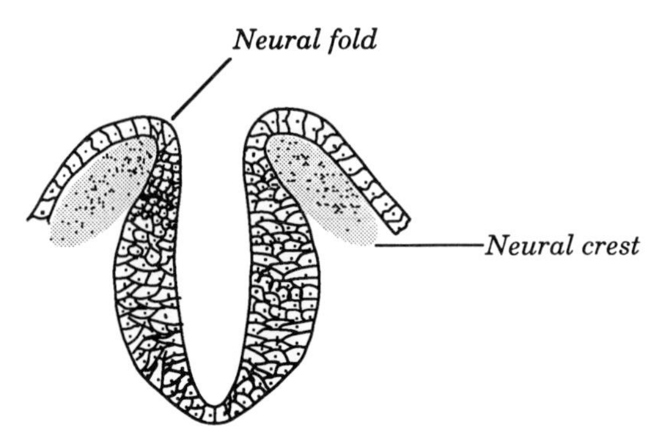

Fig. 1. Position of neural crest cells around the neural fold.

cells are also found in the leptomeninges, inner ear, pigmented retina, and the iris in birds and mammals. The pigmented cells of the retina and iris arise from neural plate ectoderm rather than the migrating neural crest cells. It has been suggested that the precursor to the pigmented retina cell is different from the precursor to the neuron or glial cell, and that the cells of the ciliary margin give rise to both neuroretina and pigment epithelium (McKay, 1989).

The determinative and differentiative events leading to the switch from a neural crest cell to a melanoblast have been investigated from several viewpoints. Niu (1947) explanted different parts of neural fold in culture medium. Hörstadius and Sellman (1946) and Hörstadius (1950) transplanted parts of the neural fold to the side of an embryo. These experiments revealed that the area from which the neural fold was taken affected the type of tissue formed. Cranial neural fold produces few melanophores but gives rise to many cells with the potential for becoming cartilage. Truncal neural folds give rise to numerous melanophores, but no precartilage cells are formed. Apparently, cartilaginous tissue develops from neural crest only when induced to do so by foregut endoderm. In contrast, trunk neural crest cultivated under the same conditions yielded only melanophores and mesenchyme with no cartilage (Okada, 1955). The exact time for the determinative events in unclear, but it has been reported that in amphibians, melanophores and guanophores are distinct by the time the cells migrate away from the site of their neural-fold origin (Stevens, 1954).

Once neural crest cells have differentiated into pigment cells, they are no longer migratory *in vitro*, but when injected into trunk somites of chick embryos, they migrate ventrally along the neural crest pathways (Bronner-Fraser, 1984). In humans, the melanocyte number over specific regions is constant (head and sexual regions are twice as populated as other regions). Szabo has shown that the distribution of melanocytes is bilaterally symmetrical and the same for males and females. Furthermore, there are no racial differences in the number of melanocytes

(Szabo, 1954, 1967), although with increasing age, there is a decline in the number of melanocytes for all races.

Factors affecting the initial migration of avian neural crest cells from the neural tube *in vitro* have been reviewed by Bronner-Fraser (1984), Le Douarin (1984) and Perris *et al.* (1989). Migration on fibronectin, laminin, collagen IV, and platelet factor 4 were examined. The 105-kDa proteolytic fragment of fibronectin, including the 65-kDa cell-binding domain, was as efficient as fibronectin in the promotion of migration. Conversely, migration was inhibited by arginine-glycine-aspartic acid (RGD) containing peptides. The 11.5-kDa fragment containing the arginine-glycine-aspartic acid-serine (RGDS) cell attachment site was able to support migration, but the 50-kDa N-terminal segment of the cell-binding domain did not. Neural-crest cells migrated on a 31-kDa peptide fragment corresponding to the C-terminal heparin binding (II) region of fibronectin. Exogenous heparin inhibited that migration. With purified laminin substrate, maximal migration occurred with low substrate concentrations, while high concentrations resulted in reduced migration. Laminin complexed with nidogen or nidogen plus collagen IV was superior to laminin alone. The migration stimulated by this complex could be antagonized with heparin, but not RGDS. The E8 domain of laminin appears to be the active site for migration. The dispersion on the β-thromboglobulin-related platelet factor 4 (PF4) substrate was comparable to that of the 31-kDa heparin-binding fibronectin fragment and 59% of that observed for the E8 laminin fragment. Heparin inhibited the neural crest cell migration on PF4.

Differentiation of melanocytes occurs in stages: synthesis of the enzyme tyrosinase; melanosome formation (premelanosomes are often called lamellar bodies); transfer of tyrosinase to melanosomes; movement of melanosomes to dendrites; and transfer through dendrites to keratinocytes. The size and shape of melanosomes exhibit racial variation (Szabo *et al.*, 1969). Tyrosinase as a regulatory point in pigmentation has been extensively reviewed (Hearing and Jiménez, 1987), so we shall not detail the biochemical events of melanin synthesis in this chapter.

Studies of melanocyte growth and differentiation have taken two approaches: (1) genetic and experimental developmental biology, and/or (2) *in vitro* or *in vivo* manipulation of melanocytes.

II. GENETIC APPROACHES TO STUDYING MELANOCYTE GROWTH AND DIFFERENTIATION

Numerous models for melanocyte growth and differentiation have been proposed. Pigment patterns in the chick, hamster, mouse, and even in humans have been characterized, and provide important insights into factors affecting melanocyte differentiation, proliferation, and function. The most extensive studies of these models have been performed in the mouse, and, therefore, we have chosen to concentrate on several of these examples. Chick and human studies, although not so complete, have also been critical for our current understanding of melanocyte growth and differentiation and will be discussed briefly in later sections. Though numerous mutants from several loci have been well characterized, we shall describe more detail for those mutations thought to affect melanocyte differentiation and proliferation. Specifically, we shall concentrate more fully on mutations that may affect migration and the initial determinative events, thereby leading to types of albinism or spotting. Discussions of mutations affecting mechanics of melanin production and release will be brief.

A. Mouse Pigment Mutants

Studies of mosaic mice developed from aggregation chimeras by Mintz (1967) led to her proposal that the entire melanocyte population results from growth of 34 clonal initiator cells. Pigment patterns then evolve from the lateral expansion of the respective clones, some of which expand more rapidly than others. For certain pigment mutants, Mintz proposes that early cell death occurs in some of the clonal initiator cells, and therefore pigment is deficient or pigmentation patterns are altered. Based on patterning studies, it would appear that clonal initiator cells arise before 8 days postcoitum (p.c.) in the neural crest cell and probably after 5 days p.c. (Mintz 1970, 1971). Therefore, the cells have become partially restricted before the time that neural-crest cells migrate away from the neural tube (approximately Day 9 p.c. in the mouse).

In contrast to Mintz's pigmentation model, Schaible (1969, 1972) has proposed that clones of pigment cells expand and merge from centers. These centers are the same as those that appear as pigment spots against a white background in white-spotted mutants. His model evokes proliferation of 14 primordial melanoblasts. These 14 melanoblasts migrate from the neural crest and locate in seven bilateral centers (nasal, temporal, aural, costal, lumbar, and sacral areas, as well as medially in the coronal and caudal areas). For animals with full pigmentation, these clones proliferate, expand, and finally merge.

Several genetic loci in the mouse control melanocyte function: nonagouti (*a*), brown (*b*), albino (*c*), piebald-spotting (*s*), white-spotting (*W*), dilute (*d*), leaden (*ln*), pink-eyed (*p*) and numerous other loci (Billingham and Silvers, 1960, Green, 1981).

1. a Locus

Genes of the *a* (nonagouti) locus affect the chemistry of the hair follicle, melanin synthesis, and the protein framework in which it is deposited. Seventeen alleles have been described for this locus on chromosome 2, and the number of reported mutations is greater than for any other locus concerned with melanin synthesis.

The genes of the agouti locus appear to influence hair color through effects on or signals from the mesoderm. The expression of these gene products alters the biology of melanocytes at a specific stage, such that yellow pigment is laid down in a subapical band (Mayer and Fishbane, 1972). An alternate explanation has been proposed by McClaren and Bowman (1969). These investigators proposed that it is the *epidermal* genotype that influences the agouti pattern, regardless of the pigment content of the melanocytes. We are therefore led to conclude that there is an intricate relationship between mesoderm and ectoderm, which can specifically affect color patterns in the coat of the mouse. The *A* alleles determine the relative amount and distribution of yellow pigment (phaeomelanin) and black or brown pigment (eumelanin) in the hairs of the coat. For example, the A^y allele affects the structure of the premelanosomes. In addition, mice homozygous for the A^y allele die *in utero*, suggesting that this gene has an important function during embryonic development. A^w mice carry the white-bellied agouti allele, which prompts or stimulates the melanocyte environment to signal synthesis of yellow, black, or brown melanin. Studies of these mutants have provided useful information about the dorsal–ventral patterning events for pigmentation. The A^w locus is sometimes referred to as A^l (light-bellied agouti). Mice exhibiting this phenotype have a unique pigmentation pattern, a subapical band of yellow on a black or brown background, that is confined to the ventrum. A^w is dominant to *A* and all lower alleles. It produces a shift in pigmentation within the hair shaft from eumelanin to phaeomelanin and back to eumelanin.

2. *b* Locus

Genes of the *b* (brown) locus on mouse chromosome 4 are involved in size and shape of the pigment granule and regulation of type of pigment (black–brown) in melanocytes. When *b* is homozygous, the gray color of the wild type is changed to a brownish color referred to as cinnamon agouti or cinnamon. The hairs are yellow-banded brown instead of black. Three other alleles at the *b* locus have been characterized: light (B^{lt}), cordovan (b^c) and white-based brown (B^w). B^{lt} and B^w appear to be dominant to *B*, but the phenotypes of mice expressing these genes is distinguishable when the allele is heterozygous with *B*. b^c is recessive to black and dominant over brown (Silvers, 1979). The genes of this locus probably affect the polymerization of brown melanin and premelanosome structure. The suggested site of action of the *b* genes is the melanoblast. A tyrosinase-related protein (TRP) maps to the *b* locus (Jackson, 1988). Jackson has speculated that TRP's wild-type function is essential for production of black rather than brown pigment.

Melanin granules from mice exhibiting the wild type *B* allele are extended ovoids and have fewer membranes, while *b* granules are spherical and include numerous membranous strands in a tangled mass. Moyer (1961) observed that when melanin is initially being deposited on the melanosome matrix, the second-order periodicity in black granules was almost 100 Å greater than that of brown granules. This increased distance could be conducive to tyrosinase activity, leading to the increased tyrosinase activity observed in brown mice. It is possible that the chemical structure of black versus brown melanin affects the structure of melanin granules, or alternatively, the proteins coded for by the *b* allele may have other functions as well.

3. *c* Locus

The *c* locus, located on chromosome 7, comprises genes that apparently affect the tyrosinase activity, number of melanosomes, and premelanosome structure. The albino allele (*c*) is recessive, and there are varying degrees of pigmentation between homozygous alleles, ranging from chinchilla (c^{ch}), himalayan (c^{h}), extreme dilution (c^{e}), and albino (*c*). Chinchilla mouse mutants (genotype a/a, c^{ch}/c^{ch}) have white hair and pink eyes as albinos do, but produce melanosomes with tyrosinase activity, even though no melanin is produced. This may result from alterations in the *c*-locus of the mouse, such that tyrosinase translation, turnover, or posttranslational processing are altered. Work by Imokawa *et al.* (1988) suggests that tyrosinase glycosylation is altered in chinchilla mutants. The tyrosinase activation process within hair-bulb melanocytes is possibly defective in chinchilla mutants in which there is a defective response to *a*-melanocyte–stimulating hormone (*a*-MSH) and dibutyryl-cyclic adenosine monophosphate (cAMP) (db-cAMP).

Two homologous cDNAs have been reported as candidates for tyrosinase. Kwon *et al.* (1987) cloned the human tyrosinase cDNA and Yamamota *et al.* (1987) cloned the mouse gene. Both map to the mouse albino (*c*) locus. More recently, Tanaka *et al.* (1990) have introduced a mouse tyrosinase minigene under the control of the authentic genomic 5′ noncoding flanking sequence into fertilized eggs of albino mice. Four of the 25 animals that developed exhibited pigment in the hair and the eyes. These results point to a defective tyrosinase gene in albinism. Similar results have been obtained in the laboratory of Paul Overbeck (Baylor University; 1990, personal communication). Takeuchi *et al.* (1988) studied the albino mouse and the black-eyed white mutant C57BL/6J-mi^{bw}/mi^{bw} mice (genotype: a/a, mi^{bw}/mi^{bw}, c/c). Absence of skin melanocytes is characteristic of this genotype. There was a 2.1 kb mRNA for tyrosinase in the skin of albino and wild-type mice but not in the skin of the black-eyed white mouse. These results suggest that albinism is due to a point mutation in the structural region of the tyrosinase gene, and that the black-eyed white mouse has a deficiency in tyrosinase expression possibly related to melanocyte differentiation.

4. *s* Locus

Piebald spotting (*s*) in mice has also been studied in detail by Grüneberg (1952). The *s* locus has been localized on chromosome 14. Several other genes that confer piebald-like phenotypes have arisen independently as distinct loci on different chromosomes. In addition, several other modifying factors (described as the "*k*" complex) seem to affect the degree of spotting and the location of spotting, such that *s*/*s* mutants may exhibit almost no pigmentation, or only a few white spots on the ventral region. In addition to "*k* genes" (some of which are more dominant–less recessive in a *k*/*k* background), there are several minor spotting genes (Dunn and Charles, 1937). Most of these minor genes exert their influence in the absence of *s*/*s*.

Mayer has hypothesized that skin (presumably the dermis) in the environment of neural-crest cells influences the expression of the piebald-spotting (*s*) gene. Transplantation experiments interchanging *s*/*s* neural tube with neural crest-free *s*/*s* skin and *s*/*s* neural crest with +/+ skin led to two possible explanations for the involvement of skin in the etiology of piebald spotting: *s* might act on skin as well as neural crest; or, alternatively, only neural crest is affected by the piebald gene and the "*k*" complex, or minor pigmentation affecting genes expressed in the *s*/*s* skin influence melanoblast differentiation. There are well-correlated effects on external and internal pigmentation.

In addition to pigmentation changes, about 10% of *s*/*s* animals exhibit megacolon, distention of the colon due to reduction in the neural crest-derived myenteric ganglion cells near the rectum. Megacolon is far more prevalent in piebald-lethal (s^l) mice than in the *s*/*s* mice. s^l can survive for approximately 1 year. However, the *s*/*s* animals usually die at approximately 15 days of age.

Two other white-spotting loci have been described, in addition to the piebald alleles. Lethal spotting (*ls*) is a recessive gene located on chromosome 2. It is very similar to piebald-lethal, and in addition to spotting, megacolon occurs as a result of deficiency in myenteric ganglion cells. Melanocyte number is reduced in the harderian gland, leg musculature, ankle skin, choroid, and membranous labyrinth of the ear. The defect in *ls*/*ls* mice is thought to result from neural-crest deficiency (Mayer and Maltby, 1964). The other white-spotting locus is belted; a recessive mutation (*bt*; chromosome 15) has spontaneously arisen at least twice (Murray and Snell, 1945; Mayer and Maltby, 1964). Belted phenotype is characterized by a transverse white patch of hair running across the lower back. The etiology of *bt* is controversial.

5. *W Locus*

Whereas albino mice have melanocytes, white spotting is characterized by absence of melanocytes. *W*/+ mice exhibit a sharply marked, white *belly spot*, while the white of the dorsal coat is somewhat variegated, producing a silvering effect. The dominant spotting gene (*W*/*W*) is lethal when homozygous (Little, 1915; Sô and Imai, 1920). *W*/*W* mice are described by Mintz as "having one big spot." Hair follicles from these animals are devoid of melanocytes. Mintz proposes that white spotting results from "preprogrammed clonal death" of a distinct population of migrating melanocytes. For white spotting, Schaible's model predicts that one of the progenitor clones failed to proliferate so that the white-spotted areas never became populated with melanoblasts.

The *W* locus of the mouse maps to chromosome 5, and mice carrying *W* locus mutant alleles exhibit varying degrees of macrocytic anemia, loss of hair pigmentation, and sterility (Russell, 1979). All of these disorders occur in cells that are migratory (germ cells, hematopoietic precursors, and neural crest-derived pigment cells). Mutations at the *W* locus result in abnormal migration events during embryogenesis and hematopoietic stem cell proliferation and/or differentiation. Just as the *k* complex can affect the level of spotting expression, in *W* heterozygotes there are modifying genes affecting *W* expression. The expression of these modifiers is such that *W* may, in some instances, act as a recessive gene (Silvers, 1979). *W* usually behaves as a semidominant on the C57BL/6 background. W^v is the viable dominant spotting mutant. Many W^v/W^v mutants survive longer than 3 weeks, and some live to be adults. Mutations at this locus result in a reduction in the number of pigment granules, and W^v mice exhibit severe macrocytic anemia and sterility.

Effects of *W* and W^v on pigmentation patterns are generally thought to result from *W*-locus allele action in melanoblasts. In addition, like the *W* allele, the W^v alleles have profound effects on hematopoiesis (particularly erythropoiesis) and gametogenesis. Mintz and Russell (1957) demonstrated that *W*/*W*, W/W^v and W^v/W^v mice exhibited defects in gametogenesis resulting from a mitotic failure that became apparent at 9 days p.c. of embryonic development. Mintz concluded that pleiotropic effects from this locus result from alterations of a single gene, and that certain kinds of cells are more vulnerable to this defect.

Recent results show that the mouse mutant W^{19H} has a deletion of the c-*kit* protooncogene (Chabot *et al.*, 1989), a gene that codes for a transmembrane receptor with tyrosine kinase activity, similar to the platelet-derived growth factor receptor and the colony stimulating factor-1 (CSF-1) receptor (Yarden *et al.*, 1987). DNA from animals carrying the W^{19H} allele lack a *Bgl*II restriction fragment that hybridizes with the c-*kit* DNA probe. The phenotype of W^{19H}/+ heterozygotes is readily distinguished by characteristic small patches of white hair. Similar results were reported with sash (W^{sh}) heterozygote backcross experiments. (W^{sh}/W^{sh} mice are devoid of pigment except around eyes and ears; sash heterozygotes have a distinctive white band about the loins.) These data are also consistent with c-*kit* mapping studies. The gene for c-*kit* maps within 0 and 1.4 cM (95% confidence interval) of the *W* locus.

Similarly, Geissler *et al.* (1988) demonstrated that the c-*kit* gene is disrupted in two spontaneous mutant *W* alleles, W^{44} and W^x. The disruption is revealed by the generation of novel *Eco*RI, *Bam*HI, and *Pst*I restriction fragments in W^{44}m/W^{44} mutants not present in wild type. There is also loss of a single *Bgl*II restriction fragment in genomic DNA prepared from W^{44}/W^{44} homozygote mice; however, the corresponding novel *Eco*RI fragment has not been identified. Genomic DNA encoding specific regions of the c-kit polypeptide (amino acids 34-40 for W^{44} or 34-791 for W^{x}) is disrupted, and when mice are homozygous for W^{44}, there is a marked reduction in c-*kit* mRNA. Mast cells from w/w^{v} mutant mice exhibit defects in the c-*kit*–associated protein kinase activity (Nocka *et al.*, 1989). Furthermore, the kinase domain of c-*kit* has a point mutation in the w^{42} mutant allele (Tan *et al.*, 1990). c-*kit* expression has been examined in developing mouse embryos by *in situ* hybridization, localizing expression of c-*kit* to numerous ectodermal and endodermal derivatives, sites of embryonic hemopoiesis, primordial germ cells, presumptive melanoblasts, the central nervous system (CNS), and in the structures of the craniofacial area (Orr-Urtreger, 1990). Based on this collection of evidence, it is strongly believed that alterations in the c-*kit* message or gene product is somehow responsible for producing the *W/W* phenotype. Some investigators have speculated that c-*kit* alone could be responsible for numerous, if not most, *w* phenotypes (macrocytic anemia, sterility and white spotting, Orr-Urtreger *et al.*, 1990). However, we cannot rule out the possibility that other genes, either within the *W* locus or elsewhere in the genome, can contribute to c-*kit*'s *in vivo* action. Since mutations at the Steel (*Sl*) locus on chromosome 10 result in a phenotype almost identical to that of *W* mutants, it has been postulated that effects of mutations on *W* and *Sl* on hematopoietic stem cells may reflect a receptor–ligand relationship between *W* and *Sl* gene products (Nocka *et al.*, 1989). This relationship has recently been confirmed with the characterization of mgf (multipotent growth factor) as the *Sl* gene product and the c-*kit* ligand (Whitte, 1990, for review).

Patch (*Ph*) and rump-white (*Rw*) are closely linked to one another and to the *W*-locus on chromosome 5. These three loci may have arisen by gene duplication. All three are lethal when homozygous. *Ph*/+ animals have sharply-marked white spots on the belly, tail, and digits, and there is usually a large white patch in the middle of the trunk. Rump-white heterozygotes (*Rw*/+) have white tails, white legs, and varying amounts of pigment loss in the sacral and lumbar regions. There are no signs of pigment dilution. *Ph/Ph* animals have some viable melanoblast clones, and their eyes have pigment. *Ph* mutants exhibit a deletion in the platelet-derived growth factor receptor α-subunit (Stephenson *et al.*, 1991). *Rw*/+ animals have a normal blood profile, and more pigment than the *Ph*/+ heterozygote. There is controversy as to whether the melanocytes present in the dermis in *Rw*/+ animals migrated in after the initial defective clones failed to survive, or whether these mutants specifically affect the ability of melanocytes to invade the hair follicle (Silvers, 1979).

Several other *W* locus alleles have been described and have been reviewed in detail by Silvers (1979), including Ames Dominant Spotting (W^{a}), Ballantyne's Spotting (W^{b}), *W*-fertile (W^{f}), Jay's Dominant Spotting (W^{j}), Panda & White (W^{PW}).

6. d, ln, and p Loci

Both the dilute (*d*) and leaden (*ln*) loci are related to changes in the morphology of melanocytes. The dilute coat-color locus maps to mouse chromosome 9. It is not associated with abnormal production of melanin, but with aberrant melanocyte morphology. Melanocytes from *d*/*d* mice are adendritic, and the melanin granules are clumped near the nucleus (Markert and Silvers, 1956). Inability to form dendritic extensions is thought to impede transfer of melanosomes to keratinocytes. The *d* mutation, as originally defined, affected only coat color and melanocyte morphology. However, other *d* mutants have also arisen with lethal effects (d^l). d^l mutants not only exhibit dilution of coat color, but also exhibit severe opisthotonus (convulsive arching upward of the head and back) at 10 days of age prior to death at 3 weeks. The neuromuscular disorder may be due to degeneration of myelin in the CNS (Kelton and Rauch, 1962). Still another variant of the dilute mutation is dilute-neurological, d^n. d^n/d^n animals are intermediate in defect between *d* and d^l. They have dilute coat color, show transient opisthotonos, and by 6 weeks of age, appear neurologically normal. It is thought that the neurological defect of d^n is caused by another gene within close proximity of dilute. The *d* mutation has been demonstrated to be dominant to d^n and d^l in that d/d^n and d/d^l heterozygotes exhibit dilute coat color but are apparently normal neurologically and live a normal lifespan. d^n/d^l heterozygotes have opisthotonus only during early life and an altered gait during adult life (Strobel *et al.*, 1988).

Like dilute and leaden, ashen (*ash*) produces a dilute coat color. *Heterozygous* or *homozygous ash* mutants also have adendritic melanocytes. Moore *et al.* (1988) have recently demonstrated that *ash* and *l* are not complementary alleles of the same gene. Additionally, a dilute semidominant suppressor gene (*dsu*) can partially suppress *ln* and *ash* by partially restoring normal melanocyte morphology. *dsu* maps to chromosome 1 and is loosely linked to leaden (*ln*).

Jenkins *et al.* (1981, and in press) have demonstrated that the original *d* allele arose as the result of integration of an ecotropic murine leukemia virus provirus EMV-3, and have thus renamed the allele d^v. When reversion from d^v to d^t occurs, most of EMV-3 is excised and lost. One viral long terminal repeat (LTR) remains, however, marking the insertion site. The mechanism for this is thought to be intrachromosomal homologous recombination mediated by the LTRs. The provirus integration site is within an intron of the *d* gene. Three splicing patterns for the >150 kb *d* gene have been described. Though it is unclear how the provirus induces the d^v mutation, it is thought to affect *d* gene transcription. d^v/d^v mice produce 11- and 8-kb transcripts, while d^t/d^t mice produce 11-, 9-, and 7-kb transcripts. Since d^v/d^v mice are normal neurologically, the 11-kb transcript may be required for normal neurological function. The 7- and 9-kb fragments are required for melanocyte function. The *d* allele is thought to encode a component of the cytoskeleton (Jenkins *et al.*, in press).

The *p* locus (pink-eye) affects melanin within the eye and hair coat. These mutants exhibit an altered premelanosome framework, probably because unit fibers fail to properly cross-link during development.

7. *Other Loci*

In addition to the pigmentation mutations described that affect the *a*, *b*, *c*, *s*, *w*, *d*, *ln*, and *p* loci, several other mutations have been described that affect mouse coat color (see Table I).

B. Chicken Pigment Mutants

Pigmentation patterns in the chick have been widely studied. Melanogenesis is thought to be the same for melanocytes of the retinal pigment epithelium, choroid melanophores of the pigmented iris epithelium, pecten, melanophores, and feathers (reviewed in Bowers, 1988). Both tyrosinase-positive and tyrosinase-negative oculocutaneous albinism also exist in chickens. Chickens also exhibit vitiligo, whereby melanocytes undergo premature degeneration in the collar area, in addition to hypomelanosis. This early melanocyte death can be caused by dominant white gene or the barring and/or blue alleles. In addition, chickens develop a disorder that mimics the postnatal amelanosis and visual defects of vitiligo. These chickens are of the autoimmune delayed-amelanotic (DAM-Smyth) line, and are characterized by an autoimmune removal of dysfunctional melanocytes (Boissy *et al.*, 1986). Another mutant that develops vitiligo-like traits is the Barred Plymouth Rock (BPR) chicken. These chickens have a feather color pattern of alternating black and white bands. The ultrastructure of these melanocytes is similar to those found in the border of vitiligo patients. The black band of BPR contains many functional melanocytes, while the white band contains no viable melanocytes. The interface between the two bands contains debilitated melanocytes (Bowers and Asano, 1984). The melanocyte defect may be associated with production of a toxic melanin precursor (Bowers, 1988).

C. Human Pigment Mutants

The autosomal recessive pigment disorder, oculocutaneous albinism, is characterized by deficiency of melanin in skin, hair, and eyes. In contrast, cutaneous albinism is a dominantly inherited trait (Stout, 1946; Sanders, 1934, 1935), while ocular albinism is inherited as an X-linked trait (Falls, 1951). Albino melanocytes have been reported to be tyrosinase negative with no tyrosinase activity *in vitro*, and/or to have the 2.1-kb mRNA for tyrosinase present, but melanin is not synthesized (Takeuchi *et al.*, 1988).

TABLE I

Mutations Affecting Mouse Coat Color

Mutation	Chromosomal location	Effects	References
Gray lethal (*gl*)	10	Yellow pigment produced but not released. Bone resorption abnormalities	Silvers (1979)
Grizzled (*gr*)	10	Melanocyte viability and phaeomelanin synthesis abnormal	Bloom and Falconer (1966)
Mocha (*mh*)	10	Reduced phaeomelanin and eumelanin behavioral disorders affecting swimming, hearing, breeding, and nervous system	Silvers (1979)
Pallid (*pa*)		Otolith development is affected	Silvers (1979)
Muted (*mu*)	13	Dilutes eumelanin and phaeomelanin and affects otolith development	Lyon and Meredith (1979)
Misty (*m*)	4	Dilutes pigmentation on nonagouti brown backgrounds	Silvers (1979)
Pearl (*pe*)	13	Dilutes pigment at the base of the hair to a greater extent than the tip	Silvers (1979)
Beige (*bg*)	13	Dilutes pigment, enlarged liposomal granules, and increased susceptibility to infection, reduced secretion of liposomal enzymes from proximal tubule cells	Brandt and Swank (1976)
Silver (*si*)	10	Melanocyte differentiation is affected, possible epithelial–melanocyte communication disturbance	Silvers (1979)
Greying with age (*ga*)	probably 10	Dominant determinate; similar to *si*; influenced by maternal effects	Kirby (1974)
Steel (*sl*)	10	Erythropoiesis, germ cell development and pigmentation are severely altered. *Sl*/*Sl* mice rarely exist through fetal development. The defect is thought to be in the environment in which the melanoblast germ cell and hematopoietic progenitors differentiate and proliferate	Bennett (1956)
Flexed-tailed (*f*)	13	Causes white spotting, tail flexion, and transient siderocytic anemia	Hunt *et al.* (1933)
Splotch (*sp*) and delayed splotch (S_{pd})	1	Homozygote state is semidominant and lethal; total inability to produce neural crest -derived melanocytes	Silvers (1979)
Varitint-Waddler (*va*) and Varitint-Waddler-J (Va^j)	1	Variegation of pigment and shaker-like syndrome	Lane (1972) and Grüneberg (1952)
Microphthalmia (*mi*)	6	Affects on internal pigment only in *mi*/*t*; *mi*/*mi* are totally without melanocytes and pigment	Silvers (1976)

The piebald trait in humans is characterized by patchy spots of hypo- and hyper-pigmentation, probably resulting from abnormal melanocyte migration and/or differentiation (Hoo *et al.*, 1986). There are few to no melanocytes in piebald hypomelanotic spots.

Minimal pigment is one of seven forms of oculocutaneous albinism (OCA). King *et al.* (1986) described these individuals as exhibiting no pigment in skin and eye (white hair/blue eyes), but during the last decade of life, some pigment began to form. Melanocytes were normal by electron microscopy (EM), and variations in premelanosome pigmentation were associated with tyrosinase activity.

D. Discussion

While genetic studies indicate that changes in certain loci affect melanocyte proliferation, viability, migration, and/or differentiation, with only a few exceptions, they do not provide a clue as to which proteins regulate those functions. *In vitro* studies have provided some information, but it is unclear how or if the factors implicated *in vitro* are altered in pigmentation mutants.

III. FACTORS REPORTED TO MODULATE MELANOCYTE GROWTH AND DIFFERENTIATION *IN VITRO*

A number of investigators have developed methods for culturing melanocytes *in vitro* (Klaus and Snell, 1967; Halaban *et al.*, 1986; Eisinger and Marko, 1982; Gilchrest *et al.*, 1984, 1986). These studies provide information about hormonal factors that control growth and differentiation of melanocytes. A variety of culture-medium additives or treatments have been reported to affect the growth of melanocytes *in vitro*: cholera toxin (CT) (Eisinger and Marko, 1982); the phorbol ester, 1-*o*-tetradecanoylphorbol-13-acetate (TPA) (Eisinger and Marko, 1982); basic fibroblast growth factor (Halaban *et al.*, 1987); insulin (Herlyn *et al.*, 1988); follicle-stimulating hormone, β-MSH, and ultraviolet light (Bertaux *et al.*, 1988); additives from bovine brain or pituitary (Wilkins *et al.*, 1985), from fibroblasts (Eisinger *et al.*, 1985); or melanoma cells (Ogata *et al.*, 1987; Bordoni *et al.*, 1990). Other factors reported to play a key role in melanocyte proliferation *in vitro* include prostaglandins E and D (Nordlund *et al.*, 1986), UVA and UVB exposure (Rosen *et al.*, 1987), and the *ras* oncogene (Wilson *et al.*, 1989).

Some of the factors reported to stimulate melanocyte proliferation also enhance melanocyte differentiation and increase melanin synthesis, including TPA, bFGF, αMSH, UVA and B, and pituitary extract (Halaban *et al.*, 1987; Herlyn *et al.*, 1988; Bertaux *et al.*, 1988; Wilkins *et al.*, 1985; Eisinger *et al.*, 1985; Ogata *et al.*, 1987). Other hormones such as vitamin D_3, androgens, and estrogens enhance melanin synthesis and tyrosinase activity without affecting proliferation (Abdel-Malek *et al.*, 1988; Ranson *et al.*, 1988).

A. TPA, Cholera Toxin, Isobutylmethylxanthine, Basic Fibroblast Growth Factor, Pituitary Extract, Insulin Effects

Eisinger and Marko (1982) improved the success rate for culturing melanocytes by including TPA, phorbol myristate acetate (PMA), and CT in the culture medium. CT increases intracellular cAMP, while TPA binds to the diacylgycerol-binding site of protein kinase C, activating the enzyme. In Eisinger and Marko's experiments, CT stimulated growth of melanocytes and increased their lifespan *in vitro.* CT also stimulated keratinocyte growth, but inhibited the growth of fibroblasts. PMA and CT stimulated the growth of newborn foreskin melanocytes much more effectively than they did adult skin melanocytes. PMA (>200 n*M*) blocked melanogenesis in some melanocytes, while lower concentrations (50–100 n*M*) delayed the onset of differentiation and decreased the rate of pigmentation.

Eisinger *et al.* (1985) reported that cell extracts from melanoma, astrocytoma, and fibroblasts contained a factor that stimulates melanocyte growth. The factor from WI-38 fibroblasts was acid- and alkaline-stable, heat-labile, and ~40,000 *M*. This factor apparently differed from the melanocyte growth factor purified from bovine brain, which was characterized as a heat- and alkaline-labile, weakly cationic mitogen, with an M_r of ~30,000. This substance stimulated growth in quiescent populations of human melanocytes *in vitro*. More recently, Ogata *et al.* (1987) demonstrated that the factor from melanoma extracts that stimulates melanocyte growth has a strong affinity for heparin sepharose and is acid- and heat-labile. The M_r of the protein is ~14,000. The growth of melanocytes was sustained by melanoma-derived melanocyte growth factor (M-McGF), TPA, and CT.

Halaban *et al.* (1986) have routinely cultured foreskin melanocytes in culture medium containing TPA, isobutylmethylxanthine, (IBMX), CT, placenta or pituitary extract, as well as serum. The growth of melanoma cells was inhibited by this mixture. More recently, Halaban *et al.* (1987) have shown that basic fibroblast growth factor (bFGF) eliminates the TPA requirement in foreskin melanocyte cultures. Furthermore, keratinocytes release a bFGF-like molecule that promotes melanocyte growth. Antibody to bFGF eliminates the melanocyte growth activity from keratinocyte culture medium. Dulbecco's modified Eagle's medium (DME) also suppresses the expression of bFGF in keratinocytes, but MCDB-153 medium does not (Halaban *et al.*, 1988). Melanocytes grown in TIP (TPA, insulin, placental extract) medium do not express bFGF, indicating that detectable levels of bFGF are not induced in response to continuous exposure to TPA. Expression of bFGF in the immortalized murine melanocyte LB10-BR cell line conferred autonomous growth in culture and resulted in a loss of differentiated function.

However, constitutive expression of bFGF alone was insufficient to render melanocytes capable of forming tumors in mice (Dotto *et al.*, 1989).

Halaban *et al.* (1988) also reported that UVB irradiation increases DNA synthesis in melanocytes cultured in TPA and increases the growth response of melanocytes to keratinocyte-conditioned medium. When keratinocytes are irradiated with UVB, there is an increase in the melanocyte growth-stimulating activity of the keratinocyte-conditioned medium. Antibody to bFGF suppressed this UVB-increased activity in keratinocyte-conditioned medium, suggesting that UVB irradiation increases bFGF availability to melanocytes. It has been proposed that UVB stimulates melanocyte growth directly by elevating cAMP and indirectly by stimulating neighboring keratinocytes to make bFGF. Though the mechanism for distribution of bFGF to the melanocyte remains mysterious, owing to the absence of a signal peptide for bFGF, bFGF accumulation on the extracellular matrix has been proposed as one possible mechanism. Still another mystery is the lack of an effect of aFGF on melanocyte growth (Halaban *et al.*, 1987) since aFGF and bFGF are reported to bind the same receptor. While bFGF affects melanocyte growth, the mechanism for this growth stimulation is unclear at present.

Herlyn *et al.* (1988) used a different basal medium for melanocyte growth, W489. They report that bFGF, TPA, and insulin are required when this basal medium is used. Medium W489 consists of a mixture of MCDB153 and L15 media (4:1 proportion) supplemented with 2 m*M* Ca^{24}. During the initial culture period, TPA ($10^{-7}M$), insulin (5 μg/ml), epidermal growth factor (EGF, 5 ηg/ml), bovine pituitary extract (40 μg/ml), and 2% fetal bovine serum (FBS) were added to supplement the melanocyte growth medium. At later times, EGF, FBS, and pituitary extract were omitted and Hepes (30 m*M*) was included in the medium. When PMA, insulin, or bFGF was removed from W489 medium, or when calcium was lowered to 0.03 m*M*, melanocyte growth decreased 40–50%. When only these agents were used to supplement the medium, doubling times were 7 and 14 days compared to 2 and 4 days with 2% FBS and 40 μg/ml pituitary extract. When factors that elevate cAMP levels were included in the culture medium (α-MSH, FSH, prostaglandin F_2a, β subunit of cholera toxin, whole cholera toxin, or forskolin), there was a significant increase in the growth of melanocytes (doubling times 5–8 days). These cAMP-inducing agents did not have additive effects on growth and would not support growth of melanocytes from freshly isolated foreskins or support clonal growth. Dibutyryl-cAMP caused an outgrowth of melanocyte dendrites, and removal of PMA led to abortive dendrite formation. Dendrite formation, induced by agents that elevate cAMP, inhibits growth of the cells. Cholera toxin had varied effects on the growth of cells, sometimes stimulating or inhibiting growth depending on experimental conditions. This result was expected since the CT receptor, monosialo-ganglioside G_{M1}, can both stimulate and inhibit growth (Spiegel and Fishman, 1987).

B. Vitamin D_3, α-MSH, β-Estradiol, and 5α-Dihydrotestosterone Effects

Abdel-Malek *et al.* (1988) reported human melanocyte cultures treated with 1,25-dihydroxyvitamin D_3 [1,25-$(OH)_2D_3$] at concentrations $\geq 10^{-8}M$, showed suppressed tyrosinase activity, with no effect on proliferation. 25-Hydroxyvitamin D_3 [25-$(OH)D_3$] effects were similar to those of 1,25-$(OH)_2D_3$. In contrast, Ranson *et al.* (1988) found that lower concentrations of 1,25-$(OH)_2D_3$ ($10^{-9}M$) produced a 100% increase in 25-hydroxyvitamin D_3-4-hydroxylase activity and a 50% increase in tyrosinase activity. The effect was apparently not signaled through an increase in cAMP, because α-MSH ($5 \times 10^{-7}M$) stimulated marked changes in cAMP (≥ seven fold), but only about a 20% increase in tyrosinase activity. β-Estradiol caused a dose-dependent increase in tyrosinase activity in their human melanocytes culture system. However, investigations by Diaz apparently give different results. Diaz *et al.* (1986) reported that melanocytes of male hamsters exhibited an increase in pigmentation in response to 5α-dihydrotestosterone *in vitro*, which was tissue specific and antagonized by estrogens. Estrogens do not appear to affect melanocyte growth (Feucht *et al.*, 1988). However, estradiol did increase tumor latency and inhibit tumor growth in athymic mice carrying an estrogen receptor–positive melanoma. Since estradiol had no effect on the growth of the melanoma *in vitro*, it was suggested that estradiol might work through an indirect mechanism.

C. Ultraviolet Light Effects

The relationship between exposure to ultraviolet light and the development of skin cancers has been examined repeatedly, and epidemiological studies show correlations between blistering sunburn and development of malignant melanoma (see Mackie and Atchison, 1982 and Glass and Hoover, 1989 for review). Because melanomas often do not develop in areas of frequent ultraviolet exposure, there was a question for many years as to the significance of these correlations. The disproportionate increase in trunkal melanomas compared with squamous cell cancers of the trunk may be related to episodes of recreational UV exposure as opposed to "chronic, long-term sun exposure" (Glass and Hoover, 1989). Recently, *in vitro* studies have demonstrated direct UV effects on melanocyte proliferation. Bertaux *et al.* (1988) developed a melanocyte explant model whereby melanocytes grow out from biopsy on epidermal sheets, thus preserving an *in vivo*–like architecture. UVB irradiation or UVA and 8-methoxypsoralen treatment stimulated melanocyte growth. However, only melanocytes obtained from donors younger than 2 years were able to grow in the skin equivalent.

Rosen *et al.* (1987) performed *in vivo* human studies to examine effects of UVA and UVB exposure on melanocyte pigment formation and growth. They reported an increase in the number of Dopa-positive cells (Dopa serves as substrate for tyrosinase) and increased tyrosinase activity after either UVA or UVB exposure. However, the responses to UVA versus UVB were different, in that the response to UVA was quicker (immediate versus 3–5 days for UVB) and was dependent on oxygen. Chedekel and Zeise (1988) described UVB and UVC irradiation converting melanin metabolites to reactive-free radicals. Certain of these free radicals could conceivably react with cellular DNA, lipids, and proteins, potentially altering cell function. Gahring *et al.* (1984) demonstrated that UV radiation (80–30 nm) enhanced release of epidermal cell thymocyte-activating factor (ETAF) from keratinocytes. ETAF is physicochemically identical to interleukin 1. Interleukin 1 is a potent inducer of the MGSA/gro gene (Wen *et al.*, 1989), linking possible indirection induction of MGSA/gro expression in response to UV radiation.

Though a number of investigators have shown that UV irradiation enhances the growth of melanocytes at the same time it enhances tyrosinase activity, the results of Friedman and Gilchrest (1987) are contradictory. These investigators demonstrated that though UV-irradiated melanocytes exhibit a dose-dependent increase in melanin content per cell and increased uptake of radiolabeled Dopa, a simultaneous inhibition of growth was induced. The UV-irradiation (A and B) did not produce detectable increases in intracellular cAMP.

D. Prostaglandin Effects

Nordlund *et al.* (1986) found that both UVB and arachidonic acid stimulate proliferation of pigment cells, and indomethacin blocks that stimulation. Since indomethacin inhibits prostaglandin (PG) synthesis, these investigators then tested the effects of PGs on the proliferation of mouse ear melanocytes *in vivo*. Daily application of PGD_2 to mouse skin caused a small increase in melanocyte density. However, PGE_2 caused a large increase in melanocyte density as well as a small increase in uptake of ^{3}H-thymidine by Dopa-positive dendritic cells. The growth regulation of malignant melanocytes is different from that of normal melanocytes with regard to PG effects. PGE_2 also enhanced melanogenesis in melanoma cells. Abdel-Malek *et al.* (1989) demonstrated that PGE_1 and PGE_2 can block the progression of Cloudman melanoma cells from the G_2 phase of the cell cycle into M or G_1. These data suggest that by blocking the cells in G_2, melanogenesis is enhanced. This is compatible with other data suggesting that melanocyte differentiation promoters act during the G_2 phase of the cell cycle.

E. α-Melanocyte–Stimulating Hormone

α-Melanocyte–stimulating hormone (α-MSH) has profound effects on tyrosinase activity of Cloudman melanoma cells (Pawelek *et al.*, 1975; Johnson and

Pastan, 1972) and can induce the establishment of neuritic processes after approximately 24 hr in culture with α-MSH. Preston *et al.* (1987) have shown that the long-term effects of α-MSH are not mimicked by cAMP. MSH-induced changes in cell shape were dependent on new protein synthesis and transcription. Cloudman S91 cells also respond to B-MSH by exhibiting changes in morphology, growth rate, and melanin production. The B-MSH stimulation of cAMP production and tyrosinase activity occurs in the G_2 phase of the cell cycle (McLane and Pawelek, 1988). This may be associated with the increased binding of MSH to synchronized populations of melanoma cells in late S and G_2.

Another murine melanoma cell line, JB/MS, responds to α-MSH with increased tyrosinase activity and increased melanin synthesis. Interferon α,B or γ significantly synergized with the α-MSH effects on tyrosinase activity (Kameyama *et al.*, 1988).

F. RAS Effects

Bennett *et al.* (1987) developed an immortalized mouse melanocyte cell line (melan-a) from embryonic skin of C-57BL mice, which is diploid and nontumorigenic in syngeneic and nude mice. These cells continuously proliferate in the presence of enriched culture medium containing 10% FCS and 200 nMTPA. The cells retain the properties of normal melanocytes except they do not exhibit senescence or a proliferative response to cholera toxin in the presence of TPA. When these cells were transfected with the v-Ha-*ras* placed under transcriptional control of the Moloney murine leukemia virus long terminal repeat (MLV-LTR), the transfectants produced rapidly growing, undifferentiated melanomas in recipient mice. v-*ras* appeared to inhibit melanin production in the v-*ras*–transformed melanocytes *in vivo* and *in vitro*. v-*ras*–transformed melan-a cells lost the requirement for TPA exhibited by the parent cell line and, in fact, became growth inhibited by TPA, suggesting that v-*ras* transformation of melan-a cells involves activation of the protein kinase C pathway.

G. MGSA Effects

Another method of elucidating the factors involved in control of melanocyte proliferation *in vivo* involves isolation of autocrine-like growth factors produced by malignant melanoma cells, followed by determination of effects of these factors on the growth or differentiative phenotype of normal melanocytes. Using this approach, we initially isolated a factor from the Hs294T human melanoma cell line that augments the growth of serum-free, low-density Hs294T cells. The activity isolated in this manner was termed MGSA, for melanoma growth–stimulatory activity. MGSA was biochemically characterized as being associated with acid- and heat-stable, trypsin- and dithiothreitol-sensitive proteins in the 16,000 and < 13,000 M_r range (Richmond and Thomas, 1986). After purification by a series of BioGel P-30 chromatography, reverse-phase, high-pressure liquid chromatography (RP-HPLC), heparin sepharose chromatography, and additional RP-HPLC steps (Thomas and Richmond, 1988), amino terminal sequence data were obtained for the first 34 amino acids of MGSA (Richmond *et al.*, 1988). This information was used to design oligonucleotide probes for screening cDNA libraries. MGSA cDNA clones were isolated from λGT10 cDNA libraries from human term placenta and from HS94T melanoma cells. The nucleotide sequence of the full-length mRNA from melanoma HS94T cells revealed that MGSA is a member of the β-thromboglobulin superfamily, including platelet factor 4, platelet basic protein, connective tissue-activating peptide III, β-thromboglobulin, and several other recently described genes (Richmond *et al.*, 1987). The nucleotide sequence of MGSA was 100% homologous to the human gro gene (Anisowicz *et al.*, 1988), a gene overexpressed in certain tumor cells. Of great interest was the finding that MGSA/gro appears to be the human homolog of the mouse PDGF-inducible competence gene referred to as KC (Oquendo *et al.*, 1989).

Determining the functional role for MGSA has been a slow process owing to the scarcity of the protein. Development of a recombinant MGSA expression system has facilitated these studies (Balentien *et al.*, 1990). Using the recombinant MGSA in collaboration with K. Matsushima, we found MGSA has neutrophil chemotactic activity and competes with ^{125}I-interleukin 8 (IL-8) for neutrophil-receptor binding (Balentien *et al.*, 1990). IL-8 shares less than 50% of nucleotide sequence homology with MGSA, though the two proteins are structurally similar. Furthermore, Schroeder and Christophers (1989) have demonstrated that human psoriatic keratinocytes overexpress both the MGSA protein and IL-8, inferring that perhaps these proteins serve as mediators of the inflammatory response in psoriasis. Expression of the MGSA/KC/gro gene is regulated by a number of modulators of the inflammatory response including lipopolysaccharide, tumor necrosis factor, IL-1, plant lectins, thrombin, platelet-derived growth factor, and EGF, further implicating the MGSA protein as a mediator of the inflammatory response (Cochran *et al.*, 1983; Matsushima *et al.*, 1988; Golds *et al.*, 1989; Larsen *et al.*, 1989; Wen *et al.*, 1989; Strieter *et al.*, 1989; Tannenbaum *et al.*, 1989; Bordoni and Richmond, 1989). Transforming growth factor beta (TGFβ) can suppress expression of the MGSA/KC/gro gene in mouse keratinocytes (Coffey *et al.*, 1988). The neutrophil chemotactic activity of MGSA is equivalent to that of IL-8. IL-8 also activates neutrophils by stimulating H_2O_2 release. MGSA recruits neutrophil migration *in vivo* and stimulates the oxidative burst in neutrophils, though it does not appear to be so potent as IL-8 in this second function (Moser *et al.*, 1990). Whether MGSA promotes neutrophil differentiation and recruits neutrophils from the bone marrow, as IL-8 does, has not been determined.

The role of MGSA in psoriatic keratinocytes is unknown. Cultured human or mouse keratinocytes do not appear to respond to MGSA with enhanced growth (Ristow, 1989 personal communication). However, normal human melanocytes respond to rMGSA with increased incorporation of ^{3}H-thymidine into DNA. Melanocytes cultured in the Clonetics melanocyte growth medium (including TPA, insulin, fibroblast growth factor, hydrocortisone, gentamicin, amphotericin B, and pituitary extract) incorporated seven times as much ^{3}H-thymidine as did melanocytes grown in the MCDB 151 basal medium without additives. Addition of either insulin, TPA, and MGSA, or insulin, TPA, and bFGF restored almost 50% of the growth-promoting activity of melanocyte-growth medium to the MCDB basal medium. The agents TPA, insulin, and either MGSA or FGF appeared to act through different mechanisms. Dibutyryl-cAMP stimulated incorporation of ^{3}H-thymidine into DNA in melanocytes, but had little effect on nevocytes. Effects of MGSA were not synergized by dibutyryl-cAMP (Bordoni *et al.*, 1990).

Based on the detection of immunoreactive MGSA in 10/15 nevi, nevocytes express fairly high levels of MGSA, as do epidermal keratinocytes of certain skin lesions (Richmond and Thomas, 1988). Though both nevi and melanomas vary considerably in their levels of expression of MGSA mRNA, there are no apparent elevations in the average MGSA expression in melanoma as compared to nevus tissue (Bordoni *et al.*, 1990). However, normal human melanocytes and immortalized mouse melanocytes do not produce levels of MGSA mRNA detectable by Northern blot analysis. These data suggest that MGSA expression is activated early during the progression from melanocyte to nevus.

CMV promoter–enhancer-driven expression of a human MGSA transgene in the immortalized murine melanocyte cell line, melan-a, results in increased ability to form colonies in soft agar as compared to controls expressing neo alone. One MGSA transgene-expressing clone produced tumors in nude mice. The histology of the tumors was compatible with that seen in melanoma, and a significant level of aneuploidy was observed in the tumors (Balentien *et al.*, submitted).

The relationship between MGSA and the piebald trait is unclear. Both genes map to the same chromosomal region (4q 13–21). One other member of the β-thromboglobulin superfamily, platelet factor 4, has been shown to affect the migration of neural crest cells *in vitro* (Perris *et al.*, 1989). The c-*kit* gene also maps to this region, and a relationship between *kit* and W phenotypes (white-spotting) has been established in the mouse. It will be interesting to determine whether recessive spotting alleles are associated with or influenced by the MGSA gene product.

IV. CONCLUDING REMARKS

Though many interesting studies have been performed to identify factors crucial to melanocyte differentiation and proliferation, the results from the *in vitro* studies are often contradictory. Genetic studies are only recently being conducted at the molecular level to allow characterization of the genes and proteins coded for by those genes regulating melanocyte growth and differentiation. A few genes clearly seem to be effective, for example, tyrosinase, bFGF, MGSA, insulin, and c-*kit*. However, based on the number of genetic loci that affect pigmentation, the potential for emergence of numerous melanocyte regulatory factors exists, suggesting that we have only scratched the surface of the black box surrounding the melanocyte-control mechanism. Certainly some of the contradictions from the *in vitro* studies result from differences in culture conditions, age of donor for the skin from which the melanocyte cultures were derived, passage number, and presence or absence of appropriate cofactors for melanocyte growth and/or differentiation. A well-defined melanocyte assay system analogous to BALB/c 3T3 cells needs to be developed. Until then, we must continue to explore the vast arena of potential melanocyte regulators, using the genetic models available, molecular biology–molecular genetics, and carefully defined *in vitro* studies.

ACKNOWLEDGMENTS

We are indebted to Brigid Hogan, Mary Dickinson, and Nancy Olashaw for their helpful discussions and comments on this manuscript, and to Carla Duley for her excellent help in word-processing the manuscript. This work was funded by the Department of Veterans Affairs and NCI CA34590.

REFERENCES

Abdel-Malek, Z. A., Ross, R., Trinkle, L., Swope, V., Pike, J. W., and Nordlund, J. J. (1988). Hormonal effects of vitamin D_3 on epidermal melanocytes. *J. Cell. Physiol.* **136**, 273–280.

Abdel-Malek, Z. A., Swope, V. B., Trinkle, L. S., and Nordlund, J. J. (1989). Stimulation of cloudman melanoma tyrosinase activity occurs predominantly in G_2 phase of the cell cycle. *Exp. Cell Res.* **180**, 198–208.

Anisowicz, A., Bardwell, L., and Sager, R. (1987). Constitutive over-expression of a growth-regulated gene in transformed Chinese hamster and human cells. *Proc. Natl. Acad. Sci. U.S.A.* **84**, 7188–7192.

Balentien, E., Mufson, B. E., Shattuck, R. L., Derynck, R., and Richmond, A. (1991). Over-expression of MGSA in normal melanocytes. *Oncogene* **6**, (in press).

Balentien, E., Han, J. H., Thomas, H. G., Wen, D., Samantha, A. K., Zachariae, C. O., Griffin, P. R., Brachmann, R., Wong, W. L., Matsushima, K., Richmond, A., and Derynck, R. (1990). Recombinant expression and biochemical characterization of the human MGSA/gro protein. *Biochemistry* **29**, 10225–10233.

Bennett, D. C., Cooper, P. J., and Hart, I. R. (1987). A line of non-tumorigenic mouse melanocytes, syngeneic with the B-16 melanoma and requiring a tumor promoter for growth. *Int. J. Cancer* **39**, 414–418.

Bertaux, B., Morliere, P., Moreno, G., Courtalon, A., Massé, J. M., and Dubertret, L. (1988). Growth of melanocytes in a skin-equivalent model *in vitro*. *Br. J. Dermatol.* **119**, 503–512.

Billingham, R. E., and Silvers, W. K. (1960). The melanocytes of mammals. *Quart. Rev. Biol.* **35**, 1–40.

Bloom, J. L., and Falconer, D. S. (1966). "Grizzled," a mutant in linkage group X of the mouse. *Genet. Res.* (Camb.) **7**, 159–167.

Boissy, R. E., Moellmann, G., Trainer, A. T., Smyth, J. R., and Lerner, A. B. (1986). Delayed amelanotic (DAM-Smyth) chicken: Melanocyte dysfunction *in vivo* and *in vitro*. *J. Invest. Dermatol.* **86**, 149–156.

Bordoni, R., Thomas, H. G., Richmond, A. (1989). Interaction of melanoma growth stimulatory activity with other growth factors and regulation of mRNA expression in melanoma cells. *J. Cell. Biochem.* **39**, 421–428.

Bordoni, R., Fine, R., Murray, D., and Richmond, A. (1990). Characterization of MGSA in normal melanocytes and malignant melanoma. *J. Cell. Biochem.* **44**, 207–219.

Bowers, R. R., and Asano, J. S. (1984). Barred plymouth rock melanocytes as a possible model for vitiligo. *Yale J. Biol. Med.* **57**, 340–341.

Bowers, R. R. (1988). The melanocyte of the chicken: A review. *In* "Advances in Pigment Cell Research" (J. T. Bagnara, ed.), Vol. XXV, pp. 49–63. Alan R. Liss, New York, New York.

Brandt, E. J., and Swank, R. T. (1976). The Chediak-Higashi (beige) mutation in two mouse strains. *Am. J. Pathol.* **82**, 573–588.

Bronner-Fraser, M. (1984). Latex beads with well-defined surface coats as probes of neural-crest migratory pathways. *In* "The role of the Extracellular Matrix in Development" (R. L. Trelstad, ed.), Vol. 42, p. 399–432. Alan R. Liss, New York, New York.

Chabot, B., Stephenson, D. A., Chapman, V. M., Besmer, T., and Bernstein, A. (1989). The protooncogene c-*kit* encoding a transmembrane tyrosine kinase receptor maps to the mouse W locus. *Nature* **335**, 88–89.

Chedekel, M. R., and Zeise, L. (1988). Sunlight, melanogenesis, and radicals of the skin. *Lipids* **23**, 587–591.

Cochran, B. H., Reffel, A. C., and Stiles, C. D. (1983). Molecular cloning of gene sequences regulated by platelet-derived growth factor. *Cell* **33**, 939–947.

Coffey, R. J., Bascom, C. C., Sipes, N. J., Graves-Deal, R., Weissman, B. E., and Moses, H. L. (1988). Selective inhibition of growth-related gene expression in murine keratinocytes by transforming growth factor β. *Mol. Cell. Biol.* **8**, 3088–3093.

Copeland, N. G., Hutchison, K. W., and Jenkins, N. A. (1983). Excision of the dba ecotropic provirus in dilute coat-color revertants of mice occurs by homologous recombination involving the viral LTRs. *Cell* **33**, 379–387.

Cruickshank, C.N.D., and Harcourt, S. A. (1964). Pigment donation in vitro. *J. Invest. Dermatol.* **42**, 183–184.

Diaz, L. C., Das Gupta, T. K., and Beattie, C. W. (1986). Effects of gonadal steroids on melanocytes in developing hamsters. *Pediatr. Dermatol.* **3**, 247–256.

Dotto, G. P., Moellmann, G., Ghosh, S., Edwards, M., and Halaban, R. (1989). Transformation of murine melanocytes by basic fibroblast growth factor cDNA and oncogenes and selective suppression of the transformed phenotype in a reconstituted cutaneous environment. *J. Cell Biol.* **109**, 3115–3128.

Dunn, L. C., and Thigpen, L. W. (1930). The silver mouse, a recessive color variation. *J. Hered.* **21**, 495–498.

Dunn, L. C., and Charles, D. R. (1937). Studies on spotting patterns. I. Analysis of quantitative variations in the pied spotting of the house mouse. *Genetics* **22**, 14–42.

Eisinger, M., and Marko, O. (1982). Selective proliferation of normal human melanocytes *in vitro* in the presence of phorbol ester and cholera toxin. *Proc. Natl. Acad. Sci. U.S.A.* **79**, 2018–2022.

Eisinger, M., Marko, O., Ogata, S. -I., and Old, L. J. (1985). Growth regulation of human melanocytes: Mitogenic factors in extracts of melanoma, astrocytoma, and fibroblast cell lines. *Science* **229**, 984–986.

Falls, H. F. (1951). Sex-linked ocular albinism displaying typical fundus changes in the female heterozygote. *Am. J. Opthalmol.* **34**, 41.

Feucht, K. A., Walker, M. J., Das Gupta, T. K., and Beattie, C. W. (1988). Effect of 17 β-estradiol on the growth of estrogen receptor–positive human melanoma *in vitro* in athymic mice. *Cancer Res.* **48**, 7093–7101.

Fitzpatrick, T. B., and Quevedo, W. C. (1960). Albinism. *In* "The Metabolic Basis of Inherited Disease" (J. B. Stanbury, J. B. Wyngaarden, D. S. Fredrickson, eds.), 2nd Ed., pp. 324–340. McGraw-Hill, New York.

Friedmann, P. S., and Gilchrest, B. A. (1987). Ultraviolet radiation directly induces pigment production by cultured human melanocytes. *J. Cell. Physiol.* **133**, 88–94.

Gahring, L., Baltz, M., Repys, M. B., and Duynes, R. (1984). Effect of ultraviolet radiation on production of epidermal cell thymocyte activating factor/interleukin 1 *in vivo* and *in vitro*. *Proc. Natl. Acad. Sci. U.S.A.* **81**, 1198–1202.

Geissler, E. N., Ryan, M. A., and Housman, D. E. (1988). The dominant-white spotting (W) locus of the mouse encodes the c-*kit* proto-oncogene. *Cell* **55**, 185–192.

Gilchrest, B. A., Vrabel, M. A., Flynn, E., Szabo, G. (1984). Selective cultivation of human melanocytes from newborn and adult epidermis. *J. Invest. Dermatol.* **83**, 370–376.

Gilchrest, B. A., Treoloar, V., Grassi, A. M., Yaar, M., Szabo, G., and Flynn, E. (1986). Characteristics of cultivated adult human nevo-cellular nevus cells. *J. Invest. Dermatol.* **87**, 102–107.

Glass, A. G., and Hoover, R. N. (1989). The emerging epidemic of melanoma and squamous cell skin cancer. *J.A.M.A.* **262**, 2097–2100.

Golds, E., Mason, P., and Nyirkos, P. (1989). Inflammatory cytokines induce synthesis and secretion of gro protein and neutrophil chemotactic factor but not β_2-microglobulin in human synovial cells and fibroblasts. *Biochem. J.* **259**, 585–588.

Green, H. (1978). Cyclic AMP in relation to proliferation of the epidermal cell: A new view. *Cell* **15**, 801–811.

Green, M. C. (1981). Catalog of mutant genes and polymorphic loci. *In* "Genetic Variants and Strains of the Laboratory Mouse" (ed. M. C. Green) p. 218. Gustav Fischer Verlag, New York.

Grüneberg, H. (1942). The anemia of flexed-tailed mice (*Mus musculus* L.) II. Siderocytes. *J. Genet.* **44**, 246–271.

Grüneberg, H. (1942). The anemia of flexed-tailed mice (*Mus musculus* L.) I. Static and dynamic hematology. *J. Genet.* **43**, 45–68.

Grüneberg, H. (1952). The Genetics of the Mouse, 2nd Ed. Nijhoff, The Hague, The Netherlands.

Halaban, R., Ghosh, S., Duray, P., Kirkwood, J. M., and Lerner, A. B. (1986). Human melanocytes cultured from nevi and melanomas. *J. Invest. Dermatol.* **87**, 95–101.

Halaban, R., Ghosh, S., and Baird, A. (1987). bFGF is the putative natural growth factor for human melanocytes. *In Vitro Cell. Devel. Biol.* **23**, 47–52.

Halaban, R., Langdon, R., Birchall, N., Cuono, C., Baird, A., Scott, G., Moellmann, G., and McGuire, J. (1988). Basic fibroblast growth factor from human keratinocytes is a natural mitogen for melanocytes. *J. Cell Biol.* **107**, 1611–1619.

Hearing, V. J., and Jiménez, M. (1987). Mammalian tyrosinase—the critical regulatory control point in melanocyte pigmentation. *Int. J. Biochem.* **19**, 1141–1147.

Herlyn, M., Mancianti, M. L., Jambrosic, J., Bolen, J. B., and Koprowski, H. (1988). Regulatory factors that determine growth and phenotype of normal human melanocytes. *Exp. Cell Res.* **179**, 322–331.

Hoo, J. J., Haslam, R.H.A., and van Orman, C. (1986). Tentative assignment of piebald trait to chromosome band 4q12. *Hum. Genet.* **73**, 230–231.

Hörstadius, S., and Sellman, S. (1946). *In* "Experimentelle Untersuchungen über die Determination des Knorpeligen Kopfshelettes bei Urodelen". Nova. Acta Soc. Scient. Uppsaliensis, Ser (4), **13**, 1–170.

Hörstadius, S. (1950). *In* "The Neural Crest" Oxford University Press, London.

Hunt, H. R. (1932). The flexed tailed mouse. *Proc. VI Int. Congr. Genet.* **2**, 91–93.

Imokawa, G., Yada, Y., and Hori, Y. (1988). Induction of melanization within hair-bulb melanocytes in chinchilla mutants by melanogenic stimulants. *J. Invest. Dermatol.* **91**, 106–113.

Jackson, I. J. (1988). A cDNA encoding tyrosinase-related protein maps to the brown locus in mouse. *Proc. Nat'l. Acad. Sci. U.S.A.* **85**, 4392–4396.

Jenkins, N. A., Copeland, N. G., Taylor, B. A., and Lee, B. K. (1981). Dilute (d) coat colour mutation of DBA/2J mice is associated with the site of integration of an ecotropic MuLV genome. *Nature* **293**, 370–374.

Jenkins, N. A., Strobel, M. C., Seperack, P. K., Kingsley, D. M., Moore, K. J., Mercer, J. A., Russell, L. B., Copeland, N. G. A retroviral insertion in the dilute (d) locus provides molecular access to this region of mouse chromosome 9. (submitted).

Johnson, G. S., and Pastan, I. (1972). N^6, $O^{2'}$-dibutyryl adenosine 3′, 5′-monophosphate induces pigment production in melanoma cells. *Nature (London)* **237**, 267–268.

Kameyama, K., Tanka, S., Ishida, Y., and Hearing, V. J. (1988). Interferons modulate the expression of hormone receptors on the surface of murine melanoma cells. *J. Clin. Invest.* **83**, 213–221.

Kelton, D. E., and Rauch, H. (1962). Myelination and myelin degeneration in the central nervous system of dilute-lethal mice. *Exp. Neurol.* **6**, 252–262.

King, R. A., Wirtschafter, J. D., Olds, D. P., and Brumbaugh, J. (1986). Minimal pigment. A new type of oculocutaneous albinism. *Clin. Genet.* **29**, 42–50.

Kirby, G. C. (1974). Greying with age: A coat-color variant in wild Australian populations of mice. *J. Hered.* **65**, 126–128.

Klaus, S. N., and Snell, R. S. (1967). The response of mammalian epidermal melanocytes in culture to hormones. *J. Invest. Dermatol.* **48**, 352–358.

Kwon, B. S., Haq, A. K., Pomerantz, S. H., and Halaban, R. (1987). Isolation and sequence of a cDNA clone for human tyrosinase that maps at the mouse c-albino locus. *Proc. Natl. Acad. Sci. U.S.A.* **84**, 7473–7477.

Lane, P. W. (1972). Two new mutations in linkage group XVI of the house mouse. Flaky tail and varitint-waddler-J. *J. Hered.* **63**, 135–140.

Larsen, C. G., Anderson, A. O., Oppenheim, J. J., and Matsushima, K. (1989). Production of interleukin-8 by human dermal fibroblasts and keratinocytes in response to interleukin-1 or tumour necrosis factor. *Immunology* **68**, 31–36.

Le Douarin, N. (1984). Migration of neural crest cells., Pigment cells. *In* "The Neural Crest" pp. 22–53, 108–133. Cambridge University Press, New York.

Little, C. C. (1915). The inheritance of black-eyed white spotting in mice. *Am. Nat.* **49**, 727–740.

Lyon, M. F., and Searle, A. G. (1989). "Genetic Variants and Strains of the Laboratory Mouse." Oxford University Press, New York.

Mackie, R. M., and Atchison, T. G. (1982). Severe sunburn and subsequent risk of primary cutaneous malignant melanoma in Scotland. *Br. J. Cancer* **46**, 955–960.

Markert, C. L., and Silvers, W. K. (1956). The effects of genotype and cell environment on melanoblast differentiation in the house mouse. *Genetics* **41**, 429–450.

Matsushima, K., Morishita, K., Yoshimura, T., Lavu, S., Kobayashi, Y., Lew, A., Appela, E., Kung, H. F., Leonard, E. J., and Oppenheim, J. J. (1988). Molecular cloning of a human monocyte-derived neutrophil chemotactic factor (MDNCF) and the induction of MDNCF mRNA by interleukin 1 and tumor necrosis factor. *J. Exp. Med.* **167**, 1883–1893.

Mayer, T. C., and Maltby, E. L. (1964). An experimental investigation of pattern development in lethal spotting and belted mouse embryos. *Develop. Biol.* **9**, 269–296.

Mayer, T. C. (1967). Pigment cell migration in piebald mice. *Dev. Biol.* **15**, 521–535.

Mayer, T. C., and Fishbane, J. L. (1972). Mesoderm–ectoderm interaction in the production of the agouti pigmentation pattern in mice. *Genetics* **71**, 297–303.

McKay, R.D.G. (1989). The origin of cellular diversity in the mammalian control nervous system. *Cell* **58**, 815–821.

McLane, J. A., and Pawelek, J. M. (1988). Receptors for β-melanocyte-stimulating hormone exhibit positive cooperativity in synchronized melanoma cells. *Biochemistry* **27**, 3743–3747.

McLaren, A., and Bowman, P. (1969). Mouse chimeras derived from fusion of embryos differing by nine genetic factors. *Nature* **224**, 238–240.

Mintz, B., and Russell, E. S. (1957). Gene-induced embryological modifications of primordial germ cells in the mouse. *J. Exp. Zool.* **134**, 207–237.

Mintz, B. (1967). Gene control of mammalian pigmentary differentiation, I. Clonal origin of melanocytes. *Proc. Nat. Acad. Sci. U.S.A.* **58**, 344–351.

Mintz, B. (1970). Gene expression in allophenic mice. *In* "Control Mechanisms of Expression Cellular Phenotypes," Symp. Int. Soc. Cell Biol., (H. Padylkula, ed.), pp. 15–42. Academic Press, New York.

Mintz, B. (1971). Clonal basis of mammalian differentiation. *In* "Control Mechanisms of Growth and Differentiation" (D. D. Davies and M. Balls, eds.), 25th Symp. Soc., pp. 345–369. University Press, Cambridge.

Moore, K. J., Swing, D. A., Rinehik, E. M., Mucenski, M. L., Buchberg, A. M., Copeland, N. G., and Jenkins, N. A. (1988). The murine dilute suppressor gene *dsu* suppresses the coat-color phenotype of three pigment mutations that alter melanocyte morphology *d*, *ash*, and *ln*. *Genetics* **119**, 933–941.

Moser, B., Clark-Lewis, I., Zwahlen, R., and Baggiolini, M. (1990). Neutrophil-activating properties of the melanoma growth stimulatory activity. *J. Exp. Med.* **171**, 1797–1802.

Moyer, F. H. (1961). Electron microscope studies on the origin, development, and genetic control of melanin granules in the mouse eye. *In* "The Structure of the Eye," C. K. Smelser (ed). pp. 469–486, Academic Press, New York.

Murray, J. M., and Snell, G. D. (1945). Belted, a new sixth chromosome mutation in the mouse. *J. Hered.* **36**, 266–268.

Niu, M. C. (1947). The axial organization of the neural crest studied with particular reference to its pigmentary component. *J. Exp. Zool.* **105**, 79–114.

Nocka, K., Majumder, S., Chabot, B., Ray, P., Cervone, M., Bernstein, A., and Besmer, P. (1989). Expression of c-*kit* gene products in known cellular targets of w mutations in normal and w-mutant mice—evidence for an impaired c-*kit* kinase in mutant mice. *Genes Devel.* **3**, 816–826.

Nordlund, J. J., Collins, C. E., and Rheins, L. A. (1986). Prostaglandin E_2 and D_2 but not MSH stimulate the proliferation of pigment cells in the pinnal epidermis of the DBA/2 mouse. *J. Invest. Dermatol.* **86**, 433–437.

Ogata, S., Furihashi, Y., and Eisinger, M. (1987). Growth stimulation of human melanocytes: Identification and characterization of melanoma-derived melanocyte growth factor (M-McGF). *Biochem. Biophys. Res. Commun.* **146**, 1204–1211.

Okada, E. W. (1955). *Mem. Coll. Sci. Kyoto Univ.* **22**, 23–28.

Oquendo, P., Alberta, J., Wen, D., Graycar, J. L., Derynck, R., and Stiles, C. D. (1989). The platelet-derived growth factor–inducible KC gene encodes a secretory protein related to platelet α-granule proteins. *J. Biol. Chem.* **264**, 4133–4137.

Orr-Urtreger, A., Avivi, A., Zimmer, Y., Givol, D., Yarden, Y., and Lonai, P. (1990). Developmental expression of c-*kit*, a proto-oncogene encoded by the w locus. *Development* **109**, 911–923.

Pawelek, J., Sansore, M., Koch, M., Christie, G., Halaban, R., Hendee, J., Lerner, A. B., and Varga, J. M. (1975). Melanoma cells resistant to inhibition of growth by melanocyte stimulating hormone. *Proc. Natl. Acad. Sci. U.S.A.* **72**, 951–955.

Perris, R., Paulsson, M., and Bronner-Fraser, M. (1989). *Develop. Biol.* **136**, 222–238.

Preston, S. F., Volpi, M., Pearson, C. M., and Berlin, R. D. (1987). Regulation of cell shape in the Cloudman melanoma cell line. *Proc. Natl. Acad. Sci. U.S.A.* **84**, 5247–5251.

Ranson, M., Posen, S., and Mason, R. (1988). Human melanocytes as a target tissue for hormones: *In vitro* studies with 1α-25, dihydroxyvitamin D_3, α-melanocyte stimulating hormone, and beta-estradiol. *J. Invest. Dermatol.* **91**, 593–598.

Rawles, M. E. (1947). Origin of pigment cells from the neural crest in the mouse embryo. *Physiol. Zool.* **20**, 248–265.

Rawles, M. E. (1948). Origin of melanophores and their role in development of the color patterns in vertebrates. *Physiol. Rev.* **28**, 383–408.

Richmond, A., and Thomas, H. G. (1986). Purification of melanoma growth stimulatory activity. *J. Cell. Physiol.* **129**, 375–384.

Richmond, A., and Thomas, H. G. (1988). Melanoma growth stimulatory activity: Isolation from human melanoma tumors and characterization of tissue distribution. *J. Cell. Biochem.* **36**, 185–198.

Richmond, A., Balentien, E., Thomas, H. G., Flaggs, G., Barton, D. E., Spiess, J., Bordoni, R., Francke, U., and Derynck, R. (1988). Molecular characterization and chromosomal mapping of melanoma growth stimulatory activity, a growth factor structurally related to β-thromboglobulin. *EMBO J.* **7**, 2025–2033.

Rosen, C. F., Seki, Y., Farinelli, W., Stern, R. S., Fitzpatrick, T. B., Pathak, M. A., and Gange, R. W. (1987). A comparison of the melanocyte response to narrow band UVA and UVB exposure *in vivo. J. Invest. Dermatol.* **88**, 774–779.

Russell, E. S. (1979). Hereditary anemias of the mouse: A review for geneticists. *Adv. Genet.* **20**, 357–459.

Sanders, J. (1934). Family with albinismus circumscriptus. *Genetica* **16**, 435.

Sanders, J. (1935). Two families with albinismus circumscriptus. *Genetica* **17**, 185.

Schaible, R. H. (1969). Clonal distribution of melanocytes in piebald-spotted and variegated mice. *J. Exp. Zool.* **1972**, 181–199.

Schaible, R. H. (1972). Comparative effects of piebald-spotting genes on clones of melanocytes in different vertebrate species. *In* "Pigmentation, Its Genesis and Biologic Control" (V. Riley, ed.), pp. 343–357. Appleton-Century-Crofts, New York.

Schroeder, J. M., and Christophers, E. (1989). Identification of a novel family of highly potent neutrophil chemotactic peptides in psoriatic scales. *J. Invest. Dermatol.* **91**, 395 (Abstr.).

Silvers, W. K. (1956). Pigment cells: Occurrence in hair follicles. *J. Morphol.* **99**, 41–55.

Silvers, W. K. (1979). The Coat Colors of Mice, A Model for Mammalian Gene Action and Interaction, W. K. Silver (ed). pp. 1–265. Springer-Verlag, New York.

Sô, M., and Imai, Y. (1920). The types of spotting in mice and their genetic behavior. *J. Genet.* **9**, 319–333.

Spiegel, S., and Fishman, P. H. (1987). Gangliosides as bimodal regulators of cell growth. *Proc. Natl. Acad. Sci. U.S.A.* **84**, 141–146.

Stephenson, D. A., Mercola, M., Anderson, E., Wang, C., Stiles, C. D., Bowen-Pope, D. F., Chapman, V. M. (1991). Platelet-derived growth factor receptor α-subunit gene (*Pdgfra*) is deleted in the mouse patch (*Ph*) mutation. *Proc. Natl. Acad. Sci. U.S.A.* **88**, 6–10.

Stevens, L. C. (1954). The origin and development of chromatophores of *Xenopus laevis* and other anurans. *J. Exp. Zool.* **125**, 221–246.

Stout, D. B. (1946). Albinism among San Blas Cuna, Panama; further notes. *Am. J. Anthropol.* **4**, 483.

Strieter, R. M., Kunkel, S. L., Showell, H. F., Remick, D. G., Phan, S. H., Ward, P. A., and Marks, R. M. (1989). Endothelial cell gene expression of a neutrophil chemotactic factor by TNF-α, LPS, and IL-1β. *Science* **243**, 1467–1469.

Strobel, M. C., Seperack, P. K., Moore, K. J., Copeland, N. G., and Jenkins, N. A. (1988). The dilute coat-color locus of mouse chromosome 9. *Adv. Pigment Cell Res.* **256**, 297–305.

Szabo, G. (1954). The number of melanocytes in human epidermis. *Br. Med. J.* **1**, 1016–1017.

Szabo, G. (1967). Regional anatomy of the human integument with special reference to distribution of hair follicles, sweat glands, and melanocytes. *Philos. Trans. R. Soc. Lond.* Series B **252**, 447–485.

Szabo, G., Gerald, A. B., Pathak, M., and Fitzpatrick, T. B. (1969). The ultrastructure of racial color differences in man. *Anat. Rec.* **163**, 342–343.

Tan, J. C., Nocka, K., Ray, P., Traktman, P., and Besmer, P. (1990). The dominant W^{42} *spotting* phenotype results from a missense mutation in the c-*kit* receptor kinase. *Science* **247**, 209–212.

Tannenbaum, C. S., Major, J., Poptic, E., DiCorleto, P. E., and Hamilton, T. A. (1989). Lipopolysaccharide-inducible macrophage early genes are induced in BALB/c 3T3 cells by platelet-derived growth factor. *J. Biol. Chem.* **264**, 4052–4057.

Takeuchi, S., Yamamoto, H., and Takeuchi, T. (1988). Expression of tyrosinase gene in amelanotic mutant mice. *Biochem. Biophys. Res. Commun.* **155**, 470–475.

Tanaka, S., Yamamoto, H., Takeuchi, S., and Takeuchi, T. (1990). Melanization in albino mice transformed by introducing cloned mouse tyrosinase gene. *Development* **108**, 223–227.

Thomas, H. G., and Richmond, A. (1988). High-yield purification of melanoma growth-stimulating activity. *Mol. Cell. Endocrinol.* **57**, 69–76.

Wen, D., Rowland, A., and Derynck, R. (1989). Expression and secretion of gro/MGSA by stimulated human endothelial cells. *EMBO J.* **8**, 1761–1766.

Wilkins, L., Gilchrest, B. A., Szabo, G., Weinstein, R., and Maciag, T. (1985). The stimulation of normal human melanocyte proliferation *in vitro* by melanocyte growth factor from bovine brain. *J. Cell. Physiol.* **122**, 350–361.

Wilson, R. E., Dooley, T. P., and Hart, I. R. (1989). Induction of tumorigenicity and lack of *in vitro* growth requirement for 12-*o*-tetradecanoylphorbol-13-acetate by transfection of murine melanocytes with v-Ha-*ras*. *Cancer Res.* **49**, 711–716.

Witte, O. N. (1990). Steel locus defines new multipotent growth factor. *Cell* **63**, 5–6.

Yarden, Y., Kuang, W.-J., Yang-Feng, T., Coussens, L., Munemitsu, S., Dull, T. J., Chen, E., Schlessinger, J., Francke, U., and Ullrich, A. (1987). Human protoocogene c-*kit*: A new cell-surface receptor tyrosine kinase for an unidentified ligand. *EMBO J.* **6**, 3341–3351.

Yamamoto, H., Takeuchi, S., Kudo, T., Makino, K., Nakata, A., Shinoda, T., and Takeuchi, T. (1987). Cloning of the mouse tyrosinase gene. *Jpn. J. Genet.* **62**, 271–277.

III

Exploring Mechanisms of Control

10

Molecular Mechanisms That Mediate a Functional Relationship between Proliferation and Differentiation

GARY S. STEIN, JANE B. LIAN, THOMAS A. OWEN, JOOST HOLTHUIS, RITA BORTELL, AND ANDRE J. VAN WIJNEN

Department of Cell Biology
University of Massachusetts Medical Center
Worcester, Massachusetts 01655

MOLECULAR AND CELLULAR APPROACHES TO THE CONTROL OF PROLIFERATION AND DIFFERENTIATION

In this chapter we shall address molecular mechanisms that mediate the relationship between cell growth and differentiation throughout the developmental sequence of events associated with cell and tissue specialization. This relationship between proliferation and differentiation traditionally has been the cornerstone of postulated models for regulation of development. However, only recently, through the combined applications of molecular, biochemical, and cellular experimental approaches, has insight been provided into the control of gene expression required for proliferation and the influence of proliferation on expression of genes that support the progressive acquisition of structural and functional properties characteristic of a fully differentiated phenotype. While we shall focus on growth control as it relates to differentiation of the osteoblast phenotype, it is with the understanding that the basic concepts and many of the experimental results that have to date been obtained are applicable to a broad spectrum of biological systems.

Throughout this volume, a description of the sequelae of events associated with growth control within the context of the onset and progression of differentiation has been provided. For expression of the bone-cell phenotype, in addition to the temporal expression of specific genes at various times during the osteoblast developmental sequence (see Chapter 6: Owen *et al.*, 1990a; Aronow *et al.*, 1990; Owen *et al.*, 1990b; Owen *et al.*, 1991; Barone *et al.*, in press; and summarized in Fig. 1), several lines of evidence are presented that support a *functionally* coupled relationship between cell growth and differentiation. This is characterized initially by proliferation supporting expression of genes for cell-cycle and cell-growth control along with those for initial biosynthesis of the extracellular matrix (e.g., transforming growth factor β, fibronectin, and collagen). Subsequently, the down-regulation of proliferation, in part mediated by the extracellular matrix, signals the expression of a series of genes associated with maturation and three-dimensional organization of the extracellular matrix in a manner that renders the osteoblast extracellular matrix competent for the ordered deposition of mineral. Equally important is the mechanism by which genes expressed following the completion of proliferative activity are suppressed, while proliferation is ongoing, and then sequentially up-regulated during maturation of the osteoblast phenotype.

We shall focus on the molecular mechanisms that support the selection expression of genes for cell growth and osteoblast phenotype expression and those that integrate the complex series of signaling mechanisms functionally coupling proliferation and differentiation. Emphasis will be placed on multiple levels of control, which are operative as a reflection of a convergence, and integration of signaling pathways, which are operative in the bidirectional exchange of regulatory macromolecules between the extracellular environment and the cytoplasm, as well as between the cytoplasm and the nucleus.

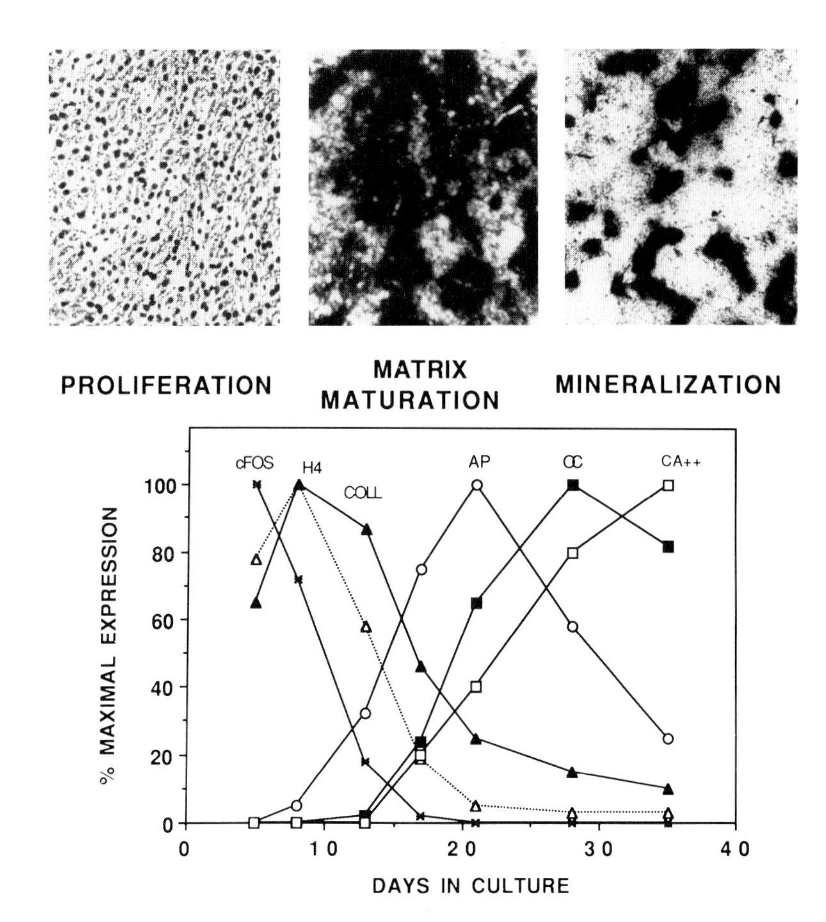

Fig. 1. Temporal expression of cell growth, extracellular matrix, and osteoblast phenotype-related genes during the development of the osteoblast phenotype *in vitro*. Isolated primary cells were cultured after confluence in BGJb medium supplemented with 10% fetal calf serum, 50 μg/ml ascorbic acid and 10 m*M* β-glycerol phosphate. Top panel: Histochemical staining of the cultures. Initially, all cells are actively proliferating, but by Day 12, multilayered regions form in which all cells have ceased proliferative activity and are alkaline phosphatase positive. The entire culture becomes alkaline phosphatase positive by Day 16 (center). The ordered deposition of mineral occurs as extracellular matrix develops within the multilayered regions (nodules) as shown by von Kossa silver staining on day 35 (right). Bottom panel: Representation of RNA transcripts of selected genes. Cellular RNA was isolated at the times indicated during the 35-day differentiation time course and assayed for the steady-state levels of various transcripts by Northern blot analysis. The resulting blots were quantitated by scanning densitometry and the data plotted relative to the maximal expression of each transcript. Cell growth-related genes shown are H4 histone (reflects DNA synthesis), and c-*fos*. Expressed during the proliferative period is Type I collagen [with fibronectin (FN) and transforming growth factor-β (TGF-β,) not shown]. Alkaline phosphatase (AP) gene expression is associated with extracellular-matrix maturation. Genes induced with extracellular-matrix mineralization (calcium) are osteopontin (not shown) and osteocalcin (OC). Note the induction of alkaline phosphatase at the end of proliferation and its down-regulation in heavily mineralized cultures (Day 35, mature osteoblasts).

I. TWO TRANSITION–RESTRICTION POINTS DURING DEVELOPMENT OF THE OSTEOBLAST PHENOTYPE CHARACTERIZE THE PROLIFERATION–DIFFERENTIATION RELATIONSHIP

A. Experimental Basis For Two Transition–Restriction Points

Both *in vivo* during bone formation (Yoon *et al.*, 1987; Weinreb *et al.*, 1990) and in primary cultures of normal diploid calvarial-derived osteoblasts (Owen *et al.*, 1990a; Aronow *et al.*, 1990; Bellows *et al.*, 1986; and Bhargava *et al.*, 1988), the progressive development of the osteoblast phenotype is associated with a developmental sequence defined by three periods. Each period is characterized by the stringently regulated expression of a set of genes (Owen *et al.*, 1990a; Aronow *et al.*, 1990; Owen *et al.*, 1990b; Owen *et al.*, 1991; and Barone *et al.*, in press), reflecting sequential periods of proliferation and extracellular-matrix biosynthesis followed by extracellular-matrix maturation and mineralization (Fig. 2). Two transition points are key components of this temporal expression of genes that support cell growth and differentiation: the first transition point, at the completion of the proliferation period when genes for cell-cycle and cell-growth control are down-regulated, and expression of genes encoding proteins for extracellular-matrix maturation and organization is initiated; and the second, at the onset of extracellular matrix mineralization (Fig. 2). These transitions have been

Fig. 2. Model of the reciprocal relationship between proliferation and differentiation during the rat osteoblast developmental sequence. (Upper panel) This relationship is schematically illustrated within the context of modifications in expression of cell cycle and cell growth-regulated genes, as well as genes associated with the maturation, development, and mineralization of the osteoblast extracellular matrix. The three principal periods of the osteoblast developmental sequence are designated within broken vertical lines (proliferation, matrix development and maturation, and mineralization). A functional relationship between the down-regulation of proliferation and the initiation of extracellular-matrix maturation and development is based on stimulation of alkaline phosphatase and osteopontin gene expression when proliferation is inhibited, but the developmental sequence is induced only to the second transition point. Growth of the osteoblast under conditions that do not support mineralization confirms the Day 20 restriction point, since the developmental sequence proceeds through the proliferation and the extracellular matrix development–maturation periods, but not further (Owen *et al.*, 1990a). AP-1, AP-1 binding activity; H4 histone; COL-I, Type αI collagen; AP, alkaline phosphatase; OP, osteopontin; OC, osteocalcin; Mineral, total accumulated hydroxyapatite (calcium + phosphate). Lower left: histone gene promoter elements. The lower panel is a diagram of the histone gene promoter indicating (1) the proximal (Site I and II) and distal regulatory elements; (2) a series of promoter-binding factors (TATA, HiNF-D, HiNF-C, HiNF-E, H4UA-1, NMP-1, H4UA-2 and H4UA-3); and (3) protein–DNA contacts at single-nucleotide resolution within both the proximal and distal regulatory elements (indicated by dots over G nts). Lower right: schematic representation of the osteocalcin gene promoter (Lian *et al.*, 1989). AP-1 sites are indicated which bind the oncogene encoded protein complex Fos–Jun within the vitamin D–responsive element and the CCAAT-containing proximal promoter element (Markose *et al.*, 1990; Owen *et al.*, 1990b).

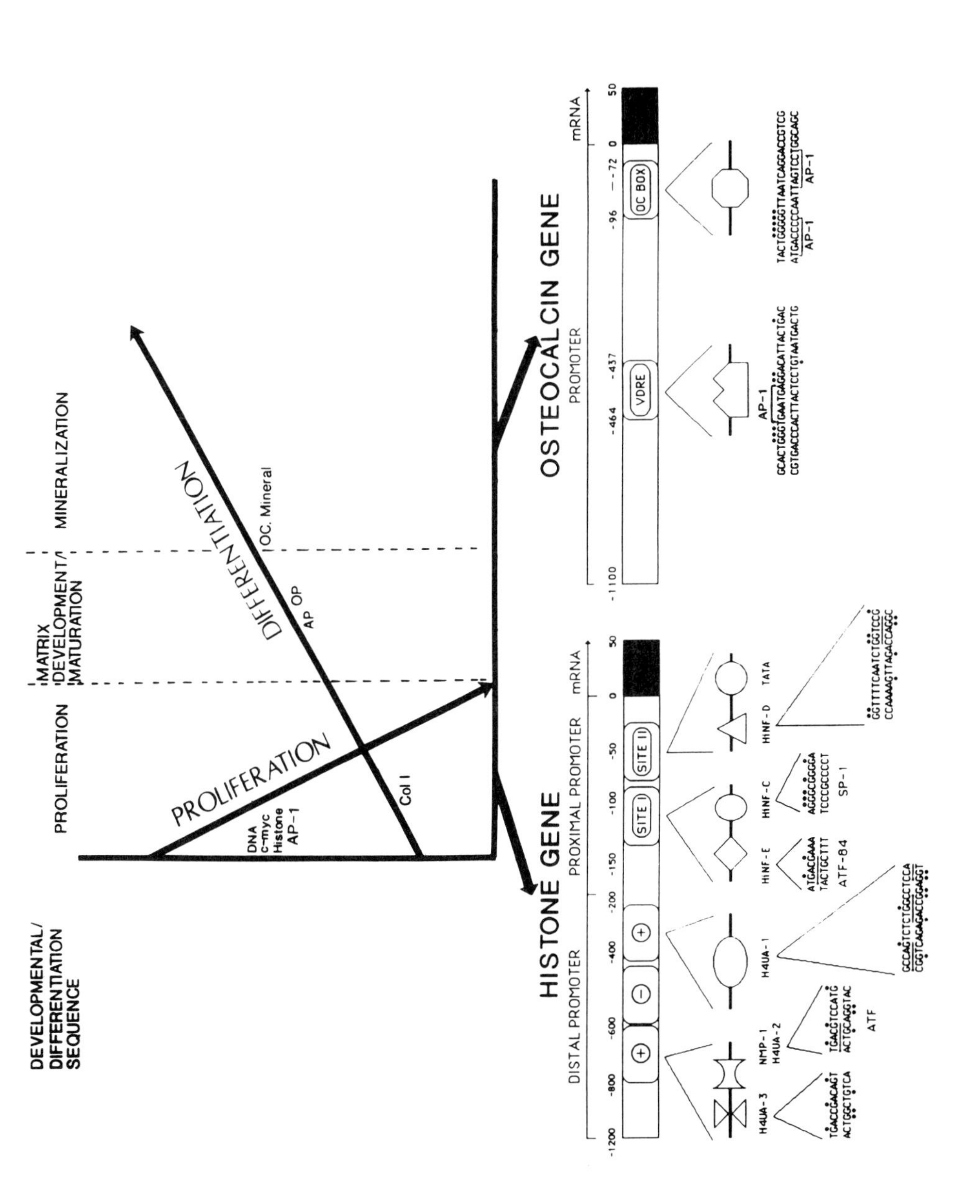
DEVELOPMENTAL/
DIFFERENTIATION
SEQUENCE
PROLIFERATION
MATRIX
DEVELOPMENT/
MATURATION
MINERALIZATION
PROLIFERATION
DIFFERENTIATION
DNA
c-myc
Histone
AP-1
Col I
AP
OP
OC, Mineral
HISTONE GENE
DISTAL PROMOTER
PROXIMAL PROMOTER
mRNA
-1200
-800
-600
-400
-200
-150
-100
-50
0
50
SITE I
SITE II
H4UA-3
NMP-1
H4UA-2
H4UA-1
HINF-E
HINF-C
HINF-D
TATA
ATF
ATF-84
SP-1
OSTEOCALCIN GENE
PROMOTER
mRNA
-1100
-464
-437
-96
-72
0
50
VDRE
OC BOX
AP-1
AP-1
AP-1

experimentally established and functionally defined as restriction points during osteoblast differentiation, to which developmental expression of genes can proceed but cannot pass, without additional cellular signaling. Thus to establish the basis of cellular competency for progression toward development of the mature osteoblast phenotype necessitates identification of the signaling pathways operative at the developmental transition points by which genes are selectively activated and/or suppressed. Additionally, the regulatory mechanisms that serve as the rate-limiting steps at these strategic points during osteoblast differentiation must be characterized.

B. Multiple Levels of Gene Regulation

Here the extent to which regulation of gene expression is mediated at the levels of transcription and/or at the multitude of posttranscriptional steps, which include processing of gene transcripts, translatability, and messenger RNA stability, must be determined. And, while the complexity and overlap of the control mechanisms and the regulatory signaling pathways is becoming increasingly apparent, this complexity provides the basis for understanding the integration of cellular responses to highly diverse physiological factors (e.g., hormonal, cyclic nucleotides, divalent cation concentrations) that modulate expression of cell growth and osteoblast-related genes.

Two striking examples of genes in which expression is controlled at multiple levels, and modified with respect to competency for responsiveness to physiological regulatory signals at various stages of osteoblast differentiation, are the cell-cycle–regulated histone genes at the proliferation–differentiation transition point (Fig. 2) (Owen *et al.*, 1990b; Sierra *et al.*, 1982; and Stein *et al.*, 1975) and the osteocalcin gene at the onset of extracellular-matrix mineralization (Fig. 2) (Owens *et al.*, 1991; Lian *et al.*, 1989; Markose *et al.*, 1990; and Owen *et al.*, 1990c). Transcriptionally, these genes are controlled by promoters with modular organizations of positive and negative regulatory elements that interact in a sequence-specific manner with a diverse series of physiological mediators (Lian *et al.*, 1989; Markose *et al.*, 1990). Additional options exercised by the osteoblast during differentiation for modulating transcriptional control of the histone and osteocalcin genes include the recruitment of transcription factors and the extent to which sequence-specific, DNA-binding proteins are phosphorylated (Roesler *et al.*, 1988). mRNA stability also accounts for changes in the extent to which the histone and osteocalcin genes are expressed during the osteoblast developmental sequence, in which cell-cycle variations during proliferation and effects of steroid hormones influence the rates of histone and osteocalcin mRNA turnover.

It should be emphasized that the rate-limiting steps of the complex and interdependent cascade of events, associated with transcriptional–posttranscriptional regulation at the key transition points during development of the osteoblast phenotype, remains to be established. However, we are beginning to gain insight into mechanisms that have the capacity to mediate rapid and long-term responses in both the activation and suppression, as well as in the extent to which genes associated with osteoblast growth and differentiation are expressed.

C. First Transition Point: Convergence of Signaling Mechanisms to Initiate Differentiation

Molecular mechanisms operative at the transition point early in the osteoblast developmental sequence, when proliferation ceases and genes associated with extracellular-matrix maturation and specialization are induced, can be addressed by examining the down-regulation mechanisms of genes expressed during the proliferation period. Examples of such genes are those encoding histone proteins, which package newly replicated DNA into chromatin. Histone gene expression is restricted to the S-phase of the cell cycle and is tightly coupled with DNA replication (Fig. 3) (Plumb *et al.*, 1983; Baumbach *et al.*, 1987) with regulation of expression controlled at both the levels of transcription and mRNA stability.

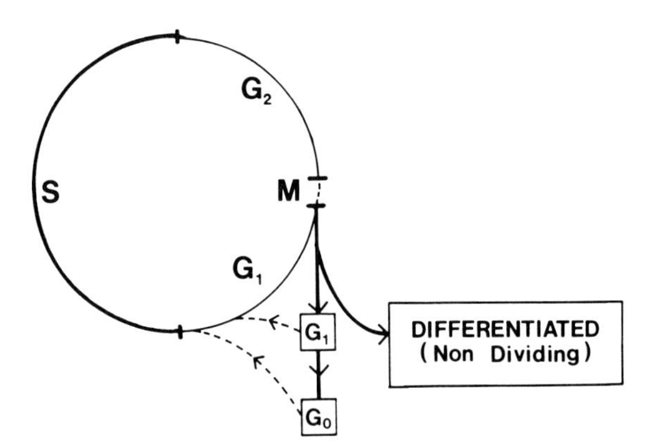

Fig. 3. Schematic representation of the osteoblast cell cycle. The doubling time of diploid osteoblasts is 20 hr with a 6-hr G_1 period, 9-hr S phase (DNA replication), 4-hr G_2, and 1-hr mitotic period. At the first transition point in the osteoblast developmental sequence with the onset of differentiation, osteoblasts exit the cell cycle following completion of mitosis as indicated. As described in the text, dramatic differences in the regulation of cell-cycle genes are operative on each side of this transition point.

1. Transcriptional and Posttranscriptional Regulation of Gene Expression

To experimentally address the extent to which regulation is transcriptionally and posttranscriptionally mediated, it is necessary to directly determine the rates of transcription and cellular levels of mRNA on both sides of the proliferation–differentiation transition point (Fig. 3), during the cell cycle in actively proliferating cells, and then following the completion of proliferative activity during the development and maturation of the osteoblast phenotype. The rationale for this approach is that a parallel relationship of cellular mRNA levels and rate of transcription is indicative that transcriptional control is operative and that mRNA turnover is coordinately regulated with the rate of transcription. Continued transcription in the absence of detectable levels of mRNA would reflect control primarily at the level of mRNA stability. Throughout the proliferation period, the level of histone gene transcription reflects the cellular level of histone mRNA and synthesis of histone protein, indicating that both transcriptional control and regulation of histone mRNA stability contribute to determining the extent to which histone protein is synthesized.

As indicated in Fig. 4, primary cultures of actively proliferating normal diploid osteoblasts can be synchronized, permitting a systematic analysis of the principal biochemical parameters (histone protein synthesis, cellular levels of histone mRNA, and rate of histone gene transcription) of histone gene expression throughout the cell cycle. Results from such an analysis of histone gene regulation in synchronized proliferating osteoblasts indicate that both transcriptional control and regulation of histone mRNA stability contribute to coupling histone gene expression with DNA replication, thereby accounting for the restriction of histone protein synthesis to the S phase of the cell cycle (Owen *et al.*, 1990a; Owen *et al.*, 1990b; and Holthuis *et al.*, 1990). These results are consistent with the absence of detectable levels of histone mRNA in non–S phase cells (Owen *et al.*, 1990b; Holthuis *et al.*, 1990) and the direct demonstration that the turnover of histone mRNA is dramatically accelerated as cells complete DNA replication (Morris *et al.*, 1991). Additionally, results from *in vitro* and *in vivo* studies have shown cell cycle–dependent modifications in the rate of transcription (Owen *et al.*, 1990b;

Fig. 4. Synchronization of rat osteoblasts. Actively proliferating rat osteoblasts (Day 3 in culture) were synchronized by two cycles of thymidine block. Panel A indicates DNA synthesis as measured by ^{3}H-thymidine labeling and determination of TCA precipitable radioactivity. These results were confirmed by autoradiography (Panel B). The lower three panels are photomicrographs of cells at three points during the cell cycle following the synchronization procedure. Note the *doublet cells* in the middle panel, which have just completed mitotic division. The twofold increase in cell density for the G_1 cells (lower right panel) compared to the S-phase cells (lower left panel), reflects completion of a round of mitotic division for most if not all of the cell population. Three vertical arrows below Panel B indicate the times during the cell cycle when the photomicrographs were taken.

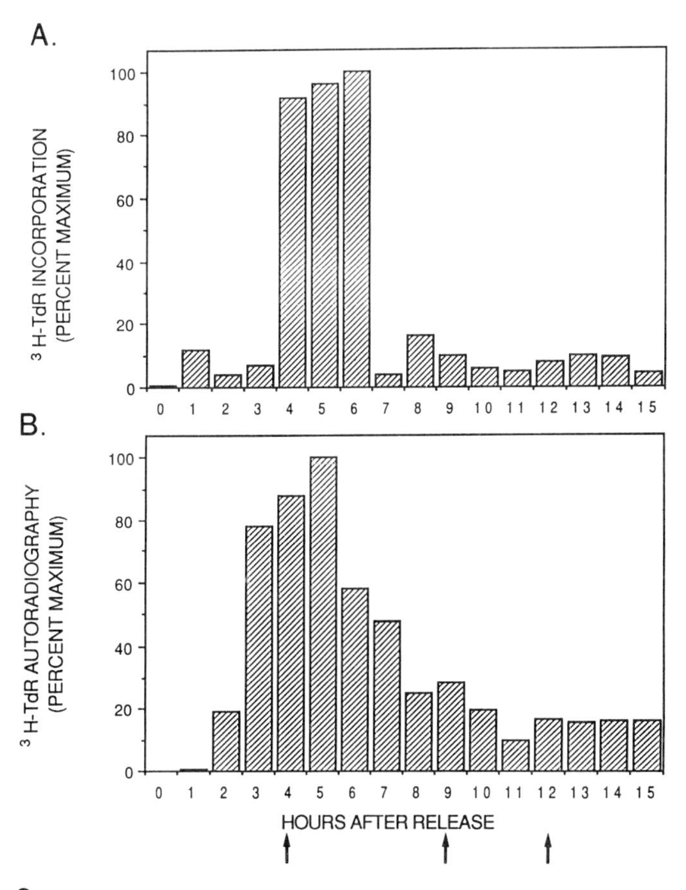
A.
100
80
60
40
20
0
3 H-TdR INCORPORATION
(PERCENT MAXIMUM)
0 1 2 3 4 5 6 7 8 9 10 11 12 13 14 15
B.
100
80
60
40
20
0
3 H-TdR AUTORADIOGRAPHY
(PERCENT MAXIMUM)
0 1 2 3 4 5 6 7 8 9 10 11 12 13 14 15
HOURS AFTER RELEASE
C.

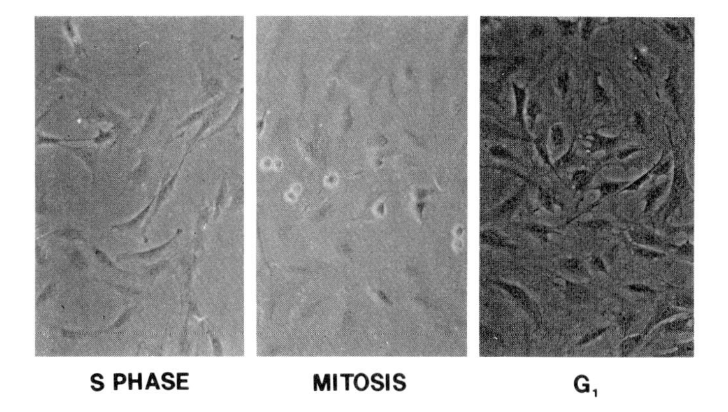
S PHASE
MITOSIS
G1

van Wijnen *et al.*, in press; and Holthuis *et al.*, 1990). The relationship of subcellular localization to histone gene transcription (Dworetzky *et al.*, 1990) and to stability properties of histone mRNAs (Lawrence *et al.*, 1988; Zambetti *et al.*, 1987) with respect to implications for modulation of histone mRNA turnover are described in Chapter 6.

2. Protein–DNA Interactions Mediating Transcription

At the completion of the proliferation period, when osteoblasts cease to traverse the cell cycle, histone gene expression is down-regulated at the transcriptional level by a selective destabilization of histone mRNA (Owen *et al.*, 1990a; Owen *et al.*, 1990b; Owen *et al.*, 1990c; Holthuis *et al.*, 1990; and Shalhoub *et al.*, 1989) (Fig. 5). A systematic examination of protein–DNA interactions in the proximal regulatory elements of cell cycle–regulated H4 and H3 histone genes both *in vivo* (Pauli *et al.*, 1987; Pauli *et al.*, 1989; and Pauli *et al.*, 1988) and *in vitro* (Owen *et al.*, 1990a; Holthuis *et al.*, 1990; van Wijnen *et al.*, 1987; van Wijnen *et al.*, 1988a; van Wijnen *et al.*, 1989; and van Wijnen *et al.*, 1988b; Kroeger *et al.*, 1987) indicates that this down-regulation of transcription is mediated by a selective loss in the binding of a transcription factor HiNF-D to sequences that influence both specificity and level of transcription (Site II) (Fig. 6A). A functional relationship between loss of HiNF-D–Site II interactions and the onset of osteoblast differentiation is further suggested by the persistence of Site II occupancy by HiNF-D when cell proliferation is inhibited, under conditions that do not promote the activation of genes associated with extracellular-matrix maturation and specialization, i.e., conditions in which cell growth is blocked but the progressive differentiation of the osteoblast does not proceed (Fig. 7) (Owen *et al.*, 1990b).

A transcriptionally mediated down-regulation of histone gene expression at the proliferation–differentiation transition point is not confined to bone-cell differentiation, since histone gene transcription has similarly been shown to cease at the completion of proliferative activity in HL-60 promyelocytic leukemia cells during monocyte differentiation (Stein *et al.*, 1989) (Fig. 6B) and during differentiation of preadipocytes. The down-regulation of histone gene transcription during monocyte and adipocyte differentiation is accompanied by and presumably functionally related to the selective loss of HiNF-D–Site II interactions, which have been established both *in vitro* and in intact cells, further supporting the general utilization of HiNF-D binding to Site II as a rate-limiting step for the down-regulation of histone gene transcription early during the differentiation process.

3. Promoter-Binding Factor Phosphorylation

Once establishing that transcriptional control is operative and a series of sequence-specific interactions of DNA-binding proteins with promoter regulatory

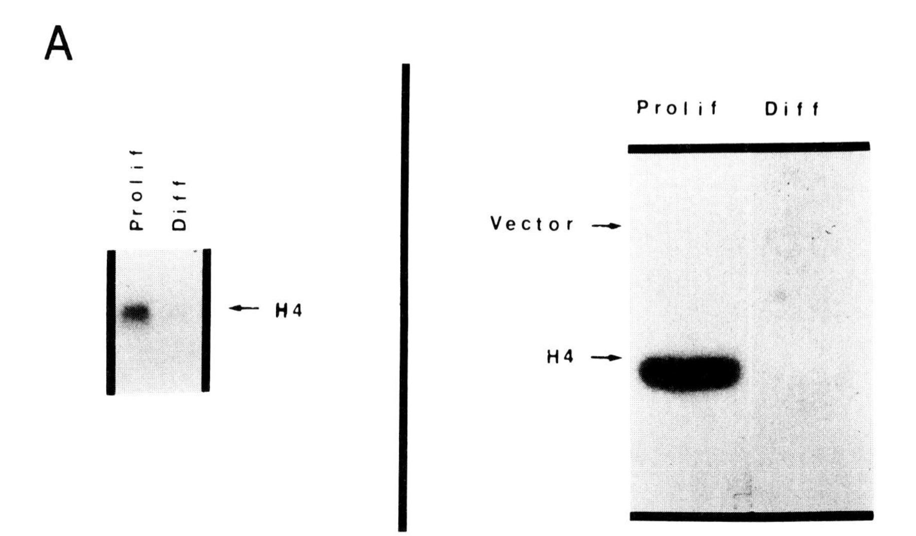

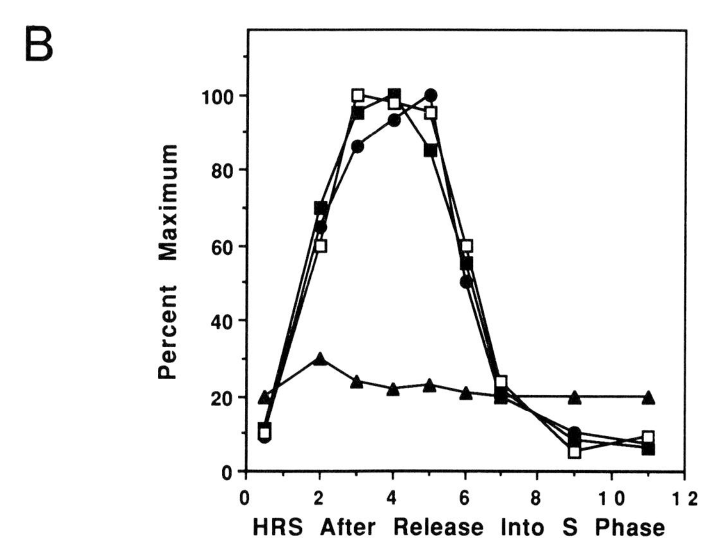

Fig. 5. (A) Postproliferative regulation of the histone gene. Down-regulation of histone expression by both mRNA destabilization and inhibition of transcription at the completion of proliferative activity with the onset of differentiation. Left panel shows detection of H4 mRNA by Northern blot analysis only in proliferating cells. Similarly, transcription of the H4 gene determined by nuclei–run on assays occurs only in proliferating cells (right panel). (B) Histone gene expression during the cell cycle in proliferating osteoblasts. Cells synchronized by two cycles of 2m*M* thymidine block and DNA synthesis (□) as well as the principal parameters of histone gene expression [histone protein synthesis (●), histone mRNA levels (■) and rate of transcription (▲)] were assayed at the indicated times. The presence of histone mRNA and histone synthesis only during S phase, while histone gene transcription is constitutive throughout the cell cycle, indicates that histone mRNA is selectively destabilized at the completion of DNA replication.

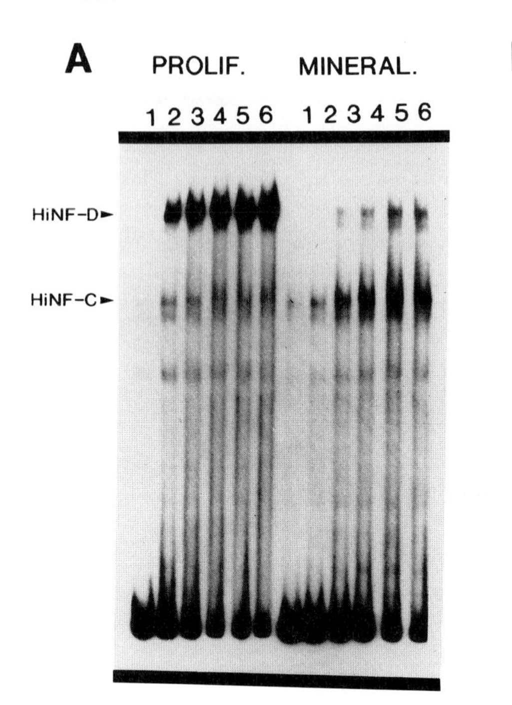
A
PROLIF.
MINERAL.
1 2 3 4 5 6
1 2 3 4 5 6
HiNF-D
HiNF-C

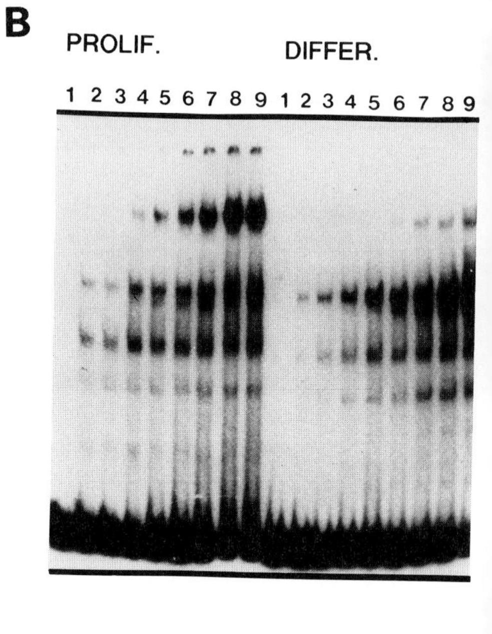
B
PROLIF.
DIFFER.
1 2 3 4 5 6 7 8 9 1 2 3 4 5 6 7 8 9

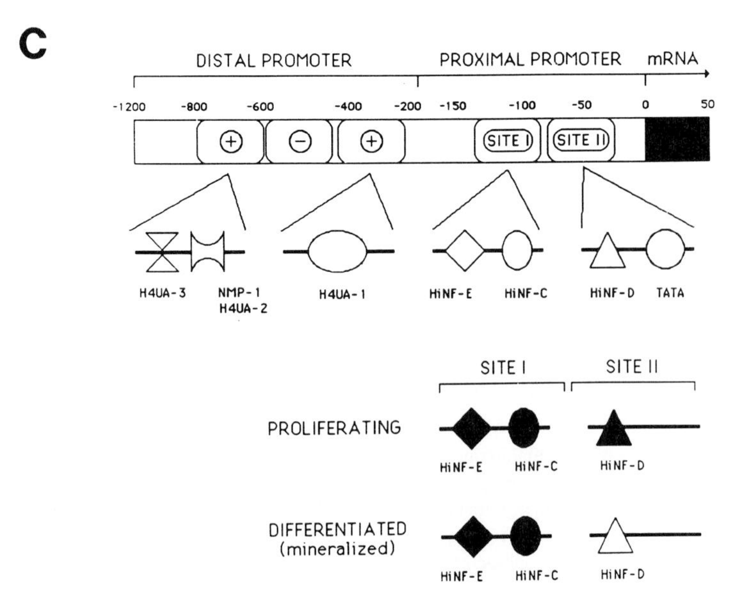
C
DISTAL PROMOTER
PROXIMAL PROMOTER
mRNA
-1200
-800
-600
-400
-200
-150
-100
-50
0
50
SITE I
SITE II
H4UA-3
NMP-1
H4UA-2
H4UA-1
HiNF-E
HiNF-C
HiNF-D
TATA
SITE I
SITE II
PROLIFERATING
HiNF-E
HiNF-C
HiNF-D
DIFFERENTIATED
(mineralized)
HiNF-E
HiNF-C
HiNF-D

Fig. 6. Selective loss of histone transcription-factor interaction at a proximal histone gene regulatory element at the onset of differentiation. (A) Osteoblast differentiation. Interactions of histone nuclear factors (HiNF-D and HiNF-C) binding to Site I and II promoter with regulatory elements of the H4 histone gene promoter. Nuclear factor extracts were prepared from proliferating (Day 9) and differentiated (Day 28 well-mineralized) osteoblasts. Binding of nuclear extracts to ^{32}P-labeled segments of the H4 histone gene promoter were carried out under low ionic strength conditions, fractionated electrophoretically, followed by autoradiography. Assays were carried out using a fixed amount of radiolabeled DNA and increasing amounts of nuclear protein fraction (lanes 1–6). Note the 79% decrease in binding of factor D to Site II in extracts of cells with a mineralized matrix. (B) HL-60 promyelocytic leukemia differentiation. Panel B demonstrates that in proliferating (S phase) HL-60 cells or during the S period of the cell cycle, there is occupancy of Sites I and II by factors C and D. In contrast, phorbal ester–induced differentiation of the promyelocytic cell to monocytes is associated with a selective loss of factor D binding to Site II with retention of Site I–HiNF-C interaction. Lanes 1–9 in each group are increased in concentrations of nuclear protein fraction. (C) Schematic illustration of the selective loss of HiNF-D–Site II interactions at the onset of differentiation. Solid symbols represent occupancy of Site I by HiNF-E and HiNF-C and Site II, by HiNF-D.

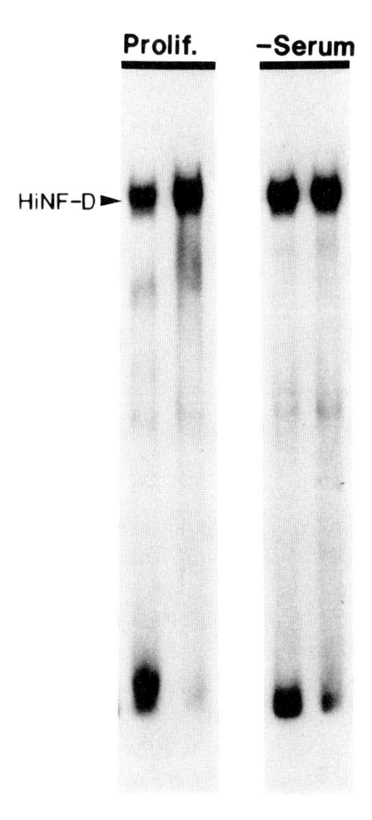

Fig. 7. Growth arrest in the absence of differentiation. When proliferation is arrested by serum deprivation (these cells retain the capacity to proliferate, but differentiation is not induced) occupancy of Sites I and II by factors A, C, and D persist.

elements has been defined (which influence the extent to which the gene is transcribed), it is necessary to pursue the mechanism by which factor–regulatory element binding is controlled. Among the critical questions to be addressed are the competency of the regulatory element to serve as a binding site for the transcription factors and the ability of the factor to be structurally competent and available for interaction with the cognate regulatory element. It is well known that many DNA-binding proteins are highly phosphorylated, primarily at serine and threonine residues, and several lines of evidence are consistent with a functional relationship between the extent to which DNA-binding proteins are phosphorylated and the capacity of these proteins to influence transcription (Roesler *et al.*, 1988) including transcription of the histone genes (Thomson *et al.*, 1976; Kleinsmith *et al.*, 1976). Recently two lines of evidence provide support for the involvement of HiNF-D phosphorylation in modulating interactions of the factor with Site II of the histone gene promoter in osteoblasts. First, enzymatic dephosphorylation of HiNF-D abrogates binding at Site II with a parallel relationship between the extent of phosphate group removal and the decline in factor-DNA interactions (Fig. 8). Second, inhibition of phosphatase activity *in vivo* in S-phase osteoblasts prevents the loss of HiNF-D–Site II binding, which occurs during passage from the S to the G_2 periods of the cell cycle. Taken together, these findings support a requirement of HiNF-D phosphorylation for mediating the up-regulation of histone gene transcription during the S phase of the cell cycle, and a requirement for HiNF-D dephosphorylation for the termination of HiNF-D–Site II interactions at the completion of S phase and the decline in transcription of histone genes. Additionally, these results provide an indication of a possible mechanism for regulating the activity of a rate-limiting histone gene transcription factor. At the same time, they extend the problem of transcriptional regulation at the proliferation–differentiation transition in the osteoblast developmental sequence to the cellular signaling pathways associated with control of protein phosphorylation.

D. Second Transition Point: Onset of Mineralization Mediates Completion of the Mature Osteoblast Phenotype

Striking modifications in gene expression are evident at the onset of and during the extracellular-matrix mineralization period of the osteoblast developmental sequence. When mineralization initiates in cultured diploid osteoblasts (Aronow *et al.*, 1990; Owen *et al.*, 1990a) and in osteocytes *in vivo*, surrounded by a fully mineralized extracellular matrix (Bruder and Caplan, 1990), there is a down-regulation of alkaline phosphatase gene expression and a dramatic increase in expression of osteocalcin. Then, as mineralization progresses, osteocalcin and osteopontin expression are partially down-regulated *in vitro* (Owen *et al.*, 1990a;

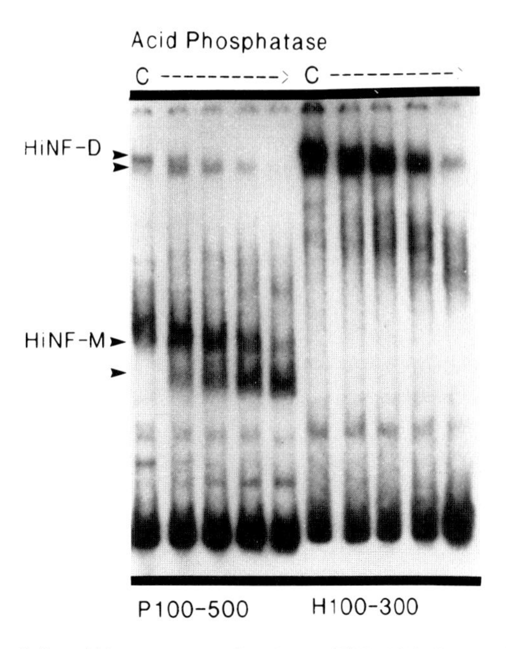

Fig. 8. Phosphorylation of histone gene nuclear factor (HiNF-D) influences interactions with the Site II promoter elements. Dephosphorylation of the histone gene-promoter factors HiNF-D and HiNF-M modify binding to Site II regulatory sequences. Fractionated nuclear proteins from proliferating cells were treated for 10 min with sweet potato acid phosphatase (SPP) before gel mobility–shift analysis. Approximately 4 μg of protein from either a phosphocellulose fraction containing both HiNF-D and HiNF-M activity (P100-500) (lanes 1–5) or a heparin-agarose fraction containing primarily HiNF-D activity (H100–300) (lanes 6–10) were added to each assay. Samples in lanes 1–5 and lanes 6–10 were each treated with, respectively, 0 (control, designated C), 0.05, 0.1, 0.2 and 0.5 units of SPP. The DNA fragment used [nt −97 to −30 (Fig. 2)] was internally labeled at an *Ava*II restriction site to prevent phosphatase cleavage of ^{32}P-label from the probe. The probe spans the entire H4–Site II, including both the HiNF-M and HiNF-D binding sites. The two phosphorylated forms of both HiNF-D and HiNF-M are indicated by arrowheads. Oligonucleotide competition analyses with partially dephosphorylated proteins showed that the lower forms of both HiNF-D and HiNF-M had similar competition behavior as the respective upper forms of each. Extended dephosphorylation abolished the binding activity of both forms of HiNF-D and HiNF-M.

Aronow *et al.*, 1990; Gerstenfeld *et al.*, 1987). These findings point to complex and minimally understood feedback mechanisms by which the genes associated with development of mature osteoblast phenotype are regulated.

1. Steroid-Hormone Responsiveness of Gene Expression in Osteoblasts

Some insight into regulatory mechanisms operative during extracellular-matrix mineralization is provided by observed modifications in responsiveness to steroid hormones of osteoblast-regulated genes involved in biosynthesis, mineralization, and turnover of extracellular matrix. The steroid hormone most extensively examined for an effect on osteoblast growth and differentiation is vitamin D, because of the well-established involvement in the regulation of calcium homeostasis *in vivo* via the intestinal absorption of calcium and the mobilization of calcium from bone (DeLuca, 1988); and additionally, on the basis of the influence of the hormone on expression of Type I collagen, alkaline phosphatase, osteopontin and osteocalcin (OC) in cultured diploid osteoblasts, all of which have been proposed to be involved in the biosynthesis, maturation, and organization of the Type I collagen extracellular matrix and related to the ordered deposition of mineral (reviewed in Rodan and Noda, 1991; Stein *et al.*, in press; Lian *et al.*, in press; Stein *et al.*, 1990). Vitamin D modulates expression of these genes via classic steroid-hormone action, with vitamin D binding to its cytosolic receptor for translocation to the nucleus, where the vitamin D–receptor complex, functioning as a transactivation factor, undergoes sequence-specific interactions with the vitamin D–responsive element (VDRE) in the promoter of the gene (DeLuca, 1988; Minghetti and Norman, 1989; Suda *et al.*, 1990). Recent studies on the expression of the vitamin D–receptor gene (Baker *et al.*, 1988; Burmester *et al.*, 1988; McDonnell *et al.*, 1987) and a potential requirement of phosphorylation for activity (Brown and DeLuca, 1990) are increasing our understanding of molecular mechanisms associated with vitamin D–mediated control of bone-cell gene expression at the level of vitamin D–receptor regulation in a manner analogous to phosphorylation-mediated activation of other transcription factors.

To address molecular mechanisms associated with vitamin D–mediated regulation of gene expression at the onset of extracellular-matrix mineralization, two types of control must be considered. First, it is necessary to address the basal expression, which may, in part, be regulated by vitamin D (Markose *et al.*, 1990; Owen *et al.*, 1991; Schüle *et al.*, 1990a). This may explain reports of both stimulation and inhibition of vitamin D–responsive genes under different biological conditions. For example, collagen regulation by vitamin D is variable (dependent on the cell origin) (Franceschi *et al.*, 1988) or *in vivo* (Hock *et al.*, 1986) versus *in vitro* (Harrison *et al.*, 1989), and alkaline phosphatase is influenced by vitamin D as a function of cell density (Majeska *et al.*, 1985) or basal level (Speiss *et al.*, 1986). Throughout the normal diploid osteoblast developmental sequence, pleiotropic effects of vitamin D are found, dependent upon the differentiated state of the cell. Fig. 9 shows inhibition of proliferation, collagen, and alkaline phosphatase expression when their basal levels are maximal, but stimulation of these parameters in mature osteocytes in the mineralized matrix when basal levels are very low. A second control is the more direct action of vitamin D, which is an enhancement rather than an activation of gene transcription.

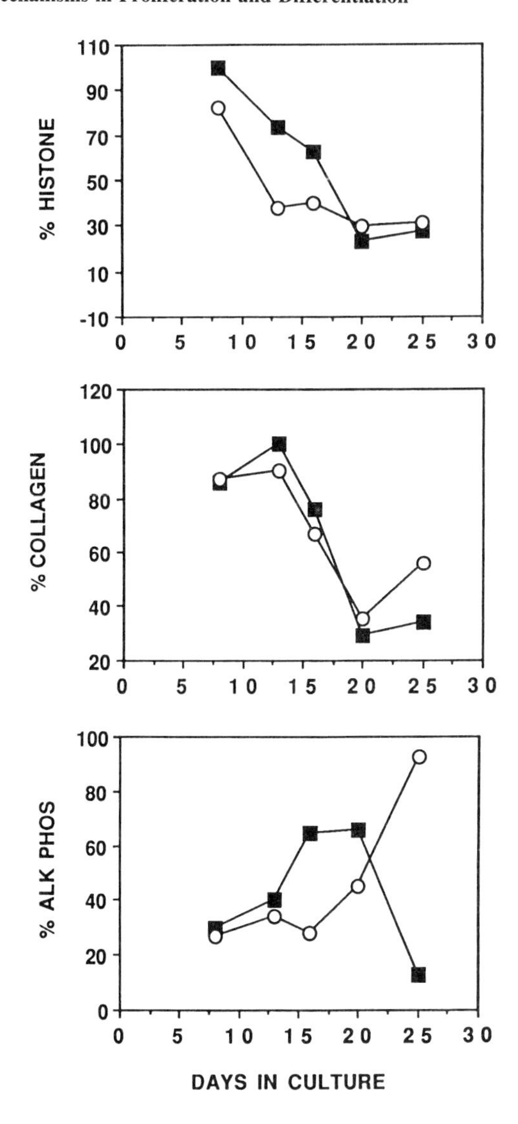

Fig. 9. Pleiotropic effects of vitamin D on cell proliferation, collagen and alkaline phosphatase expression as a function of osteoblast differentiation. At the selected times indicated throughout the osteoblast developmental sequence, $10^{-8}M$ 1,25-$(OH)_2D_3$ was added for 24 hr before harvest of the cell layer for preparation of total cellular RNA. The mRNAs for H4 histone, collagen and alkaline phosphatase (ALK PHOS) were determined from Northern or slot blot analyses for control and vitamin D–treated samples. The data for each gene are represented as percentage maximal expression for the control and vitamin D–treated samples. Note the inhibition of H4, collagen, and ALK PHOS at their peak periods of mRNA expression, but stimulation by hormone in mature cells (Day 25) for collagen and ALK PHOS.

2. *Osteocalcin Gene Expression Is Regulated at Multiple Levels*

Osteocalcin (OC) is the best-studied example of a vitamin D–responsive bone-specific protein expressed at the onset of extracellular-matrix mineralization. The OC gene (Fig. 10) encodes a 10-kD polypeptide, the OC precursor, which is processed intracellularly to a 6000-molecular-weight protein that represents one of the most abundant non-collagenous proteins accumulated in bone tissue (Gundberg *et al.*, 1984; Hauschka *et al.*, 1989). OC is a protein characterized by the presence of three γ-carboxyglutamic acid residues (resulting from the vitamin K–dependent carboxylation of glutamic acid residues 17, 21, and 24), which are necessary for the calcium-binding properties of the protein (Hauschka *et al.*, 1989). To date, OC is the only known bone-specific protein and appears to function in regulating the mineral phase of bone (Lian, 1987). Both the human

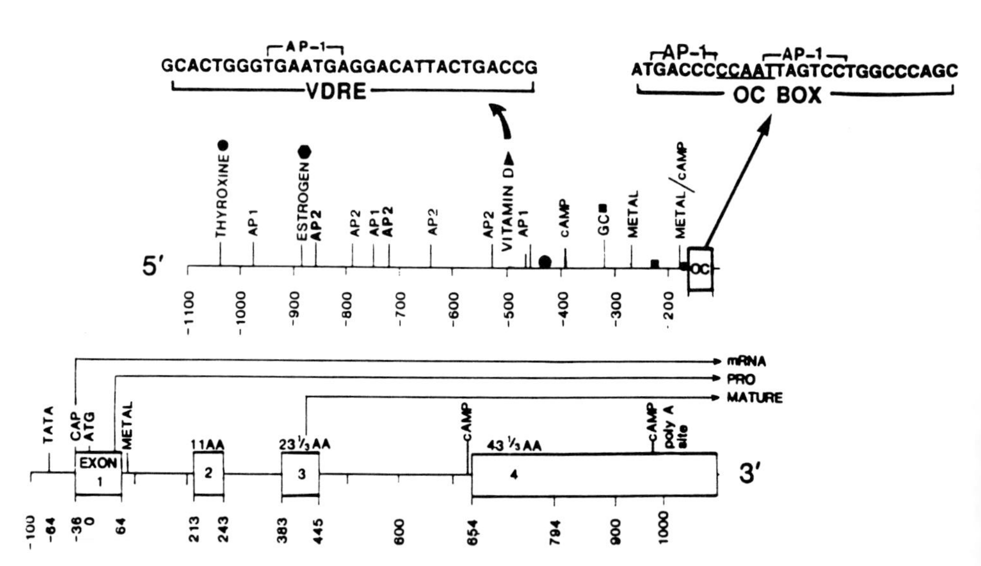

Fig. 10. Structural organization of the rat osteocalcin gene, showing intron and exon organization and location of the propeptide and mature osteocalcin. In the 1100 nucleotide, 5′ flanking sequences are indicated consensus sequences for all genes transcribed by RNA polymerase II, several steroids known to alter osteocalcin synthetic levels, and other modifiers (cAMP). Sequences of two regulatory elements that have been defined and partially characterized, the vitamin D–responsive element (VDRE) and the osteocalcin (OC) box, and the primary transcription regulatory element containing the CCAAT element as a central motif are shown. Both these regulatory elements were identified by DNase footprinting, gel retardation assays and DMS interference analysis, as described in Markose *et al.* (1990). Within these two elements are also found active AP-1 sites, which bind to the oncogene encoded Fos-Jun proteins. Since osteocalcin is not expressed in proliferating osteoblasts and cannot be induced by vitamin D, we propose these AP-1 sites serve as a regulatory mechanism to keep the gene repressed during the proliferative period until other signals activate gene transcription postproliferatively (Owen *et al.*, 1990; Lian *et al.*, 1991).

(Celeste *et al.*, 1986; Kerner *et al.*, 1989) and the rat OC genes that have been cloned (Lian *et al.*, 1989; Yoon *et al.*, 1988; Demay *et al.*, 1989) and characterized exhibit similar overall organization. Chromosome-mapping studies have assigned the human OC locus to the q region of chromosome 1 (Puchacz *et al.*, 1989).

Particularly important in addressing molecular mechanisms that contribute to the extent that the OC gene is expressed, is to account for variations in the activity and inducibility of the gene throughout the osteoblast developmental sequence, and the multiple levels of control that appear to be operative. Results summarized in Fig. 11 provide an indication of control residing at both the transcriptional and post-transcriptional levels, with the absence of expression and vitamin D regulation during the proliferation period in the osteoblast-developmental sequence. Involvement of transcriptional control is supported by a direct demonstration that OC mRNA is not synthesized in proliferating osteoblasts. The complexity of transcriptional regulation is further supported by the inability of vitamin D to induce OC gene expression before the second transition point during osteoblast differentiation, and variations in the extent to which the hormone is capable of enhancing the level of basal expression during the extracellular-matrix mineralization period (Fig. 11). This, in a more general biological context, serves as an example of differences in steroid hormone–responsive transcriptional control through binding at promoter regulatory elements as a function of the differentiated state of the cell. Not to be dismissed is the additional complexity associated with regulation at the posttranscriptional level, which is reflected by differences in the relationship between cellular mRNA levels and protein synthesis as a function of the extent to which the OC gene is expressed, implicating control of mRNA stability, translatability, or processing and secretion of OC.

The regulation of OC gene expression at multiple levels is further illustrated by the rapid induction (1 hr) of OC-gene transcription (Fig. 12A) in ROS 17/2.8 cells after $10^{-8}M$ $1,25(OH)_2D_3$ treatment. Cellular mRNA levels do not increase in a parallel manner with transcription (Fig. 12B); however, at 24 hr, increased accumulation of OC mRNA is clearly shown. The apparent inability to accumulate OC mRNA immediately following up-regulation of OC-gene transcription by vitamin D is reflected by the absence of increased synthesis of OC protein until 3 to 6 hr following hormone stimulation.

3. *Transcriptional Control of the Osteocalcin Gene*

Focusing on transcriptional regulation of OC gene expression, the VDRE of the OC gene was the first to be identified and characterized (Markose *et al.*, 1990; Kerner *et al.*, 1989; Demay *et al.*, 1990; Morrison *et al.*, 1989) (Fig. 11–13). However, to understand activation and modulation of the extent to which the gene

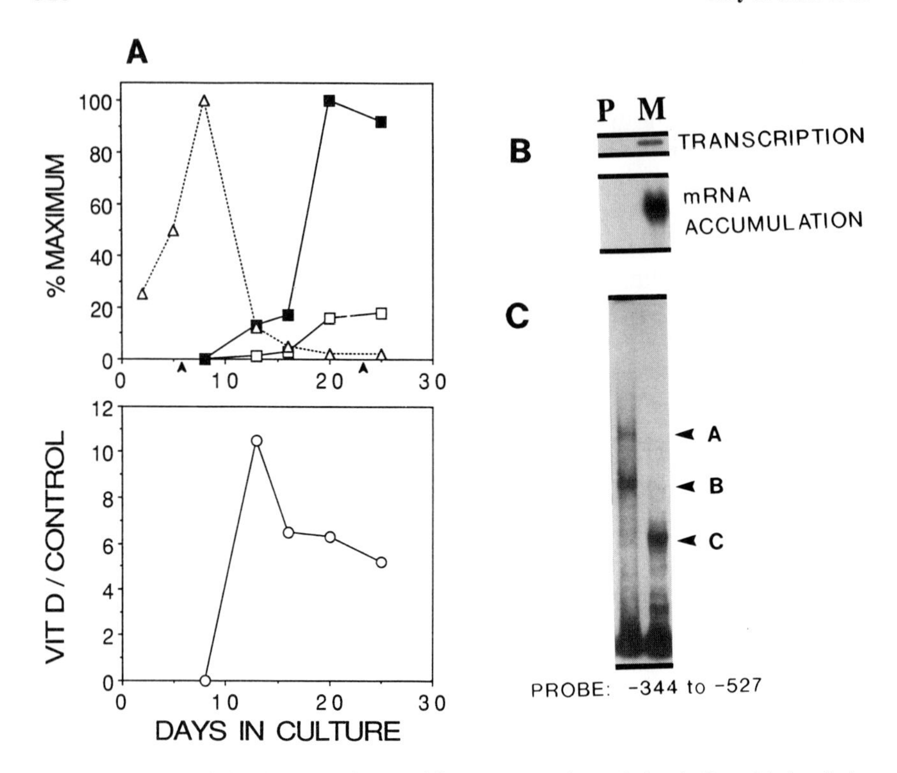

Fig. 11. Transcriptional control of osteocalcin gene expression and vitamin D modulation during the osteoblast developmental sequence. (A) Vitamin D stimulation of osteocalcin mRNA and synthesis in normal rat osteoblasts. mRNA levels were determined by Northern blot analysis. The percentage maximal values as plotted for H4 histone mRNA, an indication of proliferative activity (Δ) and percentage maximal values for OC mRNA (■,□,). Control (□) and $10^{-8}M$ $1,25(OH)_2D_3$ (■) 24-hr treated cultures before the day cells were harvested for preparation of total cellular RNA and electrophoretically fractionated, then hybridized to osteocalcin or histone H4 (Δ), which indicates the period of proliferation. (B). Osteocalcin gene transcription and mRNA levels in proliferating (P) Day 8 diploid rat osteoblasts and from cells harvested from a mineralized matrix (M) at Day 28. Transcription was assayed in isolated nuclei (run-on assays, Owen *et al.*, 1990c) and cellular mRNA levels were determined by Northern blot analysis. Note the absence of expression in proliferating osteoblasts. (C) Protein–DNA interactions in the osteocalcin promoter. A segment of the osteocalcin promoter that includes the VDRE and several AP-1 sites was used as probe for assaying the binding of proteins in nuclear extracts from P and M cells. Interactions are evident with nuclear extracts from proliferating osteoblasts reflecting (bands A and B) binding activity with negative regulatory elements that do not exhibit protein–DNA interactions in cells in mineralized cultures when the gene is transcribed. Band C clearly reflects a specific protein–DNA complex formed only when the gene is transcribed.

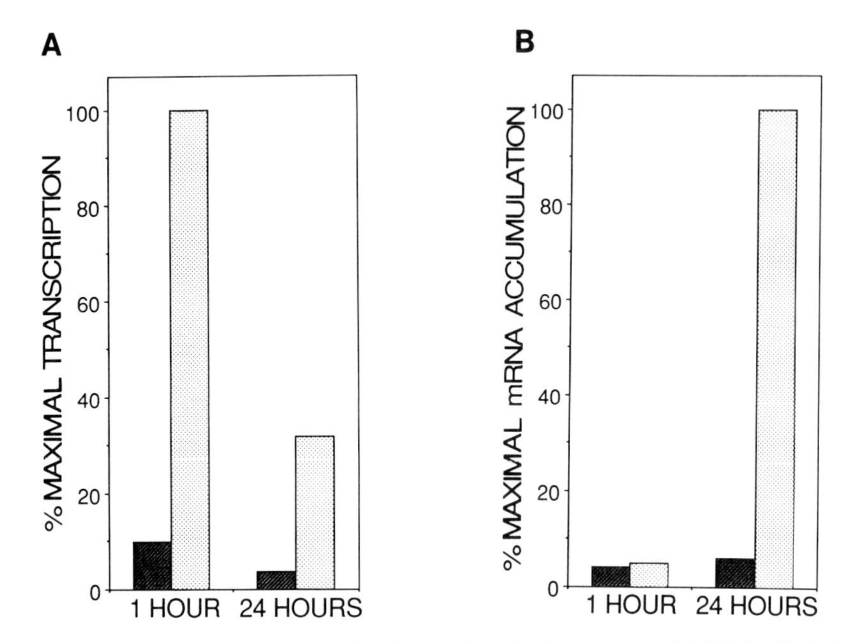

Fig. 12. Osteocalcin transcription and mRNA levels in vitamin D–treated ROS 17/2.8 cells. Cells were treated for 1 or 24 hr with $10^{-8}M$ 1,25-$(OH)_2D_3$ (light grey bar) and compared to control cells (dark hatched bar). Cells were harvested for preparation of either total cellular RNA for OC mRNA determination by Northern blot analysis or nuclei for transcription run-on assays. Note the 10-fold increase in transcription occurring at 1 hr in Panel A, whereas an increase in mRNA does not appear until the 24-hr point (Panel B).

is transcribed, it should be appreciated that the OC-gene promoter has a modular organization consisting of both positive and negative regulatory elements that are responsive *in vivo*, as well as in cultured osteoblasts, to a series of physiological mediators of OC-gene expression, including vitamin D, estrogen, glucocorticoids, retinoic acid, cyclic nucleotides, and γ-interferon (Nanes *et al.*, 1990; Lian *et al.*, 1989; Markose *et al.*, 1990; Kerner *et al.*, 1989; Demay *et al.*, 1990; Morrison *et al.*, 1989; Schüle *et al.*, 1990a). Undoubtedly it will be by understanding mechanisms for the simultaneous occupancy of a series of these regulatory elements that the synergistic and opposing actions of factors that mediate OC-gene expression will further defined. An indication of a mechanism by which vitamin D may modulate basal levels of OC-gene transcription via regulatory elements, in addition to the VDRE, is the hormone-induced changes in protein–DNA interactions in the region of the CCAAT box, only when basal levels are low.

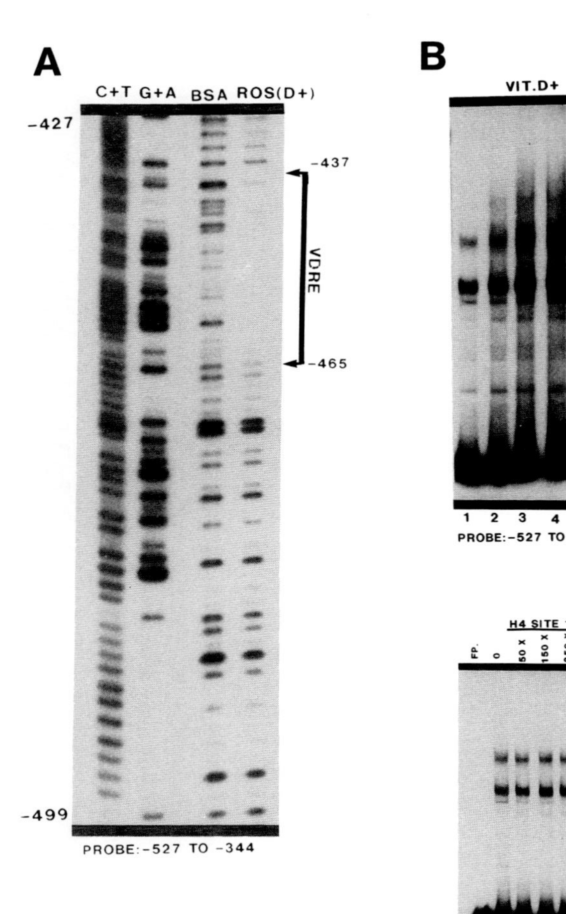

A
C+T G+A BSA ROS(D+)
-427
-437
VDRE
-465
-499
PROBE:-527 TO -344
B
VIT.D+
VIT.D-
1 2 3 4 5 6 7 8 9 10
PROBE:-527 TO -344
H4 SITE 1
VDRE
SP1
FP.
0
50 X
150 X
250 X
0
50 X
150 X
250 X
0
50 X
150 X
250 X
1 2 3 4 5 6 7 8 9 10 11 12 13
PROBE:-527 TO -344

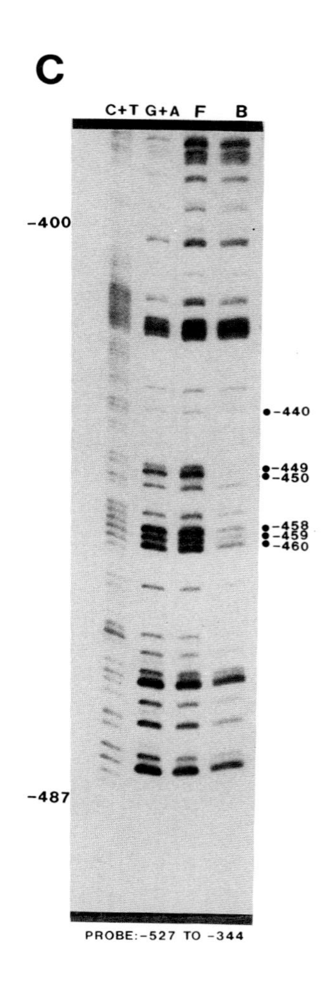

C
C+T G+A F B
-400
-440
-449
-450
-458
-459
-460
-487
PROBE:-527 TO -344

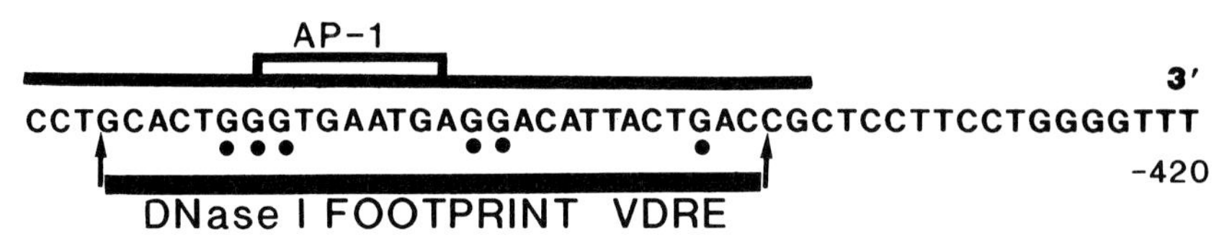

Fig. 13. Identification of the vitamin D–responsive element of the osteocalcin gene. (A) DNase I footprint analysis of protein–DNA interactions in the −527 to −344 promoter region of the rat osteocalcin gene. C + T (lane 1) and G + A (lane 2) sequencing reactions representing the coding strand were electrophoresed along with the DNase I footprint reactions. The control lane (lane 3) shows the DNase I digestion pattern of the probe incubated with 3 μg of bovine serum albumin. The pattern of digestion obtained with 13 μg of nuclear extract from 1,25$(OH)_2D_3$-treated ROS 17/2.8 cells (lane 4) shows a specific region of protection indicated by solid line and designated VDRE. (B) Vitamin D–dependent protein–DNA interactions and competition studies in two promoter elements of the osteocalcin gene, the VDRE and the OC box. (Top) Nuclear extracts from 24-hr control (vitamin D−) and $10^{-8}M$ 1,25$(OH)_2D_3$-treated ROS 17/2.8 (vitamin D+) cells were incubated with 32P-labeled segments of the osteocalcin gene promoter (−527 to −344), which included the VDRE, and analyzed by gel mobility shift assays for factors that bound to these regions. Specific protein–DNA interactions (bands) occur in vitamin D–treated cells that were absent from nuclear extracts of control cells. (Lower) Competition gel mobility shift analysis for establishing of the specificity of protein–DNA complexes. The vitamin D–dependent protein–DNA interactions can be specifically competed out using the 22-oligonucleotide vitamin D–responsive element as a competitor probe. The three lanes for each competitor represent 50-, 150-, and 250-fold excess of oligonucleotide. (C) Methylation interference analysis was performed to establish the contact sites at single nucleotide resolution of the vitamin D–induced factor. The ^{32}P-end–labeled probe (−527 to −344; pOC3.4) was partially methylated with dimethyl sulfate and used in DNA-binding reactions. Nuclear extracts from vitamin D–treated ROS 17/2.8 cells were added, and the DNA–protein complexes were resolved by native gel electrophoresis. DNA from the protein–DNA complex, indicated as V in A, and from free probe, were eluted and cleaved with piperidine. An equal number of counts of free (lane 3) and bound (lane 4) DNA were electrophoresed on 8% polyacrylamide denaturing gels. Sequencing reactions C + T (lane F) and G + A (lane B) of the coding strand were electrophoresed alongside. The G residues that interfere with the binding of the nuclear proteins are indicated by solid circles. (D) The vitamin D–responsive element, defined by the DNase I and DMS analyses. The DNase I footprint is indicated by a solid bar, and G residues exhibiting protein–DNA interactions are designated by solid circles.

4. Three-Dimensional Organization of the Osteocalcin Gene Promoter

Further insight into molecular mechanisms operative in regulating the extent to which the OC gene is transcribed can be provided by an understanding of the structural properties of the promoter. Here, the 5′ regulatory region of the gene serves as a basis for explaining observed synergistic contributions of activities at regulatory elements that are physically separated, supporting control of OC gene transcription by cooperative interactions between a complex series of sequence-specific promoter binding factors and their cognate regulatory elements. Such interactive signals between the basal transcriptional regulatory element, the OC-CCAAT containing box and the VDRE appear to modulate OC gene transcription.

We have shown specific vitamin D–dependent, vitamin D–receptor antibody sensitive protein–DNA interactions at both the VDRE and OC box using nuclear extracts from both ROS 17/2.8 cells (Markose, *et al.,* 1990) and normal diploid osteoblast cultures (Bortell, *et al.,* in press; Lian, *et al.,* in press). One must therefore consider mechanisms (1) to potentiate such interactions and (2) to account for the transcription of specific genes which are selectively inititated with a limited representation of the regulatory factors and regulatory elements.

The answers may in part reside with transcriptional regulation occurring within the three-dimensional context of nuclear architecture where the nuclear matrix, the network of polymorphic anastomosing filaments within the nucleus (Fey and Penman, 1988; Capco *et al.*, 1982; Barrack and Coffey, 1983), is potentially operative in the concentration and localization of promoter regulatory elements as well as sequence-specific transcription factors. These dominant structural components of the nucleus may relate to a number of lines of evidence that indicate transcriptional activity is associated with modifications in chromatin structure as reflected by nucleosome organization, the extent of DNA methylation, and representation of DNase I hypersensitive sites.

Participation of the nuclear matrix in transcriptional control of the progressive expression of cell growth and tissue-specific genes during osteoblast phenotype development is provided by several experimental observations (see Chapter 6 for an in-depth description). First, changes in the protein composition of the nuclear matrix by two-dimensional electrophoretic analysis (Dworetzky, *et al.,* 1990) parallels the sequential expression of genes during the three principal periods of the osteoblast developmental sequence. Secondly, there is evidence for binding of genes to the nuclear matrix only during periods of active transcription, the histone gene in proliferating osteoblasts and the OC gene postproliferatively (Dworetzky, *et al.,* submitted).

Related to the OC gene, our studies indicated the presence of nucleosomes (Bortell, *et al.,* in press), each encompassing approximately 180 bp's within the promoter sequence spanning the VDRE and the OC box region. This reduces the potential distance between the vitamin D receptors bound at the two steroid recognition sequences, thereby raising the possibility of functional cooperativity between vitamin D receptors bound at both sites resulting in an enhancement of transcriptional activity (Stein, *et al.,* in press) and schematically illustrated in Fig. 14. Indeed, another example of cooperativity between receptor–promoter binding sequences is evidence for synergism of remote glucocorticoid responsive elements (Grange, *et al.,* 1989). Alternatively, the steroid receptor binding motif in the OC box may be utilized as a low affinity site for interaction with the vitamin

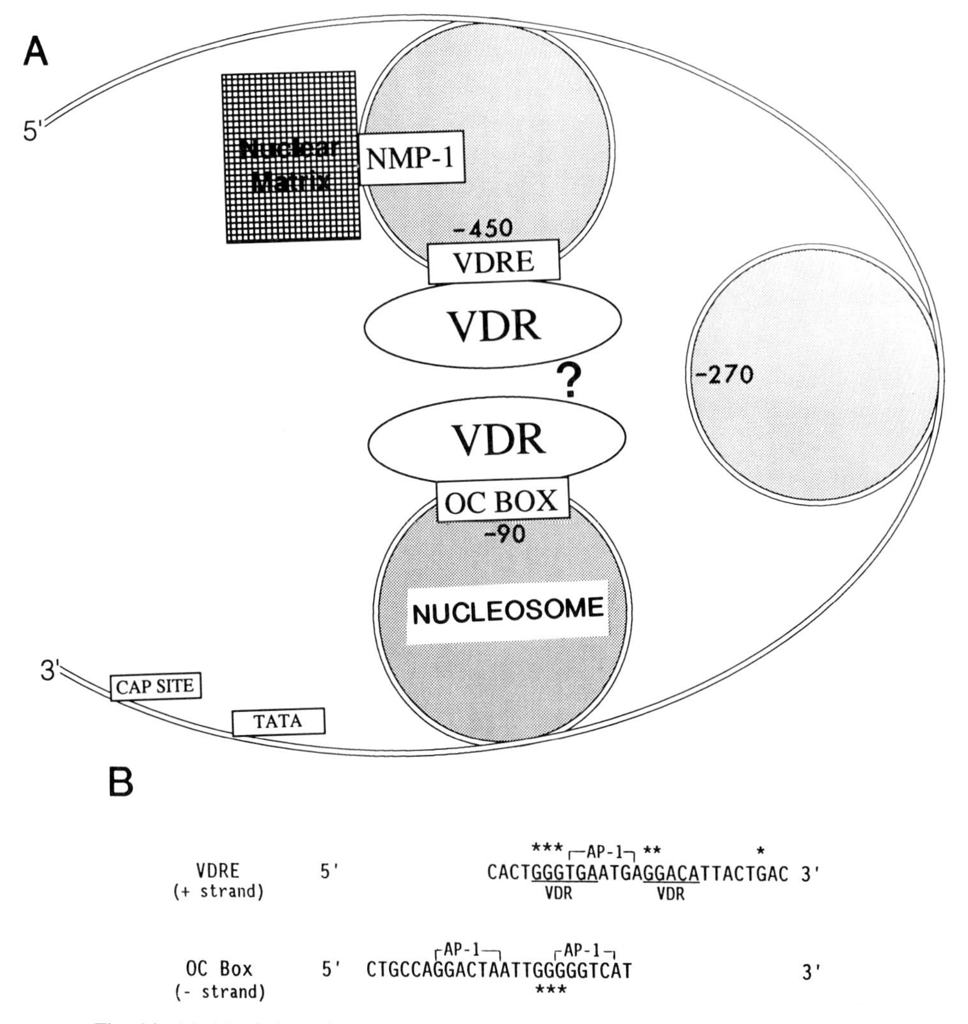

Fig. 14 Model of three-dimensional organization of the osteocalcin gene promoter. (A) The proposed model accounts for cooperative interactions between two regulatory elements in the osteocalcin gene promoter, the OC CAAT-containing box and the vitamin D–responsive element (VDRE), for precisely controlling levels of basal transcription and vitamin D modulation during development of the osteoblast phenotype. Experimental evidence for close proximity of the VDRE and the OC box includes (1) vitamin D–dependent and receptor antibody sensitive protein–DNA interactions at both the VDRE and OC box, (2) the presence of 3 nucleosomes each 180 nucleotides as detected by micrococcal nuclease digestion in the 5′ promoter of the OC gene spanning the regions of the OC box and the VDRE, thereby reducing distance between these elements, and (3) a nuclear matrix attachment consensus sequence (NMP-1) associated with the VDRE. Nuclear matrix attachment occurs when the gene is actively transcribed. A question mark signifies several possibilities for receptor interactions at the two regulatory elements, including dimerization of the vitamin D receptor or possible interaction between two receptor molecules mediated by a putative binding protein. In this model, two vitamin D–receptor molecules are each interacting with steroid receptor binding elements residing in the VDRE and OC box. These sequences are underlined in (B). Alignment of the vitamin D–receptor binding sites in the VDRE and OC box are based on homology between vitamin D-inducible protein contacts at G residues (designated by asterisks) established by dimethyl sulfate interference (Markose, *et al.,* 1990). AP-1 sites were defined by binding of recombinant Fos and Jun proteins followed by dimethyl sulfate interference analysis (Owen, *et al.,* 1990c).

D–receptor complex. Then, from the proximity of vitamin D–receptors bound at both the OC box and the VDRE, resulting from the nucleosome-associated chromatin organization of the OC gene-proximal promoter sequences, an elevated concentration and hence increased potential for interactions that account for the vitamin D enhancement of OC gene transcription may be established.

Equally important is consideration of the possibility that gene-nuclear matrix-association, together with structural properties of chromatin, may be functionally related to other protein-DNA interactions at the VDRE and OC box that influence transcriptional activity of the OC gene, possibility related to cooperativity between the two regulatory elements. The coordinate occupancy of AP-1 sites within the VDRE and OC box by the nuclear proto-oncogene-encoded fos and jun proteins which overlap hormone receptor binding domains is such an example and one where suppression appears to be the resulting activity (Owen, *et al.,* 1990c; Lian, *et al.,* 1991). To understand activation and modulation of the extent to which the gene is transcribed, the regulatory elements, together with the synergistic and opposing actions of factors that occupy these elements and mediate OC gene expression must be further defined. Here the complexity provides a basis for diversity and flexibility in the integration of signaling mechanisms that are operative in accommodating requirements for the level at which specific genes are expressed as differentiation progresses.

II. PHENOTYPE SUPPRESSION: A POSTULATED MECHANISM FOR MODULATING THE RELATIONSHIP OF PROLIFERATION AND DIFFERENTIATION BY FOS–JUN INTERACTIONS AT AP-1 SITES IN STEROID RESPONSIVE AND OTHER PROMOTER ELEMENTS

A. Developmental Activation and Suppression of Transcription

The general reciprocal relationship has been described between cell growth and the sequential expression of genes encoding proteins associated with extracellular-matrix maturation, organization, and mineralization that occurs immediately following the completion of proliferative activity. In addition, the more specific possibility should be addressed that genes expressed during proliferation may result in direct suppression of other genes that are expressed only postproliferatively at a series of stages throughout the development of the differentiated osteoblast phenotype.

Since transcriptional control of cell-growth and tissue-specific genes is an important component of the regulated development of osteoblast phenotype, a key consideration must be the mechanisms by which genes are selectively and sequentially rendered transcribable at specific stages during osteoblast differentiation. Two points must therefore be addressed to understand transcriptional control of osteoblast differentiation: (1) the identification and characterization of transcriptional regulatory elements and the promoter-binding factors that interact in a sequence-specific manner to modulate specificity and level of transcription; and (2) the mechanisms by which regulatory elements are rendered competent to bind cognate transcription factors at specific times during the differentiation process.

Several genes expressed selectively at various stages of the osteoblast developmental sequence are transcriptionally modulated by the steroid hormone vitamin D, and these genes include Type I collagen, expressed during proliferation and postproliferatively (Owen *et al.*, 1990a; Harrison *et al.*, 1989); alkaline phosphatase (Majeska *et al.*, 1985), expressed during the periods of extracellular-matrix maturation and organization; osteopontin and OC, expressed during extracellular-matrix mineralization (Owen *et al.*, 1990a; Lian *et al.*, 1989; Oldberg *et al.*, 1989; Prince and Butler, 1987). Moreover, the biological relevance of such transcriptional modulation by vitamin D is supported by the well-documented role of vitamin D as a physiologic mediator of expression of these genes *in vivo* and *in vitro* (reviewed in DeLuca, 1988; Stein *et al.*, in press; Lian *et al.*, in press; Stein *et al.*, 1990; Minghetti and Norman, 1989; Suda *et al.*, 1990). Thus, the mechanism by which the VDREs of these developmentally expressed genes exhibit occupancy by the vitamin D–receptor complex is essential to our understanding of transcriptional regulation during osteoblast differentiation.

B. Fos–Jun Suppression of Osteocalcin Transcription

An indication of a potential molecular mechanism for mediating a relationship between proliferation and transcription of a tissue-specific gene was provided several years ago, when the sequence of the bone-specific OC gene promoter was reported (Lian *et al.*, 1989; Yoon *et al.*, 1987; Celeste *et al.*, 1986; Kerner *et al.*, 1989; Demay *et al.*, 1989) and a series of AP-1 consensus sequences was identified in the 5′ regulatory region (Lian *et al.*, 1989; Markose *et al.*, 1990) (Fig. 10). The protooncogene-encoded Fos and Jun proteins form a stable heterodimeric complex via a leucine zipper that interacts in a sequence-specific manner with AP-1 sites. The presence of AP-1 consensus sequences in the OC gene promoter presented the possibility that Fos and Jun proteins in proliferating osteoblasts could suppress OC gene transcription until late in the development of the bone-cell phenotype, at which time extracellular-matrix mineralization is initiated (Figs. 1 and 2). Such a line of reasoning has recently been supported by four experimental results:

1. Identification of a vitamin D–responsive element (VDRE) in the osteocalcin gene promoter (nt −462 to −439 for the rat gene, and nt −513 to −493 for the human gene) by deletion mutagenesis (Kerner *et al.*, 1989; Demay *et al.*, 1989; Morrison *et al.*, 1989) and by direct determination of protein-DNA interactions in

the 5′ regulatory sequences (Markose *et al.*, 1990) (Fig. 13) indicated that an AP-1 consensus sequence resides within this regulatory element, which mediates vitamin D enhancement of expression (Fig. 15). Additionally, an AP-1 consensus sequence was identified in the osteocalcin box (nt −76 to −99 for the rat gene, and nt −123 to −100 for the human gene) (Fig. 15), which contains a CCAAT motif as a central element and influences tissue-specific basal levels of OC gene transcription. These results are consistent with a model in which coordinate occupancy of the AP-1 sites in the VDRE and OC box in proliferating osteoblasts may suppress both basal level and vitamin D–enhanced OC gene transcription, a phenomenon described as phenotype suppression (Fig. 16). A similar organization of the VDRE and OC box for the human and rat OC gene suggests that the functional properties of the elements with respect to the relationship of Jun–Fos interactions to vitamin D–receptor binding are conserved.

2. Sequence-specific binding of the Jun–Fos complex to the AP-1 sites within the VDRE and OC box of the OC gene promoter has been demonstrated at single-nucleotide resolution (Fig. 15) by protein–DNA interactions using purified recombinant fos and jun proteins together with competition and Dimethyl Sulfate (DMS) fingerprint analysis (Owen *et al.*, 1990c).

3. Expression of c-*fos* and c-*jun* have been shown to occur primarily during the proliferative period of the osteoblast developmental sequence (Owen *et al.*, 1990a; Shalhoub *et al.*, 1989) (Figs. 1 and 2). Similarly, electrophoretic mobility-shift analysis has indicated that AP-1 binding activity is observed primarily in proliferating osteoblasts, and dramatically decreases after the down-regulation of proliferation and the initiation of extracellular-matrix maturation and mineralization, at which time (coincident with) OC gene transcription is initiated (Owen *et al.*, 1990c) (Fig. 17).

4. Experiments in which transfection of c-*fos* and c-*jun* into cells expressing OC results in the down-regulation of OC gene transcription further support a Fos–Jun mediated suppression of the osteocalcin gene (Schüle *et al.*, 1990a).

C. Fos–Jun and Vitamin D–Mediated Transcription of Alkaline Phosphatase and Collagen

A question that must always be addressed with respect to the validity of a model proposed to explain transcriptional regulation of a gene is the extent to which the hypothesis is applicable to other genes expressed under similar or related biological circumstances. Further support for the biological relevance of postulating that Fos–Jun binding to an AP-1 site within a VDRE can suppress expression of genes normally expressed following down-regulation of proliferation comes from the organization of the alkaline phosphatase gene (Fig. 17). Similar to the observed

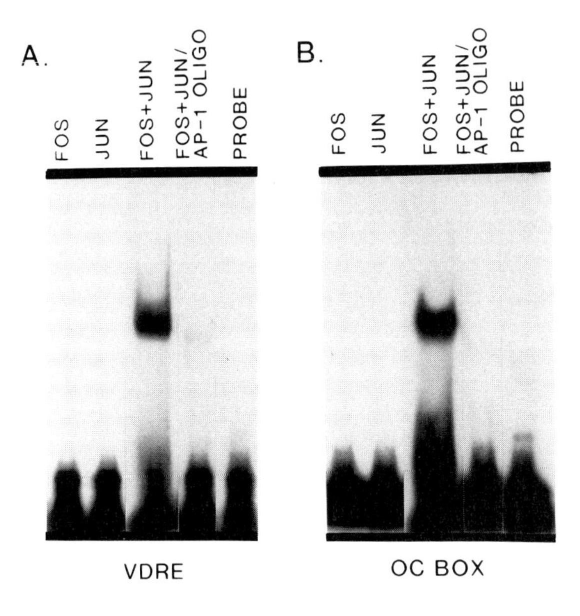

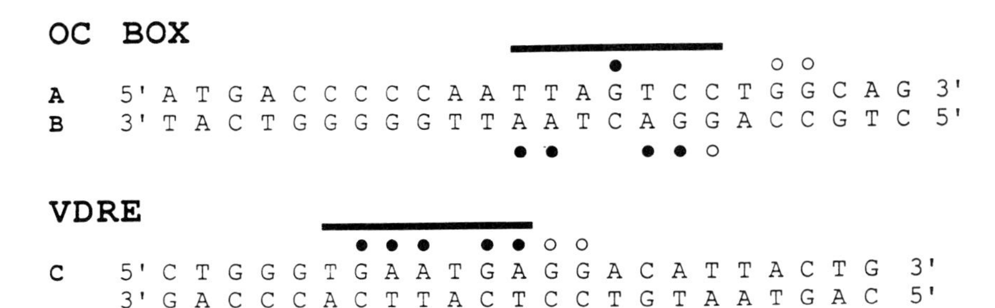

Fig. 15. Binding of purified recombinant Fos and Jun proteins to rat osteocalcin VDRE (A), OC box (B) probes as assessed by gel retardation assay. For each probe, 1 μM protein of either Fos alone (FOS), Jun alone (JUN), or Fos–Jun together (FOS + JUN) were incubated at 37°C for 15 min and 4μg of ^{32}P-labeled probe was added. Binding was allowed to occur for 15 min at room temperature. Protein–DNA complexes were resolved on 4.5% native polyacrylamide gels. For each probe, the binding of the Fos–Jun complex was specifically competed with an oligonucleotide spanning the human MT II_A gene AP-1 site. Residue contacts of purified recombinant Fos–Jun complex within the rat OC gene, OC box, and VDRE are identified by methylation interference analysis and are designated by the closed circle in the lower panel.

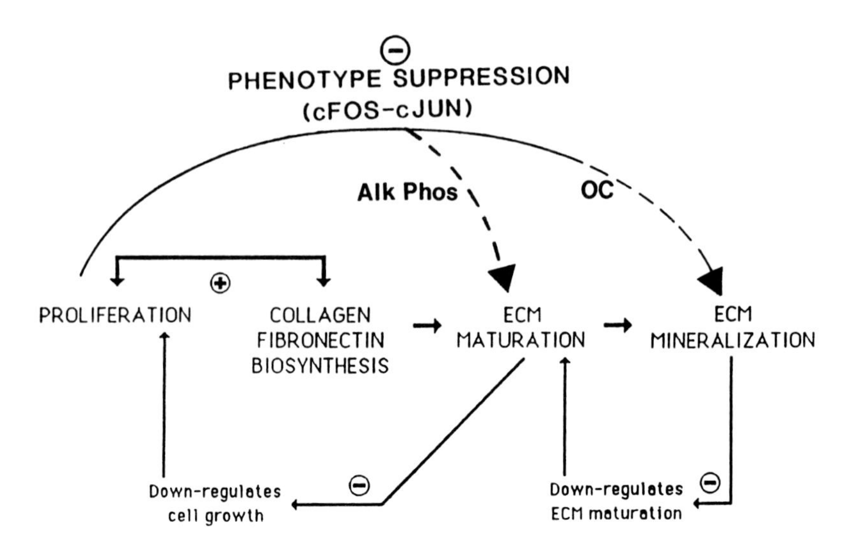

Fig. 16. A model for suppression of genes associated with expression of the mature osteoblast phenotype in actively proliferating cells and other signaling mechanisms that regulate osteoblast differentiation. The proliferation period supports the synthesis of a Type I collagen–fibronectin extracellular matrix, which continues to mature and mineralize. The formation of this matrix down-regulates proliferation, and matrix mineralization down-regulates the expression of genes associated with the matrix-maturation period. The occupancy of an AP-1 site in the vitamin D–responsive element of the alkaline phosphatase gene promoter by Fos–Jun and/or related proteins suppresses alkaline phosphatase transcription during proliferation. Similarly, occupancy of the AP-1 sites in the osteocalcin box (CCAAT-containing proximal promoter element) and the vitamin D–responsive element of the osteocalcin gene by Fos–Jun and/or related proteins suppresses the basal and vitamin D–enhanced expression of the osteocalcin gene before the initiation of osteocalcin basal expression at the onset of extracellular-matrix mineralization.

Fos–Jun interactions with the OC gene VDRE, an AP-1 motif within a putative VDRE sequence of the alkaline phosphatase gene, which is also expressed following proliferation in osteoblasts undergoing differentiation, binds the Fos–Jun complex (Owen *et al.*, 1990c). An analogous mechanism for suppression of both the alkaline phosphatase and the OC gene is particularly interesting since, while these genes are both expressed after the completion of proliferation in normal diploid osteoblasts, they are expressed sequentially; alkaline phosphatase is expressed and induced to high levels immediately following the down-regulation of cell growth, while OC is expressed only in the mature osteoblast when extracellular-matrix mineralization occurs.

Vitamin D responsiveness of the Type I collagen gene, which is actively expressed in proliferating osteoblasts, providing transcripts to support biosynthesis of the principal component of the bone-cell extracellular matrix, has been well documented. Is this incompatible with the postulated involvement of Fos–Jun and AP-1 interactions in the control of vitamin D–mediated transcriptional regulation? The sequence of the rat Type I collagen gene has recently been published by Lichtler *et al.* (1989); an examination of the promoter has revealed a VDRE sequence analogous to that found in the osteocalcin and alkaline phosphatase 5′ regulatory regions, but with an AP-1 consensus sequence contiguous to and not within the VDRE (Fig. 17) (Owen *et al.*, 1990c). Sequence-specific binding of the Fos–Jun complex to the AP-1 sites associated with the collagen promoter VDRE (Owen *et al.*, 1990c) indicates that subtle variations in the organization of the VDRE and AP-1 motifs in the osteocalcin and alkaline phosphatase genes compared to that in the Type I collagen gene promoter may contribute to their differential expression during the osteoblast developmental sequence. One other

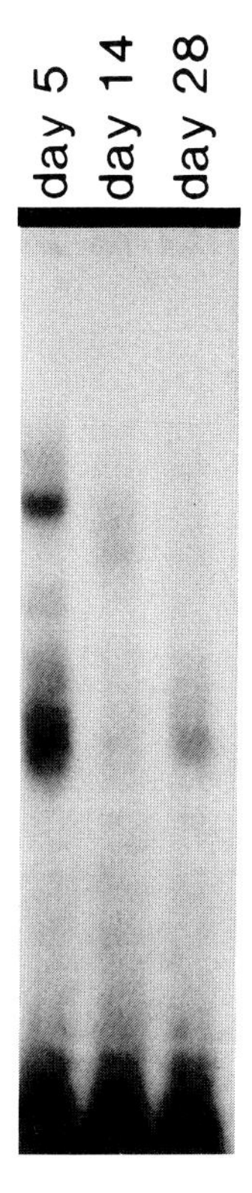

Fig. 17. AP-1 activity during the osteoblast developmental sequence. Nuclear-factor proteins were prepared from primary osteoblasts harvested during active proliferation (Day 5), postproliferatively (Day 14) during matrix development and on Day 28, from mineralized cultures. Gel retardation assays were carried out using the AP-1 metallothionein IIa consensus sequence as probe under conditions that maximize Fos–Jun protein binding. Significant protein–DNA interactions are found in Day 4 extracts.

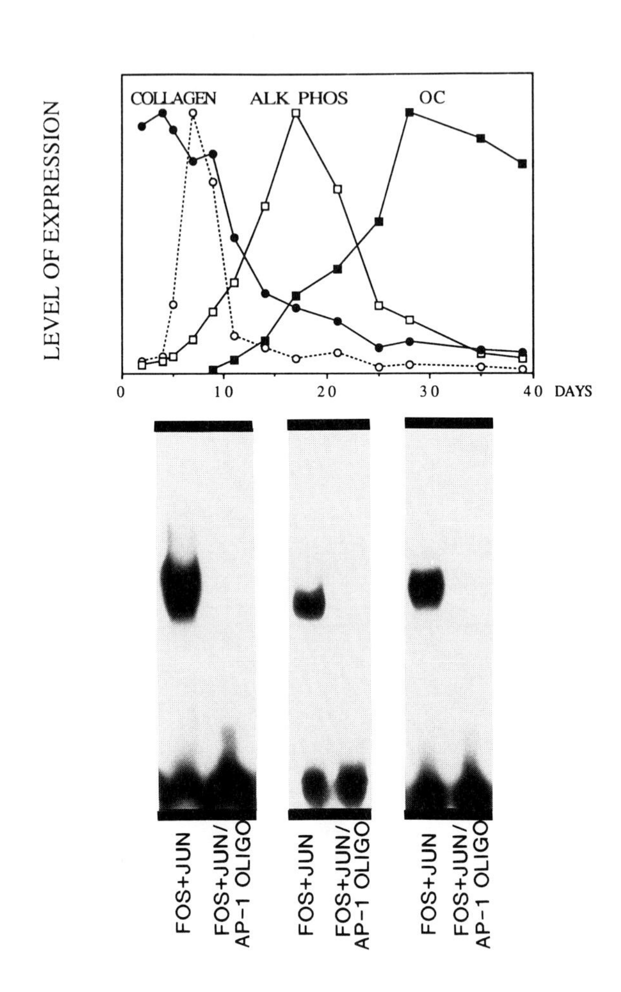

Collagen: CTGGGGGCAGAAGAACTTTCTGGAGGATTTGAGTGA

Alkaline
Phosphatase: GGGGGTGACTGATGGTAACCTGATTG

Osteocalcin: CTGGGTGAATGAGGACATTACTG

vitamin D–responsive gene, osteopontin, is expressed in both proliferating cells at low levels, and is then induced to high levels during mineralization. At both periods it is regulated by vitamin D. Of interest, recent identification of the osteopontin VDRE shows a different sequence, a 96-bp motif AGGTTCACG, compared to the 24-bp osteocalcin VDRE (Noda *et al.*, 1990).

D. A Common Mechanism for Suppression and Gene-Specific Activation of Osteoblast Genes

Yet to be addressed is the mechanism by which the OC and alkaline phosphatase genes are rendered transcribable and vitamin D–responsive at specific times during expression of the osteoblast phenotype following the down-regulation of proliferation (completion of the cell-growth period). However, the results are consistent with *a common mechanism for suppressing expression of osteoblast genes* transcribed postproliferatively by Jun–Fos binding, when the cells are actively proliferating, and *a gene specific mechanism(s) for the sequential activation of these genes* during the subsequent expression of the osteoblast phenotype. Here, the possibilities include (1) release of the Fos–Jun complex from the AP-1 sites to permit sequences to be available for occupancy by the vitamin D–receptor complex and/or by tissue-specific OC box transcription factors, or (2) modifications of the Fos–Jun complex that facilitate binding of activation–related factors. With respect to the latter possibility, binding of the Fos–Jun complex may pleiotropically play a dual positive and negative role in the regulation of transcription, suppressing transcription when proliferation is ongoing by directly or indirectly modulating sequence-specific interactions at the vitamin D–receptor binding domain, and then serving to facilitate vitamin D–receptor binding postproliferatively to facilitate the sequential up-regulation of vitamin D–responsive genes.

Fig. 18. Fos–Jun binding to AP-1 sites associated with vitamin D–responsive elements of 3 genes expressed sequentially during osteoblast differentiation, collagen (-•-), alkaline phosphatase (-□- ALK PHOS), osteocalcin (-■- OC) (top profiles). The proliferation period is indicated by histone gene expression (---O---). Lower Panel. The sequence-specific protein–DNA interactions of the oncogene encoded Fos and Jun protein complex and the VDRE sequences (used as probe) of the collagen, alkaline phosphatase and osteocalcin gene were determined. Fos–Jun binds to AP-1 sites residing within the VDREs of OC and AP, but to an AP-1 site contiguous to collagen VDRE. Sequence-specific interactions are confirmed by competition with an oligonucleotide encoding an AP-1 sequence from the human metallothionein gene promoter. We propose this organization of AP-1 sites (designated by the underline) and VDREs allows for vitamin D regulation of collagen during proliferation, but blocks hormone regulation of ALK PHOS and OC until the postproliferative period. Thus it appears that subtle variations in the organization of the VDRE and AP-1 motifs in the osteoblast gene promoters may contribute to their differential expression during osteoblast phenotype development.

Additional experimental results are necessary to determine whether the organization of steroid receptor–binding domains and AP-1 sites within promoters of genes that are hormone responsive can provide a general mechanism for the phenotype suppression and/or activation in different periods of cell and tissue differentiation. However, support for such a model is provided by the association of AP-1 sites within consensus sequences for other steroid-responsive elements, for example, glucocorticoids (Diamond *et al.*, 1990; Schüle *et al.*, 1990b). Undoubtedly the phenotype-suppression model represents a simplification of an extremely complex series of protein–DNA interactions whereby multiple physiological signals are transduced to the nucleus, resulting in modifications in transcription that progressively alter phenotypic properties of cells leading to structural and functional events associated with differentiation. However, the strength of such a model is that it provides a basis for experimentally addressing the organization of regulatory elements for phenotypic marker genes of cell differentiation within the context of genes associated with proliferation. Whether this is reminiscent of or directly reflecting a molecular parameter of the long-standing postulated association of proliferation and differentiation remains to be established.

III. DEREGULATION OF THE RELATIONSHIP BETWEEN CELL GROWTH AND TISSUE-SPECIFIC GENE EXPRESSION IN OSTEOSARCOMA CELLS

The developmental sequence associated with the differentiation of normal diploid osteoblasts, both *in vitro* and *in vivo*, supports two concepts related to the progressive development of the osteoblast phenotype. First, the proliferative period promotes collagen gene expression and initial production of a collagen extracellular matrix that subsequently supports maturation of the osteoblast phenotype. Second, a reciprocal and functionally coupled relationship exists between the down-regulation of proliferation and initiation of expression of osteoblast phenotype markers such as alkaline phosphatase and osteopontin. While there are density-dependent effects on gene expression in transformed osteoblasts and osteosarcoma cell lines (Majeska *et al.*, 1985; Speiss *et al.*, 1986), the three defined periods that characterize the developmental sequence observed in cultured normal diploid osteoblasts, with a specific pattern of gene expression and evidence for transition points where signaling mechanisms necessary for the progressive expression of the osteoblast phenotype occur, are not present (Fig. 19). In several transformed osteoblast and osteosarcoma cell lines that have been extensively examined, for example, in the ROS 17/2.8 rat osteosarcoma cell line, there is a deregulation of the sequential pattern of gene expression. This relaxation of control mechanisms (that permits sequentially expressed genes in diploid osteoblasts (Owen *et al.*, 1990a) to be expressed simultaneously) is reflected by the

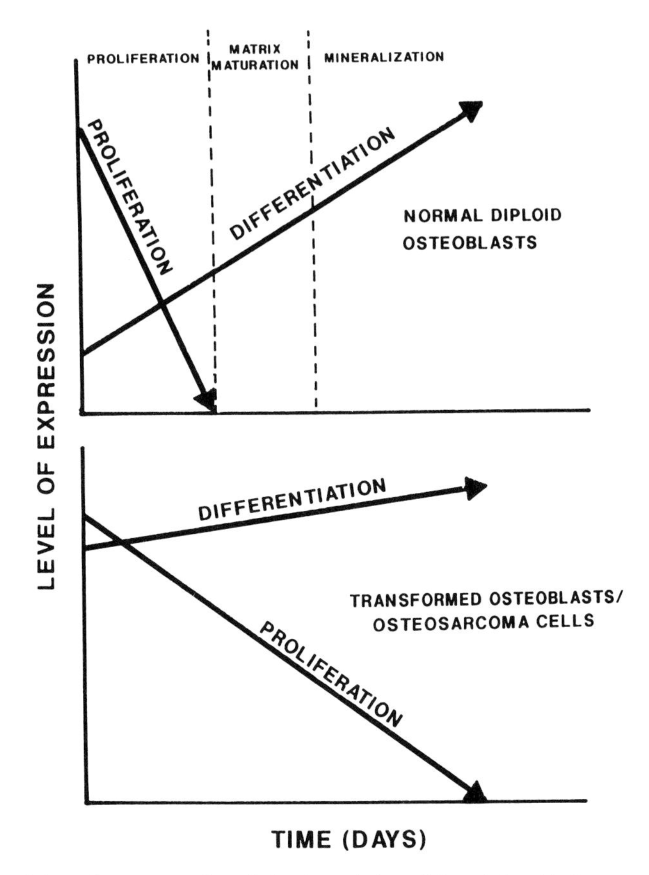

Fig. 19. Schematic representation of the deregulation of the relationship between growth and differentiation during the osteoblast developmental sequence in transformed osteoblasts or osteosarcoma cells. The proliferation vectors reflect the level of cell growth and expression of cell cycle and proliferation-related genes. The differentiation vectors reflect the expression of genes associated with the biosynthesis, maturation, and mineralization of the extracellular matrix. Top panel: In the normal diploid osteoblasts there are three distinct periods to the developmental sequence, with extracellular-matrix biosynthesis an early event occurring after the completion of proliferation. The broken vertical lines in the upper diagram indicate the two principal transition points in the developmental sequence exhibited by normal diploid osteoblasts during the progressive acquisition of the bone cell phenotype: the first at the completion of proliferation when genes associated with matrix development and maturation are up-regulated, and the second at the onset of extracellular-matrix mineralization. Lower panel: In contrast, the constitutive expression of osteoblast phenotype markers in transformed osteoblasts and osteosarcoma cells reflects the absence of these two developmentally important transition points. The relationship between cell growth and expression of genes encoding osteoblast phenotype markers is not apparent in transformed osteoblast or osteosarcoma cells; cell growth and tissue-specific gene expression occur concomitantly; thus, the relationship between growth and differentiation is deregulated.

concomitant expression of tissue-specific genes such as alkaline phosphatase, osteopontin, and osteocalcin, while the cells are actively proliferating (Rodan *et al.*, 1989; Noda and Rodan, 1987) (Fig. 19). Perturbations are implicated in the signaling mechanisms that interface the down-regulation of cell growth and induction of genes that support extracellular matrix maturation and specialization.

Consistent with such reasoning, deregulation of the cell cycle–dependent histone gene–promoter binding factor HiNF-D in osteosarcoma cells has recently been observed. In normal diploid osteoblasts, HiNF-D binding activity is cell cycle–regulated. Nuclear protein extracts prepared from these cells in S phase contain distinct and measurable levels of HiNF-D binding activity, while this activity is not detectable in G_1 phase cells (Holthuis *et al.*, 1990). In contrast, in osteosarcoma cells or transformed osteoblast cell lines, HiNF-D binding activity is constitutively elevated throughout the cell cycle and declines only with the onset of differentiation (Owen *et al.*, 1990b; Holthuis *et al.*, 1990) (Fig. 20). The change from cell cycle–mediated to constitutive interaction of HiNF-D with the promoter of a cell growth–controlled gene is consistent with, and may be functionally related to, the loss of stringent cell-growth regulation associated with neoplastic transformation.

One can speculate that the deregulation of a growth-controlled gene in transformed osteoblasts and in osteosarcoma cells may reflect modifications in the activity of tumor-suppressor genes (reviewed by Weinberg, 1989; Levine, 1990). Alternatively, or together with the loss of stringent growth control, there may be modifications in the regulatory sequences and/or in the factors that control the progressive expression of the tissue-specific genes and their response to cell growth or morphogenic regulatory factors. While the specific molecular mechanisms remain to be established, by further understanding the deregulation of growth in bone tumors, we can anticipate gaining additional insight into control of the tightly coupled relationship between proliferation and development of the osteoblast phenotype.

IV. CONCLUSIONS

Results have been presented that support a reciprocal and functionally coupled relationship between proliferation and progressive development of the osteoblast

Fig. 20. Deregulation of the cell cycle–controlled histone gene in osteosarcoma cells. Cell cycle regulation of histone gene transcription in normal diploid cells is reflected by HiNF-D–Site II interactions only during the S phase of the cell cycle. In contrast, in ROS 17/2.8 osteosarcoma cells, HiNF-D–Site II interactions are constitutive throughout the cell cycle. This abrogation of promoter factor binding activity in the histone gene promoter is schematically illustrated below. Solid boxes indicate occupancy of Sites I and II by promoter binding-factors.

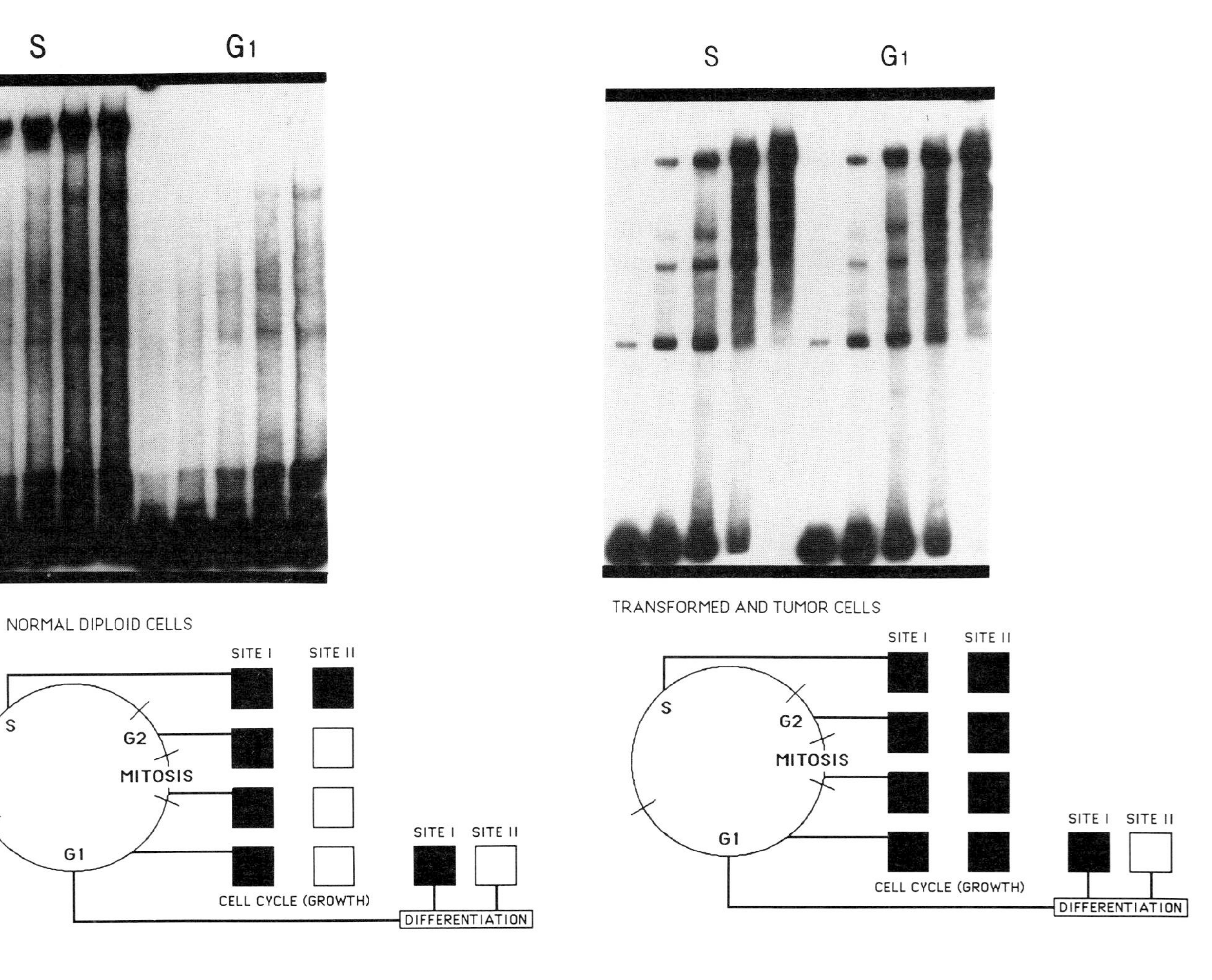
S
G1
NORMAL DIPLOID CELLS
SITE I
SITE II
S
G2
MITOSIS
G1
CELL CYCLE (GROWTH)
SITE I
SITE II
DIFFERENTIATION
S
G1
TRANSFORMED AND TUMOR CELLS
SITE I
SITE II
S
G2
MITOSIS
G1
CELL CYCLE (GROWTH)
SITE I
SITE II
DIFFERENTIATION

phenotype, mediated by the sequential expression of cell-growth and cell-cycle–regulated genes, followed by expression of genes associated with biosynthesis, maturation, and mineralization of the bone extracellular matrix. Regulation of genes during the three principal periods of the osteoblast developmental sequence, in response to physiological mediators of bone-cell growth and differentiation, are controlled at both the transcriptional and posttranscriptional levels. Complex and highly integrated signaling and gene-regulatory mechanisms are operative at two key transition points: (1) when proliferation is down-regulated, and genes associated with extracellular-matrix maturation and organization are induced; and (2) at the onset of extracellular-matrix mineralization. A model is proposed with supporting results for coordinate occupancy of AP-1 sites within steroid-responsive elements and in primary transcription-regulatory motifs by the oncogene-encoded fos and jun proteins, synthesized by actively proliferating osteoblasts and thereby suppressing both basal-level and steroid-hormone–enhanced expression of osteoblast genes transcribed postproliferatively. Interestingly, in transformed osteoblasts and in osteosarcoma cells, the loss of growth control is associated with an abrogation of the proliferation–differentiation relationship that mediates development of tissue-specific structure and function. While the complexity of regulatory mechanisms and associated signaling pathways for control of the relationship between osteoblast proliferation and differentiation is rapidly increasing, the complexity provides a basis for the convergence and integration of diverse cellular responses to a broad spectrum of physiological mediators of bone and cell growth and differentiation.

ACKNOWLEDGMENTS

Studies from the authors' laboratories described in this review were supported by grants from the National Institutes of Health (GM32010, GM32381, AR33920, AR35166, AR39122, AR39588, HD22400), the National Science Foundation (BMS 88-19989, DCB88-96116), the March of Dimes Birth Defects Foundation (1-813), the International Life Sciences Institute (Washington, D.C.), and the Northeast Osteogenesis Imperfecta Society. We thank Christine Dunshee for photographic assistance and Mary Beth Kennedy for assistance with cell cultures. The editorial assistance of Patricia Jamieson in the preparation of the manuscript is most appreciated.

REFERENCES

Aronow, M. A., Gerstenfeld, L. C., Owen, T. A., Tassinari, M. S., Stein, G. S., and Lian, J. B. (1990). Factors that promote progressive development of the osteoblast phenotype in cultured fetal rat calvaria cells. *J. Cell. Physiol.* **143**, 213–221.

Baker, A. R., McDonnell, D. P., Hughes, M., Crisp, T. M., Mangelsdorf, D. J., Haussler, M. R., Pike, J. W., Shine, J., and O'Malley, B. W. (1988). Cloning and expression of full-length cDNA encoding human vitamin D receptor. *Proc. Natl. Acad. Sci. U.S.A.* **85**, 3294–3298.

Barone, L. M., Owen, T. A., Tassinari, M. S., Bortell, R., Stein, G. S., and Lian, J. B. (1991). Developmental expression and hormonal regulation of the rat matrix Gla protein (MGP) gene in chondrogenesis and osteogenesis *J. Cell. Biochem.* (in press).

Barrack, E. R. and Coffey, D. S. (1983). Hormone receptors and the nuclear matrix. *In* "Gene Regulation by Steroid Hormones, II," (Eds., Roy, A. K. and Clark, J. H.) Springer-Verlag, pp 239–266.

Baumbach, L. L., Stein, G. S., and Stein, J. L. (1987). Regulation of human histone gene expression: Transcriptional and post-transcriptional control in the coupling of histone messenger RNA stability with DNA replication. *Biochemistry* **26**, 6178–6187.

Bellows, C. G., Aubin, J. E., Heersche, H.N.M., and Antosz, M. E. (1986). Mineralized bone nodules formed *in vitro* from enzymatically released rat calvaria cell populations. *Calcif. Tissue Int.* **38**, 143–154.

Bhargava, U., Bar-Lev, M., Bellows, C. G., and Aubin, J. E. (1988). Ultrastructural analysis of bone nodules formed *in vitro* by isolated fetal rat calvaria cells. *Bone* **9**, 155–163.

Bortell, R., Owen, T., Bidwell, J., Gavazzo, P., Breen, E., van Wijnen, A., Stein, J., Lian, J. B., and Stein, G. S. (1991). Do multiple vitamin D receptor binding elements synergistically contribute to osteocalcin gene promoter activity? (manuscript submitted).

Brown, T. A., and DeLuca, H. F. (1990). Phosphorylation of the 1,25-dihydroxyvitamin D_3 receptor: A primary event in 1,25-dihydroxyvitamin D_3 action. *J. Biol. Chem.* **265**, 10025–10029.

Bruder, S. P., and Caplan, A. I. (1990). Terminal differentiation of osteogenic cells in the embryonic chick tibia is revealed by a monoclonal antibody against osteocytes. *Bone* **11**, 189–198.

Burmester, J. K., Maeda, N., and DeLuca, H. F. (1988). Isolation and expression of rat 1,25-dihydroxyvitamin D_3 receptor cDNA. *Proc. Natl. Acad. Sci. U.S.A.* **85**, 1005–1009.

Capco, D. G., Wan, K. M., and Penman, S. (1982). The nuclear matrix: Three-dimensional architecture and protein composition. *Cell* **29**, 847–858

Celeste, A. J., Rosen, V., Buecker, J. L., Kriz, R., Wang, E. A., and Wozney, J. M. (1986). Isolation of the human gene for bone Gla protein utilizing mouse and rat cDNA clones. *EMBO J.* **5**, 1885–1890.

DeLuca, H. F. (1988). The vitamin D story: A collaborative effort of basic science and clinical medicine. *FASEB J.* **2**, 224–236.

Demay, M. B., Roth, D. A., and Kronenberg, H. M. (1989). Regions of the rat osteocalcin gene which mediate the effect of 1,25-dihydroxyvitamin D_3 on gene transcription. *J. Biol. Chem.* **264**, 2279–2282.

Demay, M. B., Gerardi, J. M., DeLuca, H. F., and Kronenberg, H. M. (1990). DNA sequences in the rat osteocalcin gene that bind the 1,25-dihydroxyvitamin D_3 receptor and confer responsiveness to 1,25-dihydroxyvitamin D_3. *Proc. Natl. Acad. Sci. U.S.A.* **87**, 369–373.

Diamond, M. I., Miner, J. N., Yoshinaga, S. K., and Yamamoto, K. R. (1990). Transcription factor interactions: Selectors of positive or negative regulation from a single DNA element. *Science* **249**, 1266–1272.

Dworetzky, S. I., Fey, E. G., Penman, S., Lian, J. B., Stein, J. L., and Stein, G. S. (1990). Progressive changes in the protein composition of the nuclear matrix during rat osteoblast differentiation. *Proc. Natl. Acad. Sci. U.S.A.* **87**, 4605–4607.

Dworetzky, S. I., Wright, K. L., Fey, E. G., Penman, S., Lian, J. B., Stein, J. L., and Stein, G. S. (1991). Sequence-specific DNA binding proteins are components of a nuclear matrix attachment site. (manuscript submitted).

Fey, E. G., and Penman, S. (1988). Nuclear matrix proteins reflect cell type of origin in cultured human cells. *Proc. Natl. Acad. Sci. U.S.A.* **85**, 121–125.

Franceschi, R. T., Romano, P. R., and Park, K. Y. (1988). Regulation of type I collagen synthesis by 1,25-dihydroxyvitamin D_3 in human osteosarcoma cells. *J. Biol. Chem.* **263**, 18938–18945.

Gerstenfeld, L. C., Chipman, S. D., Glowacki, J., and Lian, J. B. (1987). Expression of differentiated function by mineralizing cultures of chicken osteoblasts. *Dev. Biol.* **122**, 49–60.

Grange, T., Roux, J., Rigeaud, G., and Pictet, R. (1989). Two remote glucocorticoid responsive units interact cooperatively to promote glucocorticoid induction of rat tyrosine aminotransferase gene expression. *Nucl. Acids Res.* **17**, 8695–8709.

Gundberg, C. M., Hauschka, P. V., Lian, J. B., and Gallop, P. M. (1984). Osteocalcin isolation, characterization, and detection. *Methods Enzymol.* **107**, 516–544.

Harrison, J. R., Petersen, D. N., Lichtler, A. C., Mador, A. T., Rowe, D. W., and Kream, B. E. (1989). 1,25-Dihydroxyvitamin D_3 inhibits transcription of type I collagen genes in the rat osteosarcoma cell line ROS 17/2.8. *Endocrinology* **125**, 327–333.

Hauschka, P. V., Lian, J. B., Cole, D.E.C., and Gundberg, C. M. (1989). Osteocalcin and matrix Gla protein: Vitamin K-dependent proteins in bone. *Physiologic Reviews* **69**, 990–1047.

Hock, J. M., Gunness-Hey, M., Poser, J., Olson, H., Bell, N. H., and Raisz, L. G. (1986). Stimulation of undermineralized matrix formation by 1,25-dihydroxyvitamin D_3 in long bones of rats. *Calcif. Tissue Int.* **28**, 79–86.

Holthuis, J., Owen, T. A., van Wijnen, A. J., Wright, K. L., Ramsey-Ewing, A., Kennedy, M. B., Carter, R., Cosenza, S. C., Soprano, K. J., Lian, J. B., Stein, J. L., and Stein, G. S. (1990). Tumor cells exhibit deregulation of the cell-cycle histone gene-promoter factor HiNF-D. *Science* **247**, 1454–1457.

Kerner, S. A., Scott, R. A., and Pike, J. W. (1989). Sequence elements in the human osteocalcin gene confer basal activation and inducible response to hormonal vitamin D_3. *Proc. Natl. Acad. Sci. U.S.A.* **86**, 4455–4459.

Kleinsmith, L. J., Stein, J., and Stein, G. (1976). Dephosphorylation of nonhistone proteins specifically alters the pattern of gene transcription in reconstituted chromatin. *Proc. Natl. Acad. Sci. U.S.A.* **73**, 1174–1178.

Kroeger, P., Stewart, C., Schaap, T., van Wijnen, A., Hirshman, J., Helms, S., Stein, G., and Stein, J. (1987). Proximal and distal regulatory elements that influence *in vivo* expression of a cell cycle–dependent human H4 histone gene. *Proc. Natl. Acad. Sci. U.S.A.* **84**, 3982–3986.

Lawrence, J. B., Singer, R. H., Villnave, C. A., Stein, J. L., and Stein, G. S. (1988). Intracellular distribution of histone mRNAs in human fibroblasts studied by *in situ* hybridization. *Proc. Natl. Acad. Sci. U.S.A.* **85**, 463–467.

Levine, A. J. (1990). Tumor-suppressor genes. *Bioessays* **12**, 60–66.

Lian, J. B. (1987) Osteocalcin: Functional studies and postulated role in one resorption. *In* "Current Advances in Vitamin K Research" (J. W. Suttie, ed.), pp. 245–257. Elsevier Science Publishers, New York.

Lian, J., Stewart, C., Puchacz, E., Mackowiak, S., Shalhoub, V., Collart, D., Zambetti, G., and Stein, G. (1989). Structure of the rat osteocalcin gene and regulation of vitamin D–dependent expression. *Proc. Natl. Acad. Sci. U.S.A.* **86**, 1143–1147.

Lian, J. B., Stein, G. S., Bortell, R., and Owen, T. A. (1991). Phenotype suppression: A postulated molecular mechanism for mediating the relationship of proliferation and differentiation by fos/jun interactions at AP-1 sites in steroid responsive promoter elements of tissue-specific genes. *J. Cell. Biochem.* **45**, 9–14.

Lian, J. B., Stein, G. S., Owen, T. A., Aronow, M., Tassinari, M. S., Pockwinse, S., and Bortell, R. (1991). Cell structure and gene expression: Contributions of the extracellular matrix. *In* "Molecular Basis of Human Cancer" (C. A. Nicolini, ed.), Plenum Press, New York (in press).

Lian, J., and Stein, G. (1991). Vitamin D regulation of osteocalcin, collagen, and alkaline phosphatase gene expression is controlled by osteoblast growth and differentiation. *In* "Eighth Workshop on Vitamin D – 1991" (A. W. Norman, ed.), Walter de Gruyter Publishers, Paris (in press).

Lichtler, A., Stover, M. L., Angilly, J., Kream, B., and Rowe, J. (1989). Isolation and characterization of the rat α-1 (I) collagen promoter. *J. Biol. Chem.* **264**, 3072–3077.

Majeska, R. J., Nair, B. C., and Rodan, G. A. (1985). Glucocorticoid regulation of alkaline phosphatase in the osteoblastic osteosarcoma cell line ROS 17/2.8. *Endocrinology* **116**, 170–179.

Markose, E. R., Stein, J. L., Stein, G. S., and Lian, J. B. (1990). Vitamin D–mediated modifications in protein–DNA interactions of two promoter elements of the osteocalcin gene. *Proc. Natl. Acad. Sci. U.S.A.* **87**, 1701–1705.

McDonnell, D. P., Mangelsdorf, D. J., Pike, J. W., Haussler, M. R., and O'Malley, B. W. (1987). Molecular cloning of complementary DNA encoding with avian receptor for vitamin D. *Science* **235**, 1214–1217.

Minghetti, P. P., and Norman, A. W. (1989). $1,25(OH)_2$-vitamin D_3 receptors: Gene regulation and genetic circuitry. *FASEB J.* **2**, 3043–3053.

Morris, T. D., Weber, L. A., Hickey, E., Stein, G. S., and Stein, J. L. (1991) Changes in the stability of a human H3 histone mRNA during the HeLa cell cycle. *Mol. Cell. Biol.* **11**, 544–553.

Morrison, N. A., Shine, J., Fragonas, J. -C., Verkest, V., McMenemy, L., and Eisman, J. A. (1989). 1,25-Dihydroxyvitamin D–responsive element and glucocorticoid repression in the osteocalcin gene. *Science* **246**, 1158–1161.

Nanes, M. S., Rubin, J., Titus, L., Hendy, G. N., and Catherwood, B. D. (1990). Interferon-γ inhibits 1,25-dihydroxyvitamin D_3–stimulated synthesis of bone GLA protein in rat osteosarcoma cells by a pretranslational mechanism. *Endocrinology* **127**, 588–594.

Noda, M., and Rodan, G. A. (1987). Type β transforming growth factor (TGFβ) regulation of alkaline phosphatase expression and other phenotype-related mRNAs in osteoblastic rat osteosarcoma cells. *J. Cell. Physiol.* **133**, 426–437.

Noda, M., Vogel, R. L., Craig, A. M., Prahl, J., DeLuca, H. F., and Denhardt, D. T. (1990) Identification of a DNA sequence responsible for binding of the 1,25-dihydroxyvitamin D_3 receptor and 1,25-dihydroxyvitamin D_3 enhancement of mouse secreted phosphoprotein 1 (Spp-1 or osteopontin) gene expression. *Proc. Natl. Acad. Sci. U.S.A.* **1990**, 9995–9999.

Oldberg, A., Jirskog-Hed, B., Axelsson, S., and Heinegård, D. (1989). Regulation of bone sialoprotein mRNA by steroid hormones. *J. Cell Biol.* **109**, 3183–3186.

Owen, T. A., Aronow, M., Shalhoub, V., Barone, L. M., Wilming, L., Tassinari, M. S., Kennedy, M. B., Pockwinse, S., Lian, J. B., and Stein, G. S. (1990a). Progressive development of the rat osteoblast phenotype *in vitro*: Reciprocal relationships in expression of genes associated with osteoblast proliferation and differentiation during formation of the bone extracellular matrix. *J. Cell. Physiol.* **143**, 420–430.

Owen, T. A., Holthuis, J., Markose, E., van Wijnen, A. J., Wolfe, S. A., Grimes, S., Lian, J. B., and Stein, G. S. (1990b). Modifications of protein–DNA interactions in the proximal promoter of a cell growth–regulated histone gene during the onset and progression of osteoblast differentiation. *Proc. Natl. Acad. Sci. U.S.A.* **87**, 5129–5133.

Owen, T. A., Bortell, R., Yocum, S. A., Smock, S. L., Zhang, M., Abate, C., Shalhoub, V., Aronin, N., Wright, K. L., van Wijnen, A. J., Stein, J. L., Curran, T., Lian, J. B., and Stein, G. S. (1990c). Coordinate occupancy of AP-1 sites in the vitamin D–responsive and CCAAT box elements by *fos–jun* in the osteocalcin gene: A model for phenotype suppression of transcription. *Proc. Natl. Acad. Sci. U.S.A.* **87**, 9990–9994.

Owen, T. A., Aronow, M. A., Barone, L. M., Bettencourt, B., Stein, G., and Lian, J. B. (1991). Pleiotropic effects of vitamin D on osteoblast gene expression are related to the proliferative and differentiated state of the bone cell phenotype: Dependency upon basal levels of gene expression, duration of exposure, and bone-matrix competency in normal rat osteoblast cultures. *Endocrinology* **128**, 1496–1504.

Pauli, U., Chrysogelos, S., Stein, G., Stein, J., and Nick, H. (1987). Protein–DNA interactions *in vivo* upstream of a cell cycle–regulated human H4 histone gene. *Science* **236**, 1308–1311.

Pauli, U., Chrysogelos, S., Stein, J., and Stein, G. (1988). Native genomic blotting: High-resolution mapping of DNase I–hypersensitive sites and protein-DNA interactions. *Proc. Natl. Acad. Sci. U.S.A.* **85**, 16–20.

Pauli, U., Chrysogelos, S., Nick, H., Stein, G., and Stein, J. (1989). *In vivo* protein-binding sites and nuclease hypersensitivity in the promoter region of a cell cycle–regulated human H3 histone gene. *Nucleic Acids Res.* **17,** 2333–2350.

Plumb, M., Stein, J., and Stein, G. (1983). Coordinate regulation of multiple histone mRNAs during the cell cycle in HeLa cells. *Nucleic Acids Res.* **11,** 2391–2410.

Prince, C. W., and Butler, W. T. (1987). 1,25-Dihydroxyvitamin D_3 regulated the biosynthesis of osteopontin, a bond-derived cell attachment protein, in clonal osteoblast-like osteosarcoma cells. *Collagen Rel. Res.* **7,** 305–313.

Puchacz, E., Lian, J. B., Stein, G. S., Wozney, J., Huebner, K., and Croce, C. (1989). Chromosomal localization of the human osteocalcin gene. *Endocrinology* **24,** 2648–2650.

Rodan, G. A., and Noda, M. (1991). Gene expression in osteoblastic cells. *Crit. Rev. Eukaryotic Gene Expression* **1,** 85–98.

Rodan, S. B., Wesolowski, G., Yoon, K., and Rodan, G. A. (1989). Opposing effects of fibroblast growth factor and pertussis toxin on alkaline phosphatase, osteopontin, osteocalcin, and type I collagen mRNA levels in ROS 17/2.8 cells. *J. Biol. Chem.* **264,** 19934–19941.

Roesler, W. J., Vandenbark, G. R., and Hanson, R. W. (1988). Cyclic AMP and the induction of eukaryotic gene transcription. *J. Biol. Chem.* **263,** 9063–9066.

Schüle, R., Kazuhiko, U., Mangelsdorf, D. J., Bolado, J., Pike, J. W., and Evans, R. M. (1990a). Jun–Fos and receptors for vitamins A and D recognize a common response element in the human osteocalcin gene. *Cell* **61,** 497–504.

Schüle, R., Rangarajan, P., Kliewer, S., Ransome, L. J., Bolado, J., Yang, N., Verma, I. M., and Evans, R. M. (1990b). Functional antagonism between oncoprotein c-Jun and the glucocorticoid receptor. *Cell* **62,** 1217–1226.

Shalhoub, V., Gerstenfeld, L. C., Collart, D., Lian, J. B., and Stein, G. S. (1989). Down-regulation of cell growth and cell cycle–regulated genes during chick osteoblast differentiation with the reciprocal expression of histone gene variants. *Biochemistry* **28,** 5318–5322.

Sierra, F., Lichtler, A., Marashi, F., Rickles, R., Van Dyke, T., Clark, S., Wells, J., Stein, G., and Stein, J. (1982). Organization of human histone genes. *Proc. Natl. Acad. Sci. U.S.A.* **79,** 1795–1799.

Speiss, Y. H., Price, P. A., Deftos, J. L., and Manolagas, S. C. (1986). Phenotype-associated changes in the effects of 1,25-dihydroxyvitamin D_3 on alkaline phosphatase and bone Gla protein of rat osteoblastic cells. *Endocrinology* **118,** 1340–1346.

Stein, G., Park, W., Thrall, C., Mans, R., and Stein, J. (1975). Regulation of cell cycle stage–specific transcription of histone genes from chromatin by nonhistone chromosomal proteins. *Nature* **257,** 764–767.

Stein, G., Lian, J., Stein, J., Briggs, R., Shalhoub, V., Wright, K., Pauli, U., and van Wijnen, A. J. (1989). Altered binding of human histone gene transcription factors during the shutdown of proliferation and onset of differentiation in HL-60 cells. *Proc. Natl. Acad. Sci. U.S.A.* **86,** 1865–1869.

Stein, G. S., Lian, J. B., and Owen, T. A. (1990). Relationship of cell growth to the regulation of tissue-specific gene expression during osteoblast differentiation. *FASEB J.* **4,** 3111–3123.

Stein, G. S., Lian, J. B., Owen, T. A., Stein, J. L., Tassinari, M. S., van Wijnen, A., Barone, L., Shalhoub, V., Aronow, M., Zambetti, G., Dworetzky, S. I., Pockwinse, S., and Holthuis, J. (1991). Cell structure and the regulation of genes controlling proliferation and differentiation: The nuclear matrix and cytoskeleton. *In* "Molecular Basis of Human Cancer" (C. A. Nicolini, ed.). Plenum Press, New York (in press).

Stein, G. S., Lian, J. B., Dworetzky, S. I., Owen, T. A., Bortell, R., van Wijnen, A. J., and Bidwell, J. P. (1991). Regulation of transcription factor activity during growth and differentiation: Involvement of the nuclear matrix in concentration and localization of promoter binding proteins. *J. Cell. Biochem.* (in press).

Suda, T., Shinki, T., and Takahashi, N. (1990). The role of vitamin D in bone and intestinal cell differentiation. *Annu. Rev. Nutr.* **10,** 195–211.

Thomson, J. A., Stein, J. L., Kleinsmith, L. J., and Stein, G. S. (1976). Activation of histone gene transcription by nonhistone chromosomal phosphoproteins. *Science* **194**, 428–431.

van Wijnen, A. J., Stein, J. L., and Stein, G. S. (1987). A nuclear protein with affinity for the 5′ flanking region of a cell cycle dependent human H4 histone gene *in vitro. Nucleic Acids Res.* **15**, 1679–1698.

van Wijnen, A. J., Wright, K. L., Massung, R. F., Gerretsen, M., Stein, J. L., and Stein, G. S. (1988a). Two target sites for protein binding in the promoter region of a cell cycle–regulated human H1 histone gene. *Nucleic Acids Res.* **16**, 571–592.

van Wijnen, A. J., Massung, R. F., Stein, J. L., and Stein, G. S. (1988b). Human H1 histone gene-promoter CCAAT box binding protein HiNF-B is a mosaic factor. *Biochemistry* **27**, 6534–6541.

van Wijnen, A. J., Wright, K. L., Lian, J. B., Stein, J. L., and Stein, G. S. (1989). Human H4 histone gene transcription requires the proliferation-specific nuclear factor HiNF-D: Auxiliary roles for HiNF-C (Sp1-Like) and HiNF-A (high mobility group-like). *J. Biol. Chem.* **264**, 15034–15042.

van Wijnen, A. J., Choi, T. K., Owen, T. A., Wright, K. L., Lian, J. B., Jaenisch, R., Stein, J. L., and Stein, G. S. (1991). Involvement of the cell cycle–regulated nuclear factor HiNF-D in cell-growth control of a human H4 histone gene during hepatic development in transgenic mice. *Proc. Natl. Acad. Sci. U.S.A.* **88**, 2573–2577.

Weinberg, R. A. (1989). Oncogenes, antioncogenes, and the molecular bases of multistep carcinogenesis. *Cancer Res.* **49**, 3713–3721.

Weinreb, M., Shinar, D., and Rodan, G. A. (1990). Different pattern of alkaline phosphatase, osteopontin, and osteocalcin expression in developing rat bone visualized by *in situ* hybridization. *J. Bone Miner. Res.* **5**, 831–842.

Yoon, K., Buenaga, R., and Rodan, G. A. (1987). Tissue specificity and developmental expression of osteopontin. *Biochem. Biophys. Res. Commun.* **148**, 1129–1136.

Yoon, K., Rutledge, S. J., Buenaga, R. F., and Rodan, G. A. (1988). Characterization of the rat osteocalcin gene: Stimulation of promoter activity by 1,25-dihydroxyvitamin D_3. *Biochemistry* **27**, 8521–8526.

Zambetti, G., Stein, J., and Stein, G. (1987). Targeting of a chimeric human histone fusion mRNA to membrane-bound polysomes in HeLa cells. *Proc. Natl. Acad. Sci. U.S.A.* **84**, 2683–2687.

11

The Nuclear Matrix: Structure and Involvement in Gene Expression

JEFFREY A. NICKERSON AND SHELDON PENMAN

Department of Biology
Massachusetts Institute of Technology
Cambridge, Massachusetts 02139

I. INTRODUCTION

Cell structure is the basic building block of tissue and, ultimately, animal architecture. The cytoskeleton of every cell is assembled on the nucleus, which was long considered a simple repository for the cell's huge length of chromatin but

MOLECULAR AND CELLULAR APPROACHES TO THE CONTROL
OF PROLIFERATION AND DIFFERENTIATION

otherwise with little structure of interest. This view was engendered by our inability to see a nuclear structure hidden by the large mass of opaque chromatin. However, steady improvements in methods for fractionating the cell and its nucleus, coupled with better methods of imaging structural networks, have revealed a fascinating nuclear structure, which plays an important role in the most basic of cell processes.

The central structure in the nucleus is a nonchromatin armature that remains after the experimental removal of chromatin. This structure has been termed the nuclear matrix and, among its myriad functions, provides the scaffolding that organizes the great length of chromatin in the cell nucleus. In this chapter, we shall examine the morphology of the nuclear matrix exposed by *in situ* fractionation and seen by resinless section electron microscopy. We shall show how the complete nuclear matrix can be further fractionated to reveal a network of 10-nm core filaments that contain nearly all the nuclear RNA. This may give us some insight into the long known but little understood puzzle—why intact RNA is necessary for normal chromatin and nuclear architecture. We shall discuss a strategy for mapping the matrix topography using monoclonal antibodies to nuclear-matrix proteins. Finally we shall discuss how the matrix plays several roles in the regulation of gene expression.

It is increasingly apparent that the nuclear matrix participates in the critical events of nuclear metabolism. The matrix is the site of DNA replication (Berezney and Coffey, 1975; McCready *et al.*, 1980; Pardoll *et al.*, 1980), RNA synthesis and processing (Zeitlin *et al.*, 1987; Zeitlin *et al.*, 1989), and hormone binding in the nucleus (Simmen *et al.*, 1984; Barrack 1983; Barrack and Coffey, 1980; Kumara *et al.*, 1986; Rennie *et al.*, 1983; Kirsch *et al.*, 1986; Buttyan *et al.*, 1983). Perhaps most interesting is the apparent relationship of the nuclear matrix to the spatial organization of chromatin and to the selection of active chromatin regions. These relationships were uncovered by experiments showing (1) that an intact nuclear matrix is necessary for normal chromatin architecture (Nickerson *et al.*, 1988) and (2) that actively transcribed gene sequences are preferentially bound to the matrix. Together these two findings suggest the involvement of the nuclear matrix in the regulation of gene expression (Ciejek *et al.*, 1983; Hentzen *et al.*, 1984; Robinson *et al.*, 1982; Small *et al.*, 1985).

A nuclear matrix has been found in cells from a great variety of eukaryotic species. Ronald Berezney (Berezney, 1984) has suggested that the development of a nuclear matrix preceded the evolution of chromatin in eukaryotes. This view was suggested by his analysis of electron micrographs of nuclei from the dinoflagellate *Gyrodinium cohnii* (Kubai and Ris, 1969), whose DNA is arranged in highly coiled bundles and not in the histone-containing chromatin structure characteristic of higher eukaryotes. Despite the prokaryotic appearance of DNA, these unicellular organisms have a true nucleus filled with a fibrillogranular material, which is between the dense DNA bundles and resembles the nuclear matrix seen in similar micrographs from mammalian cell. The ancestors of these organisms first appeared more than 1 billion years ago; perhaps the nuclear matrix, but not chromatin, is an older structure than this. Chromatin-folding strategies might have evolved to take advantage of a preexisting nuclear structure.

II. EARLY APPROACHES TO NUCLEAR MATRIX RESEARCH

Our ideas about cell architecture have been strongly influenced by the technology available for examining that structure. Early studies of nuclear structure were made by light microscopy and provided little evidence for the existence of an internal, nonchromatin skeleton. The material in which chromatin and nucleoli were suspended appeared to be a translucent gel, and this material was called the nuclear sap or karyolymph. The application of electron microscopy to the study of nuclear organization revealed a structure of fibrillogranular material in the interchromatin space between patches of condensed heterochromatin (see, for example, Fig. 7a). The use of EDTA regressive staining (Bernard, 1969; see also Fig. 7b) showed that this material was not chromatin but did contain RNA. By 1966, Don Fawcett was forced to conclude that terms like nuclear sap were no longer appropriate (Fawcett, 1966) and to suggest nuclear matrix as a preferable substitute.

Biochemical analysis of the nucleus underwent a similar evolution. In 1942 Mayer and Gulick (Mayer and Gulick, 1942) were able to find a subfraction of nuclear proteins that resisted extraction with high-ionic-strength solutions. Unfortunately, these proteins came to be referred to as the *insoluble residue* and, although apparently of interest, insolubility proved an anathema to biochemists. Similar high-ionic-strength extractions would later be used in conjunction with nuclease digestions to isolate the nuclear matrix, largely free of chromatin. In their pioneering studies, Berezney and Coffey (Berezney and Coffey, 1974, 1975, 1977) developed a protocol employing elevated ionic strength for the isolation of the nuclear matrix. They depended primarily on biochemical analysis of the extracted and remaining material to develop their fractionation methods. In the procedure they finally adopted, chromatin was removed from isolated rat liver nuclei by DNase I digestion and 2 *M* NaCl extraction. This was a severe treatment, and it is possible that more labile structural elements of the matrix were lost.

Most early studies of the isolated nuclear matrix used the Berezney and Coffey protocol or variations of it. Despite the high ionic strength used in this procedure, a number of important features of the matrix did survive. These included the attachment sites for the chromatin loops, DNA replication complexes, binding sites for steroid hormone–receptor complexes, and most of the heterogeneous nuclear RNA (hnRNA). This original preparation method served as an important research tool, but a gentler nuclear-matrix procedure, better preserving matrix morphology, is now available (Capco *et al.*, 1982; Fey *et al.*, 1986; He *et al.*, 1990).

III. NEWER TECHNOLOGIES FOR NUCLEAR MATRIX ANALYSIS

Our ability to see nuclear-matrix structure depended on the development of two technologies: resinless-section electron microscopy (Capco *et al.*, 1984; Nickerson *et al.*, 1990) and *in situ* sequential fractionation (Capco *et al.*, 1982; Fey *et al.*, 1986; He *et al.*, 1990). The nuclear matrix is a three-dimensional structure that is poorly imaged by two-dimensional electron microscopic techniques employing metal-stained embedded sections. In such sections, only those structural components that bind metal stain at the surface of the section can be imaged. Embedment-free, whole-mount electron microscopy, while providing three-dimensional information, cannot show the nuclear interior, which is obscured by the surrounding nuclear lamina. Resinless sections, however, provide a cross-sectional view through the nuclear interior revealing all the structures in the section.

The second key technology, *in situ* sequential fractionation, allows the gentle, and yet complete, removal of chromatin. The nuclear matrix contains only a very small part (about 1%) of total cell protein, and it is immersed in and masked by the much larger amount of opaque chromatin. Visualization of the matrix structure requires the removal of chromatin, but by a procedure gentle enough to leave the matrix largely intact. The procedure developed in this laboratory omits the isolation of nuclei and extracts chromatin *in situ*, preserving the connections between the nuclear matrix and the intermediate filaments of the cytoskeleton.

A. Resinless-Section Microscopy

Traditional embedded-section electron microscopy can show many features of cell morphology. This technique excels at preserving and imaging the outlines of membrane-bound cell organelles. However, three-dimensional information is completely lost, since only that portion of the section actually at the surface forms the image. For example, filaments are seen only in cross-section unless they are both close to and parallel to the surface of the section, something that occurs rarely. Thus, heavy metal-stained embedded sections are not well suited for studying three-dimensional structures such as the cytoskeleton or nuclear matrix.

Embedded-section electron microscopy was first introduced in the 1950s to facilitate the imaging of membranes in cross section. Prior to that, specimens were imaged directly, a technique later reintroduced by Porter (Porter, 1984), who viewed entire cells using high-voltage electron microscopy. Without embedding resin, proteins form high-contrast images with no need for metal stains. If the soluble proteins are removed by suitable detergent extraction, the cytoskeleton is revealed with unprecedented clarity and is easily seen by conventional voltage (80 kV) electron microscopy.

An electron-microscopic technique for preparing resinless sections was developed in this laboratory (Capco *et al.*, 1984), based on that of Wolosewick (Wolosewick, 1980) but with a more manageable embedding material, diethylene glycol distearate. Extracted cells are prepared for electron microscopy by fixation in glutaraldehyde and embedment in diethylene glycol distearate. After sectioning, the diethylene glycol distearate embedding resin is removed by solvent extraction. Following critical point drying, the section is viewed without staining. The ultrastructure of the cytoskeleton and nuclear matrix can be seen without staining and throughout the whole section because of the absence of embedding material. This resinless-section technology is simple and permits imaging of the nuclear matrix in three dimensions with unprecedented clarity. The ideal application of this technique is with stereo electron microscopy. In this case, a micrograph is taken, the section is rotated about 15°, and a second micrograph is taken. The two electron micrographs can then be examined together with a stereo viewer to reveal the precise three-dimensional relationships between structures in the section.

More recently, we have advanced resinless-section technology by developing two techniques for staining resinless thin sections with antibodies and gold-conjugated second antibodies (Nickerson *et al.*, 1990). Immunolocalization of specific proteins is a powerful tool for cell-structure studies; when combined with resinless section microscopy, it allows the location of specific nuclear-matrix proteins in the structure to be determined with high resolution. In the first method of immunostaining, extracted cells are fixed and stained with antibodies before embedment, sectioning, removal of the embedding resin, and critical point drying. In the postembedment method, the samples are embedded and sectioned, the embedding resin is removed, and the sample is rehydrated before antibody staining.

B. Sequential *in situ* Fractionation

An ideal protocol for preparing the nuclear matrix would remove from the nucleus the large mass of tightly affixed, electron-opaque chromatin without altering the underlying nuclear-matrix structure. An important guide to developing such a procedure is the ability to visualize the resulting structures. Using resinless-section microscopy to inform our fractionation procedures, we found the gentlest conditions that effectively removed chromatin from nuclei (Capco *et al.*, 1982; Fey *et al.*, 1986). In the resulting nuclear-matrix preparation procedure, we extract whole cells first with Triton X-100 in a buffer of physiological pH and ionic strength to remove membranes and soluble proteins. Then a stronger double-detergent combination of Tween 40 and deoxycholate in a low-ionic-strength buffer removes the cytoskeleton, except for the intermediate filaments. The remaining membrane-free nuclei, still connected at their surface to the intermediate filaments, are digested with DNase I before extraction with 0.25 *M* ammonium sulfate to remove chromatin. This digestion–extraction step removes more than 97% of nuclear DNA and essentially all the histones. We call the resulting structure the *RNA–containing nuclear matrix* because it retains about 70% of nuclear RNA.

IV. MORPHOLOGY OF THE RNA-CONTAINING NUCLEAR MATRIX

The RNA-containing nuclear matrix consists of two parts: the nuclear lamina, which is a protein shell made of lamins A, B, and C (Gerace *et al.*, 1984; Krohne *et al.*, 1987; Fisher *et al.*, 1986) and the internal nuclear matrix. The extracted matrix is embedded in a network of intermediate filaments, which attach to the nuclear lamina. This can be seen most dramatically in the whole mount of Fig. 1. The nuclear matrix and intermediate filaments seem to form a single structure and, to emphasize this fact, we sometimes refer to this arrangement as the nuclear matrix–intermediate filament scaffold. The nuclear envelope, a double-membrane structure covering the lamina, has been removed by the detergent extraction, but the nuclear pores that normally transverse the envelope are not extracted, remaining embedded in the lamina. This is often referred to as the pore–lamina complex. The intermediate filaments may be joined to the lamina at pores by filamentous crossbridges (Carmo-Fonesca *et al.*, 1987). The resinless section gives a better view of the nuclear interior as shown in Fig. 2a. This RNA-containing nuclear matrix comprises a network of thick, polymorphic, knobby fibers bounded by the meshwork of the nuclear lamina.

Nuclear RNA plays a crucial role in the structure of the nuclear matrix, and preserving nuclear-matrix morphology depends on keeping nuclear RNA intact. Digesting the matrix with RNase A or pretreating cells with actinomycin D before matrix isolation causes the matrix to collapse into a few amphorous masses (Fig. 2b). We refer to the nuclear matrix after RNase A digestion as the RNA-depleted matrix.

Fig. 1. The nuclear matrix–intermediate filament scaffold of a human colon carcinoma cell. For this whole-mount view of the cell, it was grown on a Formvar covered, carbon-coated grid and extracted with 0.5% Triton X-100, digested with DNase I and RNase A, and treated with 0.25 *M* ammonium sulfate. The resulting structure was viewed by embedment-free electron microscopy after fixation, critical point drying, and carbon coating. The RNA-depleted nuclear matrix is bounded by a proteinaceous lamina which, in turn, is connected to the intermediate filaments of the cytoskeleton to form a single cell-wide structure.

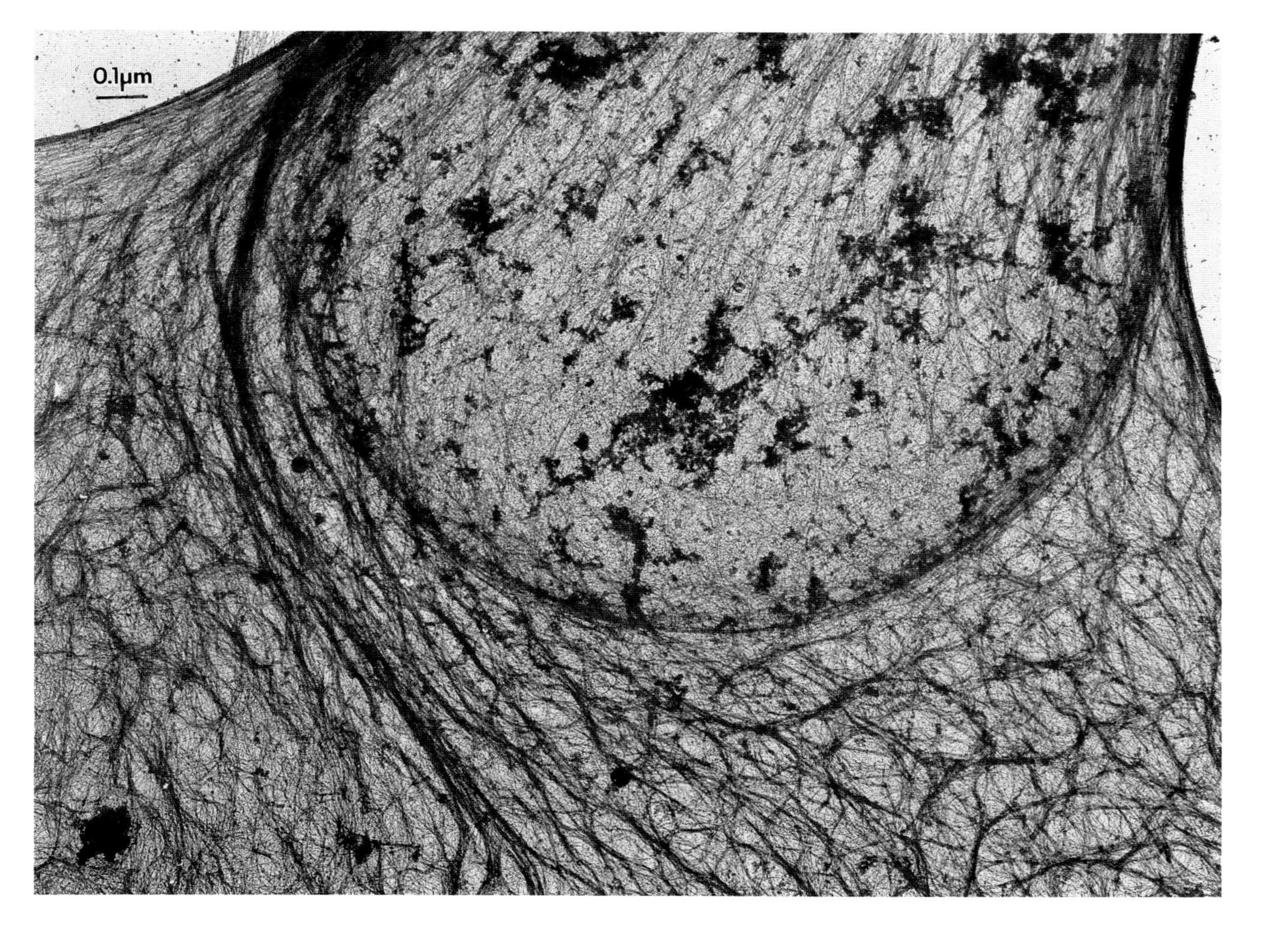
0.1µm

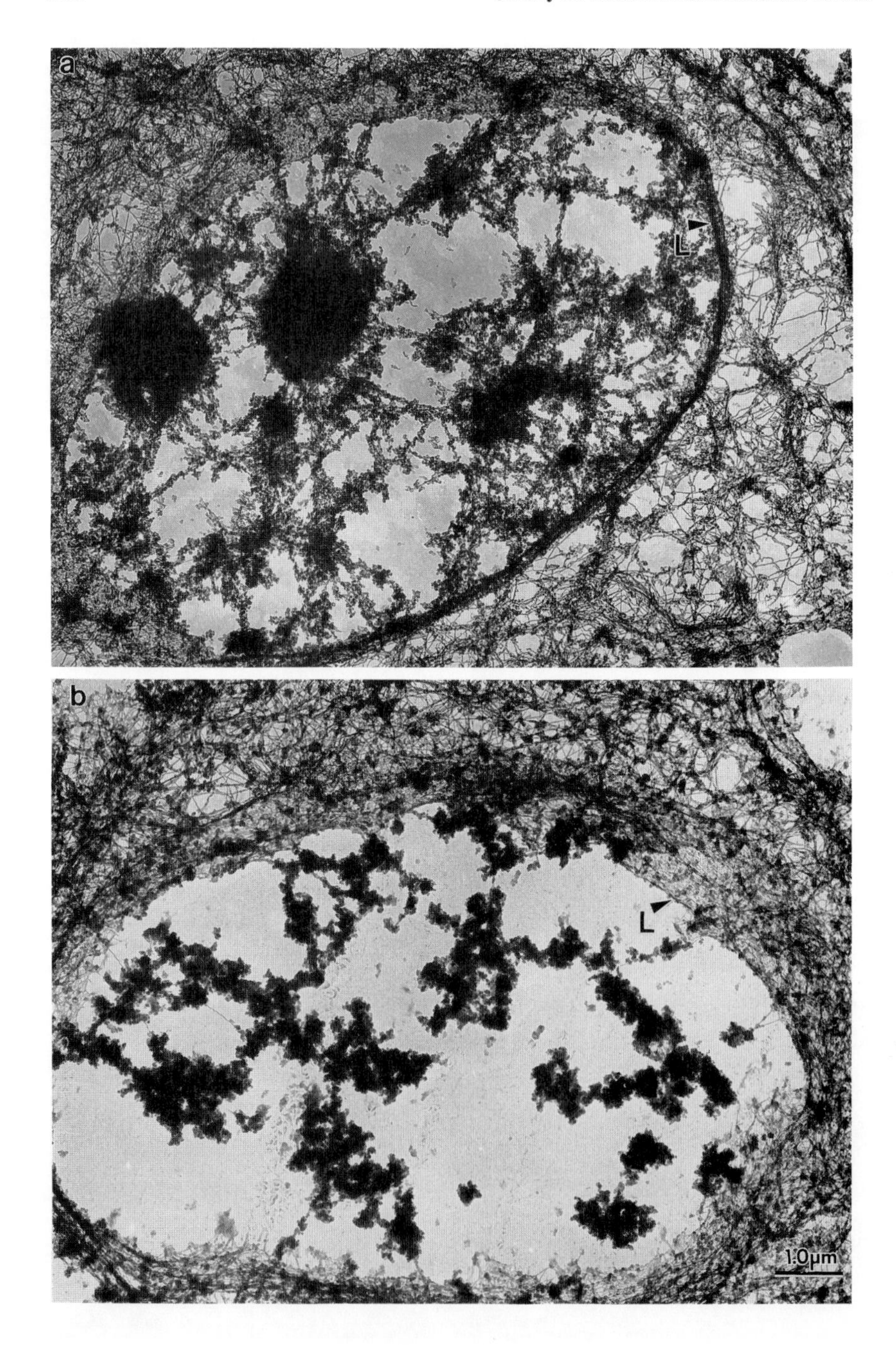
a
L
b
L
1.0μm

V. THE CORE FILAMENTS OF THE NUCLEAR MATRIX

The complete RNA-containing nuclear matrix can be further extracted to reveal an underlying network of 9 nm and 13 nm filaments, which we have called the core filaments of the nuclear matrix (He *et al.*, 1990). Further extraction of the DNase I–0.25 *M* ammonium sulfate prepared RNA–containing matrix with 2 *M* NaCl removed most of the nuclear-matrix proteins and revealed an underlying network of slender filaments (Fig. 3). We have found these core filaments in HeLa, MCF-7, and CaSki cells and believe them to be universal. They are much thinner and more uniform than the thick fibers of the complete matrix. There are two filament types distinguished by their diameters: 9 nm and 13 nm. The core filaments fill the nuclear space more densely and uniformly than the thick matrix fibers owing, in part, to the salt-induced decrease in nuclear volume. Many nuclei have large, dark masses (M) with a granular morphology consistent with their being remnant nucleoli. These presumptive nucleolar remnants are less electron dense than those seen before 2 *M* NaCl extraction (Fig. 2a). They are composed, in part, of granules ordered as though they aligned with interior filaments. Core filaments of larger-than-average diameter emanated from these masses. The smaller dark masses are very likely unextracted fragments of the nuclear matrix.

When seen at medium (Fig. 4a) and high (Fig. 4b) magnification, the nuclear-core filaments of MCF-7 human breast carcinoma cells are heterogeneous. There are two principal size classes of filaments with mean diameters of 9 nm and 13 nm. Some filaments have repeating striations similar to, but not so clear as, those in the filaments reported by Jackson and Cook (Jackson and Cook, 1988). The network is highly branched, and filaments anastomose smoothly with no obvious junction structures.

The three-dimensional organization of the nuclear-matrix core filaments and their attachments to the nuclear lamina are best seen in stereo images (Fig. 5). While the conventional embedded section images metal stains on its surface, the

Fig. 2. The RNA-containing and RNA-digested nuclear matrix of HeLa cells seen by resinless-section electron microscopy. (a) The RNA-containing nuclear matrix was revealed by sequential fractionation–detergent extraction, DNase I digestion, and 0.25 *M* ammonium sulfate extraction. After fixation, the cells were embedded in the removable resin diethylene glycol distearate. Sections of about 0.2 μm were cut, affixed to carbon-coated, Formvar-covered grids, and extracted overnight with *n*-butyl alcohol to remove the resin. The grids were transferred to ethanol, dried through the CO_2 critical point, carbon coated, and viewed. The RNA-containing nuclear matrix consists of a network of thick fibers in the nuclear interior bounded by and connected to the nuclear lamina (L). The thick fibers connect the lamina to large granular masses that may be remnant nucleoli. Intermediate filaments can be seen attached to the lamina, so the internal nuclear matrix is connected to the cytoskeleton through the lamina. (b) The complete RNA-containing nuclear matrix was revealed as described in (a) and then digested with RNase A. After fixation, the cells were processed for resinless-section electron microscopy. The nuclear matrix has collapsed into amorphous masses within the lamina.

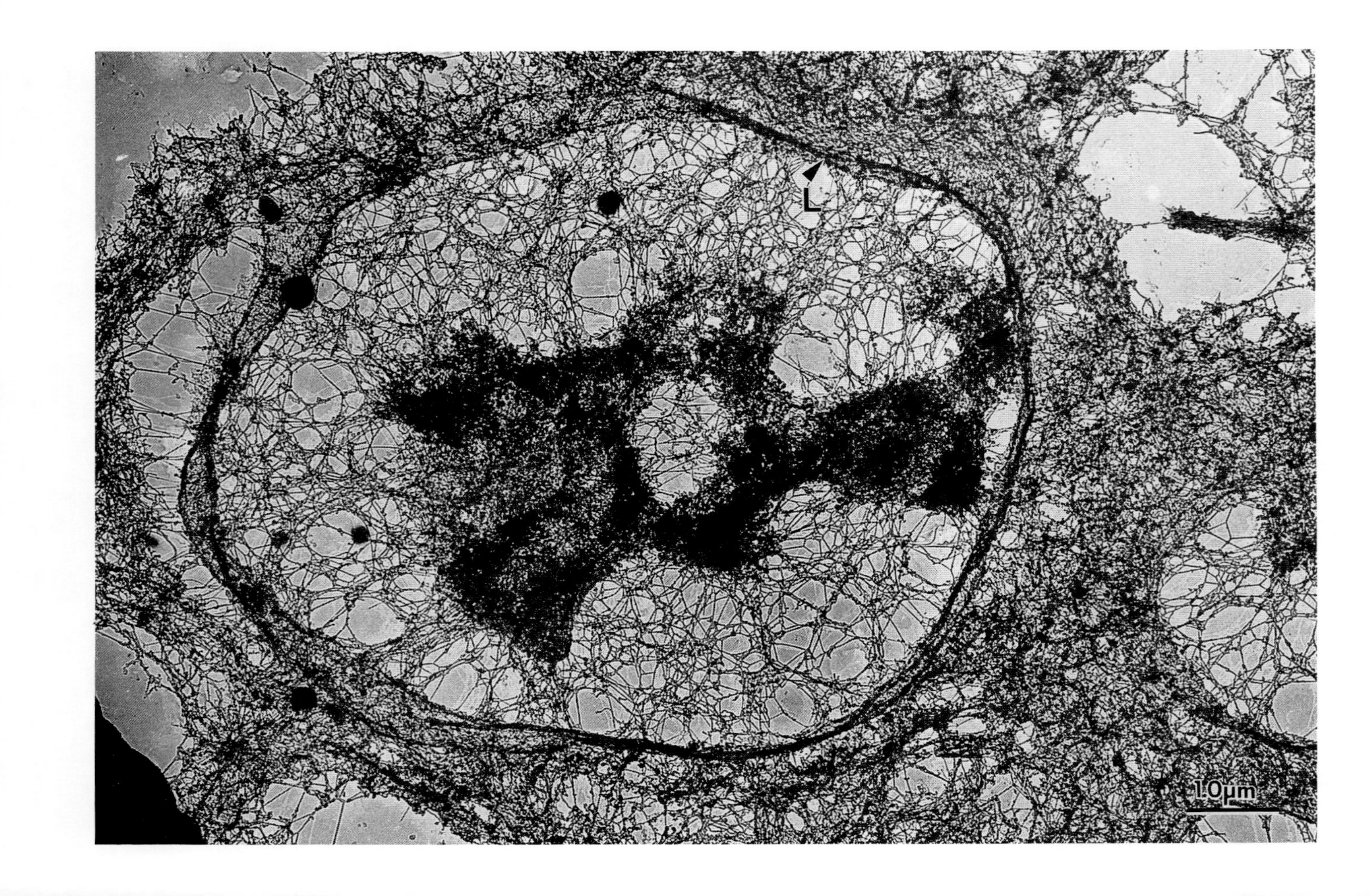
L
1.0μm

resinless section displays its entire contents. The third dimension of the image can be presented by tilted-stage stereoscopic micrographs. Figure 5a presents a stereoscopic view of the core-filament network from a monolayer MCF-7 cell. Figure 5b shows the core filaments of a suspension-grown HeLa cell in stereo at higher magnification. The three-dimensional network of core filaments is connected to the nuclear lamina, which is itself made of lamin filaments.

For reasons that are not well understood, the stepwise application of salt is necessary to reveal core filaments. Applying 2 *M* NaCl directly to the nucleus after digestion with DNase I yields the different and less well preserved structure shown in Fig. 6. With this direct application of 2 *M* NaCl, the nuclear material aggregates into amorphous masses. The two-step procedure is far more effective in removing matrix proteins, while leaving the core filaments distinct and intact. This difference is due to raising the ionic strength incrementally. Extraction with NaCl alone revealed the core filaments, provided the NaCl concentrations were increased in graded steps.

Even with a gentle, stepwise increase in salt, not every nucleus in a section is seen to be filled with core filaments. Some nuclei appear partially or completely empty. We attribute this to the incomplete preservation of structure caused, in some way not yet understood, by DNase I. When DNA is partially digested with restriction enzymes before 0.25 *M* ammonium sulfate extraction and then 2 *M* NaCl extraction, core filaments were observed in almost every nuclear section. This difference could be due either to the partial removal of filaments by DNase I or to the rearrangement of the filament network into one region of the nucleus so that some sections are empty.

It is not clear why a stepwise increase in ionic strength, first 0.25 *M* ammonium sulfate and then 2 *M* NaCl, is necessary for revealing well-preserved core filaments. We have observed that 2 *M* NaCl extraction causes a pronounced shrinkage of the nuclear structure. The nuclear radius decreases by as much as 20% with a consequent volume decrease of about 50%. When the ionic strength is increased rapidly after DNase I digestion, the nuclear matrix may be shrinking rapidly at a time when digested chromatin is breaking loose. The matrix may be mechanically damaged by this procedure, resulting in poor preservation of core-filament structure. The two-step protocol increases the ionic strength in two steps and should cause a more gradual decrease in nuclear size. This may be a mechanically more gentle procedure and should be expected to give a better preservation of nuclear-

Fig. 3. The core filaments of the nuclear matrix. HeLa cells grown in suspension were detergent extracted, digested with DNase I, and treated with 0.25 *M* ammonium sulfate to reveal the complete RNA-containing nuclear matrix, still embedded within the intermediate filaments of the cytoskeleton. This structure was further extracted with 2 *M* NaCl to remove most of the nuclear-matrix proteins and reveal an underlying network of 10-nm filaments. These core filaments were much thinner and more uniform than the original matrix fibers. L, lamina.

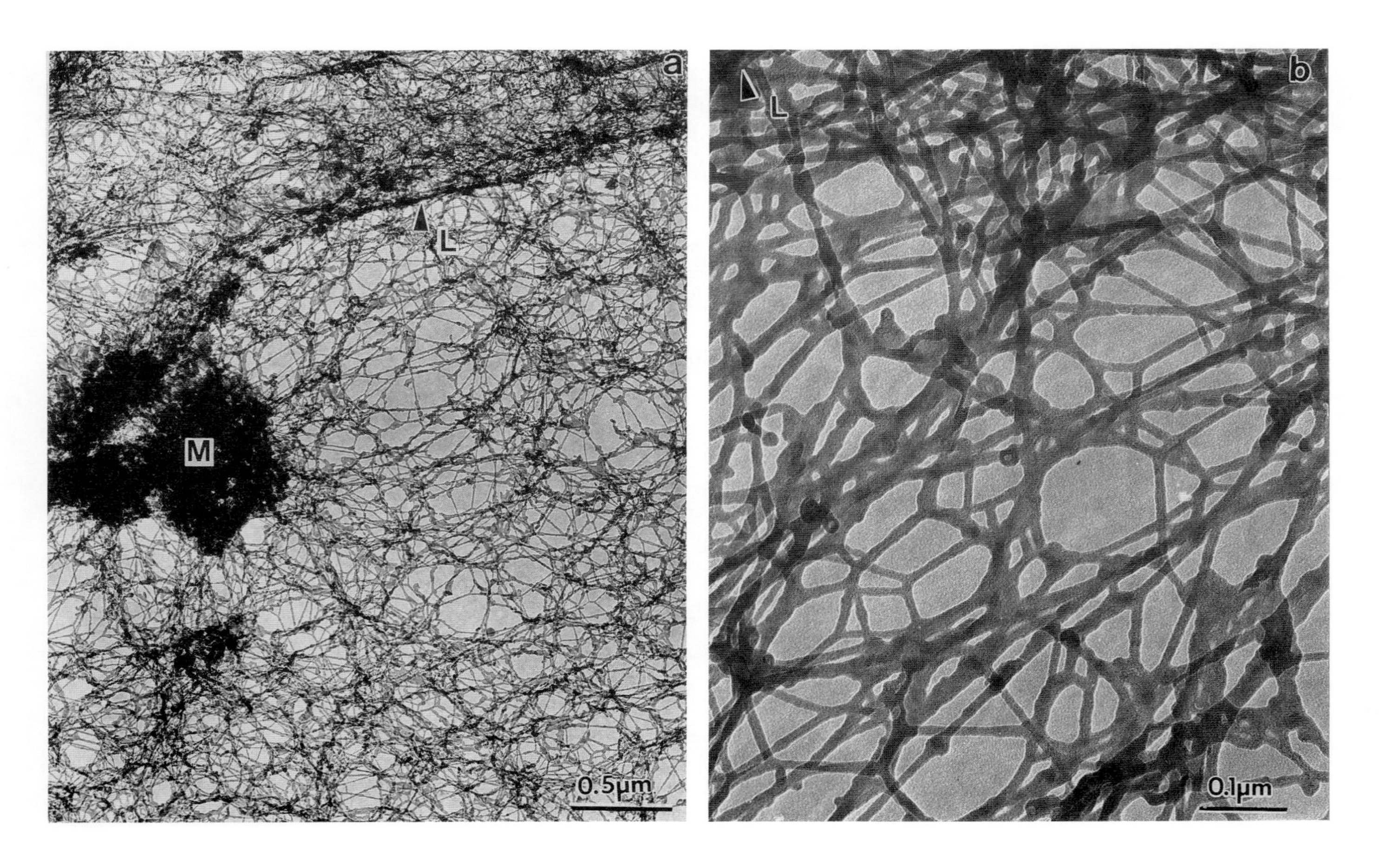
a
L
M
0.5μm
b
L
0.1μm

matrix morphology. The large shrinkage of nuclear volume may explain, in part, why the nuclear interior seems more full of filaments than would be intuitively predicted from an examination of the RNA-containing matrix.

Nuclear filaments have also been reported by Jackson and Cook (Jackson and Cook, 1988), who use a very different procedure to reveal them. They remove chromatin from agarose-encapsulated cells by *Hae*III digestion and electrophoresis. This procedure removes chromatin incompletely; about 25% of the DNA is retained compared to less than 1% in our preparation. The remaining chromatin appears to cover the core filaments, which are seen only in scattered, uncovered regions. Filaments are not seen in their procedure if DNase I is substituted for the restriction enzyme. This is, in part, consistent with our finding of fewer core filaments with DNase I than with restriction-enzyme digestion. The unusual nature of the Jackson and Cook procedure makes it difficult to compare their results directly with those reported above for the 0.25 *M* ammonium sulfate–2 *M* NaCl protocol discussed above. The filaments revealed have a similar appearance, but they have less pronounced and regular striations. The 0.25 *M* ammonium sulfate–2 *M* NaCl procedure that we use for uncovering filaments is simple, rapid, and is easily employed in any laboratory.

VI. RNA ASSOCIATION WITH THE NUCLEAR MATRIX

Many previous observations have suggested a role for RNA in chromatin architecture and nuclear organization. There is a clear relationship between RNA metabolism and nuclear-matrix structure. The matrix contains an RNA component packaged, at least in part, in nuclear ribonucleoprotein (RNP) particles (Herman *et al.*, 1978; Fey *et al.*, 1986; Miller *et al.*, 1978; van Eekelen and van Venrooij, 1981; Gallinaro *et al.*, 1983; Long and Ochs, 1983, Long and Schrier, 1983). We found that digesting nuclei with RNase A led to a collapse of nuclear-matrix structure. A similar, though less extensive, collapse was caused by treatment of cells with actinomycin D before extraction. The discovery of RNA-containing core filaments suggests a basis for this RNA requirement. The nuclear matrix appears to be constructed around the core filaments, which depend on intact RNA for their integrity. Disruption of nuclear RNA, either by digesting nuclei *in vitro* or by drug treatments *in vivo*, destroys the core filaments, disturbing the remaining matrix material.

Fig. 4. The core filaments of the MCF-7 nuclear matrix. (a) Core filaments. The nuclear lamina (L) connected the network of nuclear-core filaments (below) to the cytoplasmic intermediate-filament network (above). A residual mass (M) remained enmeshed in the network of filaments. (b) Higher-magnification view of core filaments. In this view, the tripartite junctions between core filaments could be seen more clearly. The network of core filaments connected to the lamina (L).

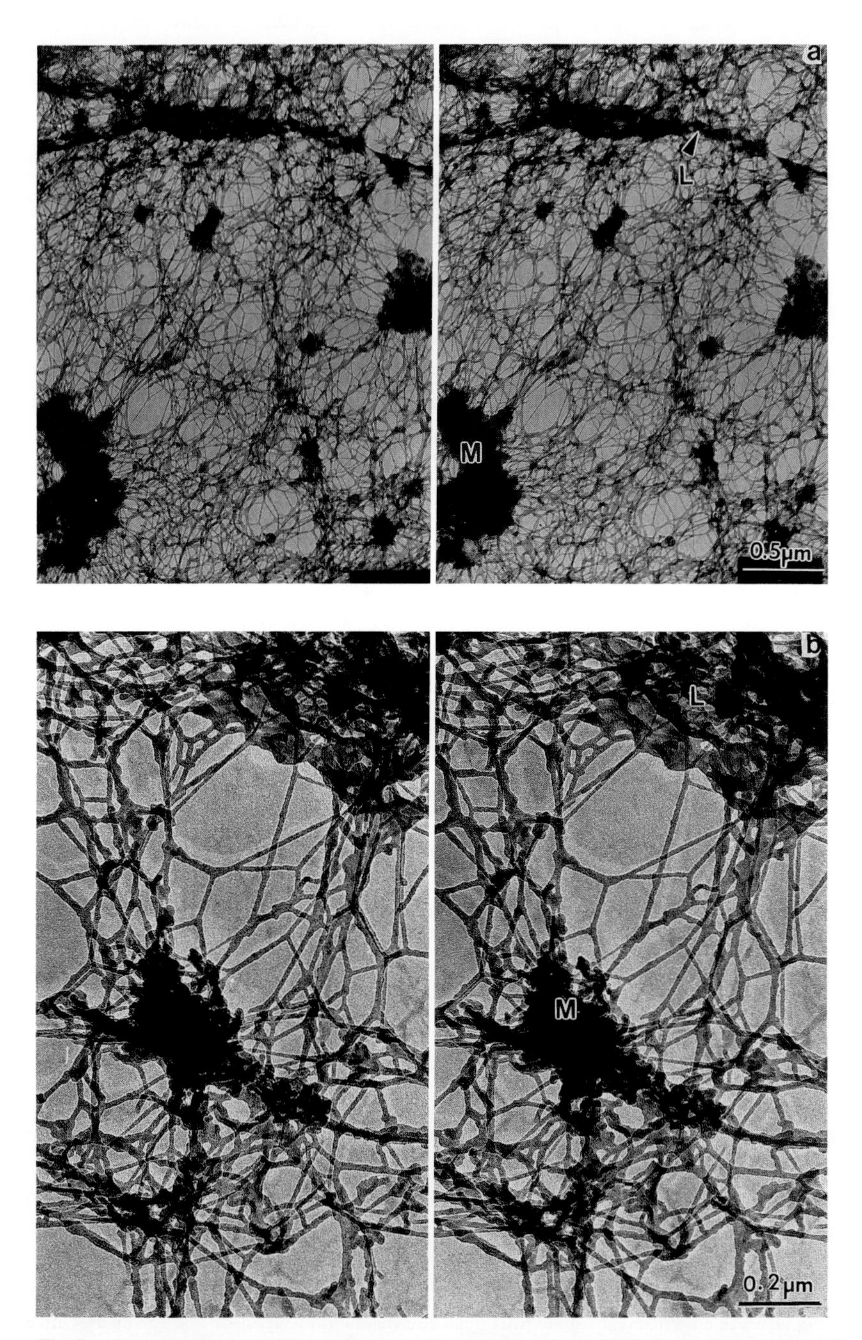

Fig. 5. Stereoscopic view of the nuclear-core filament network. (a) MCF-7 core filament network. The network of core filaments (below) is bounded by the nuclear lamina (L). A dense mass of residual material (M) remains enmeshed in the network. (b) HeLa core filament network. In this higher-magnification stereoscopic view, the core filaments can be seen connecting to filaments of the nuclear lamina (L) throughout the section. The filaments can be clearly seen entering the large, dense mass of residual material (M).

Fig. 6. The nuclear matrix made with 2 *M* NaCl applied directly after DNase I digestion. This is a section of the nuclear interior of a HeLa cell that has been extracted with Triton X-100 and digested with DNase I, just as for the HeLa cell in Fig. 4. However, instead of treatment with 0.25 *M* ammonium sulfate and then with 2 *M* NaCl, this cell was treated directly with 2 *M* NaCl. Direct application of 2 *M* NaCl results in the aggregation of core filaments. The uncovering of an intact network of 10-nm filaments requires a stepwise increase in ionic strength.

A. Chromatin Architecture Requires Intact RNA

The mammalian nucleus packages more than a yard of DNA into a 5-μm spheroid. This packaging of DNA into chromatin and of chromatin into the nucleus is highly ordered in space (Cremer *et al.*, 1982a; Cremer *et al.*, 1982b; Hens *et al.*, 1983; Rappold *et al.*, 1984; Manuelidis, 1985; Borgeois *et al.*, 1985) with the coarse features of this organization correlating with transcriptional activity. The condensed or heterochromatin is largely inactive, and transcription is localized largely to the extended, or dispersed, euchromatin. The fundamental configuration of DNA in chromatin is relatively well understood. The basic structures, the nucleosomes, are closely packed into polynucleosome chains, which are wound into 30-nm fibers (Felsenfeld and McGhee, 1986; Thomas, 1984). Much less is known about the ordering of chromatin fibers into the large masses of chromatin or about the distribution of these masses throughout the nuclear interior. We suggest that the tissue-specific architectural features of chromosomal or chromatin folding in the interphase nucleus may be imposed by an underlying structural framework.

There is a clear dependence of chromatin architecture on ongoing RNA metabolism; inhibition of RNA synthesis results in the retraction of chromatin from the nuclear lamina and its aggregation into massive clumps (Bernhard, 1971). Bouteille and co-workers (Bouvier *et al.*, 1982) found that the spatial distribution of stained DNA in nuclei extracted at high ionic strength was changed by treatment with RNase. In what may be a related phenomenon, RNA-synthesis inhibitors cause a retraction of polytene chromosome puffs (Ashburner, 1972; Lewis *et al.*, 1975; Perov and Kiknadze, 1980; Alcover *et al.*, 1984).

Extraction of nuclei *in situ* with 0.5% (v/v) Triton X-100 at physiological pH and ionic strength has little visible effect on chromatin. The structures remaining after extraction, as seen in a traditional epon-embedded section (Fig. 7a), conform closely to their preextraction morphology (Lenk *et al.*, 1977; Fulton *et al.*, 1980; Penman *et al.*, 1982; Capco *et al.*, 1982). The nuclear envelope dissolves in the detergent, leaving the nucleus bounded by the nuclear lamina. The electron-dense chromatin is more prominent against a clearer background, and numerous interchromatin granules and fibers, containing RNPs, are more conspicuous than in the unextracted cell. Most important, the distribution of chromatin appears to be unaltered by the extraction with detergent. This electron micrograph is of a

---

Fig. 7. The distribution of RNP-containing structures in the nucleus of extracted control, actinomycin D, or RNase A–treated cells. HeLa cells were treated and extracted with 0.5% Triton X-100. All the micrographs in this figure are of Epon-embedded sections. The sections shown in a, c, and e were double-stained with uranyl acetate, follwed by lead citrate, and are all shown at the same magnification to show the morphological changes to nuclei caused by actinomycin or RNase A treatment. In b, d, and f, RNP-containing structures are preferentially contrasted by the EDTA regressive-staining method. These cells are all shown at the same higher magnification in order to reveal the distribution of RNP-containing structures. (a) and (b). These were control HeLa cells. The nucleus contains a nucleolus (Nu) of uniformly distributed chromatin fibers with occasional patches of aggregated chromatin, which are perilaminar or scattered in the nuclear interior. Seen at higher magnification with EDTA regressive staining [part (b)] the RNP-containing structures (double arrowheads) within the nuclear area are interspersed with bleached chromatin (asterisk). L, Lamina. (c) and (d). These HeLa cells were treated for 2 hr with 5 μg/ml actinomycin D. The nuclear lamina (L) has become invaginated, and there is a dramatic redistribution of the chromatin with clumping and condensation into aggregates (asterisk). The nucleolus (Nu) is rearranged, the two components—fibrillar and granular—were segregated. Seen at higher magnification [part (d)] with preferential staining for RNA, actinomycin causes a massive condensation of chromatin into aggregates (asterisk). There is a redistribution of RNP-containing structures, which mostly surround the bleached aggregates of chromatin and which occasionally themselves form small aggregates (double arrowheads). (e) and (f). HeLa cells were treated with 5 μg/ml RNase A for 20 min following extraction in 0.5% Triton X-100. Treatment with RNase A causes the clumping of chromatin both onto the nuclear lamina (L) and onto the nucleolus (Nu). With preferential staining for RNPs at higher magnification [part (f)], clumps of aggregated RNPs (double arrowheads) can be seen adjacent to clumps of aggregated chromatin (asterisk).

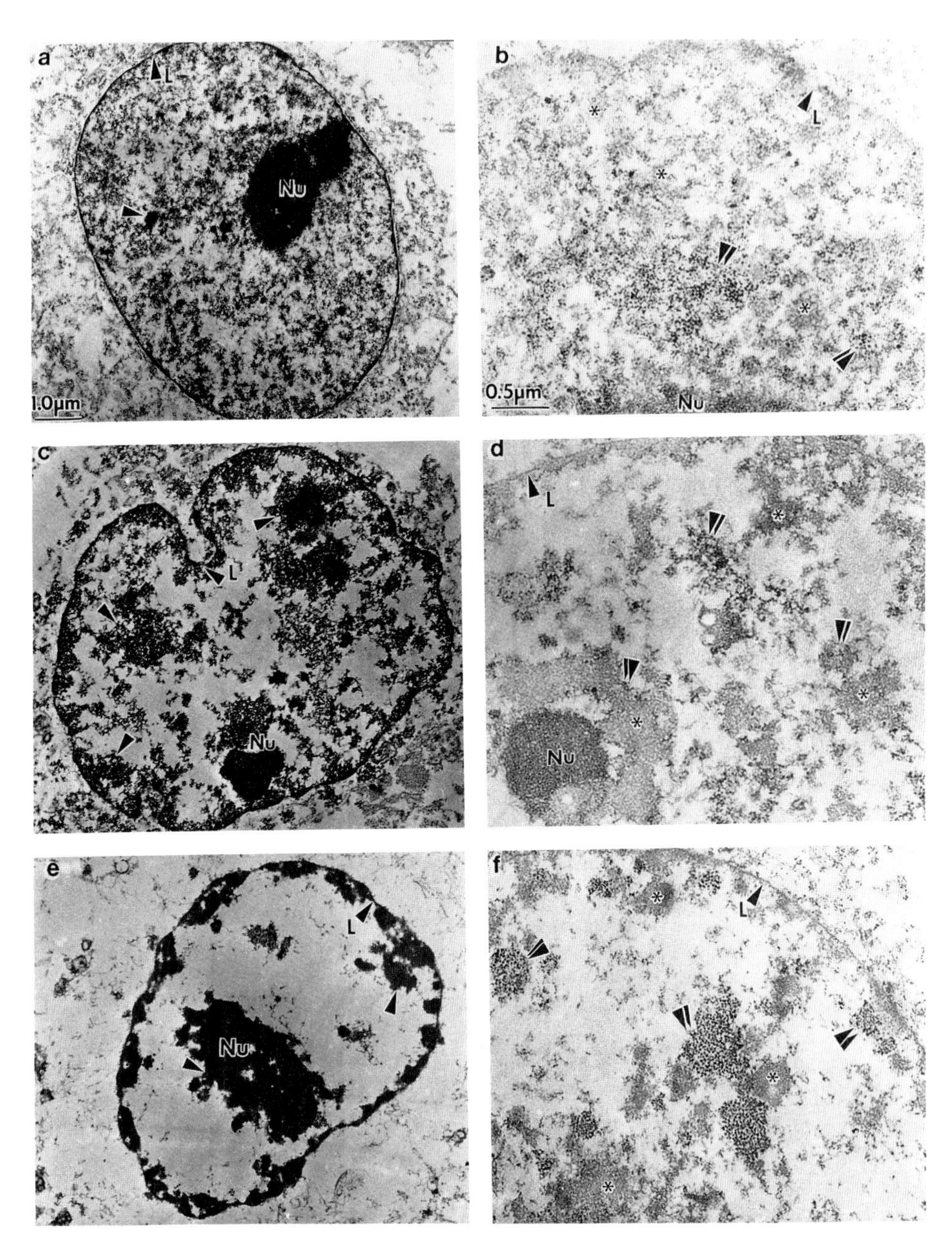
a
L
Nu
1.0μm
b
L
Nu
0.5μm
c
L
Nu
d
L
Nu
e
L
Nu
f
L

traditional epon-embedded section, which presents a simpler and two-dimensional view of chromatin architecture, but which allows selective staining for RNA.

The interchromatin granules visible in Fig. 7a were identified as RNA-containing structures by using the ethylenediaminetetraacetic acid (EDTA) regressive staining procedure of Bernhard (1969). The EDTA preferentially bleaches the chromatin after staining by removing chromatin-binding uranyl ions. This leaves chromatin as pale, unstained regions. RNA-containing structures, such as those associated with the RNA-containing nuclear matrix, destain much more slowly and remain electron dense. The result of this procedure is shown in Fig. 7b. Following this differentiation step, the RNP-containing structures appear contrasted against the bleached background. The RNP-containing interchromatin granules are distributed in the spaces between masses of bleached heterochromatin (Puvion-Dutilleul and Puvion, 1981; Wassef, 1979).

Treatment of cells for 2 hr with actinomycin caused the collapse of chromatin as well as the separation of nucleolar components (Fig. 7c). At a higher magnification (Fig. 7d), we can observe the rearrangement of the selectively stained RNP-containing structures seen mostly as filaments bordering the condensed, destained chromatin masses or as small, dense clusters of granules.

This effect of actinomycin D shows a role for RNA in chromatin organization. A more direct demonstration of the relation of RNA to chromatin organization is seen when Triton X-100–extracted nuclei are treated with RNase A (Fig. 7e). The enzyme has an effect on chromatin similar to, but even more extensive than, that of actinomycin D. Chromatin appears only as highly condensed aggregates, which are either localized in the perilaminar region or collapsed onto the modified nucleoli. The digestion-resistant, RNP-containing structures are mostly granular and coalesced into a few large clusters adjacent to the bleached, condensed chromatin (Fig. 7f).

The same treatment of cells with actinomycin that causes the collapse of chromatin also causes a breakdown of the nuclear matrix. The effect is similar to that seen after treatment with RNase A (compare Figs. 2a and 2b). Thus, the disruption of nuclear RNA, either *in situ* using nuclease or *in vivo* with actinomycin, leads to a collapse of nuclear-matrix structure. The breakdown of the nuclear matrix caused by the removal of RNA parallels the collapse of chromatin and suggests the two phenomena may be related. This correlation of effects leads us to propose that changes in chromatin architecture and in nuclear-matrix morphology are related, that RNA is an essential component of the nuclear matrix, which is in turn required for normal chromatin organization.

B. Nuclear RNA Is Associated with the Nuclear Matrix Core Filaments

Nuclear RNA consists of approximately equal amounts of ribosomal precursor RNA and hnRNA. Ribosomal precursor RNA is largely, although not entirely, located in the nucleolus (Perry, 1962; Penman *et al.*, 1969). The location of the hnRNA has been less certain, although it is known to be extranucleolar, in the *nucleoplasm* (Penman *et al.*, 1968). About 70% of total nuclear RNA remains after extraction of the nuclear matrix with 2 *M* NaCl. Thus, most hnRNA was associated with the core filaments. This hnRNA is of very high molecular weight (Fig. 8).

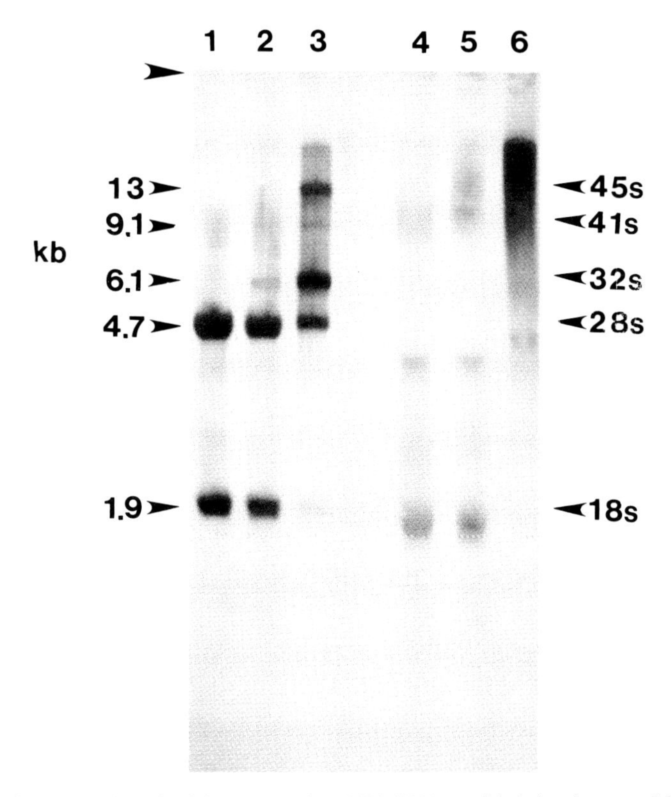

Fig. 8. Gel electrophoresis of the HeLa nuclear RNA. RNA was labeled and extracted from HeLa nuclear fractions as described in Table I and electrophoresed in a 1% (w/v) agarose gel with equal cpm of RNA loaded per lane. (Lanes 1–3) RNA from control cells. Lane 1 was the chromatin-associated RNA released by DNase I digestion and 0.2 *M* ammonium sulfate extraction. Lane 2 was RNA released by 2 *M* NaCl from the complete nuclear matrix. Lane 3 was the RNA that remained after extraction with 2 *M* NaCl. The ribosomal precursor RNAs were not removed by 2 *M* NaCl (Lane 3), but remain with the core structure. This included the 45 S (13 kb), 41 S (9.1 kb), and 32 S (6.1 kb) rRNA species. The core structure retained some of the fully spliced 28 S rRNA, but very little 18 S rRNA. (Lanes 4–6) RNA from cells treated with 0.04 μg/ml actinomycin D. In these fractions, there was little labeled rRNA and the hnRNA could be better seen. In the chromatin-associated hnRNA released by DNase I digestion and 0.25 *M* ammonium sulfate extraction (Lane 4), there was a heterogeneous population of RNA of between 2 and 5 kb. Very little high-molecular-weight hnRNA (modal size, 20 kb) was removed with the chromatin or by 2 *M* NaCl (Lane 5). This hnRNA was retained with the core structure (Lane 6).

HeLa cells were labeled for 2 hr with [^{3}H]uridine, a sufficient time for nuclear RNA to be completely labeled. A low level of actinomycin (0.04 μg/ml) was added to a second sample to selectively suppress the synthesis of ribosomal precursor RNA. After labeling, the cells were fractionated as before, and the amount of radioactive RNA in the several nuclear subfractions was determined (Table I). The RNA in each subfraction was purified and analyzed by gel electrophoresis (Fig. 8).

About 28% of nuclear RNA was released with the chromatin by DNase I digestion and 0.25 *M* ammonium sulfate extraction (Table I). An additional 4% was removed by a further 2 *M* NaCl extraction to reveal core filaments. The final 67% of nuclear RNA was retained with the core filaments. Treatment of cells with 0.04 μg/ml actinomycin D during labeling selectively inhibits ribosomal RNA synthesis and decreased the labeling in each fraction by about half.

The result of formaldehyde–agarose gel electrophoresis of these labeled RNA fractions is shown in Fig. 8. The mature 28 S ribosomal RNA was present in both the chromatin (lane 1) and 2 *M* NaCl–released fractions (lane 2), along with a smaller amount of 18 S ribosomal RNA. The 45S ribosomal RNA precursor and the 41S, 36S, and 32S processing intermediates with some 28S RNA remained with the core filaments (lane 3), perhaps in remnant nucleoli. The selective suppression of ribosomal RNA synthesis before labeling showed the hnRNA distribution in the nuclear subfractions. The RNA released by 0.25 *M* ammonium sulfate following DNase I digestion was heterogeneous but much smaller, on average, than the hnRNA of the core filaments. The modal size of this RNA was

TABLE I

The Distribution of RNA in HeLa Nuclei[a]

Fraction	CPM per cell	Percentage of nuclear RNA	Percentage of control
Control			
Chromatin	16.9	28.7	—
2 *M* NaCl	2.5	4.2	—
Core filaments	39.4	67.1	—
Total nuclear	58.8	100	—
0.04 μg/ml actinomycin D			
Chromatin	8.0	27.4	47.4
2 *M* NaCl	1.2	4.1	47.7
Core filaments	20.0	68.6	50.8
Total nuclear	29.2	100	49.7

[a]HeLa CCL2.2 cells (2.75×10^5 cells per incubation) were grown in ^{3}H-uridine (50 μCi per ml) for 2 hr at 37°C. The cells were grown with or without 0.4 μg/ml actinomycin D to inhibit ribosomal RNA synthesis and were fractionated, first by DNase I digestion and 0.25 *M* ammonium sulfate extraction to release chromatin, and then by 2 *M* NaCl extraction to uncover the core filaments of the nuclear matrix.

about 2.5 kb, and it may consist of message-sized molecules in transit to the cytoplasm. About 70% of nuclear hnRNA remained with the core filaments (Table I) and was highly enriched in very high-molecular-weight hnRNA (Fig. 8, lane 6) with a modal size of about 20 kb. Thus the core-filament network contains most hnRNA and almost all the very high-molecular-weight hnRNA.

C. The Localization of hnRNP in the Core Filaments

So far, there has been no simple way of determining the precise, spatial location of RNA in the core filament fraction. However, hnRNP proteins have been localized by immunogold staining with the FA12 monoclonal antibody (Leser *et al.*, 1984) with rather surprising results (Fig. 9). The localization of FA12 to the core-filament fraction of the interphase nucleus is not surprising, since this fraction contains essentially all the hnRNA of the cell. The FA12 antigen is one of the hnRNP proteins, and the antibody clearly decorates the granular material enmeshed in the core-filament fraction and not the filaments (Fig. 9d). This suggests strongly that the RNA itself is in the granular material rather than the filaments. The sensitivity of the entire network to RNase A would suggest that the RNA is part of a nexus that is necessary for the integrity of the filament network. However, it is possible that the FA12 protein has been separated from hnRNA by the 2 *M* NaCl extraction used to remove most of the proteins of the complete nuclear matrix.

D. Noncoding hnRNA, a Structural Component of the Nucleus?

If RNA molecules are involved in the maintenance of nuclear architecture, they may be different from any so far characterized. Both the mass and genetic complexity of the transcribed hnRNA greatly exceed what would be expected for message precursors (Scheller *et al.*, 1978; Salditt-Georgieff *et al.*, 1981; Harpold *et al.*, 1981). This is known from, admittedly, crude estimates of hnRNA complexity and from kinetic measurements showing that mRNA emerging from the nucleus is only a small percentage of the hnRNA synthesized in the nucleus. This is a much smaller conversion of nuclear RNA than can be explained by the loss of intervening sequences during splicing and can be explained either by message processing being highly nonconservative or by a large amount of hnRNA not being message precursor. Perhaps the most compelling demonstration that a major portion of hnRNA transcripts do not contain a message sequence is the experiments of Darnell and coworkers (Salditt-Georgieff *et al.*, 1981; Harpold *et al.*, 1981). These measurements showed that about 75% of hnRNA has no poly (A) and will not hybridize to cloned message sequences. Also, the number of 5′ caps

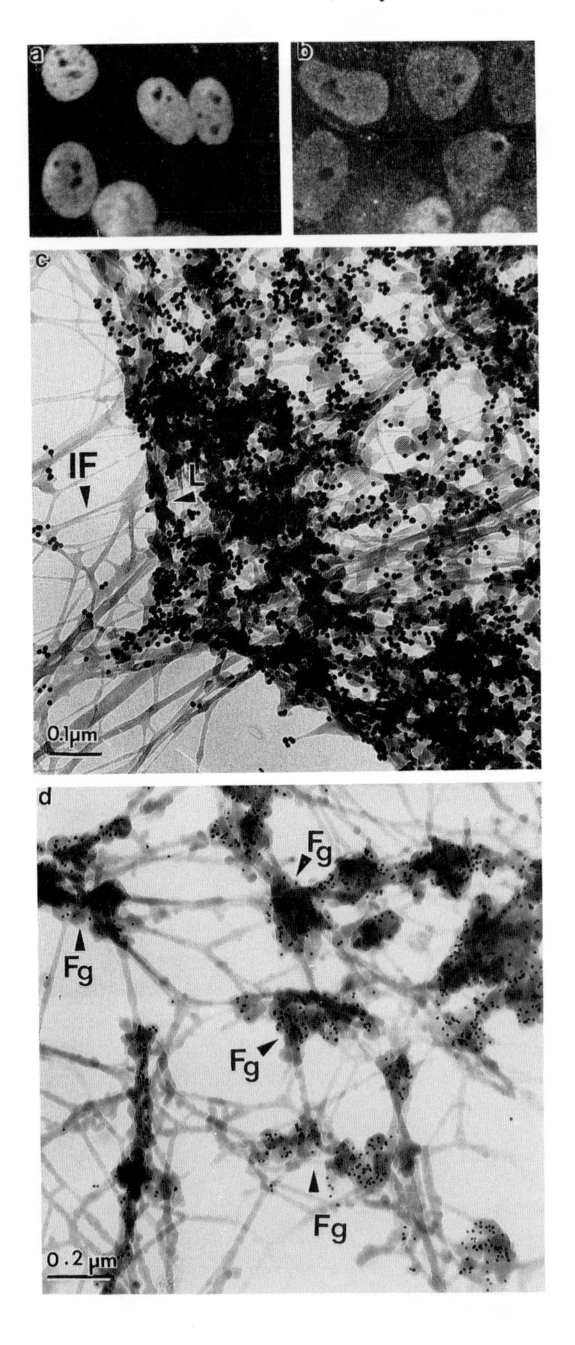
a
b
c
IF
L
0.1μm
d
Fg
Fg
Fg
Fg
0.2 μm

in RNA is much greater than the number of caps entering polyribosomes. These observations suggest that there is much more information in the hnRNA transcripts than can be accounted for by their roles as message precursors. Since most of this RNA is associated with the core nuclear-matrix filaments, it may play a role in nuclear architecture and in the organization of chromatin.

VII. ADENOVIRUS CAPSID ASSEMBLY AND THE CORE FILAMENTS OF THE NUCLEAR MATRIX

Adenovirus capsids are assembled on nuclear-matrix filaments. We have studied the rearrangements of the HeLa cell nuclear matrix at 6 to 48 hr following adenovirus infection (Zhai *et al.*, 1987). Rearrangement of the nuclear matrix is evident as early as 6 hr after infection and continues progressively until the nucleus has become one large *virus factory*, producing virions and storing them until cell lysis. This is consistent with an early nuclear-matrix paper, which reported protein changes in the matrix following adenovirus infection as well as a matrix association of the adenovirus major core protein (Hodge *et al.*, 1977).

By 28 hr after infection, many granulated areas have appeared in the nucleus (Fig. 10). In high-magnification views (Figs. 10B and 10C), these can be seen to contain viral capsids of about 75 nm in diameter. Capsids can be seen emerging from these granulated structures and bound to filaments that are very similar in appearance to the core filaments of the nuclear matrix revealed after 2 *M* NaCl extraction, and shown in Figs. 3, 4, and 5.

Fig. 9. Immunolocalization of the FA12 hnRNP antigen. The hnRNP proteins are part of the core-filament network within the nuclear matrix. These filaments are revealed by further extracting the low-salt nuclear matrix with high salt. (a) Immunofluorescence of FA12-stained Caski cells. Monolayer cells were extracted with Triton X-100 in a buffer of near-physiological pH and ionic strength. Cells were then stained with FA12 monoclonal mouse IgG followed by rhodamine-conjugated antimouse antibody. (b) Immunofluorescence of FA12 stained nuclear-matrix core filaments. The cells in (a) were further extracted with DNase I and 0.25 *M* ammonium sulfate to remove chromatin from the nuclear matrix. The nuclear matrix-containing preparation was then extracted with 2 *M* NaCl to reveal the core filaments. The core-filament preparation was then stained with FA12 antibody and rhodamine-conjugated antimouse antibody. (c) Resinless-section electron micrograph of gold-bead immunostained core filaments from Caski cells. Cells were extracted as in (b) and stained with FA12 antibody and gold-bead conjugated antimouse second antibody. Cells were then treated with preimmune goat antiserum and fixed with glutaraldehyde. The sample was then embedded in diethylene glycol distearate and ultrathin sections were cut. The embedding material was removed with *n*-butyl alcohol, and the sections then dehydrated in ethanol, dried through the CO_2 critical point, lightly carbon coated and viewed in the electron microscope. L, lamina; IF, intermediate filaments. (d) Resinless-section electron micrograph of gold-bead immunostained core filaments from MCF-7 cells. The core filaments of MCF-7 cells are more extended than those of Caski cells. MCF-7 cells were extracted, stained, and ultrathin sections were prepared as in (c). FA12 was located on granular structures (Fg) retained with the core filaments.

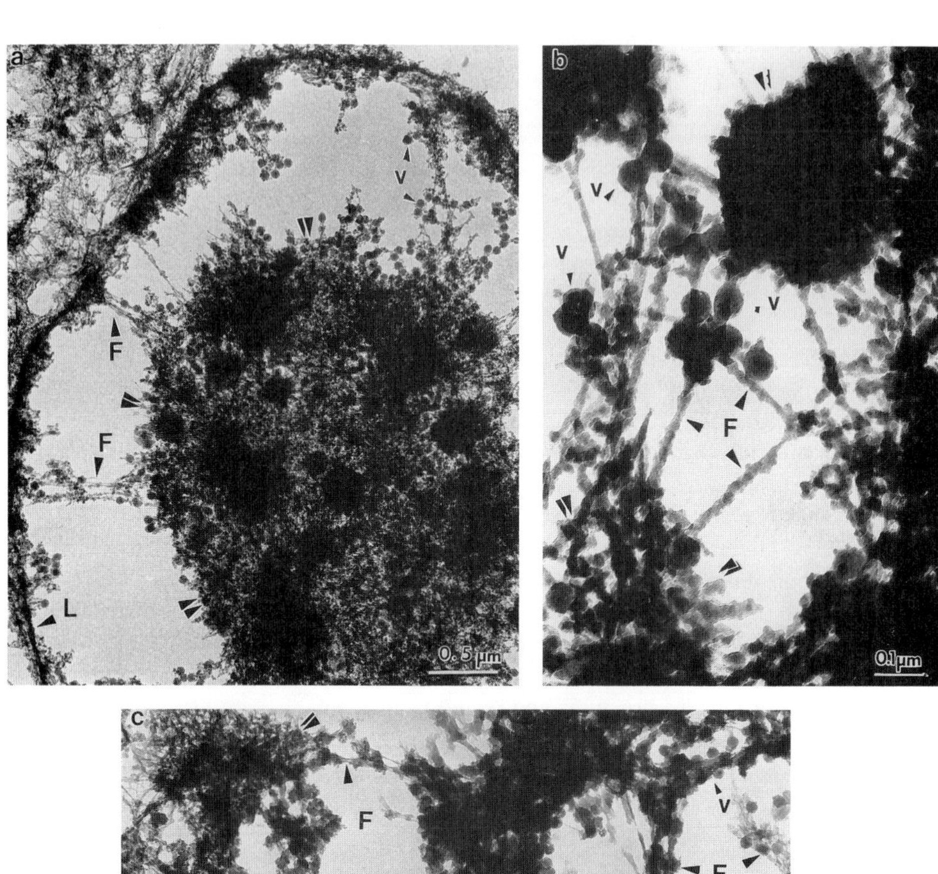
a
v
F
F
L
0.5 μm
b
v
v
v
F
0.1μm

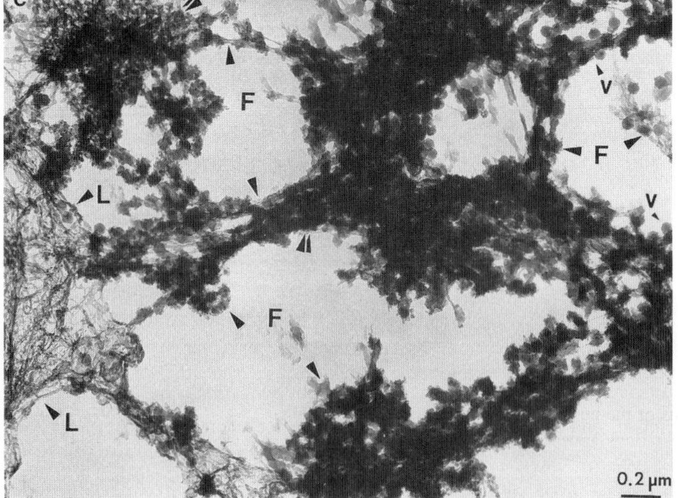
c
F
v
F
L
v
F
L
0.2 μm

The virus-related structures of the detergent-extracted nucleus are composed of at least three areas, which we see by electron microscopy as amorphous, electron-dense regions with capsid-sized granulations on their periphery, granulated regions whose granulations are composed of viral capsids, and filaments that connect these structures to each other and anchor them to the lamina. The decoration of these filaments, which so much resemble the core filaments of the nuclear matrix, with capsids suggests that they may be involved in the transport of capsids and capsid intermediates between the other structures. The granulated regions contain capsids and may be regions of capsid assembly and storage. Amorphous regions have the electron density of DNA and may be the sites of viral DNA replication. The capsid-sized granulations apparently emerging from their periphery suggests to us that DNA is packaged into empty capsid cores at this location.

VIII. CELL TYPE-SPECIFIC PROTEINS OF THE NUCLEAR MATRIX

While chromatin proteins are largely invariant from cell to cell, the protein composition of the nuclear matrix is differentiation state dependent, varying from cell type to cell type (Fey and Penman, 1989; Stuurman *et al.*, 1989, 1990) and with stage of development (Dworetzky *et al.*, 1990). A study in this laboratory has compared the protein composition of the RNA-depleted nuclear matrix, that is the matrix after RNase A digestion, from several lines of cultured cells. The nuclear-matrix proteins of these cell lines could be divided into three classes: general (found in all cells studied), class specific (unique to cells lines from similar tissue

---

Fig. 10. Adenovirus-infected cells at 28 hr after infection. These are high-magnification resinless-section electron micrographs of the cytoskeleton and nuclear matrix of a HeLa cell 28 hr after adenovirus infection. The symbols used in this figure are L, nuclear lamina; Cy, cytoplasmic region; F, filament of the nuclear matrix; double arrowheads, virus center in the nucleus; v, viral capsids. (a) This detergent-extracted, adenovirus-infected HeLa cell has a large virus center in the center of the nucleus suspended by filaments of the nuclear matrix to the nuclear lamina. Mature virions can be seen emerging from this structure, and they decorate the nuclear-matrix fibers. The large virus-specific structure of virus center consists of both amorphous dense regions and granulated regions. (b) This is a detergent-extracted cell shown at a higher magnification. We can see the intimate contact between the filaments of the nuclear matrix and virions and the anchoring of those filaments to a viral center, from which virions appear to be emerging. These filaments are identical in appearance to the core filaments of the nuclear matrix (Figs. 3,4,and 5). (c) This infected cell has been extracted with detergent, digested with DNase I, and treated with 0.25 *M* ammonium sulfate to reveal the RNA-containing nuclear matrix. This micrograph shows one region of the nuclear interior. Filaments of the nuclear matrix can be seen connecting the DNase I-resistant virus center to the nuclear lamina, which is cut in a tangential way in this section. These filaments are beginning to bundle, and virus centers are beginning to overlap onto the filaments. Virions decorate these filaments, especially in the vicinity of the virus center.

origins such as epithelia), and the completely cell-type specific. A surprisingly large number of the nuclear-matrix proteins appeared only in a single-cell type and were, thus, unique tissue markers. These markers have not been seen before, since the interior nuclear-matrix proteins are only 1% of total cell protein and are consequently masked when whole-cell or nuclear extracts are examined.

There were 37 nuclear-matrix proteins common to 3 breast ductal-carcinoma cell lines. Of these 37 proteins, 16 are observed only in the breast lines and not in any other cell lines examined. The nuclear-matrix proteins from 4 different cell types showed a strong cell type–specificity of nuclear-matrix protein composition. There are 47 nuclear-matrix proteins in the human diploid fibroblast line. Of these, only 5 nuclear-matrix proteins are specific to fibroblasts. The fibroblast line has only 7 nuclear-matrix proteins in common with the breast cell lines. Glioma, duodenal, and adrenal cortex lines had 79, 84, and 78 proteins that we analyzed, of which 30, 42, and 32, respectively, are present only in one of the five cell types examined.

The nuclear-matrix protein composition of the differentiating fetal rat osteoblast changes during development (Dworetzky *et al.*, 1990). The changes correlate with the three developmental stages of osteoblast development: proliferation, extracellular-matrix maturation, and mineralization. The temporal pattern of matrix changes is consistent with a nuclear-matrix mediation of phenotypic stage-specific gene expression.

The strong and highly reproducible cell-type and developmental-stage dependence of nuclear-matrix protein composition may have important clinical applications. It may be possible to identify the phenotype or origin of a cell, even when its morphology is altered beyond recognition by malignancy, by analyzing its matrix composition. Even more useful is the predictable change in the nuclear-matrix proteins when cells differentiate within a lineage. For example, the several types of leukemia give rise to homogeneous populations of cells, which may have characteristic patterns of nuclear-matrix proteins.

IX. NUCLEAR CARTOGRAPHY: MAPPING PROTEINS ON THE NUCLEAR MATRIX

The matrix consists of a protein shell, the lamina, surrounding an internal meshwork of protein fibers. The principal components of the lamina are the lamins A, B, and C, which have been well characterized (Gerace *et al.*, 1984; Krohne *et al.*, 1987; Fisher *et al.*, 1982, 1987). Their amino acid sequences are known and the lamins have both sequence and structural similarities to intermediate filament proteins (Fisher *et al.*, 1986; Parry *et al.*, 1986; Franke, 1987). The disassembly of the lamina at mitosis and its reassembly in the daughter cells may be regulated by phosphorylation and dephosphorylation of the lamins (Gerace and Blobel,

1980; Ottaviano and Gerace, 1985; Burke and Gerace, 1986). At mitosis, lamins A and C become soluble, while lamin B associates with a cytoplasmic vesicle.

In contrast to our rather extensive knowledge of the lamins, the portions of the nuclear matrix within the lamina have not been well characterized. A few proteins that associate with the internal matrix have been identified, most notably, topoisomerase II (Berrios *et al.*, 1985; Nelson and Coffey, 1987), which is on the matrix in interphase cells and condenses onto the chromosomes during mitosis. Several known regulatory proteins have been found in the nuclear-matrix fraction including the adenovirus E1A protein (Chatterjee and Flint, 1986), a calmodulin-stimulated protein kinase (Sahyoun *et al.*, 1984), and the transactivating protein of human T-Cell leukemia virus I (Slamon *et al.*, 1988). With these few exceptions, most basic components of the internal fibers and structures of the nuclear are uncharacterized. Little is known about the proteins involved or about how they associate with each other in the matrix structure.

In order to fill this void in our knowledge of matrix structure, we have been making monoclonal antibodies against the nuclear-matrix proteins of various human cell lines. With these antibodies we can identify particular nuclear-matrix proteins in the matrix by high-resolution immunostaining of resinless sections using gold bead–conjugated second antibodies. We can use the same antibody to characterize the antigen biochemically and to identify cDNA clones for molecular sequencing.

One example of this approach to matrix mapping is the monoclonal antibody H1B2. The antigen for this monoclonal antibody has been identified by Western blotting to be three proteins with molecular weights of 105,000, 135,000, and 260,000. Two of these, but not the 260,000 MW protein are also detected by immunoprecipitation of nuclear-matrix proteins.

The H1B2 antigen is masked in interphase cells and does not immunostain until after extraction with 0.25 *M* ammonium sulfate. The resulting immunofluorescent staining pattern is bright, punctate, and entirely nuclear. Removing chromatin with DNase I digestion and 0.25 *M* ammonium sulfate extraction left the H1B2 antigen with the nuclear matrix. Immunogold-stained resinless-section electron micrographs of the core filaments show H1B2 localized on the thick fibers of the nuclear matrix (Fig. 11).

The H1B2 antigen rearranges dramatically during mitosis. Several examples of this rearrangement can be seen in the fluorescence photomicrograph of Fig. 12. As cells approach prophase, the antigen becomes unmasked and stains without the need for salt extraction. First appearing as a bright spot, the antibody staining spreads through the nucleus and, later, around the condensing chromosomes. The antibody also brightly stains the spindle poles and, more weakly, some punctate structures in the cytoplasmic space. As the chromosomes separate at anaphase, H1B2 fluorescence separated into two distinct regions, each half remaining with a daughter chromosome set. The H1B2 antigen returns to the nucleus at telophase but leaves a bright-staining region in the midbody.

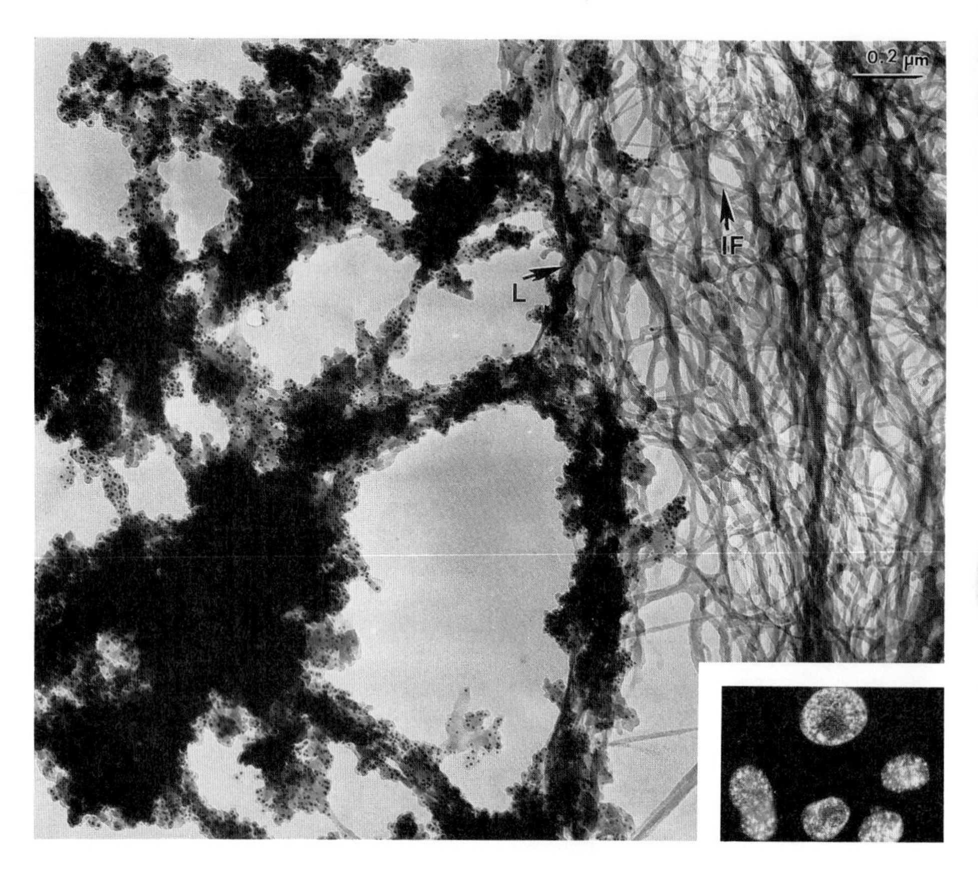

Fig. 11. Immunolocalization of the H1B2 nuclear-matrix protein in resinless sections. W12 cervical dysplasia cells were extracted, digested with DNase I, and salt-extracted with 0.25 *M* ammonuim sulfate to reveal the nuclear matrix. Following fixation, cells were stained with the H1B2 monoclonal antibody and then with either rhodamine-conjugated (inset) or gold bead–conjugated second antibodies. Samples were processed for resinless-section electron microscopy as previously described (Nickerson *et al.,* 1990). L, lamina; IF, intermediate filaments.

Fig. 12. Immunofluorescent staining of sychronized Caski cells with H1B2. These panels show the fluorescence image and the phase-contrast image of the same field. Caski cervical carcinoma cells, grown on glass slides, were synchronized by a single thymidine block. They were extracted with Triton X-100. Staining was with the H1B2 monoclonal antibody and rhodamine-conjugated second antibody. As cells approach prophase, the antigen becomes unmasked and stains without the need for salt extraction. First appearing as a bright spot, the antibody staining spreads through the nucleus and later around the condensing chromosomes. The antibody also brightly stains the spindle poles and some punctate structures in the cytoplasmic space. As the chromosomes separate at anaphase, H1B2 fluorescence separated into two distinct regions, each half remaining with a daughter chromosome set. The H1B2 antigen returns to the nucleus at telophase but leaves a bright-staining region in the midbody.

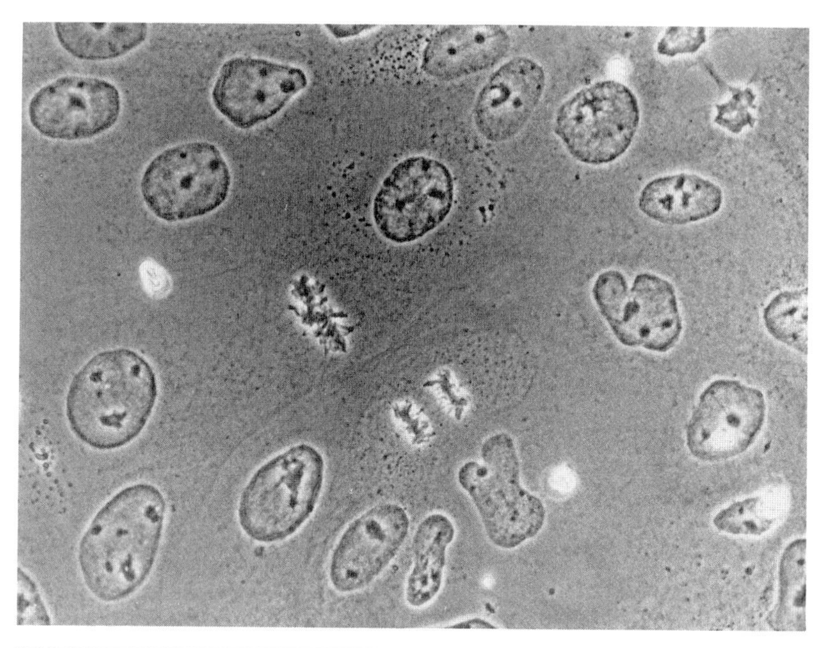

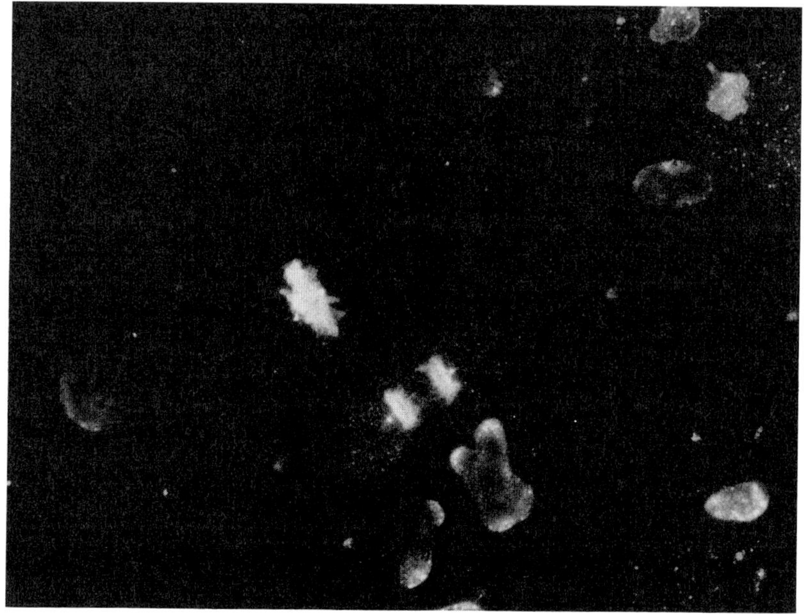

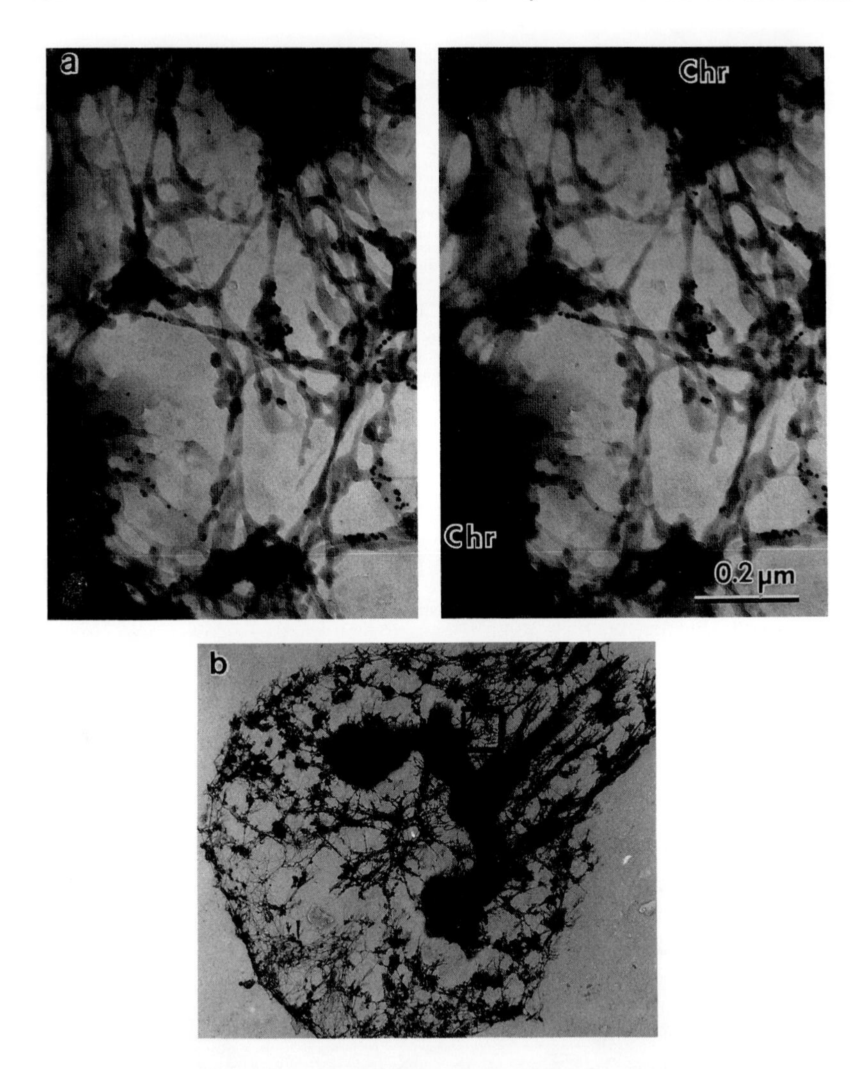

Fig. 13. Mitotic HeLa filaments decorated by H1B2 antibody and seen in three dimensions. HeLa cells were sychronized by a double thymidine block. (a) A resinless-section stereoscopic micrograph of a mitotic HeLa cell stained with H1B2 antibody. Cells were isolated by mitotic shake, extracted with Triton X-100, and processed in suspension. Cells were fixed with paraformaldehyde and stained with H1B2 and a gold bead–conjugated second antibody. The cells were processed for resinless-section electron microscopy. The micrographs were made at 10 degrees total tilt angle. H1B2 was located on a fibrogranular network connected to the Chromosomes (Chr). (b) (Bottom) The whole section from which the stereopair of part (a) was selected.

Some of the H1B2 antigen in the perichromosomal region is on a network of filaments that radiate out from and connect the chromosomes. This distribution can be seen in three dimensions in the stereo-electron micrograph of Fig. 13a.

X. THE NUCLEAR MATRIX AND REGULATION OF GENE EXPRESSION

There are three control points in gene expression at which the nuclear matrix may play a key role: transcription, RNA processing, and transport of fully processed RNA out of the nucleus. As reviewed below, there is evidence for matrix involvement with each of these processes. Additionally, there is an association of steroid-hormone receptors with the matrix, suggesting that regulation of gene expression by steroid hormones may involve matrix–chromatin interactions (Simmen *et al.*, 1984; Barrack, 1983; Barrack and Coffey, 1980; Kumara *et al.*, 1986; Rennie *et al.*, 1983; Kirsch *et al.*, 1986; Buttyan *et al.*, 1983).

Several kinds of evidence show a role for the nuclear matrix in transcription. Active RNA polymerase, nascent RNA transcripts, and actively transcribed genes are all associated with a nonchromatin nuclear structure (Jackson and Cook, 1985). While all nuclear-matrix isolation protocols remove DNA, a small fraction always remains with the matrix after isolation. The sequence content of this DNA has been studied and found to be enriched in actively transcribed sequences. Global measurements using end-labeled polyadenylated mRNA to quantify matrix-associated DNA have shown that transcribed sequences resist detachment from the matrix (Jackson and Cook, 1985; Thorburn *et al.*, 1988). The transcribed ovalbumin gene, but not the untranscribed β-globin gene, is enriched in the matrix DNA of chicken oviduct cells (Robinson *et al.*, 1982; Ciejek *et al.*, 1983). In cells producing β-globin, Hentzen *et al.* (1984) found an enrichment of this sequence in matrix-associated DNA, while Ross *et al.* (1982), using a different matrix-isolation procedure, did not. Ribosomal DNA is enriched in matrix fractions only when being actively transcribed (Pardoll and Vogelstein, 1980; Keppel, 1986). These observations may reflect the involvement of the nuclear matrix in the architecture of chromatin, organizing it into regions of transcribed euchromatin and nontranscribed condensed heterochromatin (Nickerson *et al.*, 1989). Changes in matrix composition and architecture may alter the structure of chromatin, changing the accessibility of specific genes for transcription.

Chromatin loops are constrained at their bases by the nuclear matrix. The DNA regions bound to the matrix, matrix-attachment regions (MARs), have been localized to specific sequences flanking mouse immunoglobulin genes (Cockerill and Garrard, 1986), the chicken lysosome gene (Phi-Van and Strätling, 1988), the human interferon-β gene (Bode and Maass, 1988), the human β-globin gene (Jarman and Higgs, 1988), and the Chinese hamster dihydrofolate reductase gene (Käs and Chasin, 1987) as well as various *Drosophila* genes (Mirkovitch *et al.*,

1984; Gasser and Laemmli, 1986). Most MARs (sometimes called scaffold-attachment regions) are A-T rich sequences, are conserved in evolution, and have high concentration sequences similar to topoisomerase II cleavage sites. Incorporation of MAR sequences into DNA constructs used for transfection of foreign genes increases the rates of transcription from those genes in stable transfectants by about 10-fold (Stief *et al.*, 1989; Phi-Van *et al.*, 1990). Thus, matrix attachments of genes in the regions flanking their coding sequences may be necessary for efficient transcription.

The nuclear matrix may be the site for mRNA processing from nascent transcript to mature mRNA. Pulse-labeled nascent RNA chains are associated with a nuclear structure and are not removed by DNase I digestion and removal of DNA (Jackson and Cook, 1985). Many studies have found the precursors for specific mRNAs on the matrix, suggesting that RNA splicing may occur on this structure (Ross *et al.*, 1982; Ciejek *et al.*, 1983; Ben Ze'ev and Aloni, 1983; Abulafai *et al.*, 1984; Mariman *et al.*, 1982). In most studies, more than 90% of precursor RNAs are matrix associated. Zeitlin and colleagues (Zeitlin *et al.*, 1987; Zeitlin *et al.*, 1989) have shown that splicing can occur on isolated nuclear matrices. They prepared nuclear matrices from cells producing large amounts of specific pre-mRNA. When the matrix was prepared using DNase I digestion and 0.1 *M* KCl extraction, it could perform autonomous splicing of its pre-mRNA to mature mRNA (Zeitlin *et al.*, 1989). The reaction required only ATP and Mg^{2+}, occurred without a temporal lag, and was half-complete in 5 mins. Although splicing of added pre-mRNA can occur in soluble extracts following a lag for spliceosome assembly, RNA splicing *in vivo* may occur in matrix-bound spliceosomes.

Following splicing, mature ovalbumin mRNA can be released from the isolated chicken oviduct nuclear matrix by an ATP-dependent process (Schröder *et al.*, 1987a, 1987b). Mature mRNA is not released from the matrix in order to simply diffuse out of the nucleus. Lawrence and friends (Lawrence *et al.*, 1989) have shown by *in situ* hybridization that specific mRNAs travel from gene to the nuclear periphery along single *tracks*. These transport paths are often linear and survive the removal of chromatin, showing that they are part of the underlying nuclear matrix (Xing and Lawrence, 1991).

In summary, transcription, RNA processing from precursor to mature mRNA, and mRNA transport to the nuclear periphery take place on the nuclear matrix. The matrix association of these key steps in RNA metabolism provides many points at which gene expression could be regulated by changes in nuclear-matrix composition, conformation, and architecture.

XI. FUTURE DIRECTIONS FOR NUCLEAR MATRIX RESEARCH

The structure of the nuclear matrix needs a more complete characterization than has been possible until now. Our approach of using monoclonal antibodies to specific nuclear-matrix proteins in both ultrastructural and biochemical studies will eventually allow us to *map* the matrix in much greater detail, but only the proteinaceous part of the matrix. Understanding the role that RNA may play in nuclear-matrix architecture will be a much harder task, but an exciting one. We see here, for the first time, a hint of connection between the enormous complexity of the genome, much of which is transcribed into noncoding hnRNA, and the patterns of gene regulation that serve to determine cell type and tissue architecture.

The most apparent function of the nuclear matrix is one of geometric housekeeping. The length of chromatin that must be packed into the small space of the nucleus is enormous. Such packing must be specific, in that transcriptionally active regions are extended in the euchromatic regions, while the remainder and larger portion of DNA is compacted in transcriptionally inactive heterochromatin. During DNA replication, parental and daughter strands remain organized and untangled as they are reeled through replication complexes until, at mitosis, the chromatin is condensed into chromosomes. This elaborate ballet is performed upon the matrix, and we need to understand the steps.

Nuclear-matrix research is a science in its infancy, providing many opportunities to address fundamental and difficult problems of nuclear structure and metabolism. The matrix does not exist in isolation, but must be seen as part of a tissue-wide structure of matrix, cytoskeleton, and extracellular structure (Fig. 1). Although our short-term goal will be to characterize these components individually, it may be the complex interactions between them—acting together as one larger structure—that is the most fundamental mechanism of regulation.

REFERENCES

Abulafia, R., Ben-Ze'ev, A., Hay, N., and Aloni, Y. (1984). Control of late-SV40 transcription by the attenuation mechanism and transcriptionally active ternary complexes are associated with the nuclear matrix. *J. Mol. Biol.* **172**, 467–487.

Alcover, A., Izquierdo, M., Stollar, D., Kitagawa, Y., Miranda, M., and Alonso, C. (1982). *In situ* immunofluorescent visualization of chromosomal transcripts in polytene chromosomes. *Chromosoma* **87**, 263–277.

Ashburner, M. (1972). Ecdysone induction of puffing in polytene chromosomes of *Drosophila melanogaster*. Effects of inhibitors of RNA synthesis. *Exp. Cell. Res.* **71**, 433–440.

Barrack, E. R., and Coffey, D. S. (1980). The specific binding of estrogens and androgens to the nuclear matrix of sex hormone–responsive tissues. *J. Biol. Chem.* **255**, 7265–7275.

Barrack, E. R. (1983). The nuclear matrix of the prostate contains acceptor sites for androgen receptors. *Endocrinology* **114**, 430–432.

Ben-Ze'ev, A., and Aloni, Y. (1983). Processing of SV40 RNA is associated with the nuclear matrix and is not followed by the accumulation of low-molecular-weight RNA products. *Virology* **125**, 475–479.

Berezney, R. (1984). Organization and functions of the nuclear matrix. *In* "Chromosomal Nonhistone proteins," Vol. IV (L.S. Hnilica, ed.), pp. 119–180. CRC Press, Boca Raton, Florida.

Berezney, R., and Coffey, D. S. (1974). Identification of a nuclear protein matrix. *Biochem. Biophys. Res. Commun.* **60,** 1410–1417.

Berezney, R., and Coffey, D. S. (1975). Nuclear protein matrix: Association with newly synthesized DNA. *Science* **189**, 291–292.

Berezney, R., and Coffey, D. S. (1977). Nuclear matrix: Isolation and characterization of a framework structure from rat liver nuclei. *J. Cell Biol.* **73**, 616–637.

Bernhard, W. (1969). A new procedure for electron microscopical cytology. *J. Ultrastruct. Mol. Struct.* **27**, 250–265.

Bernhard, W. (1971). Drug-induced changes in the interphase nucleus. *In* "Advances in cytopharmacology, Vol. 1.: First international symposium on cell biology and cytopharmacology" (F. Clementi and B. Ceccarelli, eds.), pp. 49–76. Raven Press, New York.

Berrios, M., Osheroff, N., and Fisher, P. A. (1985). *In situ* localization of DNA topoisomerase II, a major polypeptide component of the *Drosophila* nuclear matrix fraction. *Proc. Natl. Acad. Sci. U.S.A.* **82**, 4142–4146.

Bode, J., and Maass, K. (1988). Chromatin domain surrounding the human interferon-β gene as defined by scaffold-attached regions. *Biochemistry* **27**, 4706–4711.

Bourgeois, C. A., Laquerriere, F., Hemon, D., Hubert, J., and Bouteille, M. (1985). New data on the *in situ* position of the inactive X chromosome in the interphase nucleus of human fibroblasts. *Hum. Genet.* **69**, 122–129.

Bouvier, D., Hubert, J., Seve, A., and Bouteille, M. (1982). RNA is responsible for the three-dimensional organization of nuclear-matrix proteins in HeLa cells. *Biol. Cell.* **43**, 143–146.

Burke, B., and Gerace, L. (1986). A cell-free system to study reassembly of the nuclear envelope at the end of mitosis. *Cell* **44**, 639–652.

Buttyan, R., Olsson, C. A., Sheard, B., and Kallos, J. (1983). Steroid receptor–nuclear matrix interactions. The role of DNA. *J. Biol. Chem.* **258**, 14366–14370.

Capco, D. G., Krochmalnic, G., and Penman, S. (1984). A new method for preparing embedment-free sections for transmission electron microscopy: Applications to the cytoskeletal framework and other three-dimensional networks. *J. Cell Biol.* **98**, 1878–1885.

Capco, D. G., Wan, K. M., and Penman, S. (1982). The nuclear matrix: Three-dimensional architecture and protein composition. *Cell* **29**, 847–858.

Carmo-Fonseca, M., Cidadao, A. J., and David-Ferreira, J. F. (1987). Filamentous cross-bridges link intermediate filaments to the nuclear pore complexes. *Eur. J. Cell. Biol.* **45**, 282–290.

Chatterjee, P. K., and Flint, S. J. (1986). Partition of E1A proteins between soluble and structural fractions of adenovirus-infected and -transformed cells. *J. Virol.* **60**, 1018–1026.

Ciejek, E. M., Tsai, M.-J., and O'Malley, B. W. (1983). Actively transcribed genes are associated with the nuclear matrix. *Nature* **306**, 607–609.

Cockerill, P. N., and Garrard, W. T. (1986). Chromosomal loop anchorage of the kappa immunoglobulin gene occurs next to the enhancer in a region containing topoisomerase II sites. *Cell* **44**, 273–282.

Cremer, T., Cremer, C., Schneider, T., Baumann, H., Hens, L., and Kirsch-Volders, M. (1982a). Analysis of chromosome positions in the interphase nucleus of Chinese hamster cells by laser-UV-microirradiation experiments. *Hum. Genet. 1982.* **62**, 201–209.

Cremer, T., Cremer, C., Baumann, H., Luedtke, E. K., Sperling, K., Teuber, V., and Zorn, C. (1982b). Rabl's model of the interphase chromosome arrangement tested in Chinese hamster cells by premature chromosome condensation and laser-UV-microbeam experiments. *Hum. Genet. 1982.* **60**, 46–56.

Dworetzky, S. I., Fey, E. G., Penman, S., Lian, J. B., Stein, J. L., and Stein, G. S. (1990). Progressive changes in the protein composition of the nuclear matrix during osteoblast differentiation. *Proc. Natl. Acad. Sci. U.S.A.* **87**, 4605–4609.

Fawcett, D. W. (1966). "An Atlas of Fine Structure: The Cell, Its Organelles and Inclusions. W. B. Saunders, Philadelphia.

Felsenfeld, G., and McGhee, J. D. (1986). Structure of the 30-nm chromatin fiber. *Cell* **44**, 375–377.

Fey, E. G., Krochmalnic, G., and Penman, S. (1986). The nonchromatin substructures of the nucleus: The ribonucleoprotein (RNP)-containing and RNP-depleted matrices analyzed by sequential fractionation and resinless-section electron microscopy. *J. Cell Biol.* **102**, 1654–1665.

Fey, E. G., and Penman, S. (1988). Nuclear-matrix proteins reflect cell type of origin in cultured human cells. *Proc. Natl. Acad. Sci. U.S.A.* **85**, 121–125.

Fisher, P. A., Berrios, M., and Blobel, G. (1982). Isolation and characterization of a proteinaceous subnuclear fraction composed of nuclear matrix, peripheral lamina, and nuclear pore complexes from embryos of *Drosophila melanogaster. J. Cell. Biol.* **92**, 674–686.

Fisher, D. Z., Chaudhary, N., and Blobel, G. (1986). cDNA sequencing of nuclear lamins A and C reveals primary and secondary structural homology to intermediate filament proteins. *Proc. Natl. Acad. Sci. U.S.A.* **83**, 6450–6454.

Franke, W. W. (1987). Nuclear lamins and cytoplasmic intermediate filament proteins: A growing multigene family. *Cell* **48**, 3–4.

Fulton, A. B., Wan, K. M., and Penman, S. (1980). The spatial distribution of polyribosomes in 3T3 cells and the associated assembly of proteins into the skeletal framework. *Cell* **20**, 849–857.

Gallinaro, H., Puvion, E., Kister, L., and Jacob, M. (1983). Nuclear matrix and hnRNP share a common structural constituent with premessenger RNA. *EMBO J.* **2**, 953–960.

Gasser, S. M., and Laemmli, U. K. (1986). Cohabitation of scaffold-binding regions with upstream/ enhancer elements of three developmentally regulated genes of *D. melanogaster. Cell* **46**, 521–530.

Gerace, L., Comeau, C., and Benson, M. (1984). Organization and modulation of nuclear lamina structure. *J. Cell Sci.* (Suppl.) **1**, 137–160.

Gerace, L., and Blobel, G. (1980). The nuclear envelope lamina is reversibly depolymerized during mitosis. *Cell* **19**, 277–287.

Harpold, M. M., Wilson, M. C., and Darnell, J. E., Jr. (1981). Chinese hamster polyadenylated messenger ribonucleic acid: Relationship to nonpolyadenylated sequences and relative conservation during messenger ribonucleic acid processing. *Mol. Cell. Biol.* **1**, 188–198.

He, D., Nickerson, J. A., and Penman, S. (1990). The core filaments of the nuclear matrix. *J. Cell Biol.* **110**, 569–580.

Hens, L., Baumann, H., Cremer, T., Sutter, A., Cornelis, J. J., and Cremer, C. (1983). Immunocytochemical localization of chromatin regions UV-microirradiated in S phase or anaphase. Evidence for a territorial organization of chromosomes during cell cycle of cultured Chinese hamster cells. *Exp. Cell. Res.* **149**, 257–269.

Hentzen, P. C., Rho, J. H., and Bekhor, I. (1984). Nuclear matrix DNA from chicken erythrocytes contains b-globin gene sequences. *Proc. Natl. Acad. Sci. U.S.A.* **81**, 304–307.

Herman, R., Weymouth, L., and Penman, S. (1978). Heterogeneous nuclear RNA–protein fibers in chromatin-depleted nuclei. *J. Cell Biol.* **78**, 663–674.

Hodge, L. D., Mancini, P., Davis, F. M., and Heywood, P. (1977). Nuclear matrix of HeLa S3 cells. Polypeptide composition during adenovirus infection and in phases of the cell cycle. *J. Cell Biol.* **72**, 194–208.

Jackson, D. A., and Cook, P. R. (1985). Transcription occurs at a nucleoskeleton. *EMBO J.* **4**, 919–925.

Jackson, D. A., and Cook, P. R. (1988). Visualization of a filamentous nucleoskeleton with a 23-nm axial repeat. *EMBO J.* **7**, 3667–3678.

Jarman, A. P., and Higgs, D. R. (1988). Nuclear-scaffold attachment sites in the human globin gene complexes. *EMBO J.* **7**, 3337–3344.

Käs, E., and Chasin, L. A. (1987). Anchorage of the Chinese hamster dihydrofolate reductase gene to the nuclear scaffold occurs in an intragenic region. *J. Mol. Biol.* **198**, 677–692.

Keppel, F. (1986). Transcribed human ribosomal RNA genes are attached to the nuclear matrix. *J. Mol. Biol.* **187**, 15–21.

Kirsch, T. M., Miller, A. D., and Litwack, G. (1986). The nuclear matrix is the site of glucocorticoid receptor complex action in the nucleus. *Biochem. Biophys. Res. Commun.* **137**, 640–648.

Krohne, G., Wolin, S. L., McKeon, F. D., Franke, W. W., and Kirschner, M. W. (1987). Nuclear lamin LI of *Xenopus laevis*: cDNA cloning, amino acid sequence and binding specificity of a member of the lamin B subfamily. *EMBO J.* **6**, 3801–3808.

Kubai, D. F., and Ris, H. (1969). Division in the dinoflagellate *Gyrodinium cohnii* (Schiller). *J. Cell Biol.* **40**, 508–528.

Kumara, S.M.H., Shapiro, L. E., and Surks, M. I. (1986). Association of the 3,5,3′-triiodo-L-thyronine nuclear receptor with the nuclear matrix of cultured growth hormone–producing rat pituitary tumor cells (GC cells). *J. Biol. Chem.* **261**, 2844–2852.

Lawrence, J. B., Singer, R. H., and Marselle, L. M. (1989). Highly localized tracks of specific transcripts within interphase nuclei visualized by *in situ* hybridization. *Cell* **57**, 493–502.

Lenk, R., Ransom, L., Kaufman, Y., and Penman, S. (1977). A cytoskeletal structure with associated polyribosomes obtained from HeLa cells. *Cell* **10**, 67–78.

Leser, G. P., Escara-Wilke, J., and Martin, T. E. (1984). Monoclonal antibodies to heterogeneous nuclear RNA-protein complexes: The core proteins comprise a conserved group of related polypeptides. *J. Biol. Chem.* **259**, 1827–1833.

Lewis, M., Helmsing, P. J., and Ashburner, M. (1975). Parallel changes in puffing activity and patterns of protein synthesis in salivary glands of *Drosophila. Proc. Natl. Acad. Sci. U.S.A.* **72**, 3604–3608.

Long, B. H., and Schrier, W. H. (1983). Isolation from Friend erythroleukemia cells of an RNase-sensitive nuclear-matrix fibril fraction containing hnRNA and snRNA. *Biol. Cell.* **48**, 99–108.

Long, B. H., and Ochs, R. L. (1983). Nuclear matrix, hnRNA, and snRNA in Friend erythroleukemia nuclei depleted of chromatin by low-ionic-strength EDTA. *Biol. Cell.* **48**, 89–98.

Manuelidis, L. (1985). Individual interphase chromosome domains revealed by *in situ* hybridization. *Hum. Genet.* **71**, 288–293.

Mariman, E.C.M., van Eekelen, C.A.G., Reinders, R. J., Berns, A.J.M., and van Venrooij, W. J. (1982). Adenoviral heterogeneous nuclear RNA is associated with the host nuclear matrix during splicing. *J. Mol. Biol.* **154**, 103–119.

Mayer, D. T., and Gulick, A. (1942). The nature of the proteins of cellular nuclei. *J. Biol. Chem.* **46**, 433–440.

McCready, S. J., Godwin, J., Mason, D. W., Brazell, I. A., and Cook, P. R. (1980). DNA is replicated at the nuclear cage. *J. Cell Sci.* **46**, 365–386.

McKeon, F. D., Kirschner, M. W., and Caput, D. (1986). Homologies in both primary and secondary structure between nuclear envelope and intermediate filament proteins. *Nature (London)* **319**, 463–468.

Miller, T. E., Huang, C. Y., and Pogo, A. O. (1978). Rat liver nuclear skeleton and ribonucleoprotein complexes containing hnRNA. *J. Cell Biol.* **76**, 675–691.

Mirkovitch, J., Mirault, M-.E., and Laemmli, U. K. (1984). Organization of the higher-order chromatin loop: Specific DNA attachment sites on nuclear scaffold. *Cell* **39**, 223–232.

Nelson, W. G., and Coffey, D. S. (1987). Structural aspects of mammalian DNA replication: Topoisomerase II. *NCI Monographs* **4**, 23–29.

Nickerson, J. A., Krochmalnic, G., Wan, K. M., and Penman, S. (1989). Chromatin architecture and nuclear RNA. *Proc. Natl. Acad. Sci. U.S.A.* **86**, 177–181.

Nickerson, J. A., Krochmalnic, G., He, D., and Penman, S. (1990). Immunolocalization in three dimensions: Immunogold staining of cytoskeletal and nuclear matrix proteins in resinless electron microscopy sections. *Proc. Natl. Acad. Sci. U.S.A.* **87**, 2259–2263.

Ottaviano, Y., and Gerace, L. (1985). Phosphorylation of the nuclear lamins during interphase and mitosis. *J. Biol. Chem.* **260**, 624–632.

Pardoll, D. M., and Vogelstein, B. (1980). Sequence analysis of nuclear matrix associated DNA from rat liver. *Exp. Cell Res.* **128**, 466–470.

Pardoll, D. M., Vogelstein, B., and Coffey, D. S. (1980). A fixed site of DNA replication in eukaryotic cells. *Cell* **19**, 527–536.

Parry, D. A., Conway, J. F., and Steinert, P. M. (1986). Structural studies on lamin. Similarities and differences between lamin and intermediate-filament proteins. *Biochem. J.* **238**, 305–308.

Penman, S., Fulton, A., Capco, D., Ben-Ze'ev, A., Wittelsberger, S., and Tse, C. F. (1982). Cytoplasmic and nuclear architecture in cells and tissue: Form, functions, and mode of assembly. *Cold Spring Harb. Symp. Quant. Biol.,* **46**, 1013–1028.

Penman, S., Vesco, C., Weinberg, R., and Zylber, E. (1969). The RNA metabolism of nucleoli and mitochondria in mammalian cells. *Cold Spring Harb. Symp. Quant. Biol.* **34**, 535–546.

Penman, S., Vesco, C., and Penman, M. (1968). Localization and kinetics of formation of nuclear heterodisperse RNA, cytoplasmic heterodisperse RNA, and polyribosome-associated messenger RNA in HeLa cells. *J. Mol. Biol.* **34**, 49–69.

Perov, N. A., and Kiknadze, I. I. (1980). Differential reaction of polytene chromosome segments to high doses of actinomycin D. *Tsitologiia.* **22**, 254–259.

Perry, R. P. (1962). The cellular sites of synthesis of ribosomal and 45s RNA. *Proc. Natl. Acad. Sci. U.S.A.* **48**, 2179.

Phi-Van, L., von Kries, J. P., Ostertag, W., and Strätling, W. H. (1990). The chicken lysozyme 5′ matrix attachment region increases transcription from a heterologous promoter in heterologous cells and dampens positional effects on the expression of transfected genes. *Mol. Cell. Biol.* **10**, 2302–2307.

Phi-Van, L., and Strätling, W. H. (1988). The matrix attachment regions of the chicken lysozyme gene co-map with the boundaries of the chromatin domain. *EMBO J.* **7**, 655–664.

Porter, K. R. (1984). The cytomatrix: A short history of its study. *J. Cell Biol.* **99**, 3s–12s.

Puvion-Dutilleul, F., and Puvion, E. (1981). Relationship between chromatin and perichromatin granules in cadmium-treated isolated hepatocytes. *J. Ultrastruct. Res.* **74**, 341–350.

Rappold, G. A., Cremer, T., Hager, H. D., Davies, K. E., Muller, C. R., and Yang, T. (1984). Sex chromosome positions in human interphase nuclei as studied by *in situ* hybridization with chromosome-specific DNA probes. *Hum. Genet.* **67**, 317–325.

Rennie, P. S., Bruchovsky, N., and Cheng, H. (1983). Isolation of 3 S androgen receptors from salt-resistant fractions and nuclear matrices of prostatic nuclei after mild trypsin digestion. *J. Biol. Chem.* **258**, 7623–7630.

Reynolds, E. S. (1963). The use of lead citrate at high pH as an electron-opaque stain in electron microscopy. *J. Cell. Biol.* **17**, 208–213.

Robinson, S. I., Nelkin, B. D., and Vogelstein, B. (1982). The ovalbumin gene is associated with the nuclear matrix of chicken oviduct cells. *Cell* **28**, 99–106.

Ross, D. A., Yen, R. W., and Chae, C. B. (1982). Association of globin ribonucleic acid and its precursors with the chicken erythroblast nuclear matrix. *Biochemistry* **21**, 764–771.

Sahyoun, N., LeVine, H., and Cuatrecasas, P. (1984). Ca^{2+}/calmodulin-dependent protein kinases from the neuronal nuclear matrix and postsynaptic density are structurally related. *Proc. Natl. Acad. Sci. U.S.A.* **81**, 4311–4315.

Salditt-Georgieff, M., Harpold, M. M., Wilson, M. C., and Darnell, J. E., Jr. (1981). Large heterogeneous nuclear ribonucleic acid has three times as many 5′ caps as polyadenylic acid segments, and most caps do not enter polyribosomes. *Mol. Cell. Biol.* **1**, 179–187.

Scheller, R. H., Costantini, F. D., Kozlowski, M. R., Britten, R. J., and Davidson, E. H. (1978). Specific representation of cloned repetitive DNA sequences in sea urchin RNAs. *Cell* **15**, 189–203.

Schröder, H. C., Trölltsch, D., Friese, U., Bachmann, M., and Müller, W. E. G. (1987a). Mature mRNA is selectively released from the nuclear matrix by an ATP/dATP–dependent mechanism sensitive to topoisomerase inhibitors. *J. Biol. Chem.* **262**, 8917–8925.

Schröder, H. C., Trölltsch, D., Wenger, R., Bachmann, M., Diehl-Seifert, B., and Müller, W.E.G. (1987b). Cytochalasin B selectively releases ovalbumin mRNA precursors but not the mature ovalbumin mRNA from hen oviduct nuclear matrix. *Eur. J. Biochem.* **167**, 239–245.

Simmen, R. C., Means, A. R., and Clark, J. H. (1984). Estrogen modulation of nuclear matrix–associated steroid-hormone binding. *Endocrinology* **115**, 1197–1202.

Slamon, D. J., Boyle, W. J., Keith, D. E., Press, M. F., Golde, D. W., and Souza, L. M. (1988). Subnuclear localization of the transactivating protein of T-cell leukemia virus type I. *J. Virol.* **62**, 680–686.

Small, D., Nelkin, B., and Vogelstein, B. (1985). The association of transcribed genes with the nuclear matrix of *Drosophila* cells during heat shock. *Nucleic Acids Res.* **13**, 2413–2431.

Stief, A., Winter, D. M., Strätling, W. H., and Sippel, A. E. (1989). A nuclear-attachment element mediates elevated and position-independent gene activity. *Nature (London)* **341**, 343–345.

Stuurman, N., Meijne, A. M., van-der-Pol, A. J., de-Jong, L., van-Driel, R., and van-Renswoude, J. (1990). The nuclear matrix from cells of different origin. Evidence for a common set of matrix proteins. *J. Biol. Chem.* **265**, 5460–5465.

Stuurman, N., Van-Driel, R., De-Jong, L., Meijne, A. M., and Van-Renswoude, J. (1989). The protein composition of the nuclear matrix of murine P19 embryonal carcinoma cells is differentiation-stage dependent. *Exp. Cell Res.* **180**, 460–466.

Thomas, J. O. (1984). The higher-order structure of chromatin and histone H1. *J. Cell Sci.* (Suppl.) **1**, 1–20.

Thorburn, A., Moore, R., and Knowland, J. (1988). Attachment of transcriptionally active sequences to the nucleoskeleton under isotonic conditions. *Nucleic Acids Res.* **16**, 7183.

van Eekelen, C.A.G., and van Venrooij, W. J. (1981). hnRNA and its attachment to a nuclear-protein matrix. *J. Cell Biol.* **88**, 554–563.

Wassef, M. (1979). A cytochemical study of interchromatin granules. *J. Ultrastruct. Mol. Struct. Res.* **69**, 121–133.

Wolosewick, J. (1980). The application of polyethylene glycol (PEG) to electron microscopy. *J. Cell Biol.* **86**, 675–681.

Xing, Y., and Lawrence, J. B. (1991). Preservation of specific RNA distribution within the chromatin-depleted nuclear substructure demonstrated by *in situ* hybridization coupled with biochemical fractionation. *J. Cell Biol.* **112**, 1055–1063.

Zeitlin, S., Parent, A., Silverstein, S., and Efstratiadis, A. (1987). Pre-mRNA splicing and the nuclear matrix. *Mol. Cell. Biol.* **7**, 111–120.

Zeitlin, S., Wilson, R. C., and Efstradiadis, A. (1989). Autonomous splicing and complementation of *in vivo* assembled splicosomes. *J. Cell Biol.* **108**, 765–777.

Zhai, Z. H., Nickerson, J. A., Krochmalnic, G., and Penman, S. (1987). Alterations in nuclear-matrix structure after adenovirus infection. *J. Virol.* **61**, 1007–1018.

12

Histone Modifications Associated with Mitotic Chromosome Condensation

JOHN P.H. TH'NG*, XIAO-WEN GUO*, AND E. MORTON BRADBURY[†]*

*Department of Biological Chemistry
School of Medicine
University of California
Davis, California 95616

[†]Life Sciences Division
Los Alamos National Laboratories
Los Alamos, New Mexico 87545

I. INTRODUCTION

Each human cell contains about 2 meters of DNA that are assembled into 46 chromosomes and packed into the nucleus, which has a diameter of about 10 μm. In addition to this 6 pg of DNA, the nucleus also contains about 6 pg of RNA and 12 pg of protein, equivalent to a concentration of about 50 mg/ml. Thus, chromosome functions of replication, transcription, and mitosis are controlled within the confines of a densely packed and, as we are learning, highly organized cell nucleus. Reversible posttranslational modifications of the histones are believed to play a key role in the dynamic changes in the organization and structures of chromatin required for these functions.

II. CHROMATIN LOOPS AND THE METAPHASE CHROMOSOME

Models of chromosome organization are relevant to our understanding of the functions of histone modifications. There is increasing evidence that chromosomes have long-range order, described as chromatin loops, containing 5 to 150 kbp DNA (Marsden and Laemmli, 1979; Laemmli, 1985) that may specify both structural and genetic units. With an average DNA loop size of 50 kbp, the human genome of 3×10^9 bp would contain 60,000 loops, a number comparable with the 50,000–100,000 genes estimated to be in human cells. These DNA loops are constrained by attachments of specific DNA sequences located at the bases of the loops to nuclear-scaffold proteins (Laemmli, 1985; Lewis *et al.*, 1984) or nuclear-matrix proteins (Barrack and Coffey, 1982; Pienta and Coffey, 1984). One of the major scaffold proteins, Scl, has now been identified as topoisomerase II (Earnshaw *et al.*, 1985; Earnshaw and Heck, 1985). In eukaryotes, both topoisomerases I and II relax supercoiled DNA. However, there is as yet no clear evidence for the existence of the eukaryotic equivalent of a bacterial gyrase. Possibly the evolution of nucleosomal packaging of DNA and higher-order chromatin structures have generated alternative mechanisms for the control of DNA supercoiling in constrained DNA loops.

III. NUCLEOSOMES AND THE CHROMATIN STRUCTURE

Virtually all of the DNA in the eukaryotic genome is packaged into nucleosomes, the fundamental structural unit of chromosomes. Hence, understanding chromosome structures and functions at the molecular level requires a detailed understanding of the structures and functions of the nucleosome, and particularly, how chromatin variables associated with chromosome functions affect nucleosome structure and DNA topology in closed DNA loops.

Nucleosomes from most eukaryotic cells contain DNA with a repeat length of 195 ± 5 bp DNA, the histone octamer $[(H2A, H2B)_2 (H3_2, H4_2)]$ and one histone H1. However, the repeat lengths of the nucleosomal DNA can be shorter in some lower eukaryotes, e.g., 170 bp in *Neurospora crassa* and 165 bp in yeast, while those of some specialized cells can be longer, e.g., 212 bp in avian erythrocytes. Nuclease digestion studies of nucleosomes reveal two other defined subnucleosome particles; the chromatosome with 168 bp DNA, the histone octamer and one histone H1, and the nucleosome core particle with 146-bp DNA and the histone octamer.

The nucleosome core particle has been subjected to intense structural studies. Neutron-scatter studies showed that the low-resolution structure of the core particle in solution is a flat disc 11.0 nm in diameter, 5.5–6.0 nm thick, with 1.7 turns of DNA coiled with a pitch of 3.0 nm on the outside of the histone octamer (Fig. 1) (Pardon *et al.*, 1975, 1977a,b; Baldwin *et al.*, 1975; Bradbury *et al.*, 1975; Suau *et al.*, 1977; Braddock *et al.*, 1981). Crystallographic structure determinations have revealed the core-particle structure to a resolution of 0.7 nm (Richmond *et al.*, 1984). Within the resolution afforded by the neutron-scatter studies, the structure of the core particle in solution is very similar to the crystal structure. The crystal structure has also revealed that the path of the DNA is not a uniform coil, but follows a more irregular path of less-bent segments followed by tighter bends. Not all of the electron density of the histone octamer in the crystal structure has been identified. A probable explanation for this is that the N-terminal domains of the histones previously bound in the nucleosome have lost their DNA-binding sites in the core particle and are disordered. Controlled proteolysis can specifically remove these N-terminal domains without affecting the disc-shaped core particle (see Bohm and Crane-Robinson, 1984). It is thought that the globular domain of histone H1 seals off the two turns of the DNA that coil around the histone octamer (Crane-Robinson *et al.*, 1980). This is accomplished through the binding of the central globular domain of H1 to a *cage* of DNA, formed by the DNA strands entering and leaving the nucleosome and the central DNA segment on the pseudodyad axis of the nucleosome (see Fig. 2) (Allan *et al.*, 1980).

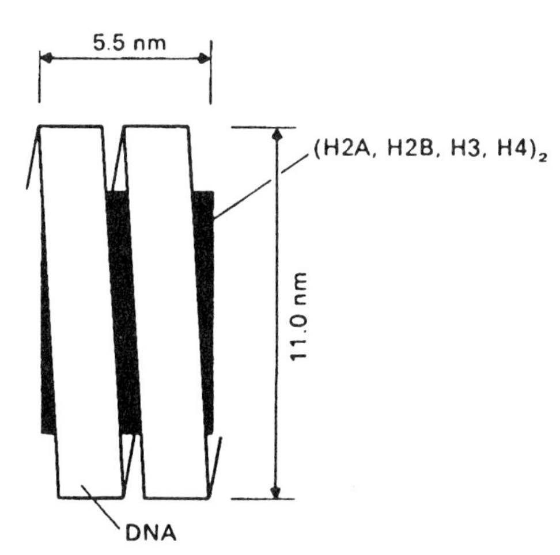

Fig. 1. The nucleosome core particle. Model of the nucleosome showing the dimensions determined from neutron-scatter studies. Reprinted with permission of E. M. Bradbury, 1981.

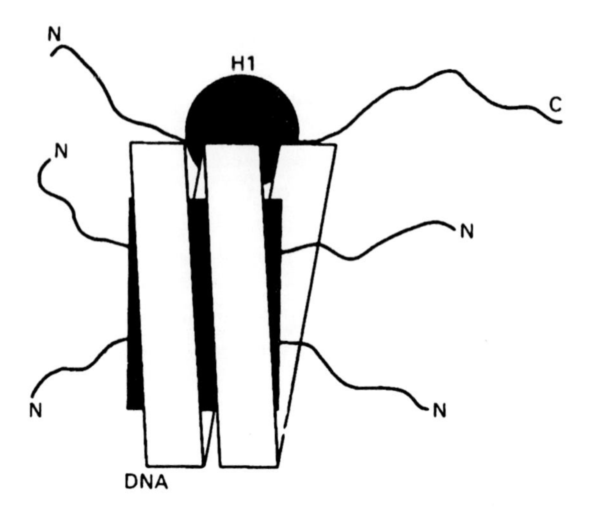

Fig. 2. Model of the nucleosome. Proposed model of the nucleosome particle showing the association of histone H1 and the relative positions of the histone tails. Reprinted with permission of E. M. Bradbury, 1981.

Higher-order chromatin structures are poorly understood because of a paucity of hard structural data which, for functional reasons, probably results from a lack of precise regularity in nucleosomal arrangements. Under conditions of low ionic strength, the chromatin is observed as a 10-nm fibril. With increasing ionic strengths, the chromatin undergoes a transition from the 10-nm fibril to the 30-nm fibril. Most of the chromatin in the cell nucleus is in this 30-nm fibril state, which then appears to be further coiled into metaphase chromosomes. Neutron-scatter studies of the 10-nm fibril gave a mass per unit length that corresponds to 1 nucleosome per 10 ± 2 nm (Suau *et al.*, 1979), and the model that best fits the neutron-scatter data is with the nucleosome discs arranged close to edge-to-edge, with the faces of the discs inclined at less than 20° to the fibril axis (Fig. 3). The DNA-packing ratio in the 10-nm fibril is 6 to 7:1. Neutron-scatter studies of long chromatin in solution further show that with increasing ionic strength, the 10-nm fibril undergoes a salt-induced transition from a loose supercoil that coils with increasing tightness until the 30-nm fibril is reached (Suau *et al.*, 1979). At this stage, the mass/unit length measurements correspond to 0.62 nucleosome/nm for calf thymus chromatin. In parallel with the increasing compaction of the chromatin fibrils, an inflection is observed in the neutron-scatter curve, which moves from 20 to 25 nm at 20 m*M* NaCl to 10 to 11 nm for the 30-nm fibril. This 10–11 nm inflection probably has the same origin as the 10–11 nm semimeridianial arc with off-meridian maxima observed in the neutron fiber-diffraction patterns (Carpenter *et al.*, 1976), and attributed to the pitch of a supercoil of nucleosomes. Electron microscopy (Thoma *et al.*, 1979), neutron and X-ray fiber-diffraction studies

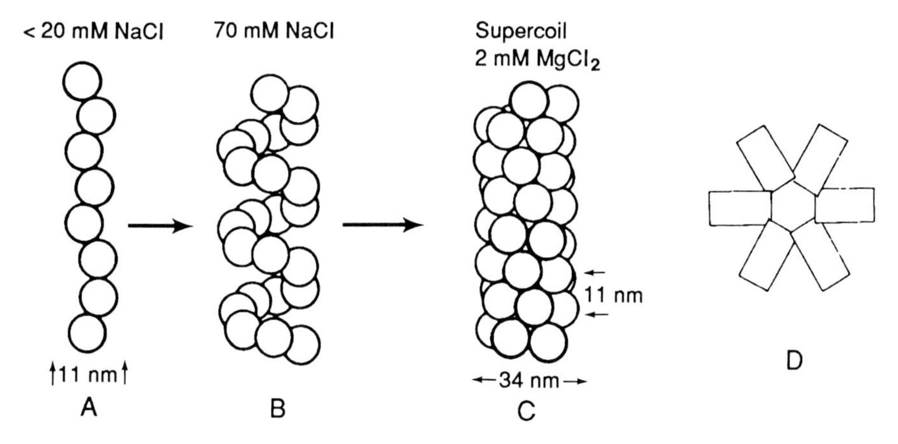

Fig. 3. Salt-induced chromatin folding. Model illustrating the effect of increasing ionic concentration on the coiling of the 11-nm chromatin (A) into higher order structures (B and C). The radial arrangement of the nucleosome discs in the 34-nm coil is shown in D. Reprinted with permission, Yasuda *et al.,* 1987.

(Carpenter *et al.*, 1976; Baldwin *et al.*, 1978) have led to the proposal of a simple model of a supercoil or solenoid of nucleosomes with a pitch of 11 nm and a diameter of 34 nm, containing radially arranged nucleosome discs at a packing of 6 to 8 nucleosomes per turn for calf thymus chromatin, corresponding to a DNA-packing ratio of about 50:1. More extensive studies of chromatin from different sources have led to the proposal of left-handed double helical structures (Williams *et al.*, 1986). Evidence for tissue and species differences in the nucleosome packing in the 30-nm supercoil comes from X-ray scatter and diffraction (Smith *et al.*, 1990). No evidence has been found for the existence of a hole down the axis of the 34-nm supercoil of nucleosomes (Suau *et al.*, 1979; Smith *et al.*, 1990). A possible explanation is that the linker DNA joining nucleosomes and the histone H1 is located inside of the supercoil.

Virtually all the DNA in the cell is packaged into the 34-nm supercoil of nucleosomes. For the formation of the condensed metaphase chromosomes, the 34-nm coil would be required to undergo only one, or possibly two, additional orders or coiling to account for the DNA-packing ratios.

IV. HISTONES AND THEIR POSTTRANSLATIONAL MODIFICATIONS

The major structural proteins that package the DNA molecules into chromosomes are the very basic histones, of which there are five types: the rigidly conserved H3 and H4, highly conserved H2A and H2B, and the conserved family of the very lysine-rich histone H1. All the histones have well-defined domains as defined by nuclear magnetic resonance spectroscopy and controlled proteolysis (reviewed in Bradbury, 1983; Bohm and Crane-Robinson, 1984). These are very flexible N- and C-terminal basic domains, extending from the central apolar globular domain, as depicted in Fig. 4. Each of these domains has been shown to have functional significance. All of the sites of reversible chemical modifications of histones—acetylations, ubiquitinations, and phosphorylations—are located in the flexible N- and C-terminal domains. The highly conserved globular domain contains the sites of interhistone interactions and the major histone–DNA interactions in the nucleosome core particle. Clearly, acetylations of the basic lysines residues and phosphorylations of the neutral serines and threonines in the flexible domains will modulate histone interactions in chromatin, and ubiquitination would be expected to perturb the packaging of nucleosomes.

Histone H1 is phosphorylated in a cell-cycle–dependent manner, with the maximal level of phosphorylation occurring at metaphase. Histone H2A is also phosphorylated, but at a level that remains constant throughout the cell cycle. About 10% of this histone is also reversibly conjugated to another extremely conserved basic protein, ubiquitin. Ubiquitinated histone H2B is found at a lower level (about 1% of the total H2B) than ubiquitinated histone H2A. Histone H3 is also found to be phosphorylated, but only on serine 10 at metaphase. Acetylation is a third modification of histones and can be detected on all the four core histones. Histone

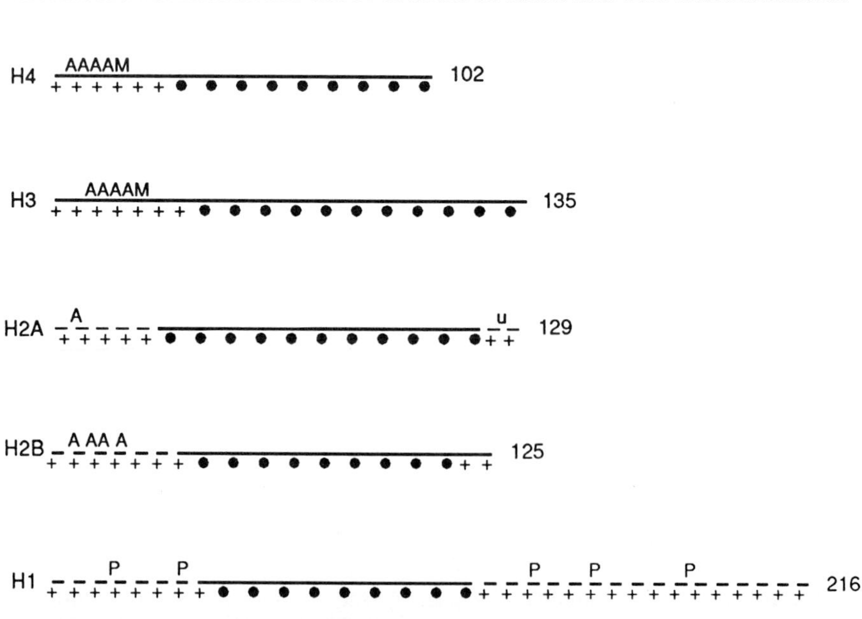

Fig. 4. Histone structure in solution. The five major histones are depicted showing their domains and the sites of posttranslational modifications. The dots, apolar globular domain; +, acidic tails; A, sites of acetylations; P, sites of phosphorylation; M, sites of methylation; u, site of ubiquitination. Reprinted with permission of E. M. Bradbury, 1983.

acetylation correlates with DNA replication and transcriptional activity. Each of these modifications and their possible functions will be discussed in greater detail.

A. Ubiquitination of Histones

Since the discovery of ubiquitin (Goldstein, 1974; Goldstein *et al.*, 1975), numerous studies have been undertaken to elucidate its function. The 76-amino acid ubiquitin is the most conserved protein identified in nature. This basic protein modifies the target protein by forming an isopeptide bond between the external carboxy terminus of ubiquitin and the ε-amino group of an internal lysine residue (Goldknopf and Busch, 1977). The resulting complex is a bifurcated protein containing two amino-termini and a single carboxy-terminus. Several functions have been attributed to the ubiquitination of proteins (recently reviewed by Jentsch *et al.*, 1990), including involvements in protein turnover, heat-shock response, and the DNA-repair pathway. This chapter will be limited to the potential role of histone ubiquitination in chromatin structure.

1. Ubiquitinated Histones in Active Chromatin

In the cell nucleus, histones H2A and H2B can be polyubiquitinated by up to 4 to 6 ubiquitin moieties per molecule (Davie and Nickel, 1987; Nickel *et al.*, 1989). Ubiquitinated H2A was originally designated as A24 in a two-dimensional gel analysis of acid-solubilized nuclear proteins (Orrick *et al.*, 1973; Yeoman *et al.*, 1973; Ballal and Busch, 1973; Goldknopf *et al.*, 1975; Goldknopf *et al.*, 1977). Protein A24 is now referred to as uH2A. The exact role of uH2A in the cell has yet to be determined. It was found to be a stable component of the nucleosome (Martinson *et al.*, 1979), and reconstitution studies showed that the nucleosome core particle can accommodate up to two uH2A molecules without having any effect on the structure of the nucleosome (Kleinschmidt and Martinson, 1981). Low-ionic-strength particle gels were initially employed to fractionate nucleosomes containing uH2A, and the DNA from these nucleosomes was found to contain actively transcribing, or potentially active, sequences (Levinger *et al.*, 1981; Levinger and Varshavsky, 1982). However, Huang *et al.* (1986) reported that nucleosomes containing a variant form of histone H2A could comigrate with uH2A-containing nucleosomes, and this raised the possibility that the observed correlations with active chromatin could be due to modifications other than ubiquitination. Indeed, Bode (1984) showed that acetylation of histones in the nucleosome could reduce the mobility of mononucleosomes in a particle gel. Similar correlations have not been made for ubiquinated histone H2B (uH2B) in these studies, probably because only about 1% of histone H2B is ubiquinated and could not be detected easily.

One physical property of chromatin is that transcriptionally inert chromatin can be selectively precipitated from active chromatin by raising salt concentrations. This insolubility of the inactive chromatin is due to the presence of higher amounts of the linker histone H1. Using such a fractionation procedure, Davie and Nickel (1987) and Nickel *et al.*, (1989) demonstrated an enrichment of ubiquitinated histones in active chromatin, with a more pronounced enrichment of uH2B than uH2A. Because the fractionation was not clean in separating active chromatin from inactive chromatin, Nickel *et al.* (1989) reexamined this correlation in the *Tetrahymena* system, in which transcriptionally active chromatin is packaged into the macronucleus, and the transcriptionally inert chromatin is found in the micronucleus. A similar enrichment of ubiquitinated histones was seen in the transcriptionally active macronucleus, with uH2B being the predominant ubiquitinated histone. These observations are consistent with the suggestion that one possible function of histone ubiquitination is in maintaining the structure of transcriptionally active chromatin.

2. *Mitosis-Dependent Deubiquitinated Histones*

The possibility that histone ubiquitination modifies structure came from the observation that metaphase chromosomes lack both uH2A and uH2B (Matsui *et al.*, 1979; Wu *et al.*, 1981). This was later studied in greater detail by Mueller *et al.* (1985) using the highly synchronous *Physarum polycephalum*. The naturally synchronous nuclear-division cycle of the macroplasmodia allowed the levels of uH2A and uH2B to be monitored through M phase with great precision. The ubiquitin moieties in these histones were found to be present up to prophase, and then absent at metaphase. The histones were reubiquitinated at anaphase. This suggested the possibility that ubiquitin labeled a set of genes that are required to be maintained in a potentially active form until the last possible moment in the cell cycle and, immediately following metaphase, are returned to this form by reubiquitination.

A possible structural role of ubiquitination in maintaining an accessible form of chromatin could also be drawn from the observations made from the mouse G_2-phase mutant, ts85, that has a temperature-sensitive lesion in the ubiquitin-activating enzyme (Finley *et al.*, 1984). Incubation of this cell line at the nonpermissive temperature results in a loss of ubiquitinated histone H2A (Matsumoto *et al.*, 1980) and a cell-cycle arrest at late S–early G_2 (Mita *et al.*, 1980). Electron microscopy of these temperature-arrested cells revealed localized condensed chromatin, particularly around the nucleolus (Mita *et al.*, 1980; Matsumoto *et al.*, 1980). This association with the nucleolus was observed in early studies, and it was suggested that uH2A could function as a repressor of the nucleolus activity (Orrick *et al.*, 1973; Goldknopf *et al.*, 1975). It is conceivable that this localized condensed chromatin occurs in specific regions of the nuclei and nucleoli of the temperature-arrested ts85 cells that have lost the ubiquitin from their histones are unable to maintain the *open conformations*.

3. Loss of uH2A during Premature Chromosome Condensation

The phenomenon of premature chromosome condensation (PCC) was first reported by Johnston and Rao (1970) when they observed that the chromatin of interphase cells could be induced to undergo condensation ahead of schedule when the cells are fused with mitotic cells. PCC can also be induced in a temperature-sensitive mutant cell line tsBN2 (Nishimoto *et al.*, 1987). When tsBN2 cells are incubated at the nonpermissive temperature, their chromatin condenses, regardless of the stage in the cell cycle. The defective gene was identified as the RCC1 and was found to encode a 45-kDa protein of unknown function (Uchida *et al.*, 1990). If ubiquitination keeps chromatin in an accessible form, the uH2A levels would be expected to decline when PCC is induced. This was indeed the situation when PCC was induced in tsBN2. Two-dimensional gel electrophoresis analyses of the histones revealed that when tsBN2 was transferred to the nonpermissive temperature, the level of uH2A declined to the level comparable to that seen in metaphase cells (Fig. 5). Thus, the results available so far are consistent with the hypothesis that ubiquitination of histones in specific regions of the chromatin function to maintain an open conformation, and that deubiquitination is required for total condensation of the chromosomes during metaphase.

B. Reversible Acetylation of Histones

Acetylation of the lysine residues occur in the N-terminal domains of core histones H2A, H2B, H3 and H4. Dynamic changes in the levels of acetylation and the specific sites of acetylation have been correlated with cellular processes such as replication, transcription, and spermiogenesis (Matthews, 1988; Csordas, 1990). Although reversible acetylation is a major modification of core histone, little is known of the effects of acetylation on chromatin structure and functions. The N-terminal domains of the core histones are not essential for the structure of nucleosome core particle, but are speculated to be necessary for the transition from the 10-nm fibril to the 30-nm coils (Lilley and Tatchell, 1977; Whitlock and Simpson, 1977; Whitlock and Stein, 1978; Allan *et al.*, 1982). Earlier efforts to examine the effects of histone acetylation on higher-order structure yielded few results (Simpson, 1978; Halleck and Gurley, 1982; Reczek *et al.*, 1982; Halleck and Schlegel, 1983; McGhee *et al.*, 1983). However, of some interest was the finding of a much enhanced DNase I sensitivity of DNA in hyperacetylated chromatin, presumably as a result of increased accessibility of unfolded chromatin (Vidali *et al.*, 1978). Consistent with this is the finding by Norton *et al.* (1989) that

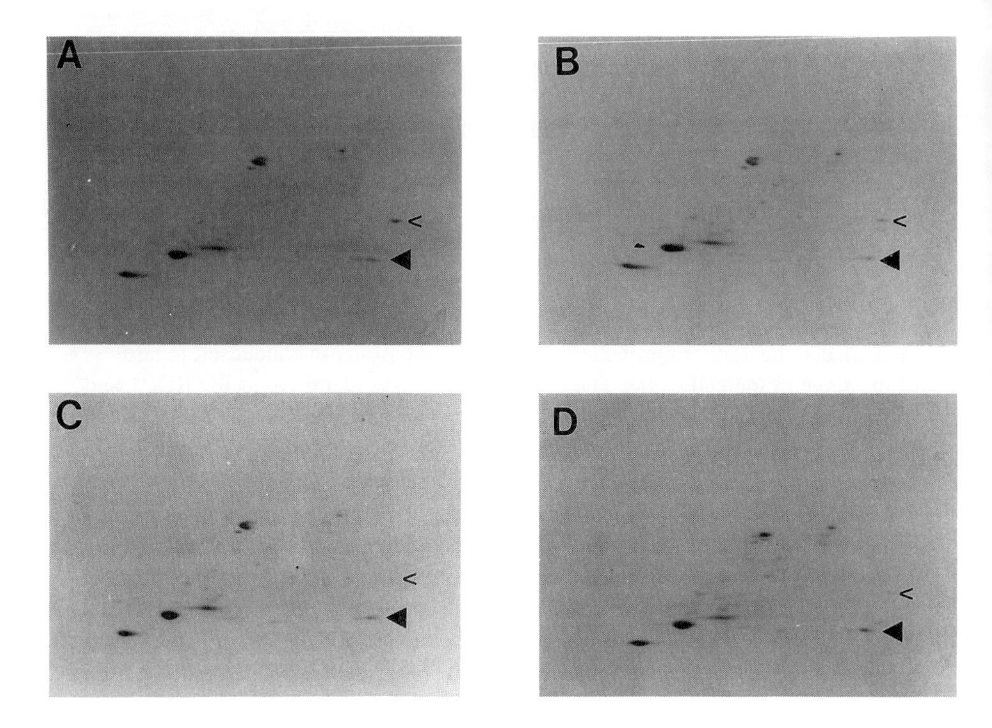

Fig. 5. Loss of uH2A during PCC. The tsBN2 was synchronized with aphidicolin and released into S phase. After 6 hr, the cells were shifted up to the nonpermissive temperature of 40.5°C for 4 hr. Histones were then extracted and analysed on a two-dimensional gel system, with the Triton-Aid-Urea in the first dimension and the SDS-polyacrylamide gel system in the second dimension. Open arrowheads indicate uH2A; closed arrowheads indicate H2A. Panel A: Histones from cells just before temperature shift-up; Panel B: Histones from cells that were maintained at 33°C for 4 hr; Panel C: Histones from cells that were shifted up to 40.5°C; Panel D: Histones from colcemid-arrested metaphase cells.

hyperacetylated nucleosome core particles have a linking number of -0.82 ± 0.05, compared to -1.04 ± 0.08 for the control. Such a change in linking number would release negative supercoiling previously constrained on nucleosomes into a chromatin loop, and thus function as a eukaryotic gyrase. How the acetylation of the core histones might affect the binding of histone H1 in chromatin is unknown.

1. Transcriptionally Active Chromatin Is Enriched in Acetylated Histones

Much circumstantial evidence has accumulated supporting a direct correlation between transcription activity and acetylation of histones (Allfrey *et al.*, 1964; Allfrey, 1980; Matthews, 1988; Csordas, 1990). In eukaryotic cells, the dominant isoforms of histone H4 were found to be nonacetylated and monoacetylated (Matthews, 1988; Csordas, 1990). The one exception was found for yeast cells in which the major isoforms are monoacetylated and tetraacetylated (Marian and Winterberger, 1982). Such differences were attributed to the highly active transcriptional states of yeast chromatin. In a recent study, the level of histone acetylation was linked with gene activity (Hebbes *et al.*, 1988). By employing antibodies specific to acetylated core histones, nucleosomes enriched in acetylated histones were isolated from bulk chromatin by immunoprecipitation. The DNA sequences associated with this hyperacetylated fraction were found to be enriched in active gene sequences by 15-fold to 30-fold.

2. *Cell Cycle-Dependent Changes in Isoforms of Histone Acetylation*

Changes in the states of acetylation of histones through the cell-division cycle have been studied in detail in the macroplasmodia of *Physarum polycephalum*, which has a naturally occurring precise nuclear division cycle (reviewed in Yasuda *et al.*, 1987). In this system, acetylation of histones was found to be minimal when the cells were at metaphase (Chahal *et al.*, 1980). Similar results have been reported for mammalian cells (D'Anna *et al.*, 1983). These changes were later confirmed by immunostaining using specific antibodies raised against acetylated forms of histone H4 (Turner, 1989; Turner and Fellows, 1989). However, deacetylation of histones may not be obligatory for the condensation of chromosomes during mitosis. Treatment of mammalian cells with sodium butyrate, an inhibitor of histone deacetylase, did not prevent chromosome condensation, but allowed the cells to traverse mitosis and arrest at G_1 (D'Anna *et al.*, 1980; Fallon and Cox, 1979; Turner, 1989).

3. *Preferential Use of Specific Lysine Residues for Acetylation*

The sites of reversible acetylation at lysine residues are rigidly conserved in histones H3 and H4. Some divergences have been found in histones H2A and H2B (Matthews, 1988; Csordas, 1990). Histone H2A is acetylated only at lysine 5, whereas the other histones can be modified by up to 4 acetyl groups; histone H2B, lysines 5, 10, 13, and 19; histone H3, lysines 9, 14, 18, and 23; and histone H4, lysines 5, 8, 12, and 16 (Matthews, 1988; Csordas, 1990; Chicoine *et al.*, 1986). Among these histones, nonrandom, cell cycle–dependent usage of acetylation sites has been reported only for histone H4. From several studies, it was concluded that Lys^5 is preferentially used in S phase, involving newly synthesized histones and the assembly of nucleosomes, whereas Lys^8 is used preferentially in the G_2 phase for transcription (Matthews, 1988; Csordas, 1990; Chicoine *et al.*, 1986). Antibodies raised against specific sites of acetylation have also been used to determine metaphase-related, site-specific usage by histone H4 (Turner and Fellows, 1989;

Turner, 1989). It was demonstrated that during interphase, Lys^{8} and Lys^{16} were more preferred, and at metaphase, the preferred sites shifted to Lys^{5} and Lys^{12} (Turner and Fellows, 1989; Turner, 1989). A slight divergence in the acetylation of histone H4 was found in the *Tetrahymena*, where the sites were on lysines 4, 7, 11, and 15. Here, Lys^{4} is preferentially acetylated during S phase, while Lys^{7} is acetylated at G_2 for transcription. Furthermore, in the macronuclei of growing cells, the diacetylated histone H4 was exclusively modified at Lys^{4} and Lys^{7}, whereas newly synthesized histone H4 was acetylated only at Lys^{4} and Lys^{11} (Chicoine *et al.*, 1986).

Analysis of the sites of acetylation in the different acetylated forms of H3 and H4 following sodium butyrate inactivation of histone deacetylase showed a specific usage of sites (Marvin *et al.*, 1990). During cellular differentiation, the nonrandom use of sites of acetylation has been found for histone H4. In the cuttlefish testis, the preferred order of acetylation is $Lys^{16} > Lys^{12} > Lys^{5} > Lys^{8}$, and this occurs just before the replacement of histones by protamine (Couppez *et al.*, 1987). In the same study, the preferred order of use during differentiation of calf thymus was found to be $Lys^{12} > Lys^{5} > Lys^{16} > Lys^{8}$. These results indicate that acetylation of different lysine residues accompanies the changes in cellular processes; however, the effects of such differential acetylation on the chromatin structure and functions still remain unclear.

4. Genetic Analysis of the N-Terminal Domains of Core Histones

To determine the importance of site-specific acetylation to cell viability, experiments were performed in yeast involving deletions of N-terminal domains of core histones that contain the sites of acetylations, and by site-directed mutagenesis of the lysine sites of acetylation. It was found that deletions in the N-terminus of histone H2B had little effect on cell viability, whereas deletion of large portions of the C-terminus was lethal (Wallis *et al.*, 1983). Intact histone H2A could complement the N-terminal deletion of H2B, but deletion of the N-termini from both histones was lethal (Schuster *et al.*, 1986). Deletion in the hydrophobic central domain of histone H4 blocked chromosome segregation and was lethal, whereas yeast-carrying deletions in either the N-terminus (residues 4–28) or the C-terminus (residues 100–102) were viable (see Table I) (Kayne *et al.*, 1988). Thus, these portions of the N- and C-termini are not absolutely required for growth. However, the cells with N-terminal deletions have much longer cell-division cycles, especially in G_2, and they exhibit a loss of mating ability. Effects on the chromatin include increased nuclease sensitivity, probably due to inefficient packing of nucleosomes (Kayne *et al.*, 1988). MeGee, *et al.* (1990) recently reported a genetic study of the role of reversible acetylation in yeast histone H4 by site-directed mutagenesis. Three classes of cell-cycle mutants were identified (Table I). In the first class, mutants with single amino acid substitution of Lys^{5}

TABLE I[a]

Genetic Analysis of Histone H4

Mutation	Phenotype, cell cycle	Mating efficiency
Deletion mutants (Kayne *et al.*, 1988)		
Δ(4-28)	Viable, extended G_2 phase	Sterile
Δ(4-34)	Lethal	—
Δ(23-34)	Lethal	—
Δ(33-44)	Lethal	—
Δ(83-88)	Lethal	—
Δ(85-86)	Lethal	—
Δ(90-95)	Lethal	—
Δ(92-93)	Lethal	—
Δ(97-102)	Lethal	—
Δ(100-102)	Viable, normal	Normal
Substitution mutants (MeGee *et al.*, 1990)		
$Lys^{5,8,12,16}$	Wild type, normal	Normal
$Lys^{5,8,12,16}$-$Arg^{5,8,12,16}$	Lethal	—
$Lys^{5,8,12,16}$-$Asp^{5,8,12,16}$	Lethal	—
$Lys^{5,8,12,16}$-$Gln^{5,8,12,16}$	Viable, extended S and G_2–M phase	Sterile
Lys^{5}-Arg^{5}	Viable, normal	Normal
His^{18}-Tyr^{18}	Viable, normal	Deficient
Lys^{16}-Arg^{16}	Viable, extended S phase	Deficient
$Lys^{5,8}$-$Arg^{5,8}$	Viable, extended S phase	Normal
$Lys^{12,16}$-$Arg^{12,16}$	Viable, extended S phase	Normal

[a]Compiled by X.W. Guo.

with arginine, or His^{18} with tyrosine, have wild-type growth characteristics. However, the latter mutant was sterile. Mutants in the second class involve substituting glutamine residues for (a) Lys^5 and Lys^8; (b) Lys^5 and Lys^{12}; or (c) Lys^{16} alone. These cells were found to have an extended S phase. Interestingly, mutants with the lone substitution at Lys^{16} also had much lower mating efficiency compared to the wild type. Thus, histone H4 contains a site that seems to be required for normal mating-type expression that spans from at least Lys^{16} to His^{18}. Both reversible acetylations of Lys^{16} and possible phosphorylation of His^{18} may be involved in the regulation of mating-type expression (Fujitaki *et al.*, 1981; Matthews and Huebner, 1984; MeGee *et al.*, 1990). Mutants with glutamine substitutions in all four lysines form the third class, and were found to have cell cycles with prolonged S and G_2–M phases. This mutant was also sterile. These results clearly demonstrated that one or more of the lysines at position 8, 12, and 16 in H4 must be functionally required for the process of replication and transcription through the cell cycle. Whereas reversible acetylation at Lys^5 seems to be important for the S and M phases in other cell types that have been studied, substitution of Lys^5 in the yeast histone H4 with arginine did not show any detectable effect on cell growth

(Matthews, 1988; Lin *et al.*, 1989; Turner and Fellows, 1989; Csordas, 1990; Turner, 1989, MeGee *et al.*, 1990). Differences between yeast and other species were also apparent when it was observed that substitutions of Lys^5 and Lys^8, or Lys^5 and Lys^{12}, with arginine residues resulted only in an extended S phase (MeGee *et al.*, 1990), whereas in *Tetrahymena*, the diacetylation patterns of the corresponding Lys^4 and Lys^7, and Lys^4 and Lys^{11} were critical for cellular functions (Chicoine *et al.*, 1986). Substitution of all four of the lysine residues with arginine or asparagine were found to be lethal, whereas substitutions with glutamine residues allowed the cells to remain viable, but would exhibit phenotypes similar to those displayed by mutants of N-terminal deletions. The results from these experiments suggest that some of the positively charged lysine residues in histone H4 have to be neutralized by reversible acetylations for cells to progress through the cell cycle.

C. Histone Phosphorylation

Early detailed studies of histone H1 phosphorylation and H1 kinase activity through the precise nuclear-division cycle of the macroplasmodium of *Physarum polycephalum* led to the first proposal that an increase in H1 kinase activity both triggered and controlled the mitotic process, and the physical process of chromosome condensation involved H1 phosphorylation. Of the posttranslational modifications associated with histones and their function in the cell, net H1 kinase activity as monitored by H1 phosphorylation was found to be the one most closely correlated with the control and regulation of the cell-division cycle. Within the last 3 years, numerous studies have been reported on the potential role of H1 kinase activity in the regulation of mitosis. This increased interest was triggered by the merging of three major lines of research on the control of the cell division cycle: (1) cell-cycle behavior of histone H1 phosphorylation and H1 kinase activity; (2) the search in yeasts for the cell-cycle control genes; (3) the role of maturation promoting factor (MPF) in inducing mitosis. Growth-associated (GA) kinase has long been proposed to regulate the onset of mitosis and to be the driving force behind the condensation of chromosomes. MPF is the activity found in metaphase cell extracts that can induce mitosis in somatic cells and meiosis in oocytes. MPF has now been demonstrated to be H1 kinase. These findings suggest very strongly that the mechanisms involved with the regulation of the cell-division cycle is universal among eukaryotees ranging from yeasts to humans.

1. Control of Chromosome Condenstion and Mitosis by GA Kinase

The very lysine-rich (vlr) histone H1 was first shown to be phosphorylated by Ord and Stocken (1966) and Stevely and Stocken (1966) in rat thymus and was believed to play a role in the S phase of the cell cycle (Panyim and Chalkley, 1969; Oliver *et al.*, 1972; Balhorn *et al.*, 1971, 1972a,b; Marks *et al.*, 1973). Histone modifications had been proposed to control the physical state of chromatin (reviewed by Allfrey, 1971), and Langan (1969) suggested that the phosphorylation of histone H1 could affect the structure of chromatin by altering the interactions between histones and DNA. In addition to the correlation with DNA replication, Marks *et al.* (1973) noted the association between histone H1 phosphorylation and mitosis in HeLa cells, and further suggested the possibility that this modification could be related to the condensation and separation of sister chromatids before and during mitosis. The major limitation in the study of histone modifications during the cell cycle at that time was the inability to synchronize mammalian cells with precision. The first direct proposal of a role of histone H1 phosphorylation in driving chromosome condensation came from Bradbury *et al.* (1973a). Physical studies on isolated calf thymus chromatin showed that histone H1 was essential for the salt-induced condensation of chromatin, and this led to the suggestion that H1 phosphorylation may drive chromosome condensation. *In vivo* evidence for this proposal came from detailed studies on the cell cycle of *Physarum polycephalum*. The major advantage of this slime mold was that in its growth stage macroplasmodium, 10^8–10^9 nuclei undergo mitotic division with exquisite precision, within 2 min of a 14-hr cycle. Using *Physarum*, a direct correlation between the G_2-phase increase in histone H1 phosphorylation and the condensation of chromosomes was observed (Bradbury *et al.*, 1973b, 1974a). Cell-cycle analysis of H1 kinase activity showed a 15-fold increase in activity from G_2 to M (Bradbury *et al.*, 1974b) and this increase resulted from an activation of the preexisting enzyme (Mitchelson *et al.*, 1978). These observations led to the proposal that histone H1 phosphorylation is the *mitotic trigger* that controls mitotic cell division. Later studies revealed that the level of phosphorylation increases from 9 to 16 phosphates per molecule of histone H1 at early S phase to 15 to 24 phosphates/H1 at prophase, and continues up to 20 to 24 phosphates/H1 at metaphase (Mueller *et al.*, 1985). Similar correlations were subsequently observed between histone H1 phosphorylation and progress through G_2 phase in CHO cells (Gurley *et al.*, 1973, 1975, 1978), rat hepatoma cells (Langan *et al.*, 1980) and HeLa cells (Ajiro *et al.*, 1981), although the maximal level of phosphorylation reached here was about 5 to 6 phosphates/molecule of H1 at metaphase. This difference in the levels of M phase phosphorylations of histone H1 probably reflects the much longer C-terminal domain of *Physarum* H1 of 200 amino acid residues, compared to 100 for mammalian H1 (Fig. 6). This universal correlation between histone H1 phosphorylation and chromosome condensation suggests that the mechanisms controlling chromosome condensation are common to all eukaryotes.

The most direct early evidence for the role of growth-associated histone H1 kinase (GA kinase) in regulating the cell-division cycle came from the observation that addition of heterologous kinase to macroplasmodia of *P. polycephalum* 3 hr

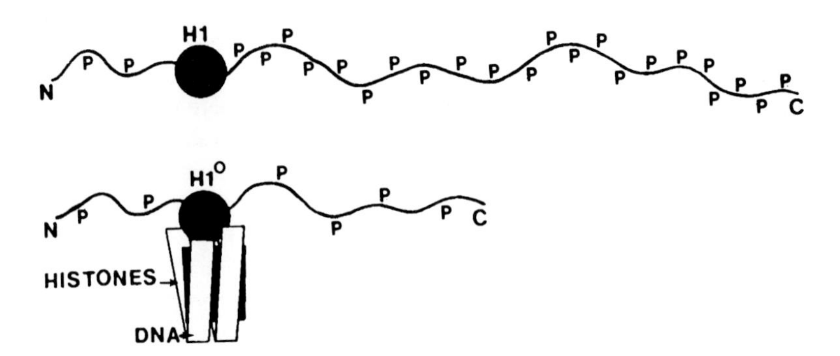

Fig. 6. Histone H1 phosphorylation sites in the nucleosome core. Model showing the positions of the phosphate groups on histone H1 when bound to the nucleosome particle. Bottom figure shows the 6 potential sites of phosphorylation on mammalian histone H1 and $H1^0$. Top figure shows the 24 potential sites of phosphorylation on histone H1 from *Physarum polycephalum* and the extra 100 amino acid residues found in the C-terminal domain. Reprinted with permission, Yasuda, *et al.,* 1987.

before M resulted in an advancement of mitosis of up to 1 hr (Bradbury *et al.*, 1974b; Inglis *et al.*, 1976). This semipurified H1 kinase was isolated from Ehrlich ascites cells, based on the assumption that the role of H1 kinase in controlling cell division is universal. Although cause and effect have yet to be demonstrated, H1 phosphorylation is thought to be the biochemical mechanism that links the physical states of chromosomes with the different functional stages of the cell cycle.

The sites of phosphorylation in mammalian histone H1 have been mapped by Langan *et al.* (1980, 1981). Among the five subtypes of histone H1, the metaphase levels of phosphorylation varied between 3 and 6 phosphates per histone molecule (also in Matsukawa *et al.*, 1985). The phosphorylations occur on either the serine or threonine residues within the consensus sequence Lys-Ser/Thr-Pro-Lys or Lys-Ser/Thr-Pro-X-Lys. These sites are located at the basic amino- and carboxy-terminal domains of H1 that interacts with DNA. Histone H1 kinases extracted from *P. polycephalum* (Chambers *et al.*, 1983) and *Xenopus laevis* (Masaracchia *et al.*, 1979) also recognize these same consensus sequences, further indicating that this mechanism of control of chromosome condensation is highly conserved through evolution.

Additional evidence for histone H1 hyperphosphorylation having a role in chromosome condensation came from studies of premature chromosome condensation (PCC). One method used for studying PCC is with the temperature-sensitive mutant, tsBN2 (Nishimoto *et al.*, 1987). When incubated at the elevated temperature of 40.5°C, the chromosomes of these cells undergo PCC, which was found to be accompanied by hyperphosphorylation of histones H1 and H3, even in S phase when the level of phosphorylation of histone H1 is normally low (Ajiro *et al.*, 1983; Nishimoto *et al.*, 1987). Another method for inducing PCC is through the use of cell fusion. Fusion between mitotic and interphase cells was found to result in the premature condensation of the chromosomes of interphase cells (Johnston and Rao, 1970). The induction of PCC in the interphase cell was also accompanied by the hyperphosphorylation of histones H1 and H3 (Hanks *et al.*, 1983). Interestingly, it was also found that the histones of the mitotic cell became partially dephosphorylated when these cells were fused to interphase cells without changing chromosome morphology. This is consistent with other observations that metaphase chromosomes could be isolated with intact morphology, but have histones that are partially dephosphorylated (D'Anna *et al.*, 1978; Paulson, 1980). Within the cell, this loss of H1 phosphate groups in condensed chromosomes was seen during PCC in the tsBN2 cell line. When these cells were transferred to the nonpermissive temperature, the level of phosphorylation on histone H1 rose with increasing condensation of the chromosomes. However, continued incubation after maximal condensation resulted in a partial dephosphorylation of the histones (Nishimoto *et al.*, 1987). A similar observation was also made during the normal cell cycle of *P. polycephalum* (Bradbury *et al.*, 1973b, 1974a). The phosphate content of histone H1 increased through G_2 with the condensation of chromosomes, but peaked just before metaphase. Right at metaphase when the chromosomes are maximally condensed, the level of phosphorylation was actually on the decline. These results indicate that histone H1 phosphorylation may be required only through the G_2 phase of the cell division cycle for initiating mitosis and driving the condensation of chromosomes.

2. The Role of MPF in Inducing Meiosis and Mitosis

The existence of mitosis-inducing factors came from studies of Johnson and Rao (1970), who reported that the fusion of interphase cells with metaphase cells resulted in the premature condensation of chromosomes in the interphase cell. This factor was later identified as MPF, or maturation-promoting factor. The term MPF was originally used by Masui and Markert (1971) to describe the activity found in cytoplasmic extracts that regulate the meiotic development of immature frog oocytes. At about the same time, Smith and Ecker (1971) described a similar maturation-inducing activity in *Rana pipiens*. Similar meiosis-inducing activities have since been found in other metaphase-arrested eukaryotic cells such as *P. polycephalum* (Adlakha *et al.*, 1988), starfish (Kishimoto and Kanatani, 1976; Kishimoto *et al.*, 1982), human (Sunkara *et al.*, 1979a,b), and hamsters (Nelkin *et al.*, 1980), suggeting that these factors are common to all eukaryotes. MPF can also induce early embryos to undergo mitotic cleavages during embryo development. Hence, MPF is now referred to as M-phase promoting factor, an activity that induces entry into M phase of the meiotic and mitotic cycles.

Early attempts at purification and characterization of MPF met with limited success primarily because of the instability of the activity with successive purification (reviewed by Adlakha and Rao, 1987). Adlakha *et al.* (1982a,b) provided evidence that MPF is chromosome bound when they showed that these factors could be released from isolated metaphase chromosomes by digestion with micrococcal nuclease or DNase II. Using its DNA-binding properties, a 200-fold purification of the MPF was achieved by affinity chromatography through a DNA-cellulose column (Adlakha and Rao, 1987). This semipurified MPF had a molecular weight of about 100,000 as determined on a Sephacryl S-200 column. SDS-polyacrylamide gel electrophoresis analyses revealed the presence of several protein bands, with the major protein mw being 50,000. The partially purified MPF had kinase activities specific for histones. Other proteins normally used in assaying for protein kinase activities, such as casein or phosvitin, were not good substrates. The kinase was also unresponsive to the presence of cAMP, calcium, calmodulin and spermine. These properties strongly indicated that this MPF activity was similar, if not identical, to the GA kinase that was partially purified by Langan (1982).

MPF from *Xenopus* has been purified to near homogeneity by Lohka *et al.* (1988) and was found to be a complex of a 32-kd and a 45-kd protein (reviewed by Lohka, 1989). The 32- to 34-kd protein ($p34^{cdc2}$) was identified as the protein kinase homolog of the *cdc2* gene product, originally identified in *S. pombe* (Arion *et al.*, 1988; Labbe *et al.*, 1988; Dunphy *et al.*, 1988). The 45- to 47-kd protein has now been identified as cyclin B (Gautier *et al.*, 1990) and is homologous to the *cdc13* gene product identified in *S. pombe* (Booher and Beach, 1988; Hagen *et al.*, 1988). There are two classes of cyclins, A and B, both of which were found to complex with $p34^{cdc2}$ (Draetta *et al.*, 1989; Minshull *et al.*, 1990). Studies of *Drosophila* (Lehner and O'Farrell, 1990) and *Xenopus* (Minshull *et al.*, 1990) cyclins suggest that cyclins A and B perform separate, yet unidentified, roles in the cell cycle. In the *Drosphila*, both cyclins are required for cells to enter mitosis. In the *Xenopus* oocyte, the cyclin A–$p34^{cdc2}$ protein kinase activity rises steadily and peaks earlier in the cell cycle than the cyclin B–$p34^{cdc2}$ protein kinase activity. The destruction of cyclin A also occurs earlier than that of cyclin B, just before the onset of mitosis. Minshull *et al.* (1990) present evidence that suggests that the cyclins could function as *targeting subunits* to guide the $p34^{cdc2}$ protein kinase to particular substrates. The complex of $p34^{cdc2}$ protein kinase and cyclin B is now believed to be the H1 GA kinase (Langan *et al.*, 1989; Chambers and Langan, 1990). In mammalian cells, cyclin B is about 62,000 and was reported to confer upon $p34^{cdc2}$ the specificity to phosphorylate histone H1 (Draetta and Beach, 1988; Brizuela *et al.*, 1989). Pines and Hunter (1989; 1990) have now cloned the genes encoding cyclins A and B from human cells.

3. The CDC2/cdc2 Gene and Its Control of the Cell Cycle

The fission yeast, *S. pombe*, and the budding yeast, *S. cerevisiae*, have been used as model systems to study the eukaryotic cell division cycle (cdc) (reviewed in

Nurse, 1985). The relative ease in isolating temperature-sensitive (ts) mutants from yeasts has provided much information on the genes that are required for the progression through the cell cycle. These *cdc* mutants have allowed identification of various points in the cell cycle at which their gene products are required, because at these points, conditionally lethal mutants arrest at the cell cycle when incubated at the restrictive temperatures. About 90 conditionally lethal mutants have been isolated from which a number of genes responsible for regulating mitosis were identified. One gene that is of major interest is the *cdc2* gene from *S. pombe*. This gene is required at two points in the cell cycle in *S. pombe* for regulating cellular growth and division (reviewed by Norbury and Nurse (1989). This first is at *Start* in late G_1, when the gene regulates entry into S phase, and the second is in G2, when the gene controls entry into mitosis. Because of this dual role, entry into M must be coordinated with the completion of the replicative phase of the cell cycle. Through the use of these conditionally lethal mutants, it was determined that *cdc2* is the gene through which other genes control mitotic division. Three of these were identified as the *wee*1, cdc25, and *nim1* (Russell and Nurse, 1986, 1987a,b) (see Fig. 7). The *wee*1 is an inhibitor of *cdc2* function. Overexpression of *wee*1 results in a delay in mitosis, and the cells grow to a larger size before undergoing division. The *cdc25* gene counteracts the action of *wee*1 by stimulating the activity of *cdc2*, thereby advancing mitosis. Overexpression of this gene in *S. pombe* causes cells to initiate mitosis prematurely, with division occurring at half the normal cell size. *nim1* is also an inducer of mitosis. It exerts its function by inhibiting the actions of *wee*1. The amino acid sequence determined from the DNA sequence of *cdc25* does not show any structural features expected of a protein kinase. *cdc2, wee*1, and *nim1* code for proteins that bear structural motifs common to protein kinases. Biochemical assays of the *cdc2*

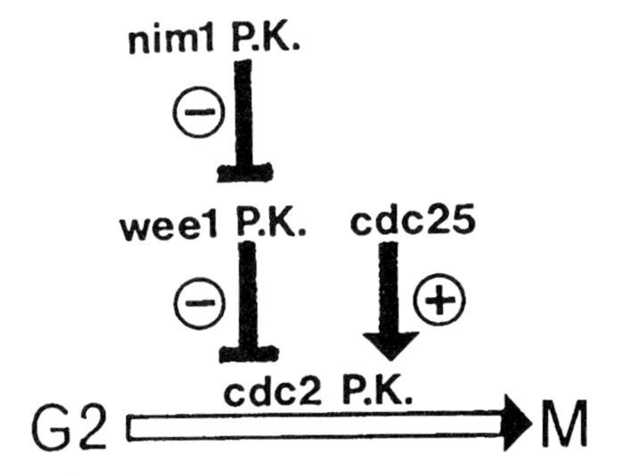

Fig. 7. Interactions between cell cycle-regulatory genes. Mitosis in *Schizosaccharomyces pombe* is regulated by the activity of the *cdc2* gene product. The *cdc2* is regulated by the activities of the activator *cdc25* gene and the inhibitor *wee*1 gene. The *wee*1 gene is in turn negatively regulated by the *nim1* gene. (From Russell and Nurse, 1987a). Reprinted with permission, P. Russel and P. Nurse, 1987.

$p34^{cdc2}$, show that it is a protein kinase (Simanis and Nurse, 1986). Interest in this gene increased greatly when Lee and Nurse (1987) demonstrated that the *cdc2* gene was also present in higher eukaryotes. They successfully cloned the human *CDC2* gene by complementation rescue of *S. pombe* cells that have a defective *cdc2* gene. Amino acid sequence analysis of the p34 gene products revealed over 60% conservation between the sequences. The fact that the human gene was able to rescue the temperature-sensitive yeast mutant suggested that the role of $p34^{cdc2}$ in regulating mitosos is universal. Since then, other genes or gene products have been found in higher eukaryotes that are homologous to those found in *S. pombe*, such as cyclin, which is homologous to the *cdc13* of *S. pombe* (Booher and Beach, 1987; Hagan *et al.*, 1988), the human p13, which is homologous to that encoded by the *suc1* gene (Draetta *et al.*, 1987), and the *string* gene which is homologous to the *cdc25* (Edgar and O'Farrell, 1989). This further supports the notion that the mechanism that regulates the onset of mitosis is common to all eukaryotes.

To date, the exact role of $p34^{CDC2/cdc2}$ in the regulation of the cell cycle remains to be identified. Histone H1 was the substrate of choice for *in vitro* studies of the enzyme, and is believed to be the major substrate in the cell for regulating mitosis. However, other cellular proteins have been reported to be potential substrates for this kinase *in vitro* (reviewed in Norbury and Nurse, 1989; Nurse, 1990). Among these are the retinoblastoma (RB) protein (Taya *et al.*, 1989), nucleolar proteins (Peter *et al.*, 1990), proteins that regulate cell shape (Verde *et al.*, 1990; Lamb *et al.*, 1990), lamins, and RNA polymerase II (Cisek and Corden, 1989).

4. The Mammalian FT210 Cell Line with a Defective CDC2 Gene Product

The genetic regulation of mitosis in mammalian cells has not been as well defined as in the yeast system. The more complex diploid genome makes the isolation of mammalian cell-cycle mutants very difficult, especially those that arrest at the G_2 phase. However, recent advances in methodology have allowed for the isolation of such mutants (reviewed by Marcus *et al.*, 1985; Eki *et al.*, 1990). One temperature sensitive G_2 phase mutant, the FT210, was found to have a defect in its ability to enter mitosis when incubated at the restrictive temperature (Mineo *et al.*, 1986). Analyses of this cell line showed that synthesis of nucleic acid and protein was not affected by a shift-up in temperature, and that the point of arrest was at the late G_2 phase of the cell cycle. Accompanying this arrest was an absence of hyperphosphorylated histone H1, which led to the proposal that the defect was in the GA kinase, and that the loss of this activity at the nonpermissive temperature prevented the mitotic phosphorylation of the histone H1.

The first clue to the defect in FT210 cells came when extracts prepared from the temperature-arrested cells showed a reduced level of $p34^{CDC2}$ (Th'ng *et al.*, 1990). The affinity-purified antibody employed for this preliminary study was raised against a highly conserved peptide region of the protein (Lee and Nurse, 1987). This 16 amino acid PSTAIR region bears the sequence EGVPSTAIREISLLKE, which was found to be perfectly conserved from yeasts to humans. The decline in the level of $p34^{CDC2}$ was seen after 6 hr incubation at the nonpermissive temperature, and prolonged incubation periods did not result in any further losses (Fig. 8). Even the level of $p34^{CDC2}$ in cycling FT210 cells appears lower than that of normal cells. Similar results were obtained with antibodies raised against the C-terminal sequence of the protein, indicating that the decline in antibody reaction was due to a loss of the protein, rather than a masking of the antigenic region. Synthesis of cellular proteins was not affected by the temperature elevation, as shown by incorporation of ^{3}H-leucine into acid-precipitable counts (Mineo *et al.*, 1986), and ^{35}S-methionine into cellular proteins as determined by SDS-PAGE (J.

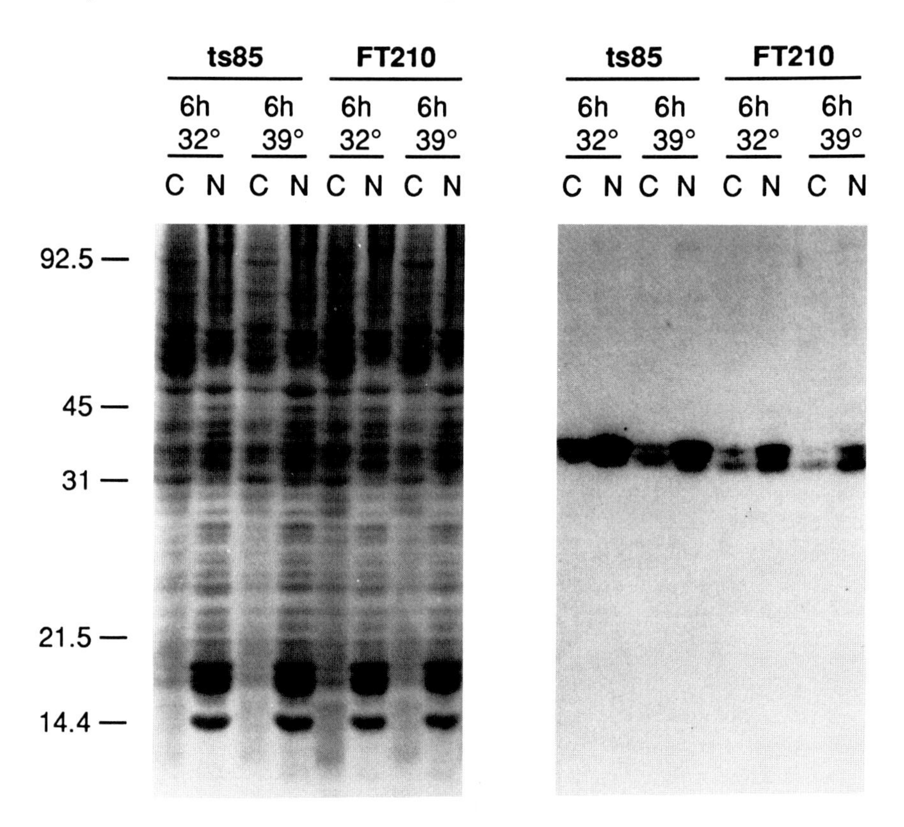

Fig. 8. Temperature-induced loss of $p34^{CDC2}$ in FT210 cells. Temperature-sensitive ts85 and FT210 cells were incubated for 6 hr at the permissive temperature of 32°C or at the restrictive temperature of 39°C. The cells were lysed, cytoplasmic and nuclear proteins were separated in SDS-PAGE, and the proteins were stained with Coomassie Blue (left panel) or immunostained for the presence of $p34^{CDC2}$ (right panel). (From Th'ng *et al.*, 1990).

Th'ng and E. M. Bradbury, unpublished observations). Hence, this decline in the relative amount of $p34^{CDC2}$ does not result from a general problem with protein synthesis. Further, Northern blot analysis performed on total cellular RNA extracted from cycling and temperature-arrested FT210 cells showed that the level of the message coding for $p34^{CDC2}$ did not diminish with prolonged incubation of cells at the elevated temperature. In fact, scans of the blot revealed a slight enhancement in FT210 in the level of the message coding for $p34^{CDC2}$ (Fig. 9). We also determined whether the $p34^{CDC2}$ protein kinase activity of the FT210 cell was affected by temperature elevation. Two methods of extracting the kinase were employed: immunoprecipitation directly from whole-cell extracts (Fig. 10), and semipurification by column chromatography of the kinase activity from cell extracts (manuscript in preparation). Both these methods yielded similar results indicating that the $p34^{CDC2}$ extracted from FT210 is inactivated more rapidly than that from wild-type cells. From these studies, the loss of $p34^{CDC2}$ in temperature-arrested FT210 cells was surmised to be caused by the temperature-induced inactivation of the enzyme activity, which leads to the enhanced rate of degradation of $p34^{CDC2}$.

To verify that the mutation is in the $p34^{CDC2}$ protein kinase, the *CDC2* gene was cloned from the FT210 cell line by the polymerase chain–reaction (PCR) method. The template used for this PCR was the cDNA prepared from the FT210 cell line. As a control, the *CDC2* gene from the parental FM3A line was similarly cloned. The DNA sequences were determined from which the amino acid sequences were derived. A comparison was made between the amino acid sequence of $p34^{CDC2FM}$ and the sequences of p34 cloned from other mammalian cells (Fig. 11). The mouse p34 that is a component of the kinase, which phosphorylates the C-terminal domain (CTD) of RNA polymerase II (Cisek and Corden, 1989), has a complete amino acid sequence identity with $p34^{CDC2FM}$ (J. Corden, 1990 private communication). Comparison between $p34^{CDC2FM}$ (Th'ng *et al.*, 1990) and $p34^{CDC2Mm}$ (Spurr *et al.*, 1990) revealed that the amino acid sequences were the same except for two residues: (1) residue 165 in which a leucine in $p34^{CDC2Mm}$ was replaced with a valine in $p34^{CDC2FM}$; (2) residue 273, in which a threonine in $p34^{CDC2Mm}$ was substituted for an alanine in $p34^{CDC2FM}$. These differences could arise from the different strains of mouse cell lines from which these genes were cloned. Ten amino acid substitutions were found between the human $p34^{CDC2Hs}$ and the mouse $p34^{CDC2FM}$. These differences are in sites that display certain degrees of variability between yeast and mammalian sequences (Cisek and Corden, 1989).

Comparison of the sequences between $p34^{CDC2FM}$ and $p34^{CDC2FT}$ revealed a surprise. Instead of the single point mutation that was expected, two mutations were found in the *CDC2* gene cloned from the FT210 cells (Fig. 12) (Th'ng *et al.*, 1990). The resulting changes in the amino acid sequence of the $p34^{CDC2FT}$ fell within regions that are highly conserved through evolution. One mutation is in the PSTAIR region, where the isoleucine residue at position 52 in $p34^{CDC2FM}$ was

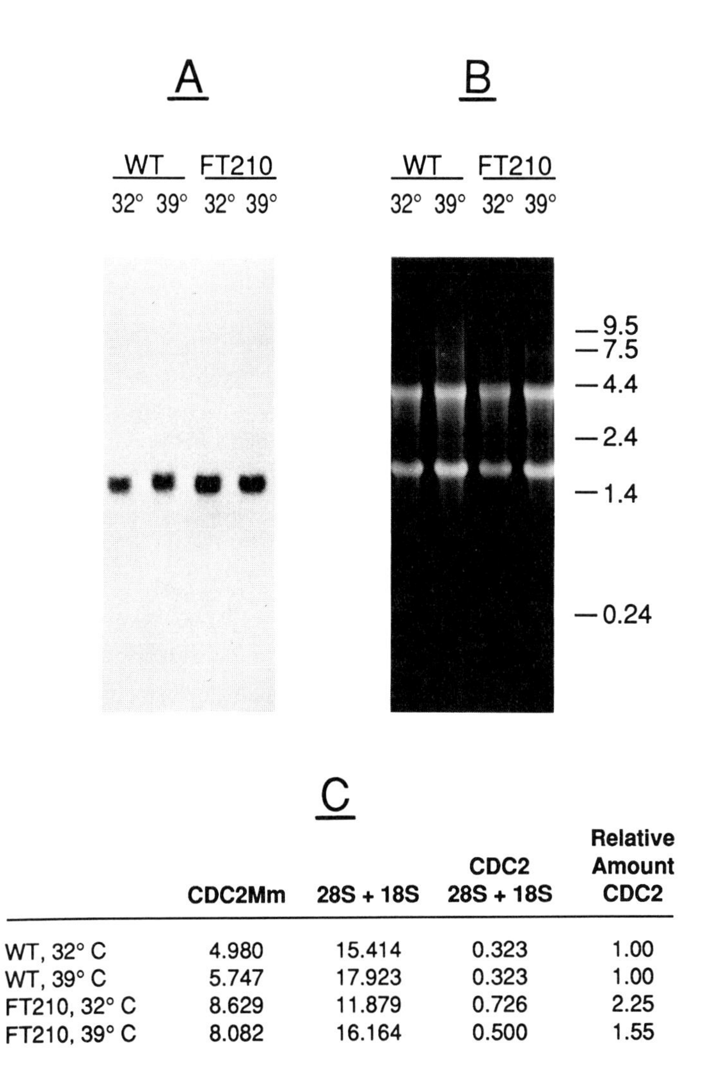

	CDC2Mm	28S + 18S	CDC2 28S + 18S	Relative Amount CDC2
WT, 32° C	4.980	15.414	0.323	1.00
WT, 39° C	5.747	17.923	0.323	1.00
FT210, 32° C	8.629	11.879	0.726	2.25
FT210, 39° C	8.082	16.164	0.500	1.55

Fig. 9. Northern blot analysis of RNA from FT210 cells. Total cellular RNA was extracted from parental cells (WT) or FT210 cells that were incubated at either 32°C or at 39°C. The RNA was fractionated and blotted for the presence of *CDC2* messages (panel A). Panel B shows the amounts of RNA loaded in each lane, as revealed by ethidium bromide staining. The levels of the *CDC2* messages relative to the amounts of RNA loaded per lane, as determined by scanning, is shown in panel C. (From Th'ng *et al.*, 1990).

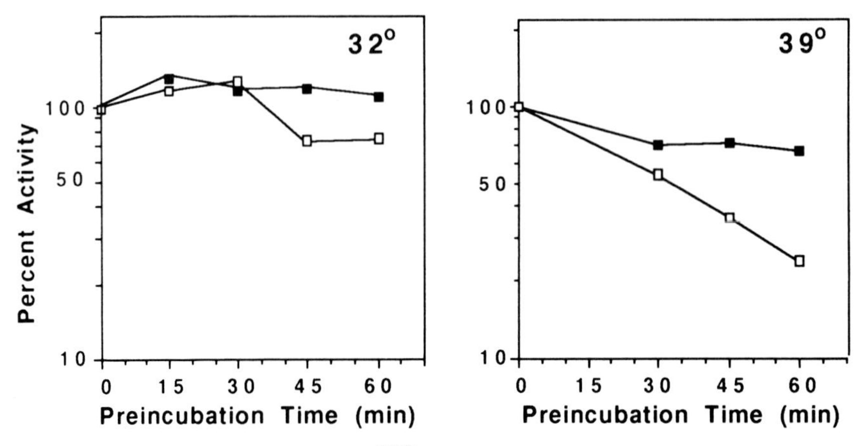

Fig. 10. Temperature-sensitivity of p34^{CDC2} protein kinase extracted from FT210 cells. Whole-cell extracts were prepared from parental FM3A cell line or from FT210 cells and the p34^{CDC2} protein kinase was immunoprecipitated. The activity of this protein kinase was then assayed after preincubating at 32°C or 39°C for the indicated period. Open squares, FT210 cell extracts; filled squares, FM3A cell extracts. (From Th'ng *et al.*, 1990).

changed to a valine in p34^{CDC2FT}. This region has been reported to have the ability to induce or accelerate the onset of MPF action (Labbe *et al.*, 1989; Gautier *et al.*, 1988). This peptide was found to also increase intracellular calcium levels in the cell (Picard *et al.*, 1990), and alteraction of a single amino acid within this peptide abolished these biological activities. The second mutation in the p34^{CDC2FT} was in the C-terminal region of the protein, where a proline at residue 272 in p34^{CDC2FM} is converted to a serine in p34^{CDC2FT}. The affected region is the tripeptide YDP that is found to be present in yeast, mouse, and human p34^{CDC2}. This change from a proline to a serine would have profound effect on the secondary structure of this region of p34. Such a change could affect both the function of the protein as well as its stability to environmental changes. Indeed, studies done on the homologous p34^{cdc28} protein kinase from *S. cerevisiae* showed that mutations within the C-terminal domain tend to give rise to temperature-sensitive kinase activity (Lorincz and Reed, 1986).

The results presented here show that the temperature-induced, cell-cycle arrest in the FT210 cells is caused by thermoinactivation of the *CDC2* gene product, p34^{CDC2}. The possibility that the cell-cycle arrest is due to a mutation in another gene that has yet to be identified can be excluded because of the observation that the FT210 cell line can be rescued by the introduction of the human *CDC2* gene cloned into an expression vector. The extremely rare double mutation that was found within the same gene posed a further question on the role of p34^{CDC2} in regulating the onset of mitosis in FT210 cells. The mutation in the PSTAIR region

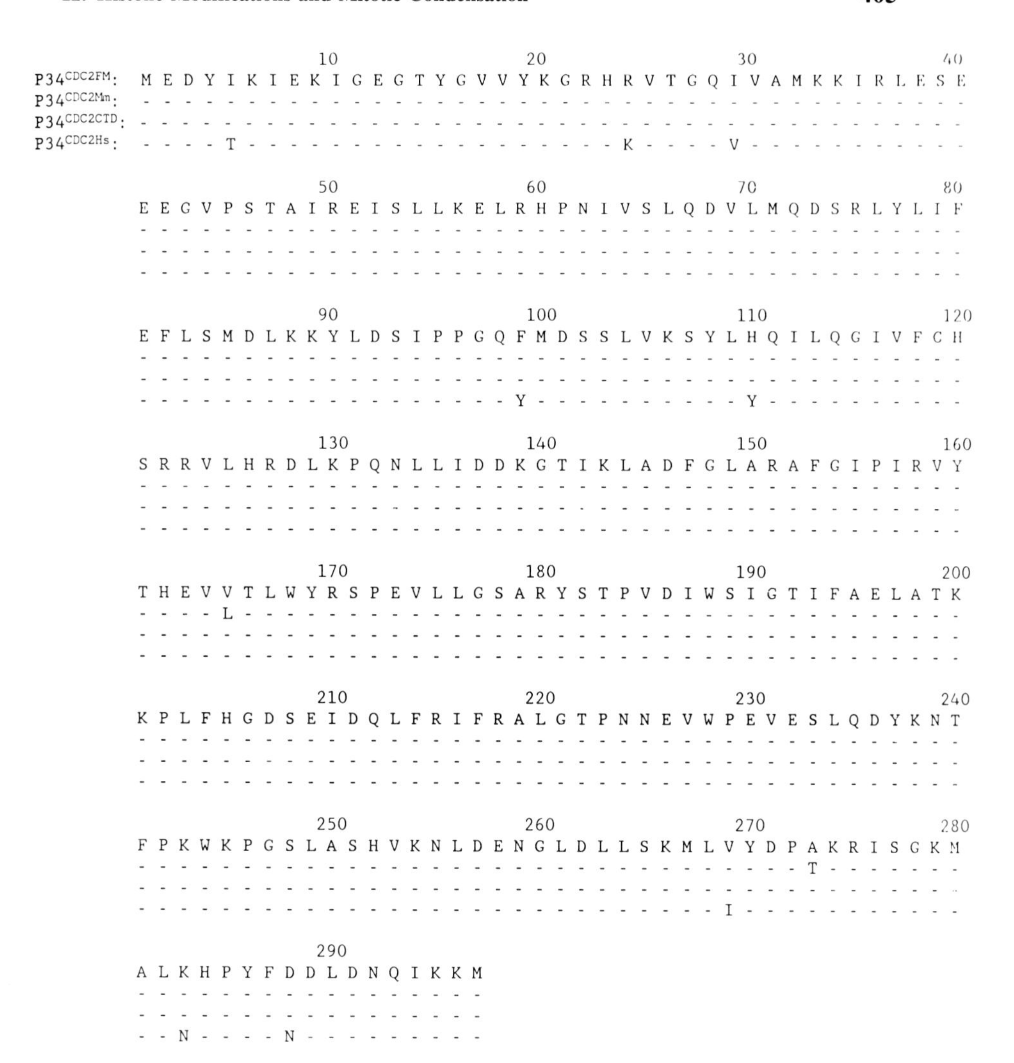

Fig. 11. Comparison of p34^{CDC2} sequences from mammalian cells. Amino acid sequences of p34^{CDC2} homologs from other mammalian cells are compared with the sequence of p34^{CDC2FM}. Identical residues are indicated by a dash, and amino acid differences are indicated by letters. (From Th'ng *et al.*, 1990).

could affect mitosis by influencing the intracellular level of calcium. Calcium fluxes have been found to be required for entry of cells into mitosis (Steinhardt and Alderton, 1988; Whitaker and Patel, 1990). Since it was shown that the PSTAIR region has to be absolutely conserved for activity (Picard *et al.*, 1990), the mutation in this region of p34^{CDC2FT} could prevent entry into mitosis by preventing

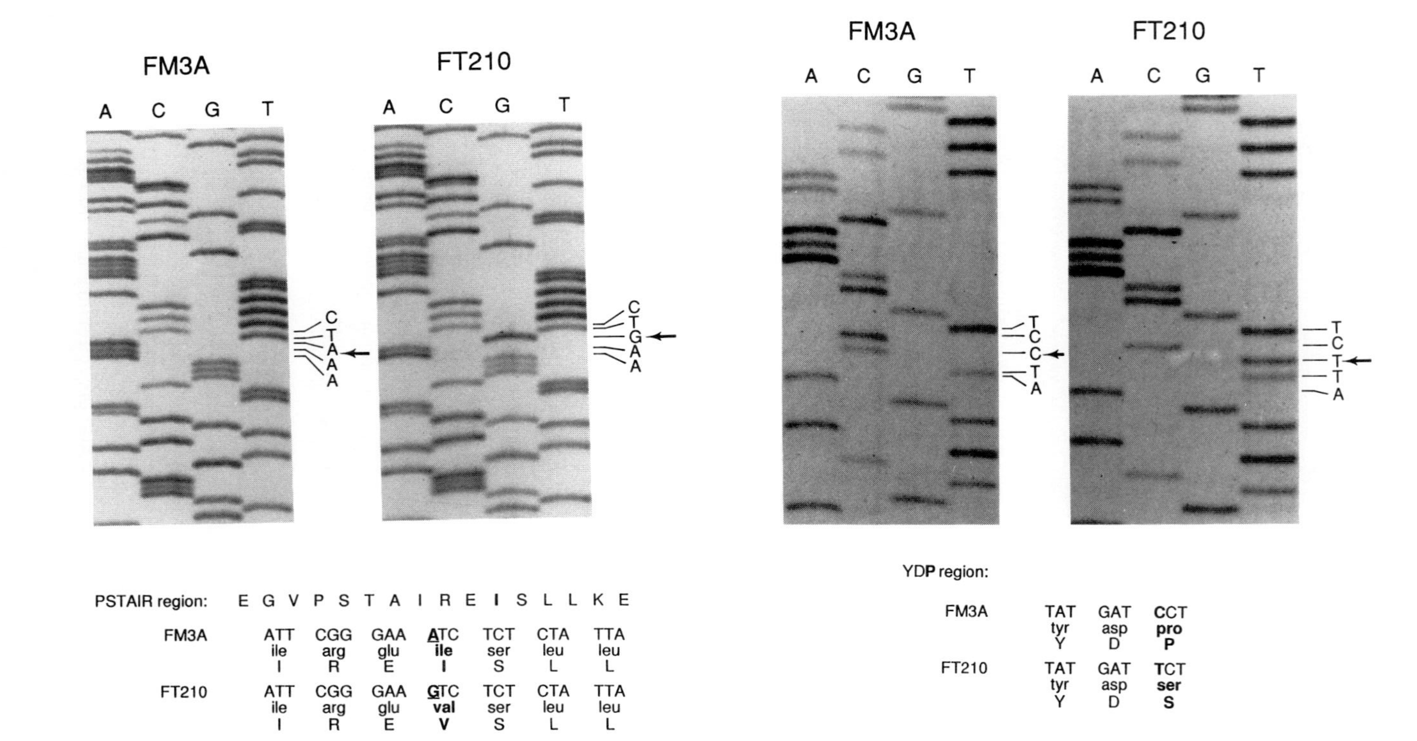

Fig. 12. Point mutations in the *CDC2*FT. the *CDC2* genes cloned from FM3A or FT210 cells were sequenced and revealed two point mutations in the *CDC2*FT: an A to a G transition in the PSTAIR region (left panel); a cytosine-to-thymine transition at the YDP region (right panel). (From Th'ng *et al.*, 1990).

increases in intracellular calcium levels. It is equally possible that the mutation in the YDP region of $p34^{CDC2FT}$ is in inactivating the GA kinase that is required to hyperphosphorylate histone H1 to drive the condensation of chromosomes and trigger the onset of mitosis. It is just as likely that both processes are required for triggering mitosis in this cell line.

V. CONCLUDING REMARKS

Until now, advances in the genetics of mitotic control have come from studies in yeasts because of the relative ease in manipulating the genome. From work done so far, it seems likely that the mechanisms in mammalian cells will turn out to be quite similar. The identification of the FT210 as a mouse cell line with a defective *CDC2* gene could provide a way to study the molecular mechanisms involved with regulating the onset of mitosis in mammalian cells.

ACKNOWLEDGMENTS

We would like to thank Dr. Takeharu Nishimoto (Department of Biology, Kyushu University, Fukuoka, Japan) for providing the tsBN2 cell line. We would also like to thank Joyce Hamaguchi for helpful comments and Becky Greer for helping in the preparation of the figures for the manuscript. This work was supported by grants from the American Cancer Society (Grant NP73D), the National Institute of Health (Grant GM26901) and the Department of Energy (Grant 88ER60673).

REFERENCES

Adlakha, R. C., Sahasrabuddhe, C. G., Wright, D. A., Lindsey, W. F., and Rao, P. N. (1982a). Localization of mitotic factors on metaphase chromosomes. *J. Cell Sci.* **54**, 193–206.

Adlakha, R. C., Sahasrabuddhe, C. G., Wright, D. A., Lindsey, W. F., Smith, M. L., and Rao, P. N. (1982b). Chromosome-bound mitotic factors: Release by endonucleases. *Nucleic Acids. Res.* **10**, 4107–4117.

Adlakha, R. C., and Rao, P. N. (1987). Regulation of mitosis by nonhistone protein factors in mammalian cells. *In* "Molecular Regulation of Nuclear Events in Mitosis and Meiosis" (R. A. Schlegel, M. S. Halleck, and P. N. Rao, eds.), pp. 179–226. Academic Press, Orlando, Florida.

Adlakha, R. C., Shipley, G. L., Zhao, J. Y., Jones, K. B., Wright, D. A., Rao, P. N., and Sauer, H. W. (1988). Amphibian oocyte maturation induced by extracts of *Physarum polycephalum* in mitosis. *J. Cell Biol.* **106**, 1445–1452.

Ajiro, K., Borun, T. W., Shulman, S. D., McFadden, G. M., and Cohen, L. H. (1981). Comparison of the structures of human histones 1A and 1B and their intramolecular phosphorylation sites during the HeLa S-3 cell cycle. *Biochemistry* **20**, 1454–1464.

Ajiro, K., Nishimoto, T., and Takahashi, T. (1983). Histone H1 and H3 phosphorylation during premature chromosome condensation in a temperature-sensitive mutant (tsBN2) of baby hamster kidney cells. *J. Biol. Chem.* **258**, 4534–4538.

Allan, J., Harborne, N., Rau, D. C., and Gould, H. (1982). Participation of core histone "tails" in the stabilization of the chromatin solenoid. *J. Cell Biol.* **93**, 285–297.

Allan, J., Hartman, P. G., Crane-Robinson, C., and Aviles, F. J. (1980). The structure of histone H1 and its location in chromatin. *Nature* **288**, 675–679.

Allfrey, V. G. (1971). Functional and metabolic aspects of DNA-associated proteins *In* "Histones and Nucleohistones" (D. M. P. Phillips, ed.), pp. 241–294. Plenum, London.

Allfrey, V. G. (1980). Molecular aspects of the regulation of eukaryotic transcription—nucleosomal proteins and their postsynthetic modifications in the control of DNA conformation and template function. *In* "Cell Biology: A Comprehensive Treatise," Vol. 3 (L. Goldstein, and D. M. Prescott, eds.), pp. 347–437. Academic Press, New York.

Allfrey, V. G., Faulkner, R., and Mirsky, A. E. (1964). Acetylation and methylation of histones and their role in the regulation of RNA synthesis. *Proc. Natl. Acad. Sci. U.S.A.* **51**, 786–794.

Arion, D., Meijer, L., Brizuela, L., and Beach, D. (1988). *cdc2* is a component of the M phase-specific histone H1 kinase: Evidence for identity with MPF. *Cell* **55**, 371–378.

Baldwin, J. P., Boseley, P. G., Bradbury, E. M., and Ibel, K. (1975). The subunit structure of the eukaryote chromosome. *Nature* **253**, 245–249.

Baldwin, J. P., Carpenter, B. G., Crespi, H., Hancock, R., Stephens, R. M., Simpson, J. K., Bradbury, E. M., and Ibel, K. (1978). Neutron-scattering from chromatin in relation to higher-order structure. *J. Appl. Cryst.* **11**, 484–486.

Balhorn, R., Balhorn, M., Morris, H. P., and Chalkley, R. (1972a). Comparative high-resolution electrophoresis of tumor histones: Variation in phosphorylation as a function of cell-replication rate. *Cancer Res.* **32**, 1775–1784.

Balhorn, R., Chalkley, R., and Granner, D. (1972b). Lysine-rich histone phosphorylation. A positive correlation with cell replication. *Biochemistry* **11**, 1094–1098.

Balhorn, R., Rieke, W. O., and Chalkley, R. (1971). Rapid electrophoretic analysis for histone phosphorylation. A reinvestigation of phosphorylation of lysine-rich histone during rat liver regeneration. *Biochemistry* **10**, 3952–3959.

Ballal, N. R., and Busch, H. (1973). Two-dimensional gel electrophoresis of acid-soluble nucleolar proteins of Walker 256 carcinosarcoma, regenerating liver, and thioacetamide-treated liver. *Cancer Res.* **33**, 2737–2743.

Barrack, E. R., and Coffey, D. S. (1982). Biological properties of the nuclear matrix. *Recent Prog. Horm. Res.* **38**, 133–195.

Bode, J. (1984). Nucleosomal conformations induced by the small HMG proteins or by histone hyperacetylation are distinct. *Arch. Biochem. Biophys.* **228**, 364–372.

Bohm, L., and Crane-Robinson, C. (1984). Proteases as structural probes for chromatin. *Biosci. Rep.* **4**, 365–386.

Booher, R., and Beach, D. (1987). Interaction between *cdc13*$^+$ and *cdc2*$^+$ in the control of mitosis in fission yeast; dissociation of the G_1 and G_2 roles of the *cdc2*$^+$ protein kinase. *EMBO J.* **6**, 3441–3447.

Bradbury, E. M. (1983). Flexibility in chromosomal proteins. *In* "Mobility and Recognition in Cell Biology" (V. Sund, C. Veeger, eds.), pp. 173–194. Walter de Gruyter & Co., Berlin.

Bradbury, E. M., Baldwin, J. P., Carpenter, B. G., Hjelm, R. P., Hancock, R., and Ibel, K. (1975). Neutron-scattering studies of chromatin. *In* "Neutron Scattering for the Analysis of Biological Structures." *Brookhaven Symposium in Biology* **27**, IV-97–IV-117. National Technical Information Service, Sringfield, Virginia.

Bradbury, E. M., Carpenter, B. G., and Rattle, H. W. E. (1973a). Magnetic resonance studies of deoxyribonucleoprotein—a proposed role and mode of action for the very lysine-rich histone F1. *Nature* **241**, 123–126.

Bradbury, E. M., Inglis, R. J., Matthews, H. R., and Sarner, N. (1973b). Phosphorylation of very lysine-rich histone in *Physarum polycephalum:* Correlation with chromosome condensation. *Eur. J. Biochem.* **33**, 131–139.

Bradbury, E. M., Inglis, R. J., and Matthews, H. R. (1974a). Control of cell division by very lysine-rich histone (f1) phosphorylation. *Nature* **247**, 257–261.

Bradbury, E. M., Inglis, R. J., Matthews, H. R., and Langan, T. A. (1974b). Molecular basis of control of mitotic cell division in eukaryotes. *Nature* **249**, 553–556.

Braddock, G. W., Baldwin, J. P., and Bradbury, E. M. (1981). Neutron-scattering studies of the structure of chromatin core particles in solution. *Biopolymers* **20**, 327–343.

Brizuela, L., Draetta, G., and Beach, D. (1989). Activation of human *CDC2* protein as a histone H1 kinase is associated with complex formation with the p62 subunit. *Proc. Natl. Acad. Sci. U.S.A.* **86**, 4362–4366.

Carpenter, B. G., Baldwin, J. P., Bradbury, E. M., and Ibel, K. (1976). Organization of subunits in chromatin. *Nucleic Acids Res.* **3**, 1739–1746.

Chahal, S. S., Matthews, H. R., and Bradbury, E. M. (1980). Acetylation of histone H4 and its role in chromatin structure and function. *Nature* **287**, 76–79.

Chambers, T. C., and Langan, T. A. (1990). Purification and characterization of growth-associated H1 histone kinase from Novikoff hepatoma cells. *J. Biol. Chem.* **265**, 16940–16947.

Chambers, T. C., Langan, T. A., Matthews, H. R., and Bradbury, E. M. (1983). H1 histone kinase from nuclei of *Physarum polycephalum. Biochemistry* **22**, 30–37.

Chicoine, L. G., Schulman, I. G., Riehman, R., Cook, R. G., and Allis, C. D. (1986). Nonrandom utilization of acetylation sites in histones isolated from *Tetrahymena. J. Biol. Chem.* **261**, 1071–1076.

Cisek, L. J., and Corden, J. L. (1989). Phosphorylation of RNA polymerase by the murine homologue of the cell-cycle-control protein *cdc2. Nature* **339**, 679–684.

Couppez, M., Martin-Ponthieu, A., and Santiere, P. (1987). Histone H4 from cuttlefish testis is sequentially actylated; comparison with acetylation of calf thymus histone H4. *J. Biol. Chem.* **262**, 2854–2860.

Crane-Robinson, C., Bohm, L., Puigdomenech, P., Cary, P. D., Hartman, P. G., and Bradbury, E. M. (1980). Structural domains in histones. *In* "(FEBS) DNA–Recombination Interactions and Repair" (G. Zadrazil and J. Sponar, eds.), pp. 293–300. Pergamon Press, Oxford and New York.

Csordas, A. (1990). On the biological role of histone acetylation. *Biochem. J.* **265**, 23–38.

D'Anna, J. A., Gurley, L. R., and Deaven, L. L. (1978). Dephosphorylation of histones H1 and H3 during the isolation of metaphase chromosomes. *Nucleic Acids Res.* **5**, 3195–3207.

D'Anna, J. A., Gurley, L. R., and Tobey, R. A. (1983). Extent of histone modifications and $H1^0$ content during cell-cycle progression in the presence of butyrate. *Exp. Cell Res.* **147**, 407–417.

D'Anna, Tobey, R. A., and Gurley, L. R. (1980). Concentration-dependent effects of sodium butyrate in Chinese hamster cells: Cell-cycle progression, inner-histone acetylation, histone H1 dephosphorylation, and induction of an H1-like protein. *Biochemistry* **19**, 2656–2671.

Davie, J. R., and Nickel, B. E. (1987). The ubiquitinated histone species are enriched in histone H1-depleted chromatin regions. *Biochim. Biophys. Acta* **909**, 183–189.

Draetta, G., and Beach, D. (1988). Activation of *cdc2* protein kinase during mitosis in human cells: Cell cycle–dependent phosphorylation and subunit rearrangement. *Cell* **54**, 17–26.

Draetta, G., Brizuela, L., Potashkin, J., and Beach, D. (1987). Identification of p34 and p13, human homologs of the cell-cycle regulators of the fission yeast encoded by $cdc2^+$ and $suc1^+$. *Cell* **50**, 319–325.

Draetta, G., Luca, F., Westendorf, J., Brizuela, L., Ruderman, J., and Beach, D. (1989). *cdc2* protein kinase is complexed with both cyclin A and B: Evidence for proteolytic inactivation of MPF. *Cell* **56**, 829–838.

Dunphy, W., Brizuela, L., Beach, D., and Newport, J. (1988). The *Xenopus cdc2* protein is a component of MPF, a cytoplasmic regulator of mitosis. *Cell* **54**, 423–431.

Earnshaw, W. C., Halligan, B., Cooke, C. A., Heck, M. M., and Liu, F. (1985). Topoisomerase II is a structural component of mitotic chromosome scaffolds. *J. Cell Biol.* **100**, 1706–1715.

Earnshaw, W. C., and Heck, M. M. (1985). Localization of topoisomerase II in mitotic chromosomes. *J. Cell Biol.* **100**, 1716–1725.

Edgar, B. A., and O'Farrell, P. H. (1989). Genetic control of cell-division patterns in the *Drosophila* embryo. *Cell* **57**, 177–187.

Eki, T., Enomoto, T., Miyajima, A., Miyazawa, H., Murakami, Y., Hanaoka, F., Yamada, M., and Ui, M. (1990). Isolation of temperature-sensitive cell-cycle mutants from mouse FM3A cells. *J. Biol. Chem.* **265**, 26–33.

Fallon, R. J., and Cox, R. P. (1979). Cell-cycle analysis of sodium butyrate and hydroxyurea, inducers of ectopic hormone production in HeLa cells. *J. Cell. Physiol.* **100**, 251–261.

Finley, D., Ciechanover, A., and Varshavsky, A. (1984). Thermolability of ubiquitin-activating enzyme from the mammalian cell-cycle mutant ts85. *Cell* **37**, 43–55.

Fujitaki, J. M., Fung, G., Oh, E. Y., and Smith, R. A. (1981). Characterization of chemical and enzymatic acid labile phosphorylation of histone H4 using phosphorus-31 nuclear magnetic resonance. *Biochemistry* **20**, 3658–3664.

Gautier, J., Minshull, J., Lohka, M., Glotzer, M., Hunt, T., and Maller, J. L. (1990). Cyclin is a component of maturation-promoting factor from *Xenopus. Cell* **60**, 487–494.

Gautier, J., Norbury, C., Lohka, M., Nurse, P., and Maller, J. (1988). Purified maturation-promoting factor contains the product of a *Xenopus* homolog of the fission yeast cell-cycle control gene *cdc2*$^{+}$. *Cell* **54**, 433–439.

Goldknopf, I. L., and Busch, H. (1977). Isopeptide linkage between nonhistone and histone 2A polypeptides of chromosomal conjugate-protein A24. *Proc. Natl. Acad. Sci. U.S.A.* **74**, 864–868.

Goldknopf, I. L., French, M. F., Musso, R., and Busch, H. (1977). Presence of protein A24 in rat liver nucleosomes. *Proc. Natl. Acad. Sci. U.S.A.* **74**, 5492–5495.

Goldknopf, I. L., Taylor, C. W., Baum, R. M., Yeoman, L. C., Olson, M. O. J., Prestayko, A. W., and Busch, H. (1975). Isolation and characterization of protein A24, a "histone-like" non-histone chromosomal protein. *J. Biol. Chem.* **250**, 7182–7187.

Goldstein, G. (1974). Isolation of bovine thymin: A polypeptide hormone of the thymus. *Nature* **247**, 11–14.

Goldstein, G., Scheid, M. S., Hammerling, V., Boyse, E. A., Schlesinger, D. H., and Niall, H. D. (1975). Isolation of a polypeptide that has lymphocyte-differentiating properties and is probably represented universally in living cells. *Proc. Natl. Acad. Sci. U.S.A.* **72**, 11–15.

Gurley, L. R., Walters, R. A., and Tobey, R. A. (1973). The metabolism of histone fractions VI. Differences in the phosphorylation of histone fractions during the cell cycle. *Arch. Biohem. Biophys.* **154**, 212–218.

Gurley, L. R., Walters, R. A., and Tobey, R. A. (1975). Sequential phosphorylation of histone subfractions in the Chinese hamster cell cycle. *J. Biol. Chem.* **250**, 3936–3944.

Gurley, L. R., D'Anna, J. A., Barham, S. S., Deaven, L. L., and Tobey, R. A. (1978). Histone phosphorylation and chromatin structure during mitosis in Chinese hamster cells. *Eur. J. Biochem.* **84**, 1–15.

Hagan, I., Hayles, J., and Nurse, P. (1988). Cloning and sequencing of the cyclin-related *cdc13*$^{+}$ gene and a cytological study of its role in fission yeast mitosis. *J. Cell Sci.* **91**, 587–595.

Halleck, M. S., and Gurley, L. R. (1982). Histone variants and histone modifications in chromatin fractions from heterochromatin-rich *Peromyscus* cells *Exp. Cell Res.* **138**, 271–285.

Halleck, M. S., and Schlegel, R. A. (1983). C-banding of *Peromyscus* constitutive heterochromatin persists following histone hyperacetylation. *Exp. Cell Res.* **147**, 269–279.

Hanks, S. K., Rodriguez, L. V., and Rao, R. N. (1983). Relationship between histone phosphorylation and premature chromosome condensation. *Exp. Cell Res.* **148**, 293–302.

Hebbes, T. R., Thorne, A. W., and Crane-Robinson, C. (1988). A direct link between core histone acetylation and transcriptionally active chromatin. *EMBO J.* **7**, 1395–1402.

Huang, S.-Y., Barnard, M. B., Xu, M., Matsui, S.-I., Rose, S., and Garrard, W. T. (1986). The active immunoglobulin κ chain gene is packaged by non–ubiquitin-conjugated nucleosomes. *Proc. Natl. Acad. Sci. U.S.A.* **83**, 3738–3742.

Inglis, R. J., Langan, T. A., Matthews, H. R., Hardie, D. G., and Bradbury, E. M. (1976). Advance of mitosis by histone phosphokinase. *Exp. Cell Res.* **97**, 418–425.

Jentsch, S., Seufert, W., Sommer, T., and Reins, H.-A. (1990). Ubiquitin-conjugating enzymes: Novel regulators of eukaryotic cells. *Trends Biochem. Sci.* **15**, 195–198.

Johnson, R. T., and Rao, P. N. (1970). Mammalian cell fusion: Induction of premature chromosome condensation in interphase cells. *Nature* **226**, 717–722.

Kayne, P. S., Kim, U.-J., Han, M., Mullen, J. R., Yoshizaki, F., and Grunstein, M. (1988). Extremely conserved histone H4 N-terminus is dispensible for growth but essential for repressing the silent mating loci of yeast. *Cell* **55**, 27–39.

Kishimoto, T., and Kanatani, H. (1976). Cytoplasmic factor responsible for germinal-vesicle breakdown and meiotic maturation in starfish oocyte. *Nature* **260**, 321–322.

Kishimoto, T., Kuriyama, R., Kondo, H., and Kanatani, H. (1982). Generality of the action of various maturation-promoting factors. *Exp. Cell Res.* **137**, 121–126.

Kleinschmidt, A. M., and Martinson, H. G. (1981). Structure of nucleosome core particles containing uH2A (A24). *Nucleic Acid Res.* **9**, 2423–2431.

Labbe, J. C., Lee, M. G., Nurse, P., Picard, A., and Doree, M. (1988). Activation at M phase of a protein kinase encoded by a starfish homologue of the cell cycle–control gene $cdc2^+$. *Nature* **335**, 251–254.

Labbe, J. C., Picard, A., Peaucellier, G., Cavadore, J. C., Nurse, P., and Doree, M. (1989). Purification of MPF from starfish: Identification as the H1 histone kinase $p34^{cdc2}$ and a possible mechanism for its periodic activation. *Cell* **57**, 253–263.

Laemmli, U. K. (1985). "Higher-Order Chromatin Loops; Evidence for Cell-Cycle and Differentiation-Dependent Changes." 10th Edward de Rothschild Sch. Mol. Biophys., Weizmann Institute of Science, Jerusalem.

Lamb, N. J. C., Fernandez, A., Watrin, A., Labbe, J. C., and Cavadore, J. C. (1990). Microinjection of $p34^{cdc2}$ kinase induces marked changes in cell shape, cytoskeletal organization, and chromatin structure in mammalian fibroblasts. *Cell* **60**, 151–165.

Langan, T. A. (1969). Phosphorylation of liver histone following the administration of glucagon and insulin. *Proc. Natl. Acad. Sci. U.S.A.* **64**, 1276–1283.

Langan, T. A. (1982). Characterization of highly phosphorylated subcomponents of rat thymus H1 histone. *J. Biol. Chem.* **257**, 14835–14846.

Langan, T. A., Gautier, J., Lohka, M., Hollingsworth, R., Moreno, S., Nurse, P., Maller, J., and Sclafani, R. A. (1989). Mammalian growth-associated H1 histone kinase: A homolog of $cdc2^+/CDC28$ protein kinases controlling mitotic entry in yeast and frog cells. *Mol. Cell. Biol.* **9**, 3860–3868.

Langan, T. A., Zeilig, C. E., and Leichtling, B. (1980). Analysis of multiple-site phosphorylation of H1 histone. *In* "Protein Phosphorylation and Bioregulation" (G. Thomas, E. G., Podesta, and J. Gordon, eds.), pp. 70–82. S. Karger, Basel.

Langan, T. A., Zeilig, C. E., and Leichtling, B. (1981). Characterization of multiple-site phosphorylation of H1 histone in proliferating cells. *In* "Cold Spring Harbor Conferences on Cell Proliferation, Vol. 8: Protein Phosphorylation" pp. 1039–1052. Cold Spring Harbor Laboratory, New York.

Lee, M. G., and Nurse, P. (1987). Complementation used to clone a human homologue of the fission yeast cell cycle–control gene *cdc2*. *Nature* **327**, 31–35.

Lehner, C. F., and O'Farrell, P. H. (1990). The roles of *Drosophila* cyclins A and B in mitotic control. *Cell* **61**, 535–547.

Levinger, L., Barsoum, J., and Varshavsky, A. (1981). Two-dimensional hybridization mapping of nucleosomes: Comparison of DNA and protein patterns. *J. Mol. Biol.* **146**, 287–304.

Levinger, L., and Varshavsky, A. (1982). Selective arrangement of ubiquitinated and D1 protein-containing nucleosomes within the *Drosophila* genome. *Cell* **28**, 375–385.

Lewis, C. D., Lebkowski, J. S., Daly, A., and Laemmli, U. K. (1984). Interphase nuclear matrix and metaphase scaffolding structure: A comparison of protein components. *J. Cell Sci. (Suppl.)***1**, 103–122.

Lilley, D. M., and Tatchell, K. (1977). Chromatin core particle unfolding induced by tryptic cleavage of histones. *Nucleic Acids Res.* **4**, 2039–2055.

Lin, R. L., Leone, J. W., Cook, R. G., and Allis, C. D. (1989). Antibodies specific to acetylated histones document the existence of deposition- and transcription-related histone acetylation in *Tetrahymena. J. Cell Biol.* **108**, 1577–1588.

Lohka, M., Hayes, M. K., and Maller, J. L. (1988). Purification of maturation-promoting factor, an intracellular regulator of early mitotic events. *Proc. Natl. Acad. Sci. U.S.A.* **85**, 3009–3013.

Lohka, M. J. (1989). Mitotic control by metaphase-promoting factor and *cdc* proteins. *J. Cell Sci.* **92**, 131–135.

Lorincz, A. T., and Reed, S. I. (1986). Sequence analysis of temperature-sensitive mutations in the *Saccharomyces cerevisiae* gene *CDC28. Mol. Cell. Biol.* **6**, 4099–4103.

Marcus, M., Fainsod, A., and Diamond, G. (1985). The genetic analysis of mammalian cell-cycle mutants. *Ann. Rev. Genetics* **19**, 389–421.

Marian, B., and Winterberger, U. (1982). Modification of histones during the mitotic and meiotic cycle of yeast. *FEBS Lett.* **139**, 72–76.

Marks, D. B., Paik, W. K., and Borun, T. W. (1973). The relationship of histone phosphorylation to deoxyribonucleic acid replication and mitosis during the HeLa S-3 cell cycle. *J. Biol. Chem.* **248**, 5660–5667.

Marsden, M., and Laemmli, U. K. (1979). Metaphase chromosome structure: Evidence for a radial-loop model. *Cell* **17**, 849–853.

Martinson, H. G., True, R., Burch, J. B. E., and Kunkel, G. (1979). Semihistone protein A24 replaces H2A as an integral component of the nucleosome histone core. *Proc. Natl. Acad. Sci. U.S.A.* **76**, 1030–1034.

Marvin, K. W., Yau, P., and Bradbury, E. M. (1990). Isolation and characterization of acetylated histones H3 and H4 and their assembly into nucleosomes. *J. Biol. Chem.* **265**, 19839–19847.

Masaracchia, U. A., Maller, J. L., and Walsh, D. A. (1979). Histone H1 phosphotransferase activities during the maturation of oocytes of *Xenopus laevis. Arch. Biochem. Biophys.* **194**, 1–12.

Masui, Y., and Markert, C. L. (1971). Cytoplasmic control of nuclear behavior during meiotic maturation of frog oocytes. *J. Exp. Zool.* **177**, 129–146.

Matsui, S. I., Seon, B. K., and Sandberg, A. A. (1979). Disappearance of a structural chromatin protein A24 in mitosis: Implications for molecular basis of chromatin condensation. *Proc. Natl. Acad. Sci. U.S.A.* **76**, 6386–6390.

Matsukawa, T., Adachi, H., Kurashina, Y., and Ohba, Y. (1985). Phosphorylation of five histone H1 subtypes of L5178Y cells at the exponential growth and mitotic phases. *J. Biochem.* **98**, 695–704.

Matsumoto, Y., Yasuda, H., Mita, S., Marunouchi, T., and Yamada, M. (1980). Evidence for the involvement of H1 histone phosphorylation in chromosome condensation. *Nature* **284**, 181–183.

Matthews, H. R. (1988). Histone modification and chromatin structure. *In* "Chromosomes and Chromatin" (K. W. Aldoph, ed.), Vol. 1, pp. 3–32. CRC Press, Boca Raton, Florida.

Matthews, H. R., and Heubner, V. D. (1984). Nuclear protein kinases. *Mol. Cell. Biochem.* **59**, 81–99.

McGhee, J. D., Nickol, J. M., Felsenfeld, G., and Rau, D. C. (1983). Histone hyperacetylation has little effect on the higher-order folding of chromatin. *Nucleic Acids Res.* **11**, 4065–4075.

MeGee, P. C., Morgan, B. A., Mittman, B. A., and Smith, M. M. (1990). Genetic analysis of histone h4: Essential role of lysine subject to reversible acetylation. *Science* **247**, 841–845.

Mineo, C., Murakami, Y., Ishimi, Y., Hanaoka, F., and Yamada, M. (1986). Isolation and analysis of a mammalian temperature-sensitive mutant defective in G_2 functions. *Exp. Cell Res.* **167**, 53–62.

Minshull, J., Goldsteyn, R., Hill, C. S., and Hunt, T. (1990). The A- and B-cyclin associated cdc2 kinases in *Xenopus* turn on and off at different times in the cell cycle. *EMBO J.* **9**, 2865–2875.

Mita, S., Yasuda, H., Marunouchi, T., Ishiko, S., and Yamada, M. (1980). A temperature-sensitive mutant of cultured mouse cells defective in chromosome condensation. *Exp. Cell Res.* **126**, 407–416.

Mitchelson, K., Chambers, T., Bradbury, E. M., and Matthews, H. R. (1978). Activation of histone kinase in G_2 phase of the cell cycle in *Physarum polycephalum. FEBS Lett.* **92**, 339–342.

Mueller, R. D., Yasuda, H., Hatch, C. L., Bonner, W. M., and Bradbury, E. M. (1985). Identification of ubiquitinated histones 2A and 2B in *Physarum polycephalum: Disappearance of these proteins at metaphase and reappearance at anaphase. J. Biol. Chem.* **260**, 5147–5153.

Nelkin, B., Nicholas, C., and Vogelstein, B. (1980). Protein factor(s) from mitotic CHO cells induce meiotic maturation in *Xenopus laevis* oocytes. *FEBS Lett.* **109**, 233–238.

Nickel, B. E., Allis, C. D., and Davie, J. R. (1989). Ubiquitinated histone H2B is preferentially located in transcriptionally active chromatin. *Biochemistry* **28**, 958–963.

Nickel, B. E., and Davie, J. R. (1989). Structure of polyubiquitinated histone H2A. *Biochemistry* **28**, 964–968.

Nishimoto, T., Ajiro, K., Davis, F. M., Yamashita, K., Kai, R., Rao, P. N., and Sekiguchi, M. (1987). Mitosis-specific protein phosphorylation associated with premature chromosome condensation in a *ts* cell cycle mutant. *In* "Molecular Regulation of Nuclear Events in Mitosis and Meiosis" (R. A. Schlegel, M. S., Halleck, and P. N. Rao, eds.), pp. 295–318. Academic Press, Orlando, Florida.

Norbury, C. J., and Nurse, P. (1989). Control of the higher eukaryotic cell cycle by $p34^{cdc2}$ homologues. *Biochem. Biophys. Acta* **989**, 85–95.

Norton, V. G., Imai, B. S., Yau, P., and Bradbury, E. M. (1989). Histone acetylation reduces nucleosome core particle linking number change. *Cell* **57**, 449–457.

Nurse, P. (1985). Cell cycle-control genes in yeast. *Trends Genet.* **1**, 51–55.

Nurse, P. (1990). Universal control mechanism regulating onset of M phase. *Nature* **344**, 503–507.

Nurse, P., and Bissett, Y. (1981). Gene required in G_1 for commitment to cell cycle and in G_2 for control of mitosis in fission yeast. *Nature* **292**, 558–560.

Oliver. D., Balhorn, R., Granner, D., and Chalkley, R. (1972). Molecular nature of F_1 histone phosphorylation in cultured hepatoma cells. *Biochemistry* **11**, 3921–3925.

Ord, M. G., and Stocken, L. A. (1966). Metabolic properties of histones from rat liver and thymus gland. *Biochem. J.* **98**, 888–897.

Orrick, L. R., Olson, M. O. J., and Burch, H. (1973). Comparison of nucleolar proteins of normal rat liver and Novikoff hepatoma ascites cells by two-dimensional polyacrylamide gel electrophoresis. *Proc. Natl. Acad. Sci. U.S.A.* **70**, 1316–1320.

Panyim, S., and Chalkley, R. (1969). The heterogeneity of histones. I. A quantitative analysis of calf histones in very long polyacrylamide gels. *Biochemistry* **8**, 3972–3979.

Pardon, J. F., Worcester, D. L., Wooley, J. C., Tatchell, K., van Holde, K. E., and Richards, B. M. (1975). Low-angle neutron scattering from chromatin subunit particles. *Nucleic Acids Res.* **2**, 2163–2175.

Pardon, J. F., Worcester, D. L., Wooley, J. C., Cotter, R. I., Lilley, D. M. J., and Richards, B. M. (1977a). The structure of the chromatin core particle in solution. *Nucleic Acids Res.* **3**, 3199–3214.

Pardon, J. F., Cotter, R. I., Lilley, D. M. J., Worcester, D. L., Campbell, A. M., Wooley, J. C., and Richards, B. M. (1977b). Scattering studies of chromatin subunits. *Cold Spring Harbor Symp. Quant. Biol.* **44**, 11–22.

Paulson, J. R. (1980). Sulfhydryl reagents prevent dephosphorylation and proteolysis of histones in isolated HeLa metaphase chromosomes. *Eur. J. Biochem.* **111**, 189–197.

Peter, M. Nakagawa, J., Doree, M., Labbe, J.-C., and Nigg, E. A. (1990). Identification of major nucleolar proteins as candidate mitotic substrates of *cdc2* kinase. *Cell* **60**, 791–801.

Picard, A., Cavadore, J.-C., Lory, P., Bernango, J.-C., Ojeda, C., and Doree, M. (1990). Microinjection of a conserved peptide sequence of $p34^{cdc2}$ induces a Ca^{2+} transient in oocytes. *Science* **247**, 327–329.

Pienta, K. J., and Coffey, D. S. (1984). A structural analysis of the role of the nuclear matrix and DNA loops in the organization of the nucleus and chromosome. *J. Cell. Sci. (Suppl.)***1**, 123–135.

Pines, J., and Hunter, T. (1989). Isolation of a human cyclin cDNA: Evidence for cyclin mRNA and protein regulation in the cell cycle and for interaction with $p34^{cdc2}$. *Cell* **58**, 833–846.

Pines, J., and Hunter, T. (1990). Human cyclin A is adenovirus E1A–associated protein p60 and behaves differently from cyclin B. *Nature* **292**, 558–560.

Reczek, P. R., Weissman, D., Huvoc, P. E., and Fasman, G. D. (1982). Sodium butyrate–induced structural changes in HeLa cell chromatin. *Biochemistry* **21**, 993–1002.

Richmond, T. J., Finch, J. T., Rushton, B., Rhodes, D., and Klug, A. (1984). Structure of the nucleosome core particle at 7 Å resolution. *Nature* **311**, 532–537.

Russell, P. and Nurse, P. (1986). *cdc25*$^+$ functions as an inducer in the mitotic control of fission yeast. *Cell* **45**, 145–153.

Russell, P., and Nurse, P. (1987a). The mitotic inducer *nim1*$^+$ function in a regulatory network of protein kinase homologs controlling the initiation of mitosis. *Cell* **49**, 569–576.

Russell, P., and Nurse, P. (1987b). Negative regulation of mitosis by *wee1*$^+$, a gene encoding a protein kinase homolog. *Cell* **49**, 559–567.

Schuster, T., Han, M., and Grunstein, M. (1986). Yeast histone H2A and H2B have interchangeable functions. *Cell* **45**, 445–451.

Simanis, V., and Nurse, P. (1986). The cell cycle–control gene *cdc2*$^+$ of fission yeast encodes a protein kinase potentially regulated by phosphorylation. *Cell* **45**, 261–268.

Simpson, R. T. (1978). Structure of chromatin containing extensively acetylated H3 and H4. *Cell* **13**, 691–699.

Smith, M. F., Athey, B. D., Williams, S. P., and Langmore, J. P. (1990). Radial density distribution of chromatin. *J. Cell Biol.* **110**, 245–254.

Smith, L. D., and Ecker, R. E. (1971). The interaction of steroids with *Rana pipiens* oocytes in the induction of maturation. *Dev. Biol.* **25**, 232–247.

Spurr, N., Gough, A., and Lee, M. (1990). Cloning of the mouse homologue of the yeast cell cycle control gene *cdc2*. *Sequence* (in press).

Steinhardt, R. A., and Alderton, J. (1988). Intracellular free-calcium rise triggers nuclear envelope breakdown in the sea urchin embryo. *Nature* **332**, 364–369.

Stevely, W. S., and Stocken, L. A. (1966). Phosphorylation of rat-thymus histone. *Biochem. J.* **100**, 20c–21c.

Suau, P., Bradbury, E. M., and Baldwin, J. P. (1979). Higher-order structures of chromatin in solution. *Eur. J. Biochem.* **97**, 593–602.

Suau, P., Kneale, G. G., Braddock, G. W., Baldwin, J. P., and Bradbury, E. M. (1977). A low-resolution model for the chromatin core particle by neutron scattering. *Nucleic Acids Res.* **4**, 3769–3786.

Sunkara, P. S., Wright, D. A., and Rao, P. N. (1979a). Mitotic factors from mammalian cells induce germinal vesicle breakdown and chromosomal condensation in amphibian oocytes. *Proc. Natl. Acad. Sci. U.S.A.* **76**, 2799–2802.

Sunkara, P. S., Wright, D. A., and Rao, P. N. (1979b). Mitotic factors from mammalian cells: A preliminary characterization. *J. Supramol. Struct.* **11**, 189–195.

Taya, Y., Yasuda, H., Kamijo, M., Nakaya, K., Nakamura, Y., Ohba, Y., and Nishimura, S. (1989). *In vitro* phosphorylation of the tumor-suppressor gene RB protein by mitosis-specific histone H1 kinase. *Biochem. Biophys, Res. Commun.* **164**, 580–586.

Th'ng, J. P. H., Wright, P. S., Hamaguchi, J., Lee, M. G., Norbury, C. J., Nurse, P., and Bradbury, E. M. (1990). The FT210 cell line is a mouse G_2 phase mutant with a temperature-sensitive *CDC2* gene product. *Cell* **63**, 313–324.

Thoma, F., Koller, T., and Klug, A. (1979). Involvement of histone H1 in the organization of the nucleosome and of the salt-dependent superstructures of chromatin. *J. Cell Biol.* **83**, 403–427.

Tuomikoski, T., Felix, M.-A., Doree, M., and Gruenberg, J. (1989). Inhibition of endocytic vesicle fusion *in vitro* by the cell-cycle–control protein kinase *cdc2*. *Nature* **342**, 942–945.

Turner, B. M. (1989). Acetylation and deacetylation of histone H4 continue through metaphase with depletion of more acetylated isoforms and altered site usage. *Exp. Cell Res.* **182**, 206–214.

Turner, B. M., and Fellows, G. (1989). Specific antibodies reveal ordered and cell cycle–related use of histone H4 acetylation sites in mammalian cells. *Eur. J. Biochem.* **179**, 131–139.

Uchida, S., Sekiguchi, T., Nishitani, H., Miyauchi, K., Ohtsubo, M., and Nishimoto, T. (1990). Premature chromosome condensation is induced by a point mutation in the hamster RCC1 gene. *Mol. Cell. Biol.* **10**, 577–584.

Verde, F., Labbe, J.-C., Doree, M., and Karsenti, E. (1990). Regulation of microtubule dynamics by *cdc2* protein kinase in cell-free extracts of *Xenopus* eggs. *Nature* **343**, 233–238.

Vidali, G., Boffa, L. C., Bradbury, E. M., and Allfrey, V. G. (1978). Suppression of histone deacetylation leads to accumulation of multiacetylated forms of histone H3 and H4 and increased DNase I sensitivity of associated DNA sequences. *Proc. Natl. Acad. Sci. U.S.A.* **75**, 2239–2244.

Wallis, J. W., Rykowski, M., and Grunstein, M. (1983). Yeast histone H2B containing large aminoterminus deletion can function *in vivo. Cell* **35**, 711–719.

Whitaker, M., and Patel, R. (1990). Calcium and cell-cycle control. *Development* 108, 525–542.

Whitlock, J. P., and Simpson, R. T. (1977). Localization of the sites along nucleosome DNA which interact with NH_2-terminal histone regions. *J. Biol. Chem.* **252**, 6516–6520.

Whitlock, J. P., and Stein, A. (1978). Folding of DNA by histones which lack their NH_2-terminal regions. *J. Biol. Chem.* **253**, 3857–3861.

Williams, S. P., Athey, B. D., Muglia, L. J., Schappe, R. S., Gough, A. H., and Langmore, J. P. (1986). Chromatin fibers are left-handed double helices with diameter and mass per unit length that depend on linker length. *Biophys. J.* **49**, 233–248.

Wu, R. S., Kohn, K. W., and Bonner, W. M. (1981). Metabolism of ubiquitinated histones. *J. Biol. Chem.* **256**, 5916–5920.

Yasuda, H., Mueller, R. D., and Bradbury, E. M. (1987). Chromatin structure and histone modifications through mitosis in plasmodia of *Physarum polycephalum. In* "Molecular Regulation of Nuclear Events in Mitosis and Meiosis" (R. A. Schlegel, M. S. Halleck, and P. N. Rao, eds.), pp. 319–361. Academic Press, Orlando, Florida.

Yeoman, L. C., Taylor, C. W., and Busch, H. (1973). Two-dimensional polyacrylamide gel electrophoresis of acid-extractable nuclear proteins of normal rat liver and Novikoff hepatoma ascites cells. *Biochem. Biophys. Res. Commun.* **51**, 956–966.

Index

I

J

K

L

M